LabVIEW for Engineers

LabVIEW for Engineers

Ronald W. Larsen
Montana State University

Prentice Hall
Boston Columbus Indianapolis New York San Francisco Upper Saddle River
Amsterdam Cape Town Dubai London Madrid Milan Munich Paris Montreal Toronto
Delhi Mexico City Sao Paulo Sydney Hong Kong Seoul Singapore Taipei Tokyo

VP/Editorial Director, Engineering/Computer Science: Marcia J. Horton
Assistant/Supervisor: Dolores Mars
Senior Editor: Holly Stark
Associate Editor: Dee Bernhard
Editorial Assistant: Keri Rand
Director of Marketing: Margaret Waples
Senior Marketing Manager: Tim Galligan
Marketing Assistant: Mack Patterson
Vice-President, Production: Vince O'Brien
Senior Managing Editor: Scott Disanno
Project Manager: Greg Dulles
Senior Operations Supervisor: Alan Fischer
Operations Specialist: Lisa McDowell
Senior Art Director: Jayne Conte
Art Director: Kenny Beck
Cover Designer: Bruce Kenselaar
Media Editor: Daniel Sandin
Composition: Integra
Printer/Binder: Hamilton Printing Co.
Cover Printer: Lehigh-Phoenix Color

Library of Congress Cataloging-in-Publication Data

Larsen, Ronald W.
LabVIEW for Engineers / Ronald W. Larsen.
p. cm.
Includes index.
ISBN-13: 978-0-13-609429-6 (alk. paper)
ISBN-10: 0-13-609429-5 (alk. paper)
1. LabVIEW. 2. Engineering—Data processing. 3. Engineering—Computer programs.
4. Scientific apparatus and instruments—Data processing. I. Title.
TA345.5.L33L37 2011
620.00285—dc22

2009052015

10 9 8 7 6 5 4 3 2 1

Prentice Hall
is an imprint of

www.pearsonhighered.com

ISBN 10: 0-13-609429-5
ISBN 13: 978-0-13-609429-6

Contents

ESource Reviewers

We would like to thank everyone who helped us with or has reviewed texts in this series.

Naeem Abdurrahman, *University of Texas, Austin*
Sharon Ahlers, *Cornell University*
David G. Alciatore, *Colorado State University*
Stephen Allan, *Utah State University*
Anil Bajaj, *Purdue University*
Grant Baker, *University of Alaska–Anchorage*
William Bard, *University of Texas*
William Beckwith, *Clemson University*
Haym Benaroya, *Rutgers University*
John Biddle, *California State Polytechnic University*
Ray Biswajit, *Bloomsburg University of PA*
Donald Blackmon, *UNC Charlotte*
Tom Bledsaw, *ITT Technical Institute*
Fred Boadu, *Duke University*
Gregory Boardman, *Virginia Tech*
Stuart Brand, *The Ohio State University*
Jerald Brevick, *The Ohio State University*
Tom Bryson, *University of Missouri, Rolla*
Ramzi Bualuan, *University of Notre Dame*
Dan Budny, *Purdue University*
Betty Burr, *University of Houston*
Fernando Cadena, *New Mexico State University*
Joel Cahoon, *Montana State University*
Dale Calkins, *University of Washington*
Monica Cardella, *Purdue University*
Linda Chattin, *Arizona State University*
Harish Cherukuri, *University of North Carolina–Charlotte*
Vanessa Clark, *Washington University in St. Louis*
Arthur Clausing, *University of Illinois*
Barry Crittendon, *Virginia Polytechnic and State University*
Donald Dabdub, *University of CA Irvine*
Richard Davis, *University of Minnesota Duluth*
Kurt DeGoede, *Elizabethtown College*
John Demel, *Ohio State University*
James Devine, *University of South Florida*
Heidi A. Diefes-Dux, *Purdue University*
Jeffrey A. Doughty, *Northeastern University*
Jerry Dunn, *Texas Tech University*
Ron Eaglin, *University of Central Florida*
Dale Elifrits, *University of Missouri, Rolla*
Timothy Ellis, *Iowa State University*

Nurgun Erdol, *Florida Atlantic University*
Christopher Fields, *Drexel University*
Patrick Fitzhorn, *Colorado State University*
Julie Dyke Ford, *New Mexico Tech*
Susan Freeman, *Northeastern University*
Howard M. Fulmer, *Villanova University*
Frank Gerlitz, *Washtenaw Community College*
John Glover, *University of Houston*
Richard Gonzales, *Purdue Calumet*
John Graham, *University of North Carolina–Charlotte*
Hayden Griffin, *Virginia Tech*
Laura Grossenbacher, *University of Wisconsin Madison*
Ashish Gupta, *SUNY at Buffalo*
Otto Gygax, *Oregon State University*
Malcom Heimer, *Florida International University*
Robin A. M. Hensel, *West Virginia University*
Donald Herling, *Oregon State University*
Orlando Hernandez, *The College of New Jersey*
David Herrin, *University of Kentucky*
Thomas Hill, *SUNY at Buffalo*
A. S. Hodel, *Auburn University*
Susan L. Holl, *California St. U. Sacramento*
Kathryn Holliday-Darr, *Penn State U Behrend College, Erie*
Tom Horton, *University of Virginia*
David Icove, *University of Tennessee*
James N. Jensen, *SUNY at Buffalo*
Mary Johnson, *Texas A & M Commerce*
Vern Johnson, *University of Arizona*
Jean C. Malzahn Kampe, *Virginia Polytechnic Institute and State University*
Moses Karakouzian, *University of Nevada Las Vegas*
Autar Kaw, *University of South Florida*
Kathleen Kitto, *Western Washington University*
Kenneth Klika, *University of Akron*
Harold Knickle, *University of Rhode Island*
Terry L. Kohutek, *Texas A&M University*
Thomas Koon, Binghamton University
Reza Langari, *Texas A&M*
Bill Leahy, *Georgia Institute of Technology*
John Lumkes, *Purdue University*
Mary C. Lynch, *University of Florida*
Melvin J. Maron, *University of Louisville*
Christopher McDaniel, *UNC Charlotte*
Khanjan Mehta, *Penn State University Park*
F. Scott Miller, *University of Missouri-Rolla*
James Mitchell, *Drexel University*
Robert Montgomery, *Purdue University*
Naji Mounsef, *Arizona State University*
Nikos Mourtos, *San Jose State University*
Mark Nagurka, *Marquette University*
Romarathnam Narasimhan, *University of Miami*
Shahnam Navee, *Georgia Southern University*

James D. Nelson, *Louisiana Tech University*
Soronadi Nnaji, *Florida A&M University*
Sheila O'Connor, *Wichita State University*
Matt Ohland, *Clemson University*
Paily P. Paily, *Tennessee State University*
Kevin Passino, *Ohio State University*
Ted Pawlicki, *University of Rochester*
Ernesto Penado, *Northern Arizona University*
Michael Peshkin, *Northwestern University*
Ralph Pike, *Louisiana State University*
Andrew Randall, *University of Central Florida*
Dr. John Ray, *University of Memphis*
Marcella Reekie, *Kansas State University*
Stanley Reeves, *Auburn University*
Larry Richards, *University of Virginia*
Marc H. Richman, *Brown University*
Jeffrey Ringenberg, *University of Michigan*
Paul Ronney, *University of Southern California*
Christopher Rowe, *Vanderbilt University*
Blair Rowley, *Wright State University*
Liz Rozell, *Bakersfield College*
Mohammad Saed, *Texas Tech University*
Tabb Schreder, *University of Toledo*
Heshem Shaalem, *Georgia Southern University*
Randy Shih, *Oregon Institute of Technology*
Howard Silver, *Fairleigh Dickenson University*
Avi Singhal, *Arizona State University*
Greg Sun, *University of Massachusetts Boston*
John Sustersic, *The Penn State University*
Tim Sykes, *Houston Community College*
Murat Tanyel, *Geneva College*
Toby Teorey, *University of Michigan*
Scott Thomas, *Wright State University*
Virgil A.Thomason, *University of TN at Chattanooga*
Neil R.Thompson, *University of Waterloo*
Dennis Truax, *Mississippi State University*
Raman Menon Unnikrishnan, *Rochester Institute of Technology*
Thomas Walker, *Virginia Tech*
Michael S.Wells, *Tennessee Tech University*
Ed Wheeler, *University of Tennessee at Martin*
Joseph Wujek, *University of California, Berkeley*
Edward Young, *University of South Carolina*
Garry Young, *Oklahoma State University*
Steve Yurgartis, *Clarkson University*
Mandochehr Zoghi, *University of Dayton*

LabVIEW for Engineers

CHAPTER 1

Introduction

Objectives

After reading this chapter, you will know:

- what LabVIEW is and how it can be used to acquire, process, and analyze data
- what a LabVIEW VI is, and how front panel and block diagrams are used
- how to start LabVIEW and create a blank VI
- how to use LabVIEW menus to open and save VIs

1.1 WHAT IS LABVIEW?

In the past, *LabVIEW* was just a graphical programming language that was developed to make it easier to collect data from laboratory instruments using data acquisition systems. LabVIEW was always easy to use once you got used to wiring connectors to write your computer programs, and it definitely makes data acquisition an easier task than without LabVIEW, but LabVIEW is not just for data acquisition any more.

LabVIEW can be used to perform the following:

- acquire data from instruments
- process data (e.g., filtering, transforms)
- analyze data
- control instruments and equipment

For engineers, LabVIEW makes it possible to bring information from the outside world into a computer, make decisions based on the acquired data, and send computed results back into the world to control the way a piece of equipment operates.

As an example, the LabVIEW program (front panel) shown in Figure 1.1 reads a process measurement (a temperature value) from a piece of equipment, compares the measured process temperature with the desired temperature (called a *setpoint*), and outputs a signal to a controller to try to control the temperature at the setpoint value. You can see in Figure 1.1 that when the temperature went above setpoint, the controller output decreased. This causes a valve on a heat source to close (partially) to bring the temperature back to setpoint.

In this brief example:

- A temperature value was read from an external device.
- The desired setpoint temperature was entered on a control on the front panel.
- A controller output was calculated using a PI Controller algorithm.
- The controller output was written to an external device.

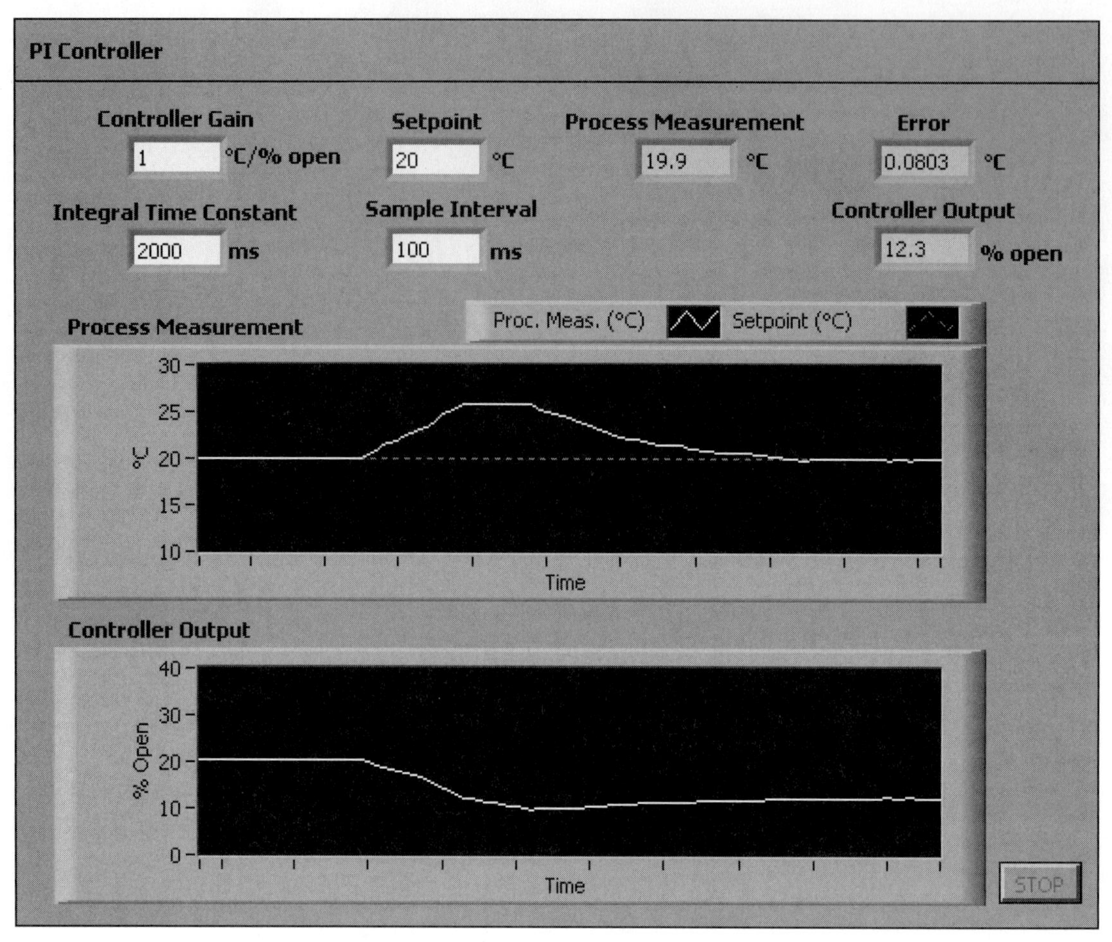

Figure 1.1
LabVIEW VI for PI Controller (front panel).

LabVIEW's ability to get data from outside the real world, use the data inside a program, and send results back out to the real world allows engineers to interact with and control events in the real world, not just inside computers. Using LabVIEW programs is a fast and efficient way to develop a new device or prototype a new instrument. And LabVIEW is becoming such an industrial standard that the LabVIEW program used to create the prototype may soon be the program used in the commercial version as well.

1.2 ASSUMPTIONS

The author is making a few assumptions about the reader and about the version of LabVIEW that you have available.

1.2.1 Target Audience

As part of the Pearson-Prentice Hall E-Source series, this text is targeted at first- and second-year engineering students. As such, the reader is assumed to have some mathematical ability, but very little experience with LabVIEW. And while LabVIEW is often used for data acquisition, that is not the primary focus of this text. Instead, we will focus on using the mathematical power of LabVIEW to tackle the analysis of data sets, whether they are acquired from an experimental system or not.

1.2.2 LabVIEW Versions

LabVIEW is a well-developed program, and the changes from one version to another are small. The author has used LabVIEW 8.5 and LabVIEW 2009 Full versions with no added bells or whistles in developing the examples in this text. For the material covered in this text, users of earlier versions of LabVIEW will see very few differences.

LabVIEW is sold in the following packages:

- **Base Package**—reduced mathematics functionality
- **Full Package**—complete set of math functions
- **Student Edition**—full package with a watermark in the lower right corner of each front panel
- **Professional Package**—can create stand-alone applications
- **NI Developer Suite**—includes extra add-ons and toolkits

In this text, we will make use of many of the math functions that are available only in the Full, Student, Professional, and Developer packages. Some of the more advanced analysis techniques illustrated in this text will be unavailable in the Base package. For example,

Base	Full	Student	Pro	Dev	Topic
✓	✓	✓	✓	✓	Trig Functions
✓	✓	✓	✓	✓	Boolean Functions
✓	✓	✓	✓	✓	Matrix Math
No*	✓	✓	✓	✓	Simultaneous Equations Function
✓	✓	✓	✓	✓	File I/O
✓	✓	✓	✓	✓	Graphs
✓	✓	✓	✓	✓	Basic Statistics
No	✓	✓	✓	✓	Interpolation
No	✓	✓	✓	✓	Curve Fitting
No	✓	✓	✓	✓	Regression
No	✓	✓	✓	✓	Integration
No	✓	✓	✓	✓	Differentiation
No	✓	✓	✓	✓	Differential Equations

*An easy workaround is presented in the text.

It is assumed that the reader has access to at least the Student LabVIEW package.

1.3 CONVENTIONS IN THE TEXT

The following conventions are used in this text:

- *Keywords*—shown in italics the first time they appear.
- **Literals**—items meant to be typed exactly as they appear in the text are shown in bold font.

- **Function** and **Control names**—the functions to be selected from the Functions Palette and the Controls to be selected from the Controls Palette will be shown in bold font. The location within the palette structure is indicated using slashes, as **Main Palette / Sub-Pallet / Group / Function**.
- **Menu Selections**—when actions are initiated from a menu, the menu and submenu choices are indicated, separated by slashes as **Menu Option / Submenu Option**.

1.4 LABVIEW VIs

LabVIEW programs are called VIs. Originally, VI stood for *virtual instrument*, but LabVIEW is now used for many more applications than just creating a computer simulation of an instrument, and LabVIEW programs are typically referred to simply as VIs.

A LabVIEW VI has two parts:

- *Front Panel*—Displays the *controls* (knobs, buttons, graphs, etc.) and represents the graphical interface for the VI. An example of a VI front panel is shown in Figure 1.1.
- *Block Diagram*—Holds the programming elements (called *blocks*, *functions*, or sometimes *subVIs*) that are wired together to build the graphical program. The block diagram for the PI Controller VI is shown in Figure 1.2.

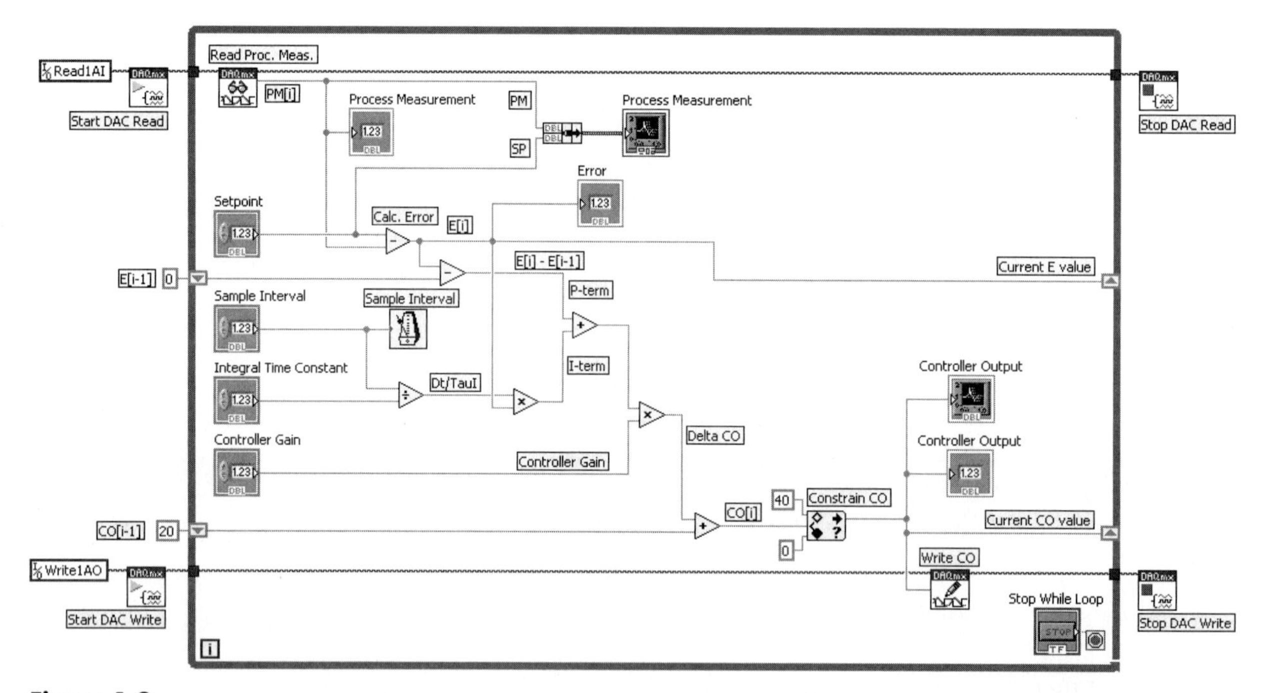

Figure 1.2
PI Controller VI, block diagram.

This text is intended for students who are new to LabVIEW, so Figure 1.2 is presented as a preview only. For students who want to know a little more about how the LabVIEW program works, Figure 1.3 shows the major program sections in the PI Controller. (If you are not interested in the program details, you can skip ahead to Section 1.5.)

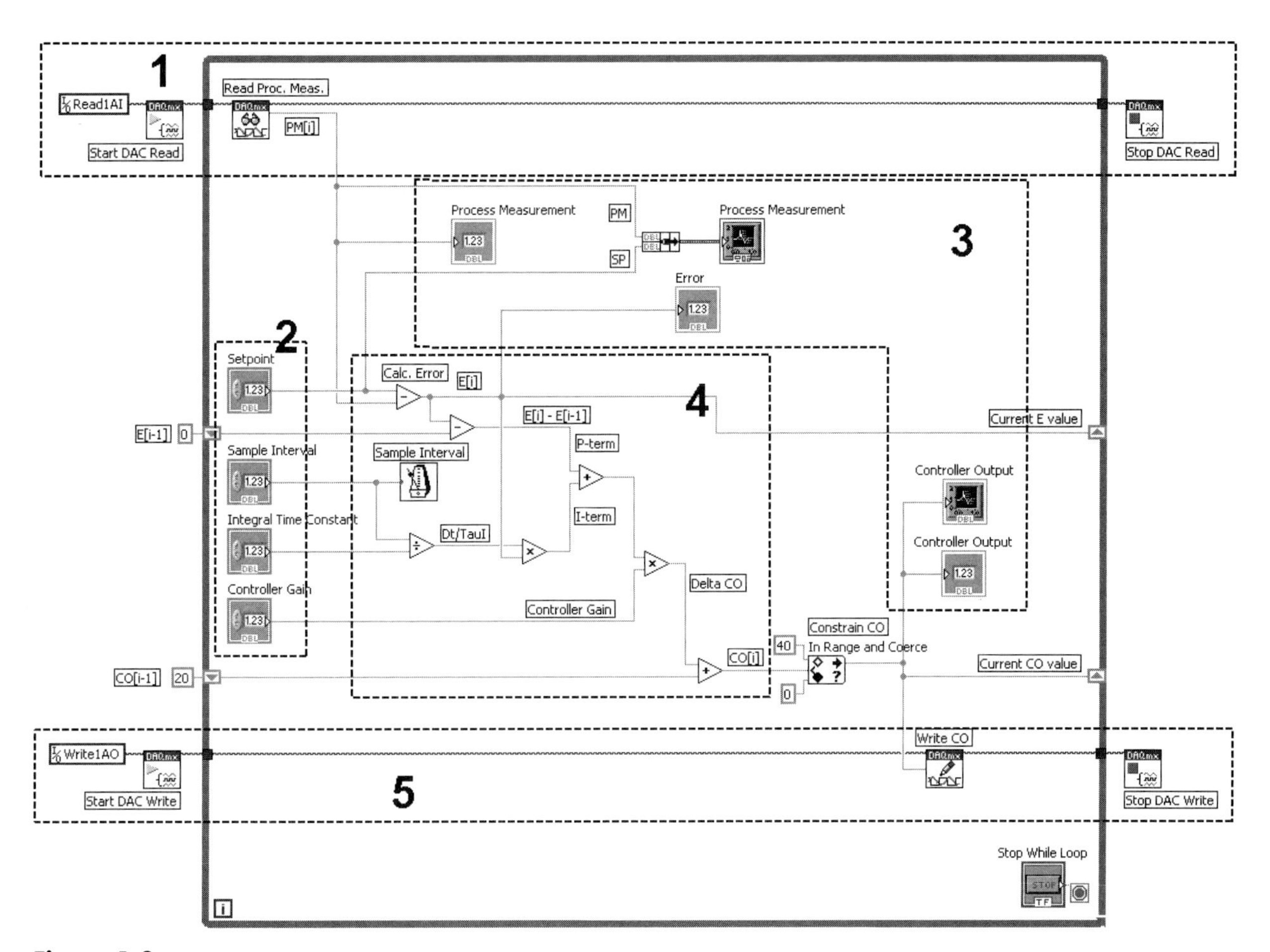

Figure 1.3
The parts of a LabVIEW program.

1. Read an analog voltage (the process measurement) from the data acquisition system.
2. Get parameter values from the controls on the front panel.
3. Display values on the front panel using numeric indicators and graphs.
4. Calculate the controller output value.
5. Write an analog voltage (the controller output) to the data acquisition system.

1.5 STARTING LABVIEW

The learning approach that is used in this text is to try to get the reader creating LabVIEW programs as quickly as possible. To accomplish this, some features may be presented briefly at first, with just enough information to allow an example to be developed. The details will be presented later in the chapter.

That said; let's start LabVIEW.

LabVIEW is started from the Windows Start menu as illustrated in Figure 1.4.

Start Menu / All Programs / National Instruments LabVIEW

If LabVIEW has been used recently, there will be an icon in the left panel of the Windows Start menu (marked with (1) in Figure 1.4). Otherwise, use the **All**

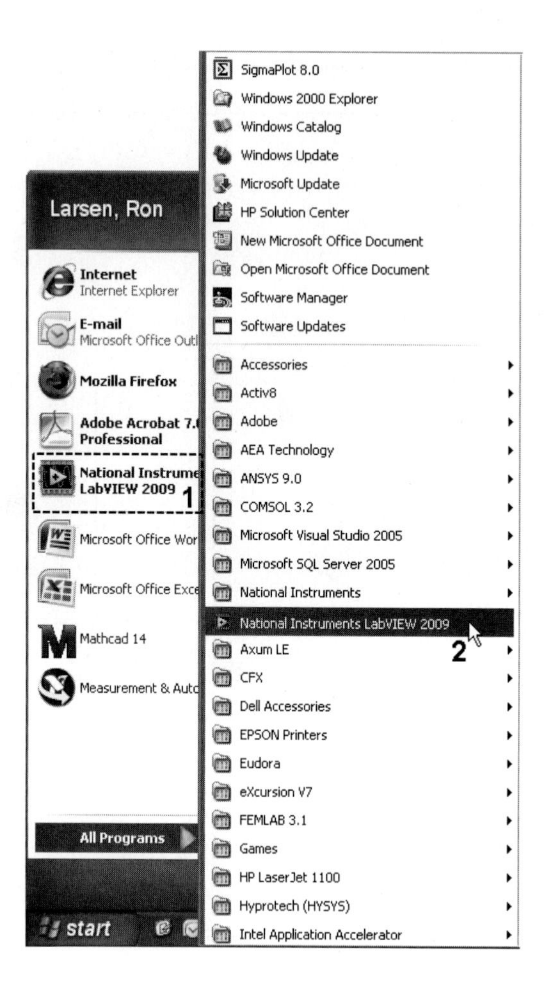

Figure 1.4
Start menu showing two options for starting LabVIEW.

Figure 1.5
Desktop shortcut icon for LabVIEW.

Programs button and find the **National Instruments LabVIEW** icon in the list of installed programs (marked with (2) in Figure 1.4).

Alternatively, there might be a shortcut to LabVIEW on the computer desktop, as shown in Figure 1.5. If your computer does not have a desktop shortcut for LabVIEW, you can create one by right-clicking on the **National Instruments LabVIEW** icon (marked with (2) in Figure 1.4) and selecting **Create Shortcut** from the pop-up menu.

As LabVIEW loads, the title screen shown in Figure 1.6 is displayed. Once the program has loaded into memory, the title screen disappears, and the *Getting Started window* (shown in Figure 1.7) is displayed.

The Getting Started window performs the following:

- provides access to online support for LabVIEW
- provides access to the LabVIEW Help system
- allows you to create a blank VI or an empty project
- allows you to open a recently used VI or project
- allows you to search for LabVIEW examples

Note: The Getting Started window is displayed by default, but that can be changed by using menu options **Tools / Options** to open the Options dialog, then selecting the **Environment** category, and checking or clearing the box before **Skip Getting Started window on launch**.

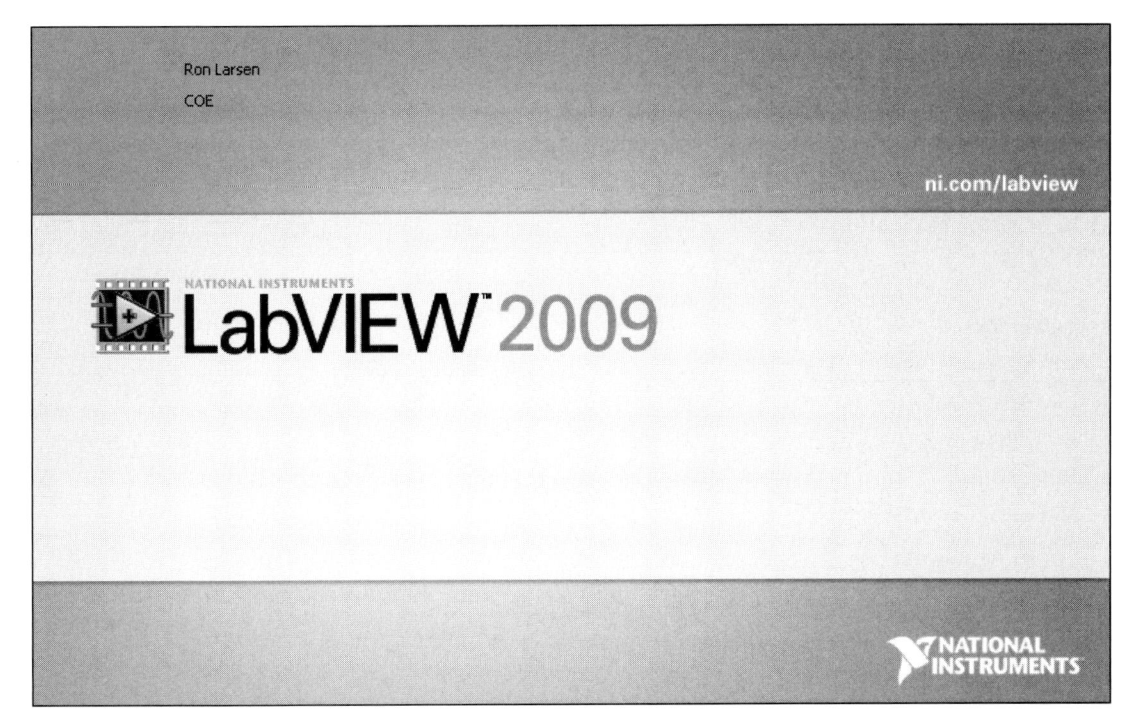

Figure 1.6
LabVIEW title screen, shown as program is loading.

LabVIEW Nomenclature:

- *VI* is synonymous with *LabVIEW program*. LabVIEW programs are stored as files with .vi extensions. LabVIEW VIs include a graphical user interface (*front panel*), and a *block diagram* that contains the programming elements.
- A *project* is a collection of related program elements that are intended to work together. A project can contain multiple VIs plus additional program elements.

1.5.1 The LabVIEW Editing Environment

LabVIEW VIs can be created quickly, can be modified as needed, and give scientists and engineers the ability to collect and analyze the data they need in order to accomplish their goals. LabVIEW provides an editing environment that makes it easy to create, modify, and run VIs.

Creating a LabVIEW VI is easy. First, you open a blank VI, then you add controls to the front panel and programming functions to the block diagram, and wire them together to create a functioning program. For now, we will create a blank VI just so we can look around at the LabVIEW workspace.

To create a blank VI in LabVIEW,

- Start LabVIEW (**Start / All Programs > / National Instruments LabVIEW**). Wait for the Getting Started screen to be displayed.
- Click **Blank VI** on the Getting Started window.

Note: If your version of LabVIEW has been set to skip the Getting Started window, starting LabVIEW should automatically open a blank VI.

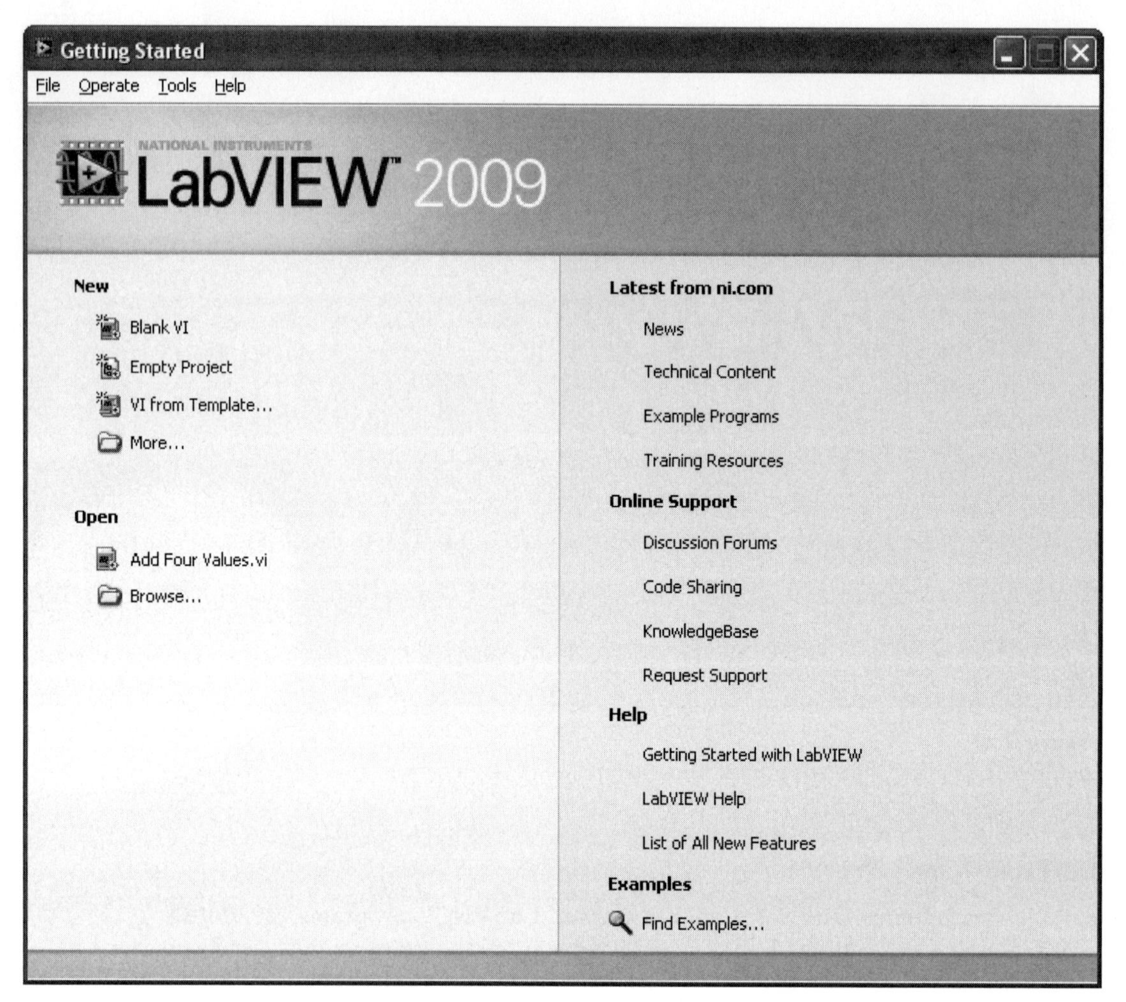

Figure 1.7
LabVIEW Getting Started window.

The blank VI will be displayed in two windows:

- front Panel, labeled Untitled 1 Front Panel
- block Diagram, labeled Untitled 1 Block Diagram

Controls Palette

When the front panel is displayed, the *Controls Palette* (Figure 1.8) is opened as well. The Controls Palette provides access to the objects (controls, indicators, knobs, and graphs) that are placed on the front panel.

Note: By default, the Controls Palette is displayed any time a front panel is being edited, but the default can be changed. If the Controls Palette is not visible, use menu options **View / Controls Palette** from the front panel to display the Controls Palette.

Since there are a large number of controls available, they are collected into a number of categories and each category can be expanded or collapsed. In Figure 1.8 only the *Express* category is shown expanded.

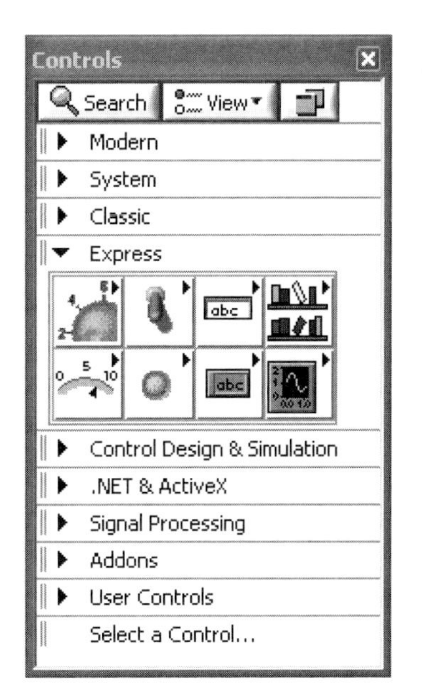

Figure 1.8
The Controls Palette is available when editing the Front Panel of a VI.

- The *Classic* set of tools includes the switches and knobs that originally came with LabVIEW, while the *Modern* set provides controls with a more updated appearance.
- The *Express* set of controls collects the most commonly used tools in one place, which can be very handy when developing a front panel.

Functions Palette

When editing a VI's block diagram, the *Functions Palette* is shown. In Figure 1.9 the *Programming* and *Express* categories are shown expanded.

Note: By default, the Functions Palette is displayed any time a block diagram is being edited, but the default can be changed. If the Functions Palette is not visible, use menu options **View / Functions Palette** from the block diagram to display the Functions Palette.

Programming in LabVIEW is all about selecting objects from the Controls and Functions Palettes and placing them on either the front panel (controls) or block diagram (functions). Then, the objects must be connected (wired) appropriately on the block diagram. We will demonstrate this process many times throughout the rest of this book.

Note: The Functions Palette contains *functions*, *VIs*, and *Express VIs*. All of these can be placed on a block diagram to create your graphical programs—they can be used in the same fashion. In this text we use the term *function* loosely, applying the term to most of the programming elements on the Functions Palette. The more specific definitions are as follows:

- **Function**—a program element that does not have a front panel or block diagram, but does have a *connector pane* indicating how the function should be wired. Functions appear on the Functions Palette with a pale yellow background.
- **VI**—a VI is a LabVIEW program. A VI can be used within another VI. When this is done, it is called a *SubVI*. VIs appear on the Functions Palette with a pale yellow strip across the top of the icon (or yellow border when expanded) and ".vi" in the name.

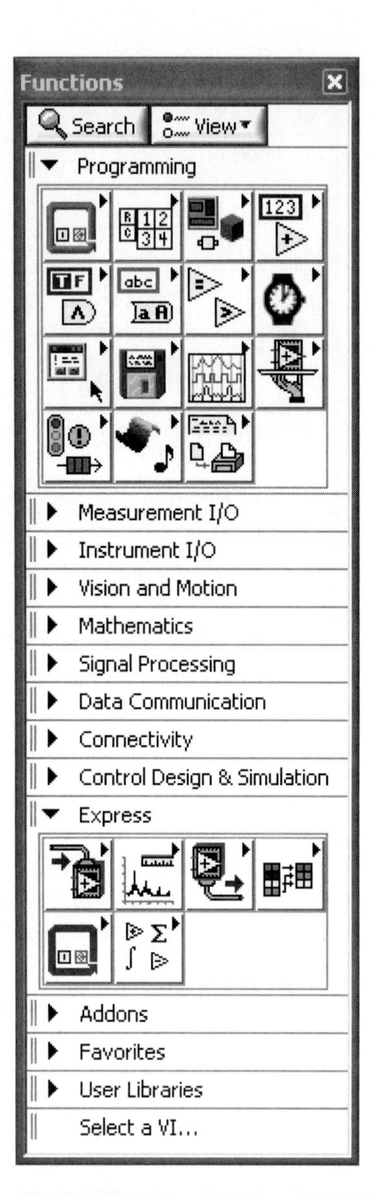

Figure 1.9
The Functions Palette is available when editing the Block Diagram of a VI.

- **Express VI**—an Express VI is a more sophisticated VI that can be configured using a dialog box. The dialog box automatically opens when the Express VI is placed on the block diagram. Double-click the VI's icon to re-open the dialog when needed. Express VIs appear on the Functions Palette with a blue strip across the top of the icon (or blue border when expanded).

PRACTICE!

Look in the Functions Palette to find the groups containing the following functions:

- Add
- Wait (look for a wristwatch icon)

Look in the Controls Palette for the groups containing:

- Dial Numeric Control
- Toggle Switch

Solution

Add function:

From the block diagram:

- functions Palette/Mathematics Group/Numeric Group/Add function
- functions Palette/Express Group/Arithmetic & Comparison Group/Express Numeric Group/Add function

Wait function:

From the block diagram:

- functions Palette/Programming Group/Timing Group/Wait (ms) function

Dial Numeric Control:

From the front panel:

- controls Palette/Express Group/Numeric Controls Group/Dial
- controls Palette/Modern Group/Numeric Group/Dial
- controls Palette/Classic Group/Classic Numeric Group/Dial

Toggle Switch:

From the front panel:

- controls Palette/Express Group/Buttons & Switches Group/Toggle Switch
- controls Palette/Modern Group/Boolean Group/Vertical (or Horizontal) Toggle Switch
- controls Palette/Classic Boolean Group/Vertical (or Horizontal) Toggle Switch

---

1.5.2 The Tools Palette

LabVIEW provides a third palette, called the *Tools Palette* (see Figure 1.10), but it is not automatically displayed. By default, *automatic tool selection* is activated and the Tools Palette is not shown.

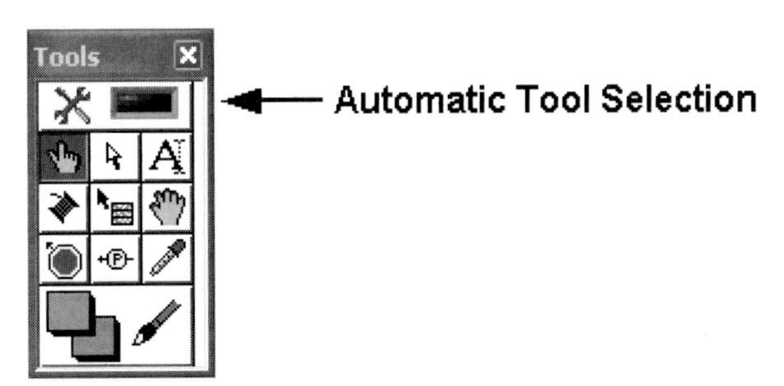

Figure 1.10
The Tools Palette.

The Tools Palette is not needed for routine tasks, but it can be displayed using menu options **View / Tools Palette** from either the front panel or the block diagram.

The Tools Palette options provide a good overview of the various tasks that must be accomplished to program in LabVIEW. In the following list, the usage of each of the tools in the Tools Palette are described, along with the way to accomplish the same task using automatic tool selection.

- **Automatic Tool Selection Button** (top of Tools Palette)—this is a toggle button with a green LED display that indicates when automatic tool selection is activated.
- **Operate Value Tool** (finger)—used to push buttons (to toggle a Boolean value, to select a menu item, etc). If automatic tool selection is activated, moving the mouse over a control that can be operated selects this tool.
- **Position / Size / Select Tool** (arrow)—used to relocate and resize controls. If automatic tool selection is activated, moving the mouse near the border of a control selects this control.
- **Edit Text Tool** (A with cursor)—used to enter text (on labels and string constants). If automatic tool selection is activated, double-click in a text field to select this control.
- **Connect Wire Tool** (Spool)—used to connect wires between block outputs and inputs. If automatic tool selection is activated, positioning the mouse near a connector or a wire selects this tool.
- **Object Shortcut Menu Tool** (Menu icon)—opens a pop-up menu of options for controls and programming blocks. Right-clicking on any object also opens the pop-up menu.
- **Scroll Window Tool** (Cupped Hand)—used to drag a window (e.g., to scroll to a hidden portion of a large block diagram). If automatic tool selection is activated, you must use the scroll bars at the edges of the windows to scroll.
- **Set / Clear Breakpoint Tool** (Stop sign)—Breakpoints are used when debugging programs to freeze execution so you can see what is happening within the program. Breakpoints are always set on the block diagram. If automatic tool selection is activated, you can right-click on a function or wire and select **Set Breakpoint** from the pop-up menu.
- **Probe Data Tool** (Probe symbol: yellow circle, arrow, P character)—Probes can be placed on wires to show the value in the wire when the program is run. Probes are placed on the block diagram, but they are visible over the front panel as well. If automatic tool selection is activated, you can right-click on a wire and select **Probe** from the pop-up menu.
- **Get Color Tool** (Dropper)—used to set the current foreground and background colors. Click the dropper on a colored object, and the foreground and background colors shown at the bottom of the Tools Palette will be set to the colors of the selected object.

  Note: You can also click on the foreground and background colors shown at the bottom of the Tools Palette and select colors from a color selection palette.
- **Set Color Tool** (Paintbrush)—sets the foreground and background colors of a colored object to the colors shown at the bottom of the Tools Palette. This is predominantly used on the front panel, although there are a few objects that can be colored on the block diagram as well (e.g., labels).

For most common tasks, the automatic tool selection mechanism works very well and eliminates the need to keep changing the currently selected tool.

1.6 CREATING A VI

We will demonstrate how to build a LabVIEW programs VIs with an example. In the example, we will build a very simple VI that has a toggle switch and an LED indicator that illuminates when the switch is "on" (Figure 1.11).

The example is intended to be about as simple as possible. It contains two controls (toggle switch and stop button) and one indicator (LED). It takes one wire on the block diagram to complete the programming.

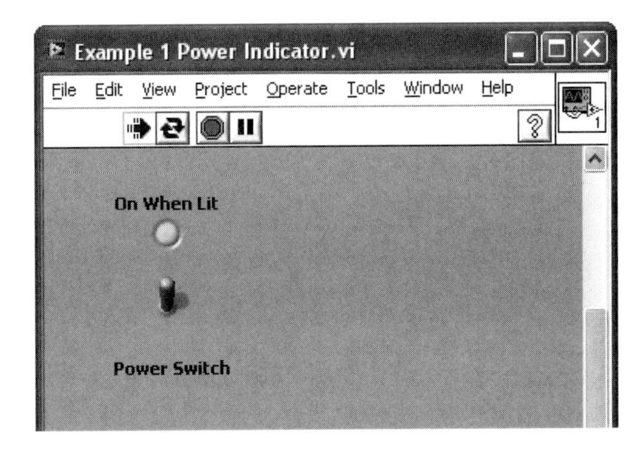

Figure 1.11
Power Indicator VI developed as Example 1.

We will work through the example quickly, with the intent of providing an overview of the VI development process. The best way to learn LabVIEW is to create the VIs on your own computer as they are presented in the text.

1.6.1 Developing a Simple Virtual Instrument—Example 1

If you click the **Blank VI** link on the Getting Started window, LabVIEW will close the Getting Started window and create a blank VI with the temporary name "Untitled 1". (If you have multiple unsaved VIs being edited, they will be named "Untitled 1", Untitled 2", and so on.) Standard practice is to assign more descriptive names the first time you save the VI.

The new VI's front panel and block diagram are shown in two new windows as illustrated in Figure 1.12.

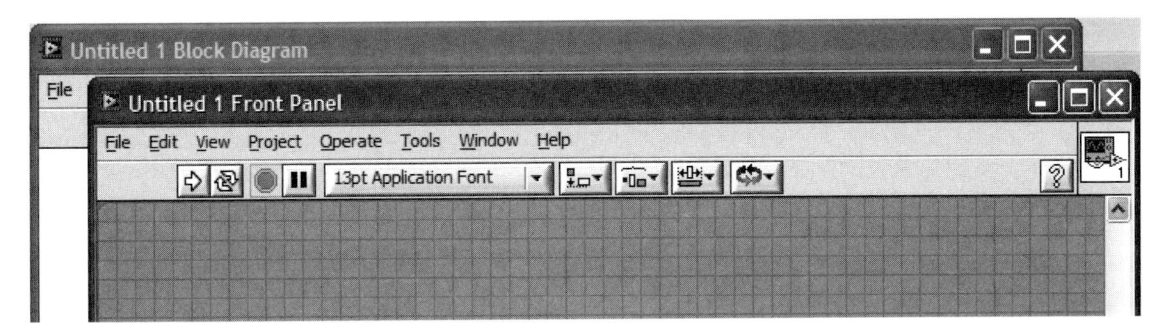

Figure 1.12
The front panel and block diagram for a new VI (temporarily named "Untitled 1").

By default, the front panel is shown on top of the block diagram because LabVIEW assumes that the "standard" way to build a VI is as follows:

- First, add controls to the front panel.
- Second, wire the *nodes* (the back side of the controls) on the block diagram.

The standard approach works in most instances.

More specifically, we will use the following steps to build the power indicator VI.

1. Create a blank VI.
2. Add a toggle switch to the front panel.
3. Add an LED indicator to the front panel.

4. Wire the toggle switch node to the LED indicator node on the block diagram.
5. Save the VI with a descriptive name.
6. Run and test the VI.

Step 1. Create a blank VI

There are two ways to create a new, blank VI, depending on whether or not the LabVIEW editor is already running.

- If LabVIEW is already running, create a new blank VI using menu options **File / New VI** from either a front panel or a block diagram.
- If LabVIEW is not already running, start LabVIEW and click on **Blank VI** on the Getting Started window as shown in Figure 1.13. (You can also use menu options **File / New VI** from the Getting Started window.)

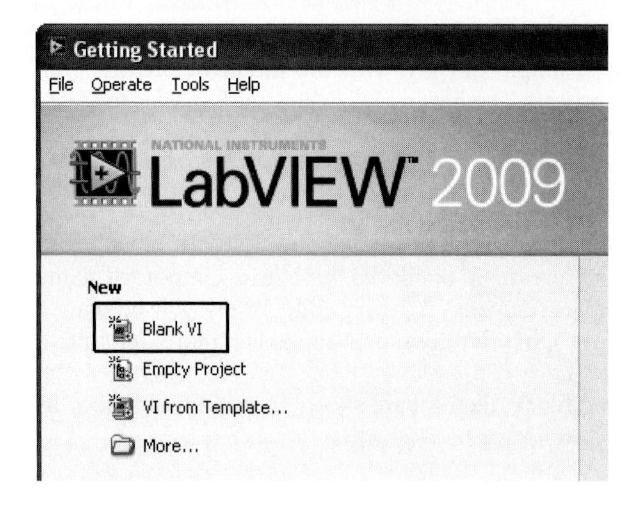

Figure 1.13
Creating a blank VI from the Getting Started window.

Whichever method you use, a new, blank VI front panel and block diagram will be displayed.

Note: The Getting Started window is displayed whenever LabVIEW is running and no VIs are being edited. As soon as the blank VI is displayed, the Getting Started window will be hidden.

Step 2. Add a toggle switch control to the front panel

When you select (click on) the front panel, the *Controls Palette* (Figure 1.14) will be displayed. Drag a Vertical Toggle Switch control from the Controls Palette to the front panel.

> **Controls Palette / Express Group / Buttons and Switches Group / Vertical Toggle Switch**

The Controls Palette is a graphical menu of all of the controls that you can place on the front panel. The controls are collected into groups to make the control you want to use easier to find. For this example, all of the controls we will need will be in the Express group.

Notes:

- The most commonly used controls have been gathered into the Express group. This keeps most of the controls you will need together in one location.
- The LabVIEW Palettes can be resized. In this text they are typically shown resized to save space.

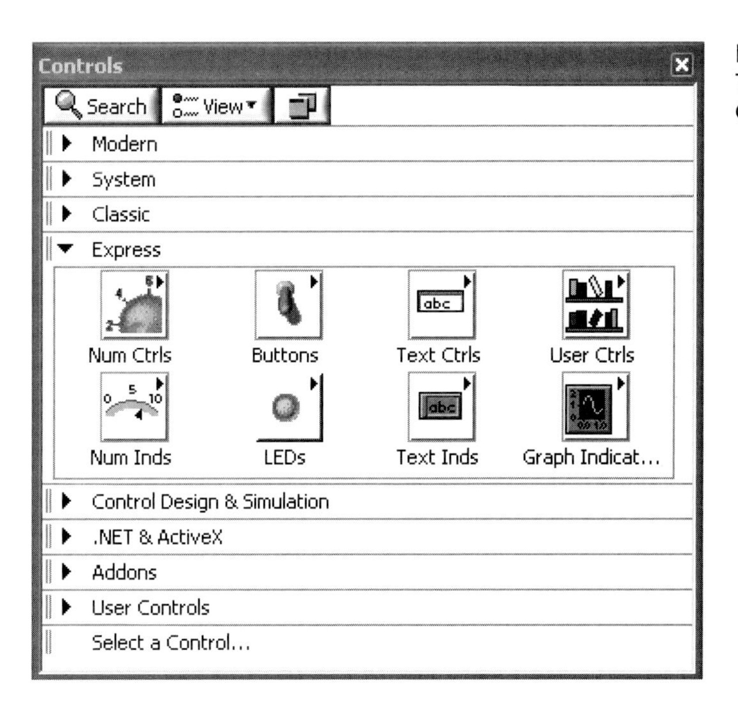

Figure 1.14
The Express group in the Controls Palette.

The following mouse clicks will get you to the toggle switch:

1. On the Controls Palette, click on the Express group. This causes the various types of controls and indicators available in the Express group to be shown, as illustrated in Figure 1.14.
2. Click on the Buttons group within the Express group. This will cause all of the various types of buttons and switches to be displayed, as illustrated in Figure 1.15.

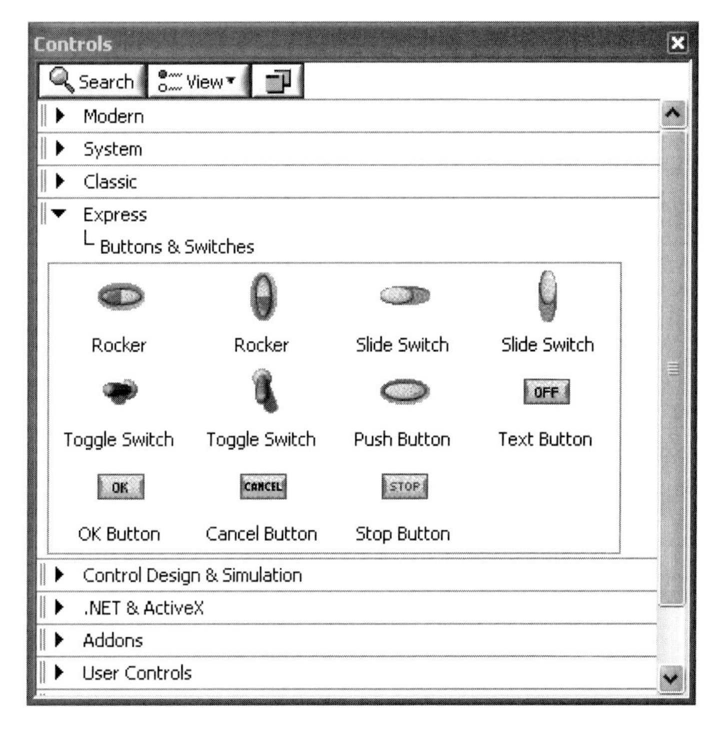

Figure 1.15
The buttons and switches available through the Express group.

3. Click on a toggle switch icon and drag it onto the front panel, as illustrated in Figure 1.16.

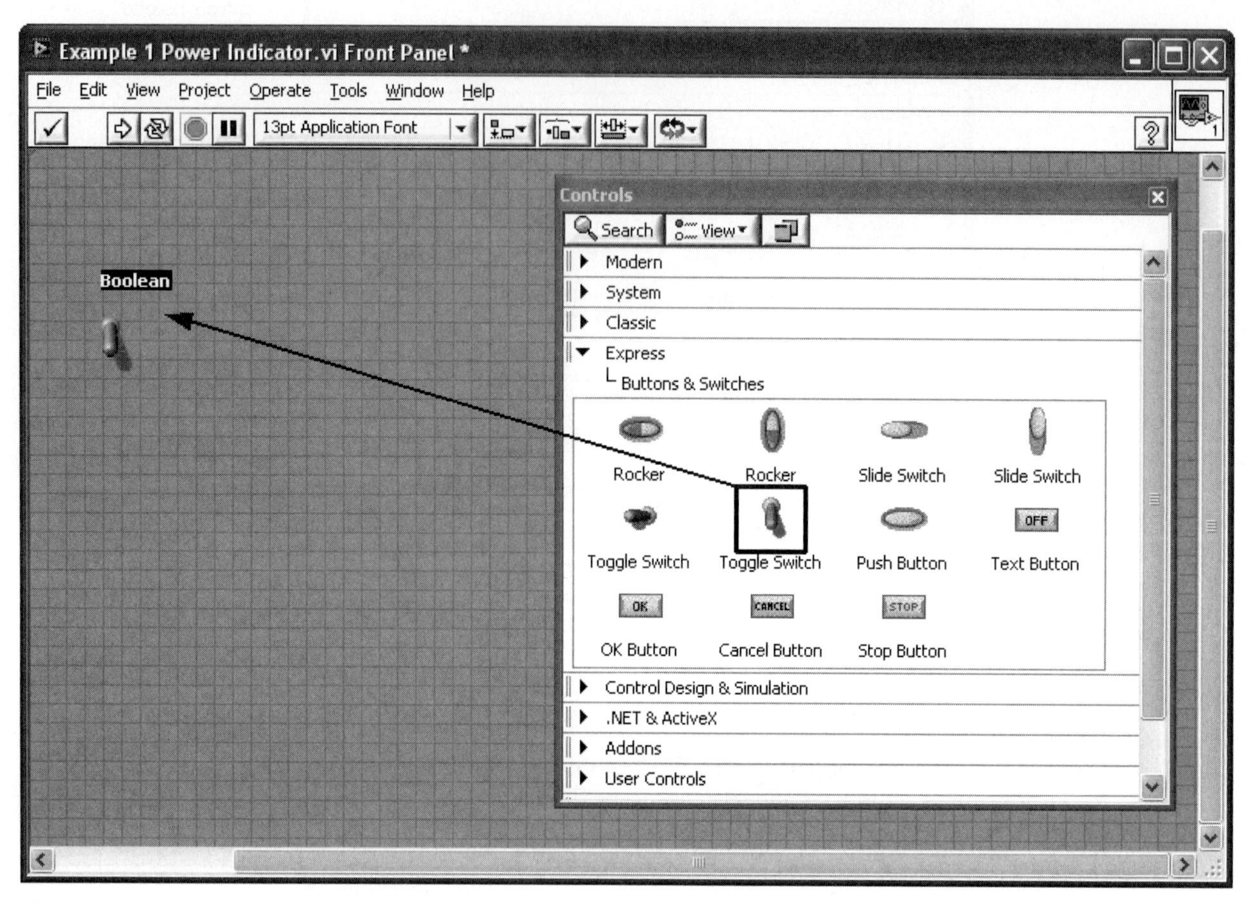

Figure 1.16
Dragging the toggle switch to the front panel.

By default, the switch is labeled "Boolean" but the label is selected (white letters on black background) so that it can be changed easily. We'll rename the control "Power Switch" and move the label below the switch.

Step 3. Add an LED Indicator to the front panel
Drag a round LED Indicator from the Controls Palette to the front panel. To find the round LED indicator, start with the Controls Palette, select the Express group, then select the LED's group, and finally drag the round LED Indicator to the front panel as illustrated in Figure 1.17. This is summarized in the following command sequence:

Controls Palette / Express Group / LED's Group / Round LED

Again, the indicator is initially labeled "Boolean" and selected so that it can be renamed. We will change the label to "On When Lit" and center the label over the LED.

Step 4. Wire the toggle switch output to the LED indicator input
When controls and indicators are placed on the front panel, a *node* for each is automatically placed on the block diagram. A node is an icon that

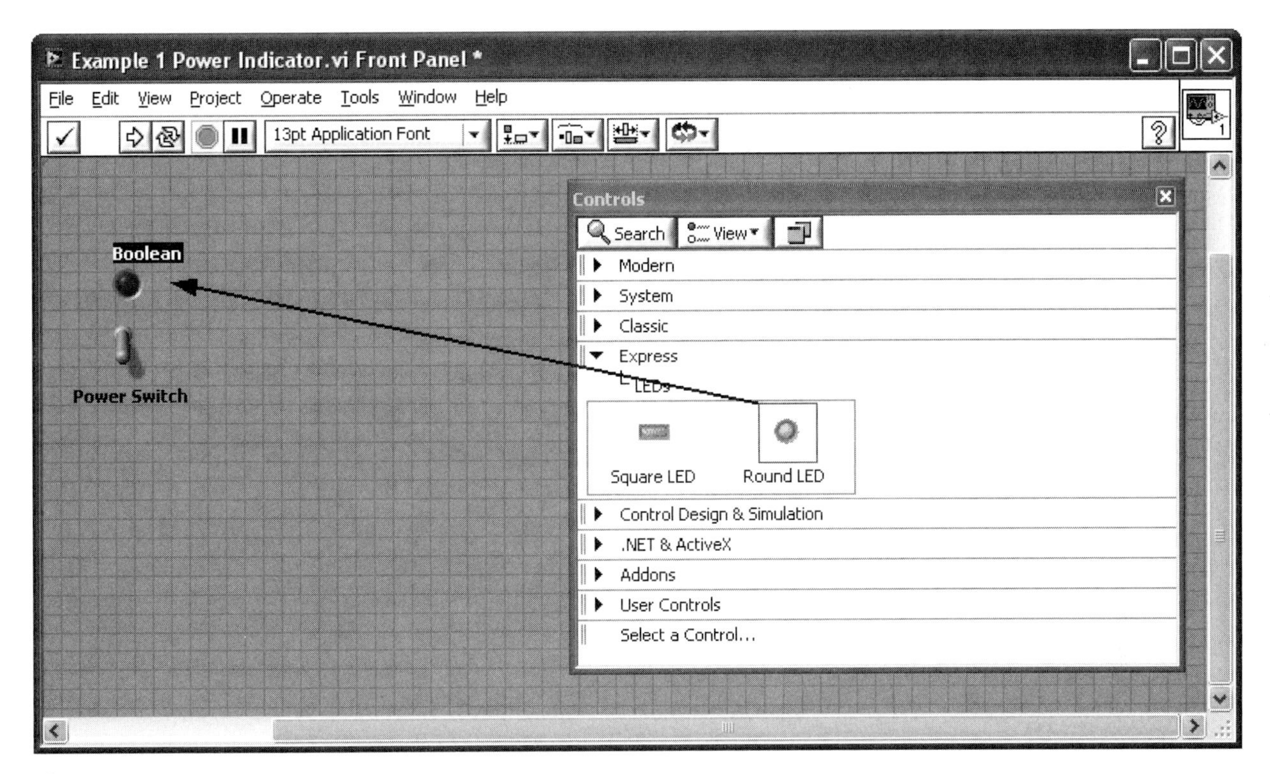

Figure 1.17
Adding a Round LED indicator to the front panel.

describes the control (or indicator, or function) and holds the terminals for the node's inputs and outputs.

In this example, the LED indicator node has to be connected (wired) to the switch node to function correctly. Wiring is done on the block diagram, which is illustrated in Figure 1.18. Specifically, the Power Switch control output terminal needs to be wired to the LED indicator input terminal.

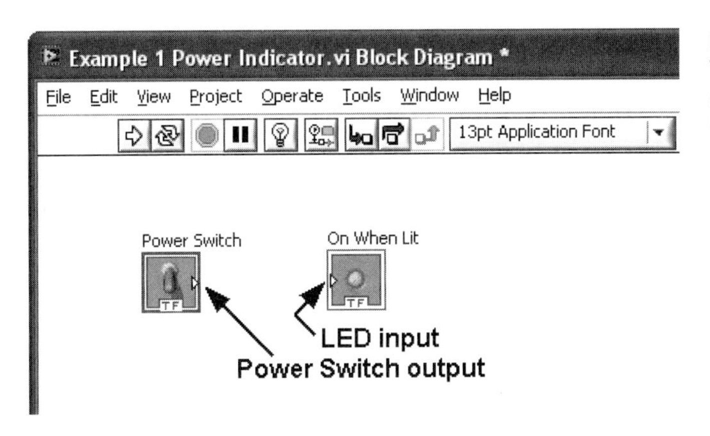

Figure 1.18
The Power Switch output needs to be wired to the LED input.

Wiring the two terminals together can be done in two ways. You can use whichever seems to work best; the dragging method is used in these examples.

- Drag the mouse from the output terminal on the Power Switch control node to the input terminal on the LED indicator node.
- Click on the output terminal on the Power Switch to begin wiring, and then click on the input terminal on the LED indicator to complete the connection.

As you move the mouse over the Power Switch output terminal, the mouse icon changes from the usual arrow to something that looks like a spool (supposedly a spool of wire) as illustrated in Figure 1.19.

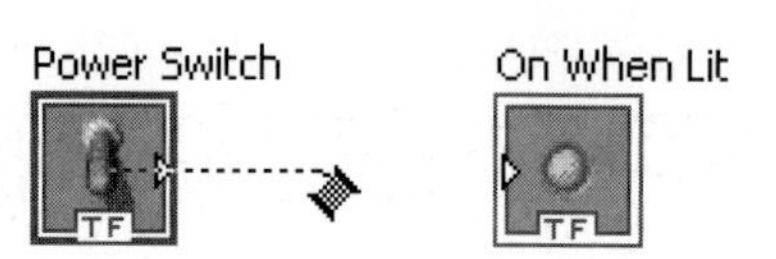

Figure 1.19
Drag the "spool" mouse icon from the output terminal to the input terminal.

Note: The changing mouse icon assumes that LabVIEW is operating in *Automatic Tool Selection mode*, which is the default. If the mouse icon does not change when moved over a terminal on the block diagram, you need to activate either Automatic Tool Selection mode or Connect Wire mode. The mode is selected on the Tools Palette, shown in Figure 1.20. Click the top button on the Tools Palette to toggle Automatic Tool Selection mode.

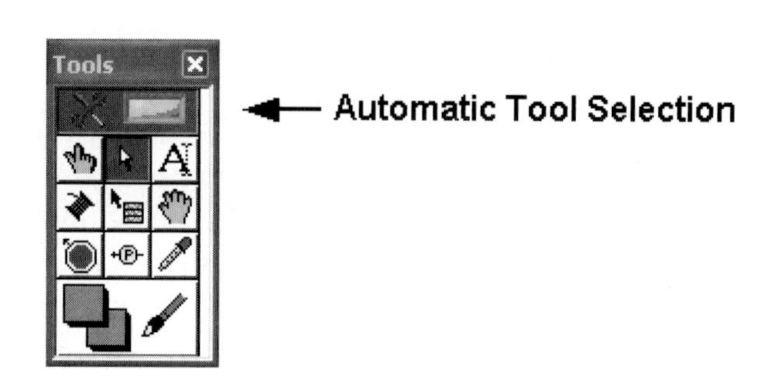

Figure 1.20
The LED indicator is bright when Automatic Tool Selection is active.

Note: To display the Tools Palette, use menu options **View / Tools Palette** from either the front panel or the block diagram.

To complete the wiring, drag the "spool" from the Power Switch output terminal to the input terminal on the LED indicator and release the mouse button. The completed wire is shown in Figure 1.21.

Figure 1.21
The Power Switch has been wired to the LED indicator.

Note: The mouse icon changes to a spool when the mouse is over a terminal so that you can wire the terminal. The mouse icon also changes to a spool when the mouse is near (but not over) a wire so that you can connect from a wire to another indicator, or a function.

When all required terminals have been wired, the VI is ready to run. LabVIEW indicates whether or not the VI is ready to run by the icon on the ***Run*** *button*, as illustrated in Figure 1.22.

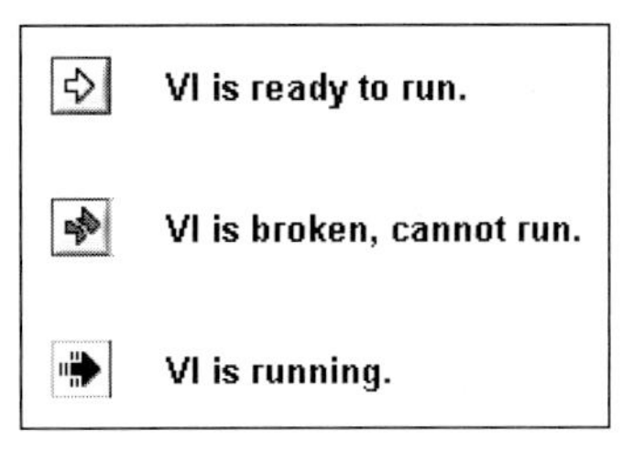

Figure 1.22
The Run button icon changes to indicate the VI status.

The VI shown in Figure 1.21 would run, but it would only evaluate the position of the power switch once and, if the switch is closed, illuminate the LED indicator momentarily before the program stops executing. To allow time to flip the switch a few times to see what happens to the LED, we want the program to continue running until we give the command to stop it. To accomplish this, we will use the **Run Continuously** button, just to the right of the **Run** button (see Figure 1.23). The **Run Continuously** button causes the VI to restart over and over again, until the **Abort Execution** button is clicked.

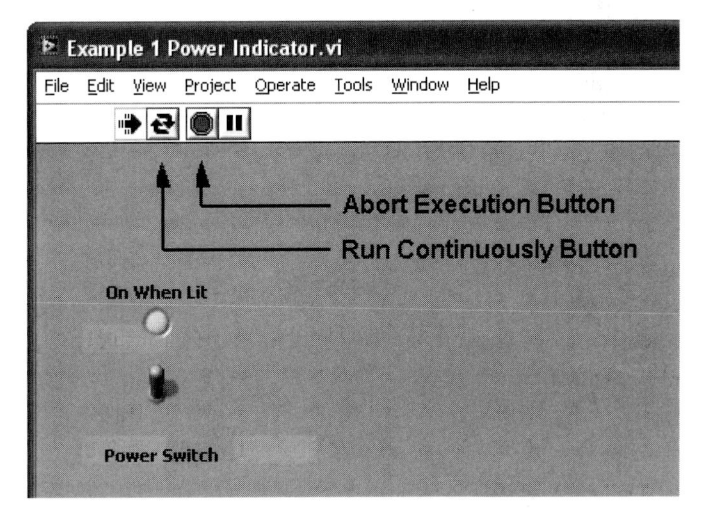

Figure 1.23
Running the VI continuously.

Step 5. Save the VI
You will want to save your VIs with a descriptive name so that you can find them again later, if needed. The first time you save a VI you will be given the opportunity to assign a descriptive name to the VI. In this example, we have saved the VI with the name "Example 1 Power Indicator". The ".vi" file extension is added automatically.

Note: When you save your VI from either the front panel or the block diagram, both the front panel and the block diagram are saved—you don't need to save them separately.

Step 6. Test the VI
Return to the front panel and click the **Run Continuously** button (indicated in Figure 1.23.). Once the VI is running, you can click on the toggle switch to flip the switch and turn the LED indicator on and off. In Figure 1.23 the

switch is in the up position and the LED is illuminated. Use the **Abort Execution** button to stop the VI.

When a toggle switch is thrown, it stays in the new position. Officially, its *mechanical action* is termed "Switch When Pressed". There are a variety of mechanical actions for switches:

- **Switch When Pressed**—toggle switches
- **Switch When Released**—mouse buttons, usually
- **Switch Until Released**—doorbell buzzer

There are also three *latch* actions.

- **Latch When Pressed**—like a starting pistol
- **Latch When Released**
- **Latch Until Released**—behaves like the doorbell buzzer

The "Latch When Pressed" action causes a momentary switch signal (latch signal), and then the signal reverts to the default value and stays at the default value for the duration of the VI's execution.

The "Latch When Released" action is similar except that the starting pistol would be fired when the trigger was released. After the momentary latch signal, the switch reverts to the default value and stays at the default value for the duration of the VI's execution.

You can change the mechanical action of a toggle switch by right-clicking on the switch and selecting **Mechanical Action** from the pop-up menu as shown in Figure 1.24.

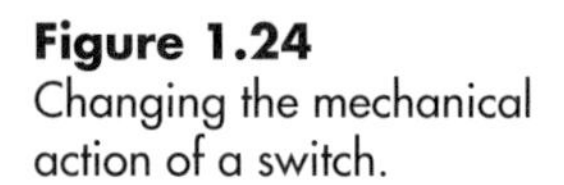

Figure 1.24
Changing the mechanical action of a switch.

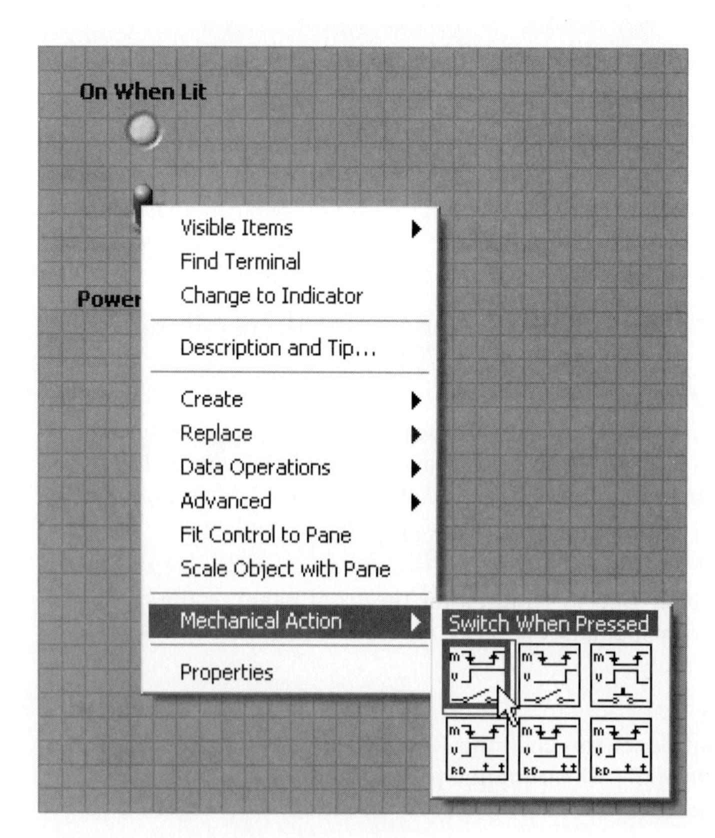

PRACTICE: USING A MOMENTARY "ON" SWITCH

Replace the toggle switch with a pushbutton and set the mechanical action to "Switch Until Released". Run the VI continuously to observe how the pushbutton now lights the LED only when it is being pressed.

- Pushbutton Control: **Controls Palette / Express Group / Buttons Group / Pushbutton**
- Change Mechanical Action: Right-click on pushbutton (see Figure 1.21), then select **Mechanical Action / Switch Until Released**

Solution: Once the mechanical action has been changed to "Switch Until Released", the LED will light only while the pushbutton is pressed (as illustrated in Figure 1.25).

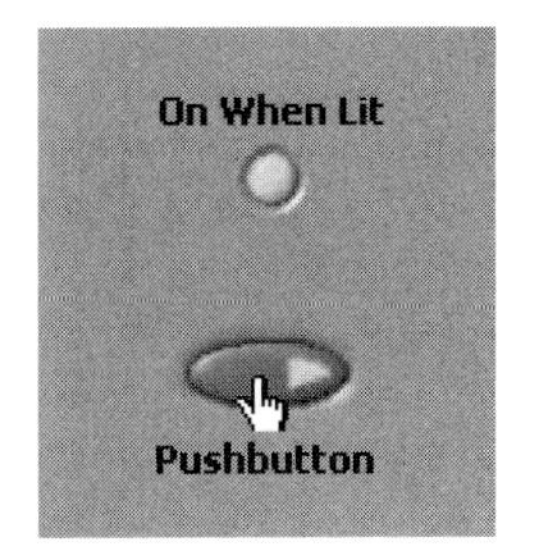

Figure 1.25
Using a pushbutton with the action set to "Switch Until Released".

1.7 LABVIEW MENUS

We conclude this chapter with a brief look at the LabVIEW menus to begin to develop a sense of how to accomplish particular tasks and to see what LabVIEW can do. The following is not a complete list of LabVIEW menu options, but a selected list of useful features.

1.7.1 File Menu

The File menu contains options that will look very familiar to most people, plus a few LabVIEW-specific options.

- **New VI** Opens a new, blank VI.
- **Open . . .** Opens the Select a File to Open dialog to allow you to find an existing VI for editing or running.
- **Close** Closes the VI being edited. If the VI has not been saved, you will be asked if you want to save the file before closing (see Figure 1.26).

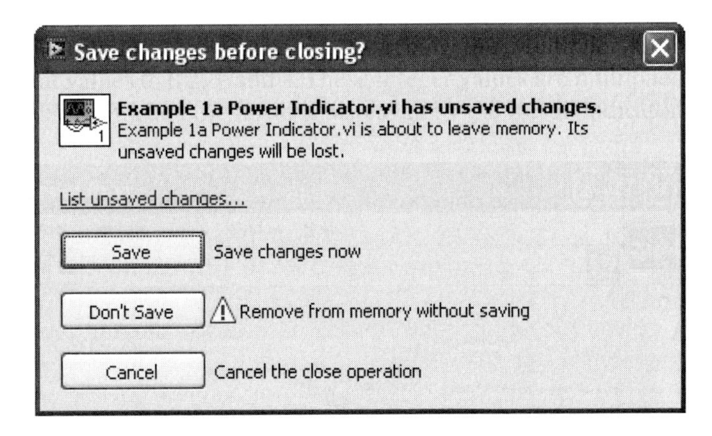

Figure 1.26
LabVIEW will prompt to save a VI before closing the file.

- **Close All** Closes all open VIs. If any VIs have been changed, LabVIEW will prompt you to save the file(s) before closing.
- **Save** Saves the current VI. The first time a VI is saved, you will be asked to give it a file name.

- **Save As . . .** Opens the Save As dialog that gives several save options.
- **Save All** Saves all open VIs.
- **Save for Previous Version . . .** Opens the Save for Previous Version dialog that allows you to select the desired version of LabVIEW (e.g., 2009, 8.6, 8.5, . . .).
- **Revert . . .** Discards all changes made since the VI was last saved.
- **New Project** Creates a new LabVIEW project (collection of related VIs and associated files).
- **Print . . .** Opens the Print dialog.
- **VI Properties** Opens the VI Properties dialog (see Figure 1.27) to allow you to observe and set a variety of property values.

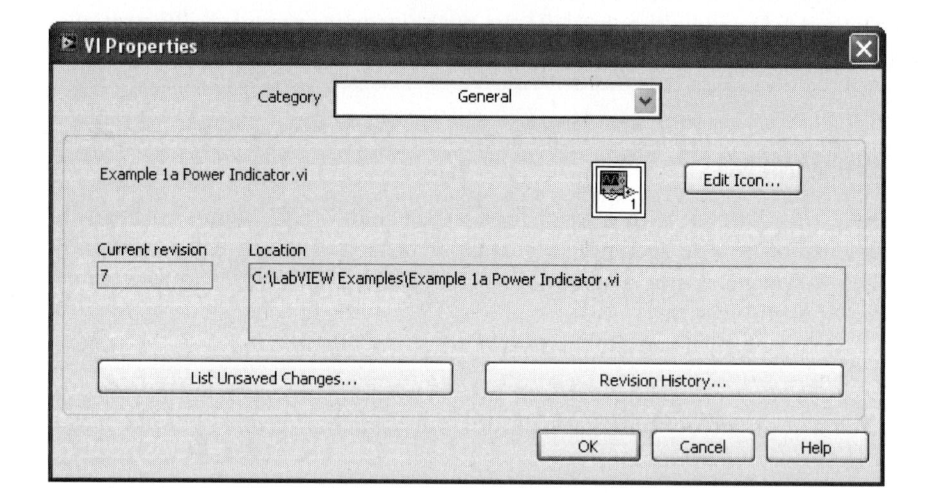

Figure 1.27
VI Properties dialog, General category.

- **Recent Files** Opens a menu listing the most recently edited VIs.
- **Exit** Shuts down LabVIEW. If there are unsaved VIs, you will be prompted to save them before exiting.

1.7.2 Edit Menu

The Edit menu collects menu options related to making various changes to the VI.

- **Undo** Reverses the last editing action. You can use the option repeatedly to back out of a series of edit steps.
- **Redo** Reapplies an edit step that was undone.
- **Cut** Copies an object to the Windows clipboard and removes the object from the VI.
- **Copy** Copies an object to the Windows clipboard and leaves the object in the VI.
- **Paste** Pastes an object previously copied to the Windows clipboard into the VI.
- **Select All** Selects all objects. Useful for moving all existing objects to make space for something new.
- **Make Current Values Default** Any values that are entered into controls in an edit session will be lost when the VI is saved and reloaded unless the values are made the default values for the control. This menu option is used to make all values currently in controls the default values for the controls.
- **Reinitialize Values to Defaults** Resets values in controls back to the default values.
- **Remove Broken Wires** Removes all broken wires on the block diagram. Ctrl-B is the keyboard shortcut.

- **Create VI Snippet from Selection** A *snippet* is a piece of program code that you want to reuse. This menu option allows you to select a section of a VI and save it for reuse in the future.
- **Create SubVI** A *SubVI* is a VI that will be used within another VI. It is easy to create a SubVI by selecting a group of program elements and using this menu option.

1.7.3 View Menu

The View menu provides access to various palettes and windows.

- **Controls Palette** Toggles the display of the Controls Palette. Only active from the front panel.
- **Functions Palette** Toggles the display of the Functions Palette. Only active from the block diagram.
- **Tools Palette** Toggles the display of the Tools Palette.
- **Quick Drop** Opens the Quick Drop dialog, which is used to quickly locate functions and controls by name.
- **Breakpoint Manager** Opens the Breakpoint Manager dialog to allow you to enable, disable, and delete breakpoints.
- **Probe Watch Window** Opens the Probe Watch Window, which can be used to monitor the values of all probes on a block diagram.
- **Error List** Opens the Error List window, which lists all errors that are preventing the VI from running. The Error List is also displayed when you click the **Broken Run** button.
- **Getting Started Window** Opens the Getting Started Window.

1.7.4 Project Menu

The Project menu is used to manage LabVIEW projects. A LabVIEW project is a collection of related VIs and associated files. Keeping related files collected makes creating a complex program easier, and simplifies creating a run-time version of the program.

- **New Project** Creates a new (empty) LabVIEW project.
- **Open Project . . .** Opens an existing LabVIEW project selected using a dialog.
- **Save Project** Saves the current LabVIEW project.
- **Close Project** Closes the current LabVIEW project.
- **Add to Project >** Allows you to add a file to an open project.

1.7.5 Operate Menu

The Operate menu allows you to control how a VI runs.

- **Run** Starts the VI. Equivalent to clicking the **Run** button on the front panel or block diagram window.
- **Stop** Stops a running VI. Equivalent to clicking the **Abort Execution** button on the front panel or block diagram window.
- **Step Into** Starts the VI, but runs only the first step of the VI. Use **Step Over** to continue step by step and **Step Out** to finish program execution.

1.7.6 Tools Menu

The Tools menu offers some fairly advanced options, only a few are mentioned here.

- **Measurement & Automation Explorer** The Measurement & Automation Explorer is used to keep track of the various data acquisition devices on a computer, and to create data acquisition tasks.

- **Build Application (EXE) from VI . . .** Used to create a standalone application that can run outside of LabVIEW.
- **Options . . .** Provides access to the default options used by LabVIEW.

1.7.7 Window Menu

The Window menu is used to control the way the LabVIEW windows are displayed. The bottom section of the Window menu shows a menu of all currently open LabVIEW windows. This allows you to access any open LabVIEW window from any other.

- **Show Block Diagram** Available from the Front Panel, the Show Block Diagram option displays the block diagram associated with the current front panel.
- **Show Front Panel** Available from the Block Diagram, the Show Front Panel option displays the front panel associated with the current block diagram.
- **Show Project** Displays the project window associated with the current VI (if any).
- **Tile Left and Right** Fills the screen with the front panel and block diagram side by side.
- **Tile Up and Down** Fills the screen with the front panel and block diagram one over the other.
- **Full Size** Expands the current window to the full screen size.

1.7.8 Help Menu

The Help menu provides access to the LabVIEW Help system.

- **Show Context Help** Opens a small Context Help window (see Figure 1.28), which provides a brief description for any object that the mouse hovers over. Most descriptions include a link to more detailed Help information.
- **Search the LabVIEW Help . . .** Opens the full LabVIEW Help system.

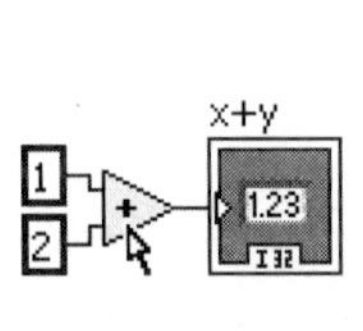

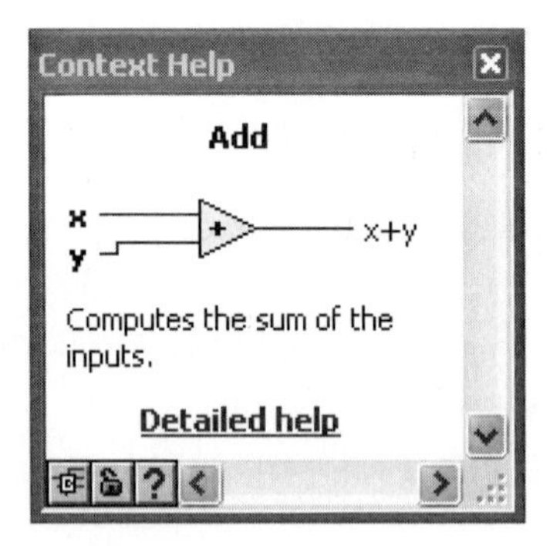

Figure 1.28
Context Help shows brief descriptions and a link to additional information.

KEY TERMS

Automatic Tool Selection
Automatic Tool Selection mode
Block Diagram
Connector pane
Control
Controls Palette
Express VIs
Front Panel
Function
Functions Palette
LabVIEW
LabVIEW program
Latch
Mechanical Action
Node
Project
Setpoint
Snippet
SubVIs
Tools Palette
Virtual instrument (VI)

SUMMARY

LabVIEW—software for data acquisition and analysis
VI—a LabVIEW computer program
Front Panel—the graphical user interface—holds controls
Block Diagram—the graphical program—holds functions

Controls Palette
The Controls Palette contains the switches, knobs, dials, and indicators that are used to set and display values on the front panel.

Functions Palette
The Functions Palette contains:

- **Functions** (pale yellow background)
- **VIs** (pale yellow strip across the top of the icon, or yellow border when expanded and ".vi" in the name)
- **Express VIs** (a blue strip across the top of the icon, or blue border when expanded)

SELF-ASSESSMENT

1. What is LabVIEW designed to accomplish?
   ANS: LabVIEW may originally have been designed to make data acquisition with National instruments hardware easier, but LabVIEW is a full-fledged programming language and math application in its own right.
2. The user (as opposed to the programmer) of a virtual instrument primarily uses which part of the virtual instrument?
   a. Front Panel
   b. Block Diagram
   ANS: The front panel is the graphical interface used by the user.
3. If you wanted to add a virtual on–off switch to a virtual instrument, the switches would be found on which palette?
   a. Functions Palette
   b. Controls Palette
   ANS: Switches are controls that are found on the Controls Palette and placed on the front panel.
4. If you wanted to add a virtual on–off switch to a virtual instrument, where would the control be placed?
   a. Front Panel
   b. Block Diagram
   ANS: Switches are controls that are found on the Controls Palette and placed on the front panel.
5. If you had developed a LabVIEW program that created a data set, and you wanted to create a graph, where would the graph be placed?
   a. Front Panel
   b. Block Diagram
   ANS: On the front panel.

CHAPTER 2

LabVIEW Basics

Objectives

After reading this chapter, you will know:

- two ways to open a VI in LabVIEW for editing
- how to use LabVIEW functions to perform basic math operations
- how to use a While Loop to keep a VI running continuously
- what dataflow programming is, and how it works
- several of the data types used for LabVIEW variables, and how to use them
- how to document VIs to make them easier to understand
- how to print and save your VIs

2.1 OPENING A VI

There are two ways to open a VI for editing in LabVIEW:

- Double-click on a VI file in a file browser such as Windows Explorer or My Computer to start LabVIEW and open the VI front panel. The VI block diagram does not open by default, but can be opened from the front panel using menu options: **Window / Show Block Diagram**.
- Start LabVIEW (**Start / All Programs > / National Instruments LabVIEW**) and open the VI file for editing from the LabVIEW environment.

When you start LabVIEW, the Getting Started window is opened (Figure 2.1). The **Open panel** provides a list of recently edited VIs and a **Browse . . .** option to find previously created VIs on your computer. To open one of the VIs in the recent files list, simply click on the file name in the list.

Once a VI is open for editing, the Getting Started window is hidden. You can open additional VIs from an open VI (either front panel or block diagram) using the menu options **File / Open**. You can see a list of recently edited files using the menu options **File / Recent Files**.

When you open an existing VI, LabVIEW opens the front panel only. You can open the block diagram from the front panel using the menu options **Window / Show Block Diagram**.

2.2 BASIC MATH IN LABVIEW—USING FUNCTIONS

Most of this text focuses on using LabVIEW for engineering data analysis, so we need to know how to do math in LabVIEW. Math is very straightforward in LabVIEW, but it is all based on *functions*—there are no math operators in LabVIEW.

A LabVIEW function is a piece of program code that works as a unit. There are many math functions in LabVIEW, including

- Basic Math Functions
  - Add, Subtract, Multiply, Divide
  - Increment, Decrement

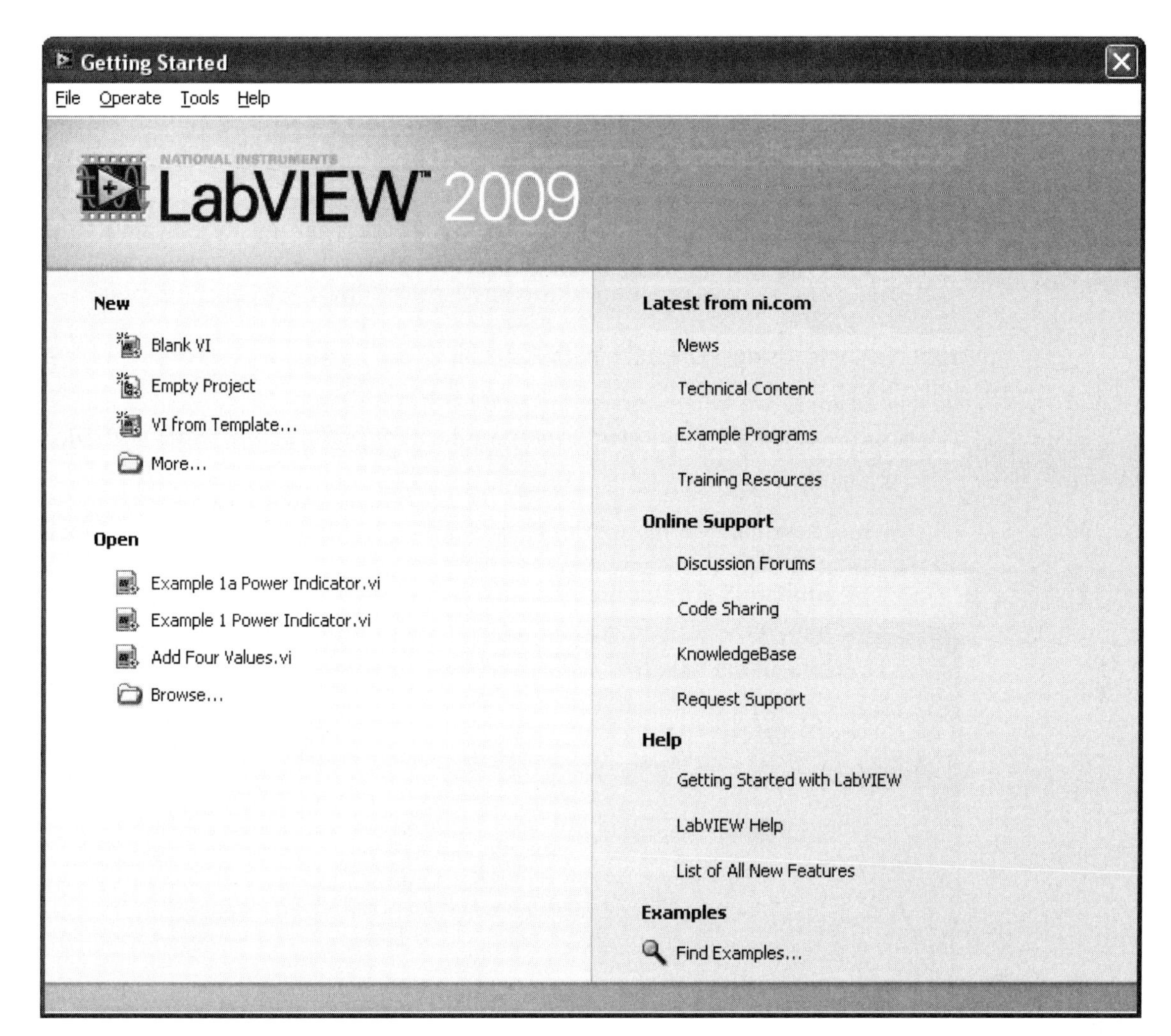

Figure 2.1
LabVIEW's Getting Started window.

  - Absolute Value
  - Square, Square Root
  - Reciprocal
- Trig and Hyperbolic Trig Functions
- Log and Exponential Functions
- Matrix Functions
- Optimization Functions
- Differential Equations Functions

Functions typically accept one or more values as inputs and return a result.

Because LabVIEW is a graphical programming language, the functions are placed on the block diagram as nodes. For example, the node for the **Add** function is illustrated in Figure 2.2. The Add function requires two inputs (the values to be added) and has one output (the sum).

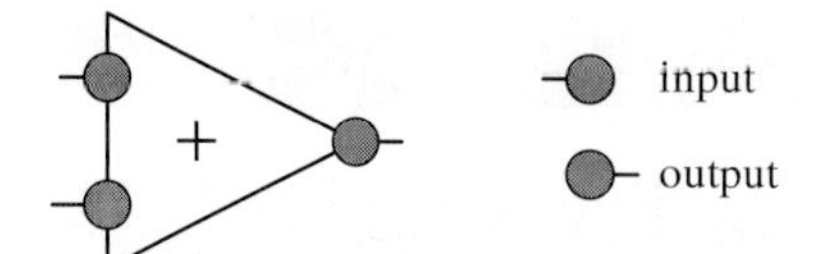

Figure 2.2
The Add function node showing terminals.

The nodes for math functions have terminals for inputs (required values) and outputs (results). The function nodes are wired to control nodes and indicator nodes to create the LabVIEW program. This is easiest to see by means of an example.

2.2.1 Example: Using a LabVIEW Math Function

In this example we will subtract one number from another. The process can be described as subtracting the *subtrahend* from the *minuend* to compute the *difference*.

Minuend − Subtrahend = Difference

While these math terms are rarely used nowadays, they will work fine as labels on the front panel as illustrated in Figure 2.3.

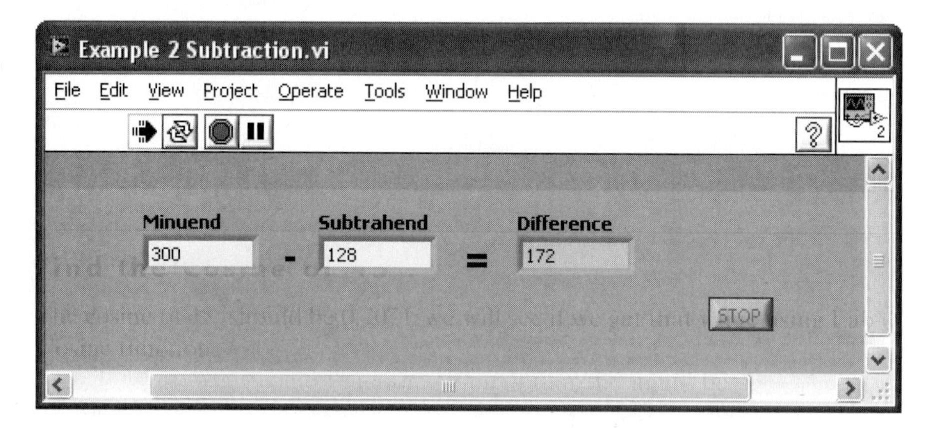

Figure 2.3
Subtraction VI developed as Example 2.

We need to understand the difference between controls and indicators in LabVIEW:

- A *control* is used to set a value.
- An *indicator* is used to display a value.

Note: The term *control* is actually used two ways in LabVIEW.

- Control is used in a generic sense to mean any object that can be placed on the front panel. The Controls Palette uses the term in this way.
- Control is also used in a more specific sense to indicate an object on the front panel that is used to set a value. This is contrasted with the term indicator, which is an object used to display a value.

We will use controls to set input values, and indicators to display results. In Figure 2.3 the **Minuend** and **Subtrahend** boxes are controls and, when the program is running, values can be typed into the boxes (technically, they're called *input fields*). The **Difference** box is an indicator that is being used to display the calculated result. Notice in Figure 2.3 that the background of the indicator is shown in gray. This is a visual clue to the user that they should not be typing a value into that box. (LabVIEW will not allow a value to be typed into an indicator.)

The steps required to create the Subtraction VI include the following:

1. Create a blank VI.

On the front panel . . .

2. Add two numeric controls (input fields for Minuend and Subtrahend), and set their properties.
3. Add one numeric indicator (Difference box), and set its properties.
4. Add two labels to show the subtraction operator and equal sign.

On the block diagram . . .

5. Add a Subtract function.
6. Wire the Subtract function to the controls and indicator:
   a. The Minuend and Subtrahend output terminals to the Subtract Function inputs.
   b. The Subtract Function output terminal to the input terminal on the Difference indicator.
7. Save the VI.
8. Run and test the VI.

Step 1. Create a blank VI
To create a blank VI, use one of these two approaches:

- From the LabVIEW Getting Started window, click on the **Blank VI** link shown in Figure 2.1. This causes a new, blank VI front panel and block diagram to be displayed.
- From another VI, use the menu options **File / New VI**.

Either way, a new front panel and block diagram will be displayed.

Step 2. Add two numeric controls to the front panel, and set their properties
Find the Numeric Control icon (labeled "Num Ctrl" in Figure 2.4) using the following commands:

Controls Palette / Express Group / Numeric Controls Group / Num Ctrl

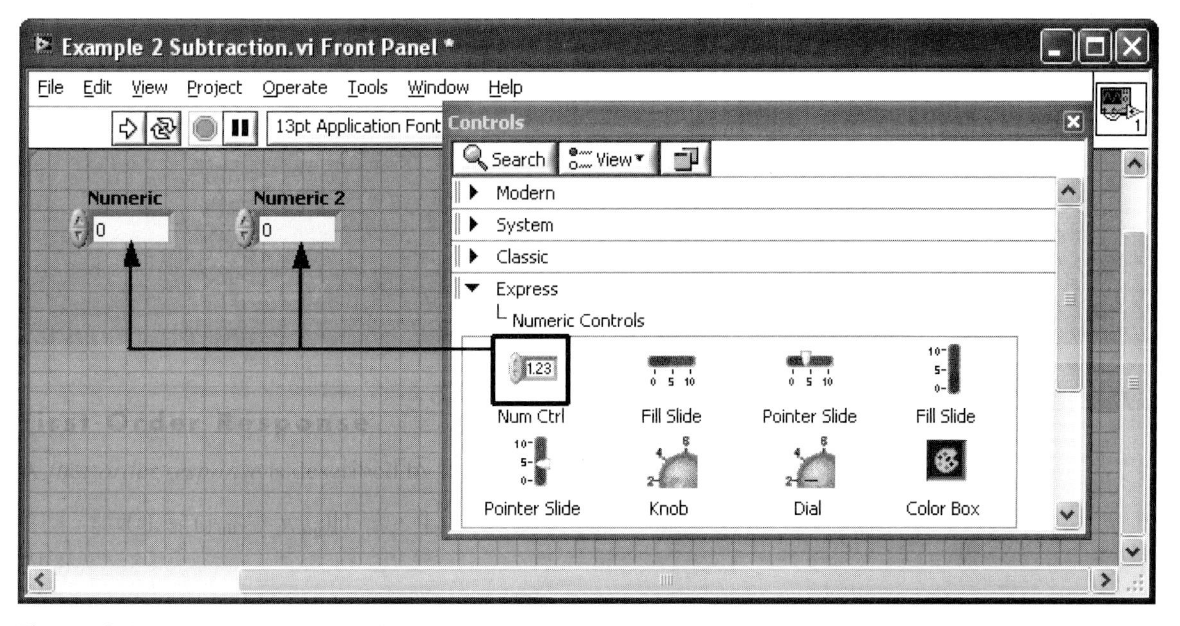

Figure 2.4
Adding two numeric controls to the front panel.

Drag the icon to the front panel to create the control labeled "Numeric" as illustrated in Figure 2.4. Repeat the process to create the control labeled "Numeric 2".

Note: When multiple instances of the same control are placed on the front panel, LabVIEW automatically adds numbers to the control labels so that each control is uniquely identified.

The numeric controls on the front panel are now fully functional, but we will change a few of the controls' properties to better suit our needs. Specifically, we will

- Rename the controls to call them **Minuend** (instead of "Numeric") and **Subtrahend** (instead of "Numeric 2").
- Hide the increment/decrement controls at the left side of each input field.

Renaming the controls

The control's name is called the *label*. To change the text displayed in a control's label, you simply double-click on the label to select the label and enter text edit mode. The label text will be highlighted (white letters on black background). Once the text is highlighted, simply type in the desired name for the control. The result of renaming the numeric controls is shown in Figure 2.5.

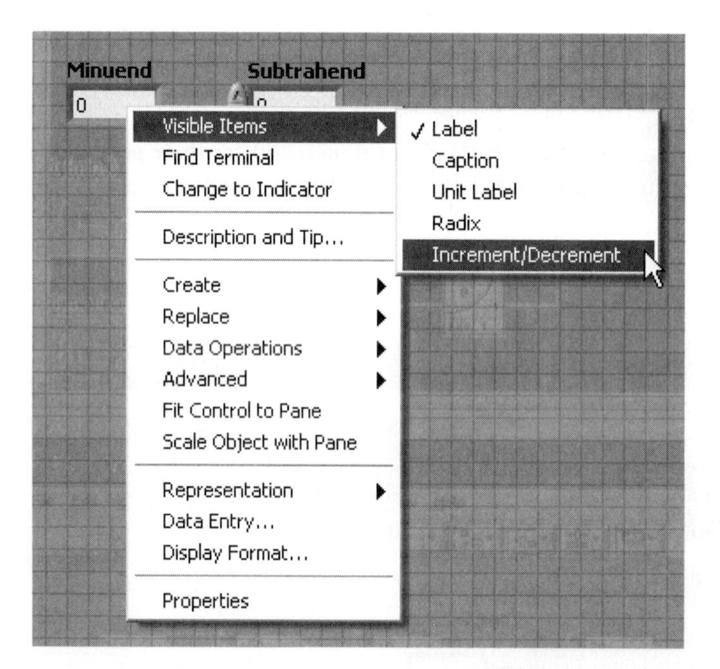

Figure 2.5 Hiding the Increment / Decrement controls on a Numeric Control.

Hiding the increment/decrement controls

The increment/decrement controls allow the value displayed in the numeric control's input field to be increased or decreased using the mouse. If you hide the increment/decrement controls, the user must type a value into the input field.

To hide the increment/decrement controls, right-click on the Minuend numeric control, then select **Visible Items** from the pop-up menu. Click on **Increment / Decrement** to clear the check mark indicating that the increment/ decrement controls are visible. Items that are unchecked are hidden, so clearing the check mark in front of the **Increment / Decrement** menu option causes the increment/decrement controls to be hidden (as shown in Figure 2.5).

Repeat the process to hide the increment/decrement controls on the Subtrahend control.

- Right-click on the Subtrahend numeric control (opens the pop-up menu).
- Select **Visible Items** (opens the submenu).
- Click on **Increment / Decrement** (clears the check mark).
- Click on the gray grid on the front panel to close the menus.

Step 3. Add a numeric indicator to the front panel, and set its properties
Find the Numeric Control icon (labeled "Num Ind" in Figure 2.6) using the following commands:

Controls Palette / Express Group / Numeric Controls Group / Num Ctrl

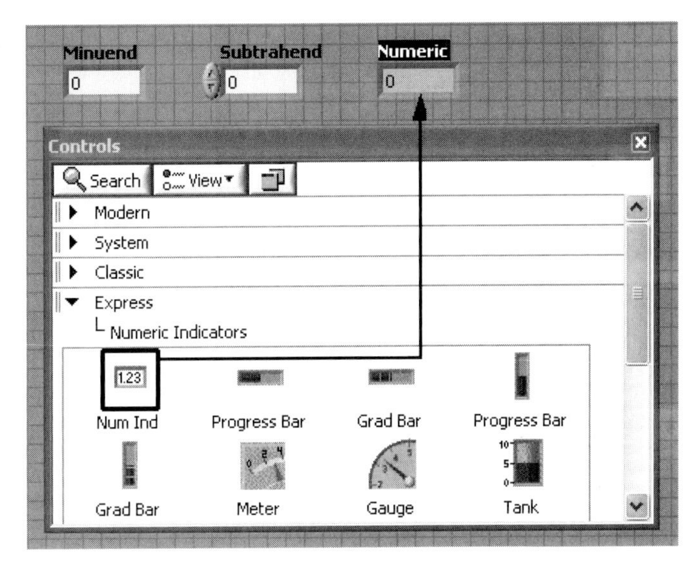

Figure 2.6
Add a numeric indicator to the front panel.

Drag the icon to the front panel to create the control labeled "Numeric" as illustrated in Figure 2.6.

The control's label is (by default) selected (white letters on black background) right after it is placed on the front panel.

- If the label is selected, just type in the control's new name, "Difference".
- If the label is not selected, double-click on the label to select it, and then type in the new name.

The front panel with all of the required controls is shown in Figure 2.7.

Step 4. Add two labels to show the subtraction operator and equal sign
To make it clear what the VI is designed to do, we want to add a subtraction operator and an equal sign as shown in Figure 2.8.

Adding labels to the front panel (or the block diagram) is easy in LabVIEW; just double-click where you want the label to be placed. LabVIEW will place an empty label at that location and leave the label in edit mode so that you can type in the text that the label should display. In Figure 2.8 the text size in the labels containing the subtraction operator and

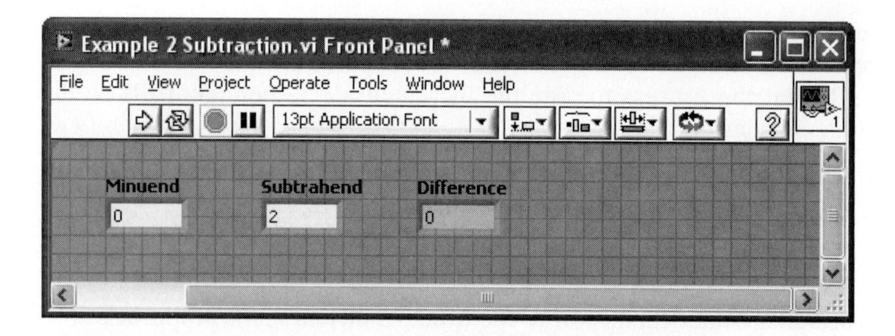

Figure 2.7
The front panel with all required controls.

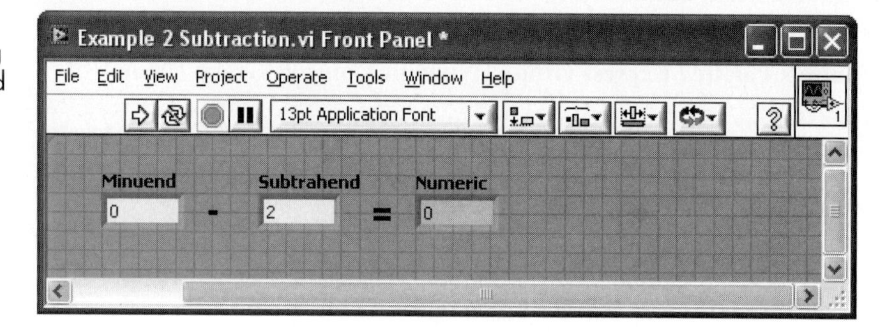

Figure 2.8
The front panel after adding the subtraction operator and equal sign labels.

the equal sign has been increased. To change the displayed text size, select the label and then press

- [Ctrl –] to decrease the text size
- [Ctrl =] to increase the text size

Note: The notation [Ctrl –] means that you must hold down the [Ctrl] key while you press the [–] key.

Alternatively, you can change the text size using the **Text Settings** button (labeled "13pt Application Font" in Figure 2.8).

The front panel is now complete, and the VI will run—but it won't actually calculate the difference between the values in the Minuend and Subtrahend fields until we finish the VI by placing a Subtract function on the block diagram and wiring the controls to the function, and the function to the indicator.

Step 5. Place a Subtract function on the block diagram

Before we make any changes, let's take a look at what LabVIEW was doing to the block diagram as we were adding controls and indicators to the front panel. The block diagram created by LabVIEW is shown in Figure 2.9.

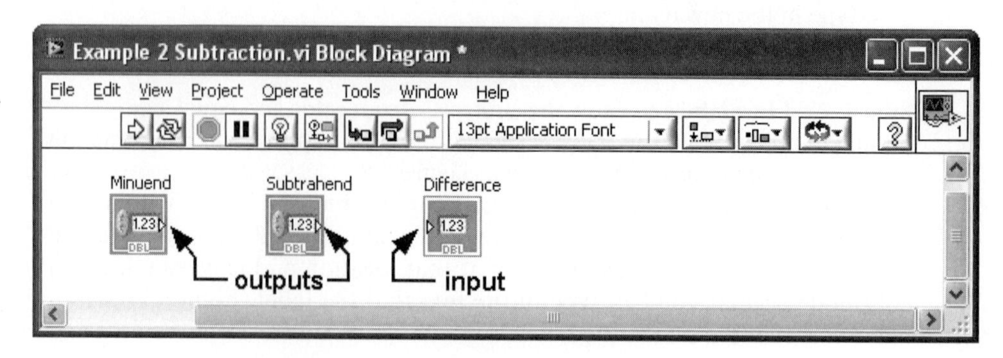

Figure 2.9
The block diagram after completing the front panel.

Notice that the labels that were defined on the front panel controls and indicator are shown on the block diagram too. This makes it easy to identify each control and indicator. Also, notice that each control has an output terminal and the indicator has an input terminal. We will use those terminals later when we wire up the VI.

First, we are going to rearrange the existing controls to stack the Minuend and Subtrahend numeric controls, and place the Difference indicator to the right and in between—leaving some space for the Subtract function, which will be placed on the block diagram later. The blocks (called *nodes*) representing the controls can be moved by dragging the icons with the mouse. After rearranging the existing controls, the updated block diagram is shown in Figure 2.10.

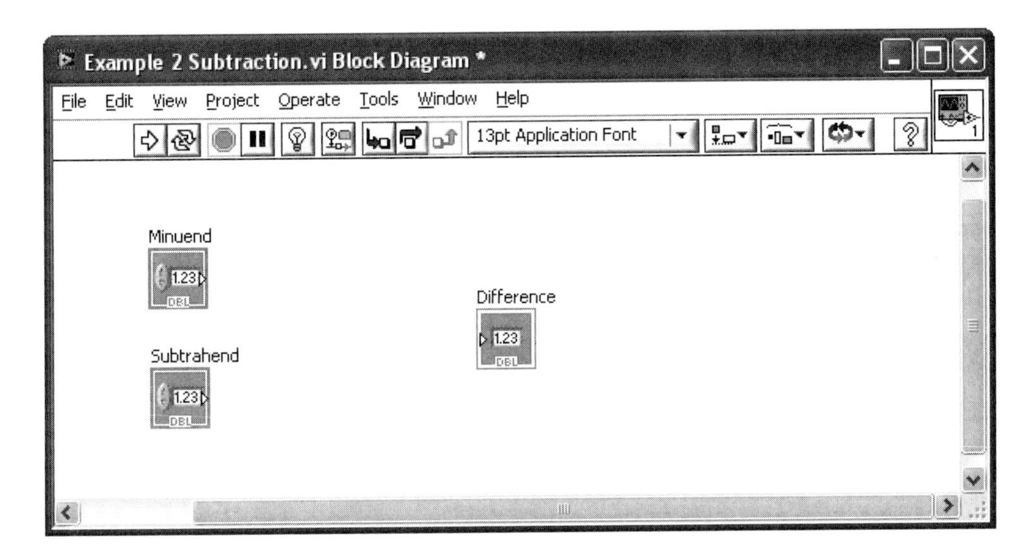

Figure 2.10
The block diagram after rearranging the controls.

Note: Rearranging the blocks on the block diagram does not move the controls on the front panel.

Next, add a Subtract function to the block diagram from the Functions Palette using the following commands:

Functions Palette / Mathematics Group / Numeric Group / Subtract Function

This process is illustrated in Figure 2.11.

Step 6. Wire the Subtract function to the controls and indicator

The next step is to wire the block diagram as follows:

- Connect the Minuend and Subtrahend outputs to the Subtract function inputs.
- Connect the Subtract function output to the Difference indicator input.

If you position the mouse over the Subtract function icon, the terminals are displayed, as shown in Figure 2.12.

Since order matters in subtraction, we have to wire the correct input value to each input terminal on the Subtract node. LabVIEW provides *context sensitive help*, which is very useful for figuring out how the various functions need to be wired.

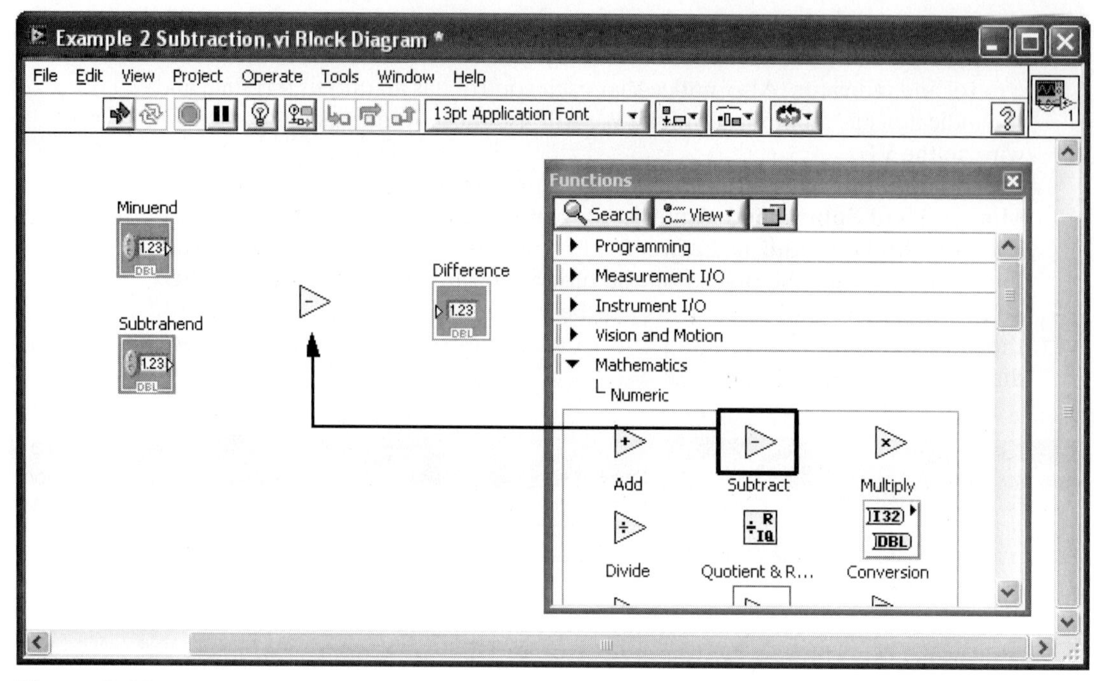

Figure 2.11
Adding a Subtract function to the block diagram.

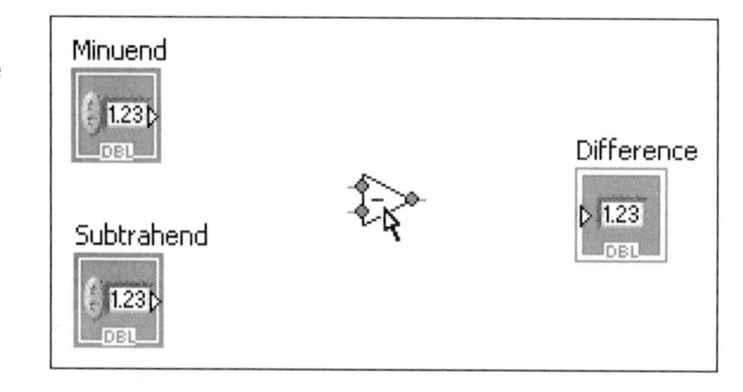

Figure 2.12
When the mouse is over the Subtract function, the terminals are displayed.

To activate context sensitive help, use the following menu options:

Help / Show Context Help (or, press [Ctrl H])

Once the Context Help window is displayed, click on the function of interest to learn about the inputs and outputs for that function. Figure 2.13 shows the context help for the Subtract function.

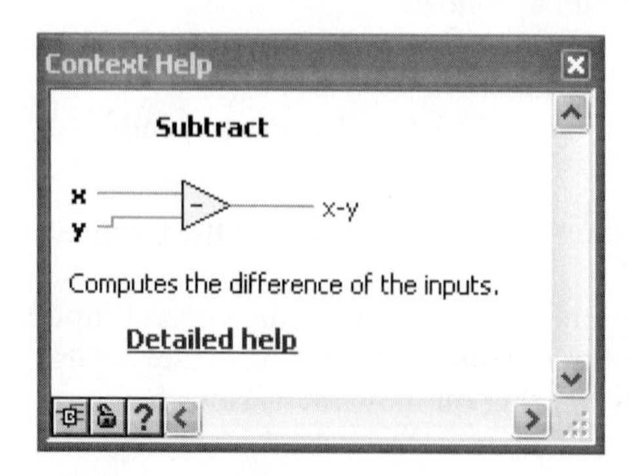

Figure 2.13
Context Help for the Subtract function.

From the context help, we see that the top input terminal on the left is called "x" and the bottom input terminal on the left is called "y". The output is labeled "x–y", which tells us how we need to wire the terminals.

> *Note:* When order is important, we have to be careful to write the variables in the correct order from left to right. For example, for subtraction, writing $x–y$ implies that the y value is to be subtracted from the x value. LabVIEW uses the same variable order with input terminals on math functions, but they are arranged from top to bottom. In LabVIEW, the bottom input value will always be subtracted from the top input value.

We want Minuend − Subtrahend = Difference, so

- The **Minuend** output needs to be wired to the "x" input on the Subtract function.
- The **Subtrahend** output needs to be wired to the "y" input on the Subtract function.
- The Subtract function output needs to be wired to the input on the **Difference** indicator.

The wired block diagram is shown in Figure 2.14.

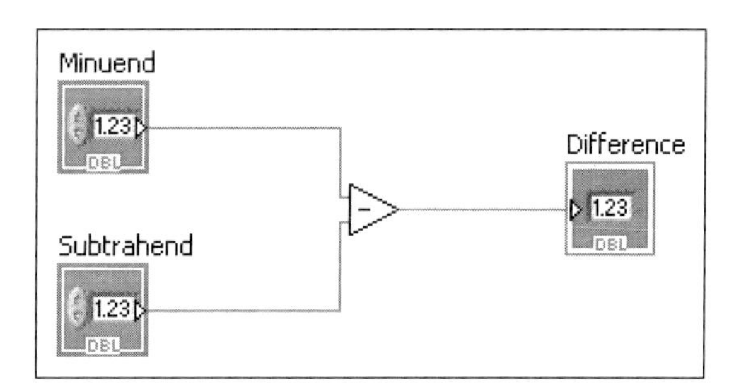

Figure 2.14
The wired block diagram.

The VI will now run (once), but we need it to keep running so that we can enter values into the Minuend and Subtrahend fields. To keep the VI running, use the **Run Continuously** button on the front panel. To stop the VI, use the **Abort Execution** button. These buttons are indicated in Figure 2.15.

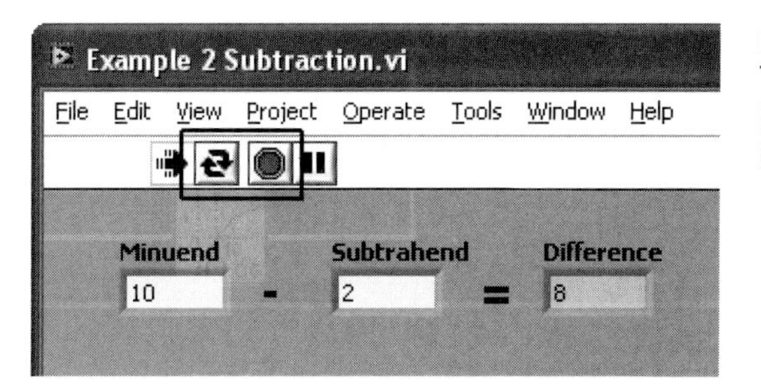

Figure 2.15
The Subtraction VI with the Run Continuously and Abort Execution buttons indicated.

Step 7. Save the VI
Be sure to save the VI occasionally to protect your work. To save the VI, use the following menu commands from either the front panel or the block diagram:

File / Save

The first time you save the VI you will have the opportunity to assign a name and select a folder location.

Step 8. Run and test the VI
The VI is ready to run. Return to the front panel and click the **Run Continuously** button (see Figure 2.15).

While the VI is running, the grid on the front panel is not displayed, the **Run** button (now inactive) changes to an arrow with a vapor trail, and the **Abort Execution** button is active as shown in Figure 2.15.

While the VI is running, you can change the values in the Minuend and Subtrahend fields, and the difference will be displayed in the Difference field (see Figure 2.16). You can continue to change input values as long as the VI is running, and LabVIEW will continue to calculate and display the difference.

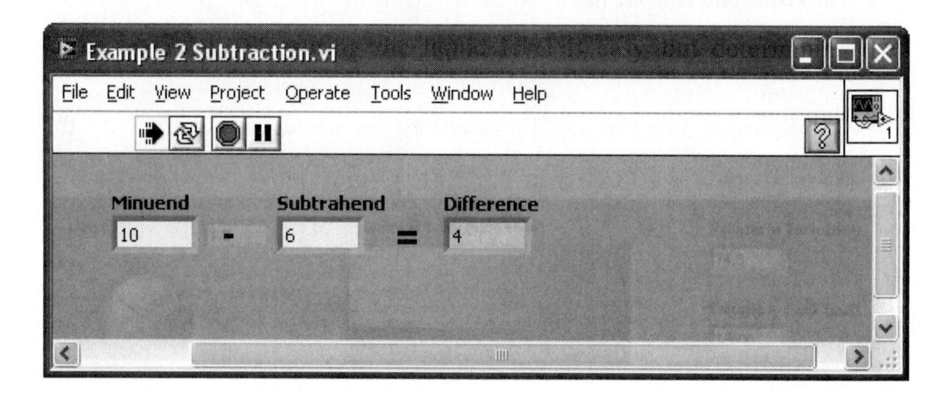

Figure 2.16
Change input values to solve multiple problems.

PRACTICE

Create the Division VI shown in Figure 2.17. What does LabVIEW do if you try to divide by zero?

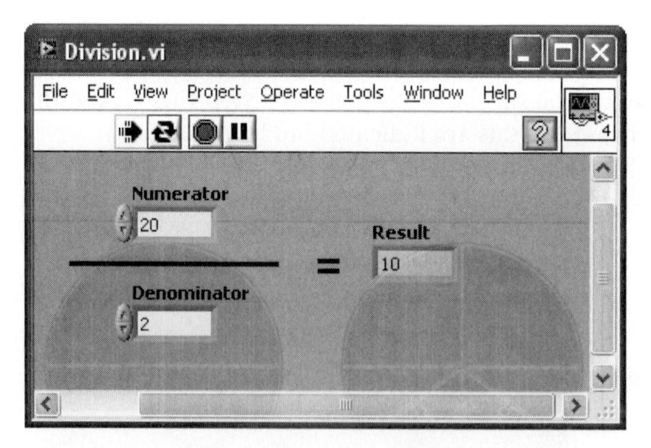

Figure 2.17
Division VI.

The required front panel items can be found in the following locations:

- Numeric controls (for numerator and denominator): **Controls Palette / Express Group / Numeric Controls Group / Num Ctrl**
- Numeric Indicator (for result): **Controls Palette / Express Group / Numeric Indicators Group / Num Ind**
- Thick Line: **Controls Palette / Modern Group / Decorations Group / Thick Line**

The block diagram is wired as shown in Figure 2.18.

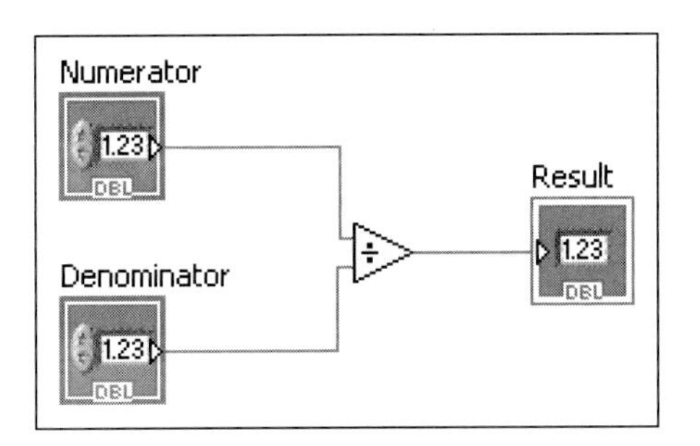

Figure 2.18
Block diagram for Division VI.

If you divide by zero, LabVIEW displays "Inf" (infinity) in the Result indicator.

2.3 PROGRAMMING PREVIEW: WHILE LOOPS

Most of the programming features of LabVIEW will be covered much later in the book, but occasionally some useful programming features will be previewed. Here, we look at an alternative to using the **Run Continuously** button: building a *While Loop* into our LabVIEW program.

A While Loop is a programming structure that causes some program elements to be evaluated repeatedly until some condition is satisfied. Since LabVIEW is a graphical programming language, a While Loop looks like a container (see Figure 2.19) and the program elements within the container are evaluated each time the While Loop cycles.

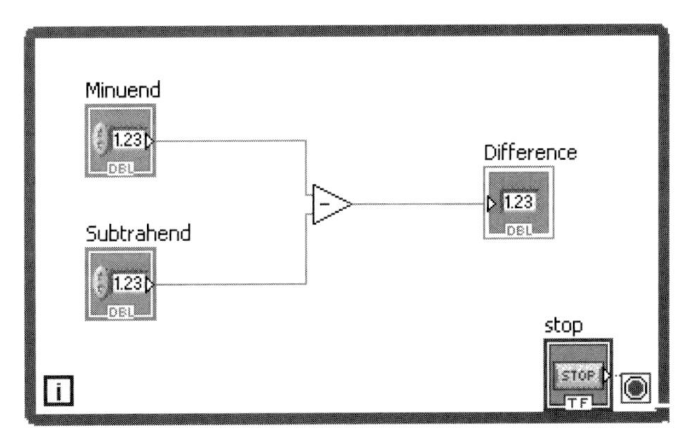

Figure 2.19
Adding a While Loop to the Subtraction VI.

To add a While Loop to the Subtraction VI, select a While Loop from the Functions Palette, and then draw a box around all of the nodes on the block diagram, as illustrated in Figure 2.19.

While Loops can be found in two locations on the Functions Palette:

Functions Palette / Programming Group / Structures Group / While Loop

Functions Palette / Express Group / Execution Control Group / While Loop

The two options are not identical. When the While Loop from the Express Group is used, a **STOP** button is automatically connected to the While

Loop Condition (bottom-right corner of the While Loop). When the While Loop from the Programming Group is used, you must wire a switch to the While Loop Condition manually. In Figure 2.19 the While Loop from the Express Group is used.

In this example, the While Loop causes the following actions to be performed each time the loop cycles:

- The value of the Minuend control is evaluated.
- The value of the Subtrahend control is evaluated.
- The Minuend and Subtrahend values are subtracted.
- The Difference indicator value is updated.
- The position of the **STOP** button is evaluated.
- The While Loop Condition is evaluated.

By default, as long as the While Loop Condition is set to False, the loop continues to cycle. When the **STOP** button is clicked, the While Loop Condition is set to True and the loop stops.

Note: The While Loop stop condition can be changed to "Continue if True". To change the stop condition, right-click on the While Loop Condition icon and select "Continue if True" from the pop-up menu.

While Loops are very commonly used to create VIs in which part or all of the VI loops until a **STOP** button is pressed. While it appears that the Run Continuously button does the same thing as a While Loop around the entire program, there are subtle differences. In the next section, we will look at an example that shows how the extra control available using a While Loop can be useful.

2.4 DATAFLOW PROGRAMMING

Sometimes you want something to happen when the While Loop ends. For example, if you use a loop to build values into an array, you might want to wait until the loop ends to compute the average of the values in the array. This can be accomplished by placing the calculations that are dependent on the loop results outside of the loop.

The word "dependent" is the key, because LabVIEW is a *dataflow programming language*. Dataflow programming means that a node (or block) on a block diagram executes as soon as all of the inputs have values. Placing a calculation outside of the While Loop does not guarantee that it will wait until the loop finishes before executing; the calculation will only wait until the While Loop has completed if the calculation is dependent on the While Loop results.

To demonstrate this, we will add a couple of calculations outside of the While Loop in the Subtract VI, as shown in Figure 2.20.

Note: The Extended Subtract VI shown in Figure 2.20 sends output values across the While Loop using two *tunnels* (the markers on the right While Loop boundaries). Tunnels can have *indexing* enabled or disabled—in this example indexing is disabled.

- Indexing Disabled—When the loop stops cycling, the value in the wire going through the tunnel is available for subsequent calculations. This is how the tunnels in Figure 2.20 are set up.

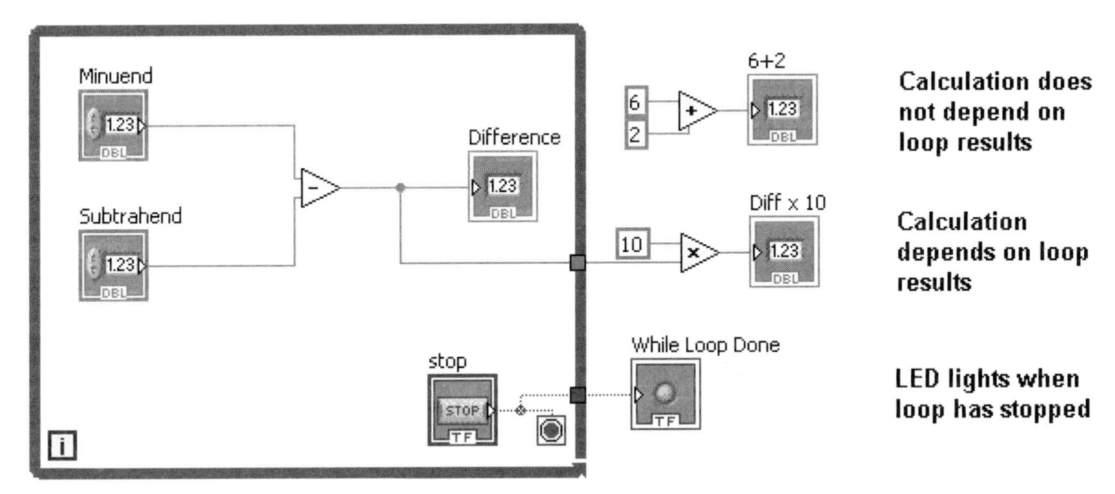

Figure 2.20
Extended Subtract VI, calculations outside While Loop.

- Indexing Enabled—The values in the wire going through the tunnel are built into an indexed array (one element is added to the array each time the loop cycles), and the entire array is available for subsequent calculations after the loop stops cycling. (Building arrays using tunnels with indexing enabled is covered in a later chapter.)

With dataflow programming, as soon as a node (a block on the block diagram) has values for all inputs, the programming code for that node executes.

- The Add function at the top-right corner of Figure 2.20 has all of the information it needs as soon as the VI is run, so the result of that calculation should appear virtually instantly after starting the VI.
- The Multiply function on the right side of Figure 2.20 depends on the results of the While Loop, so the Multiply function will not execute until the While Loop has stopped.
- The While Loop Done LED indicator gets its Boolean input from inside the While Loop, so that input will not be available until the While Loop stops. As soon as the **STOP** button is clicked, the While Loop will stop and a True is sent to the LED, causing it to light up. So the LED is an indicator that the While Loop has stopped.

When the Subtract VI is run, the "6 + 2" indicator immediately shows the calculated result (see Figure 2.21), because all inputs on the Add function were specified.

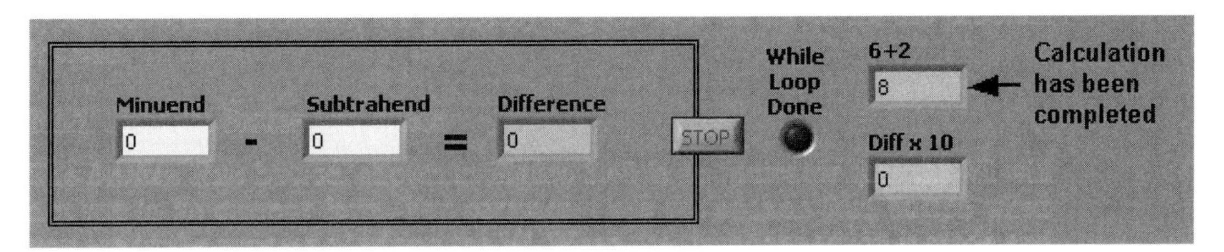

Figure 2.21
Front panel immediately after starting the Subtract VI.

The While Loop is running and the inputs to the While Loop Done LED and Multiply function have not been assigned values.

When the user changes the Minuend and/or Subtrahend values, the calculations inside the While Loop are performed (see Figure 2.22), but no values are sent out of the While Loop until the loop stops.

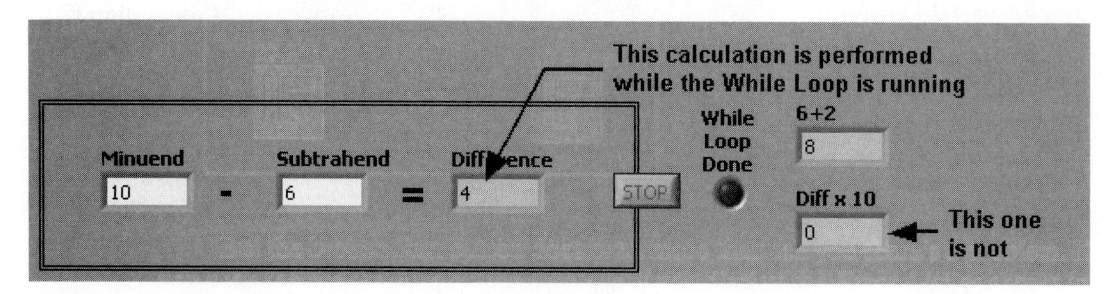

Figure 2.22
Front panel while the While Loop is running.

When the user clicks the **STOP** button, the While Loop stops, and the results from inside the While Loop flow to the Multiply function and the LED outside the loop. The "Diff x 10" indicator receives a value, and the "While Loop Done" LED lights up, as shown in Figure 2.23. Then, since there are no other nodes to evaluate, the VI terminates.

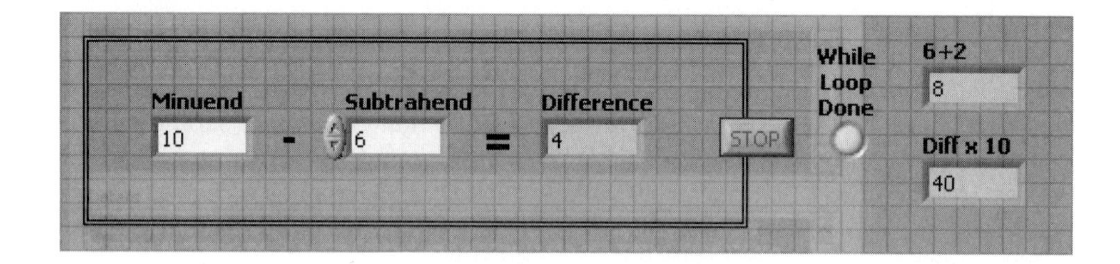

Figure 2.23
Front panel after stopping the While Loop.

This example has illustrated how LabVIEW decides when to perform various calculations (when all of the inputs are specified). Dataflow programming is very convenient, and most people are (possibly unknowingly) familiar with it from using spreadsheet programs, but it may still take a little practice to become comfortable with its use.

2.5 DATA TYPES AND CONVERSIONS

You cannot work with LabVIEW for very long without learning about *data types* because the functions and controls in LabVIEW require data in certain forms.

So far we have actually used two data types:

- **DBL**—*double-precision real values*
- **TF**—True/False or *Boolean values*

You can see how LabVIEW indicates data type by looking at the block diagram for the Subtraction VI, shown in Figure 2.24.

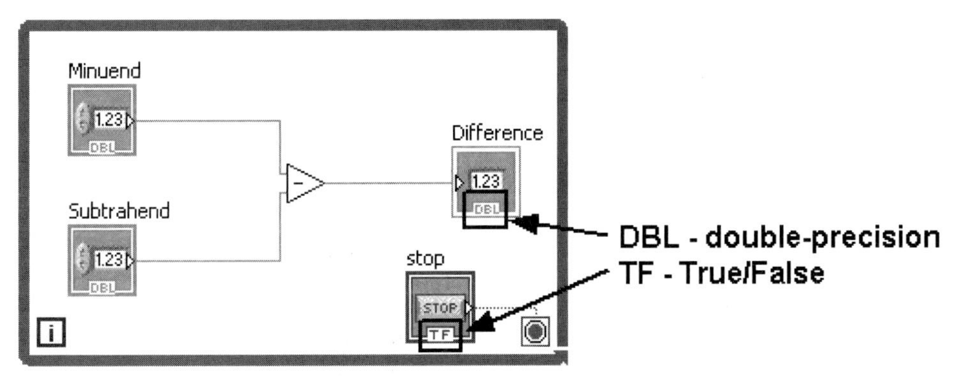

Figure 2.24
The required data type is indicated on the control blocks.

What you cannot tell in the figures in this text is that LabVIEW uses color coding to indicate data type. The most commonly used data types in LabVIEW are listed in Table 2.1.

Table 2.1 Common LabVIEW data types

Symbol	Data type	Color	Range	Default value	Comment
DBL	Double-precision floating point numeric	Orange	4.94e–324 to 1.79e+308	0.0	Default data type for floating point numeric values
I32	32-bit signed integer numeric	Blue	–2,147,483,648 to 2,147,483,647	0	Default data type for integer numeric values
TF	Boolean	Green	True or False	False	
[DBL]	Matrix of double-precision numbers				Brackets indicate array or matrix. Color indicates data type of matrix elements. Wires carrying matrices are displayed with thick lines.
abc	String	Magenta		Empty string	
	Path	Gray		Empty path	Holds a file path
	128-bit (64.64) Time stamp	Brown	01/01/1600.00:00:00 to 01/01/3001.00:00:00	12:00:00.000 AM 1/1/1904	Format is Date.Time.
	Cluster	Pink or Brown (see comment)			Clusters are collections of (potentially) multiple data types. Clusters are shown in pink if all elements are of the same data type, and brown if multiple data types are clustered.
	Waveform	Brown			Holds the start time, time step, and data of a waveform. Wires carrying waveforms are displayed with thick lines.

Table 2.1 shows only the most commonly used data types in LabVIEW. For example, there are single- (SGL), double- (DBL), and extended-precision (EXT) data types for floating point numbers, plus several data types for complex numbers.

There are also 8-, 16-, 32-, and 64-bit signed integer types, plus four more unsigned (positive only) integer data types.

If you try to wire together two different data types (not allowed), you will see two wires from two different colored blocks coming together with an "X" in the middle, as illustrated in Figure 2.25.

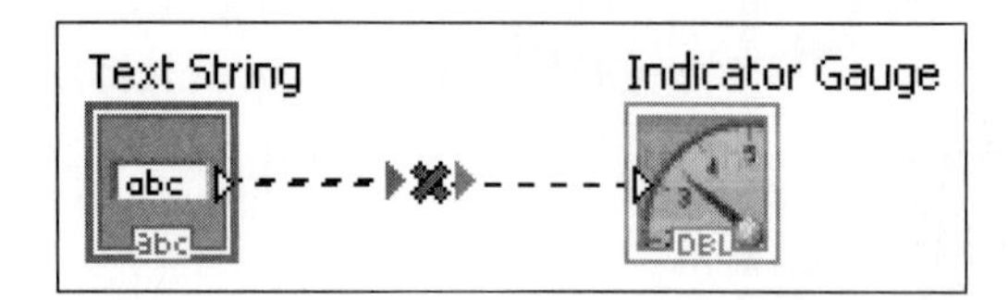

Figure 2.25
Connecting a String (abc) output to a Double (DBL) input is not allowed.

Those "X" marks indicate programming errors (called *broken wires*) and they must be fixed before the VI will run. Sometimes you can change the data type of the control or indicator. For example, you can change a control from the default DBL (double-precision) data type to an integer data type. But you cannot change a string data type (such as the Text String control shown in Figure 2.25) to a numeric data type. The solution for the error indicated in Figure 2.25 is to delete the Text String control and replace it with a Numeric Control and rewire the connections (see Figure 2.26).

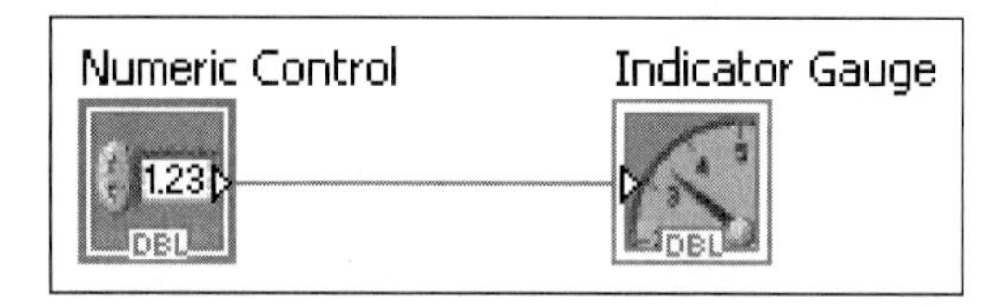

Figure 2.26
A numeric control can be connected to a numeric indicator.

The absence of an "X" in the wire in Figure 2.26 shows that there is no programming error in that connection. The VI is ready to run and the value entered in the Numeric Control will be displayed on the Indicator Gauge. The front panel is shown in Figure 2.27.

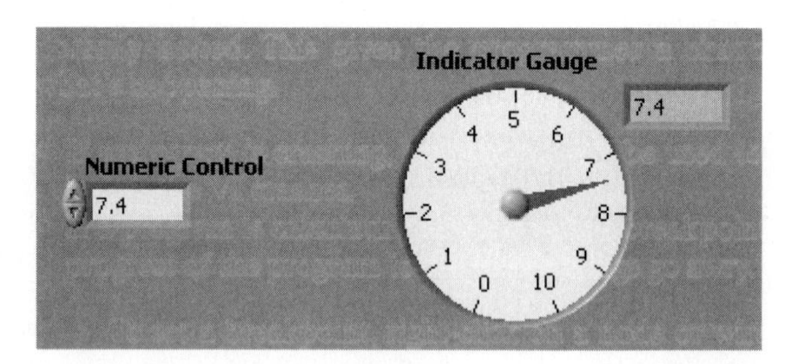

Figure 2.27
Front panel showing the Numeric Control and the Indicator Gauge.

Note: LabVIEW will allow you to display a *digital indicator* next to many of the more graphical indicators, such as the Indicator Gauge in Figure 2.27. To show the digital indicator, right-click on the indicator and select **Visible Items / Digital Display** from the pop-up menu.

An alternative to replacing the Text String control with a Numeric Control is to specifically convert the text string to a numeric value. One way to do this is to use LabVIEW's Scan Value function. The Scan Value function receives two inputs:

- The text string that should contain the number (Text String control output).
- A format string ("%#g" tells LabVIEW to use a general numeric format).

The output from the Scan Value function is the numeric value (DBL data type) read from the string. This output can be connected to the Indicator Gauge input as shown in Figure 2.28.

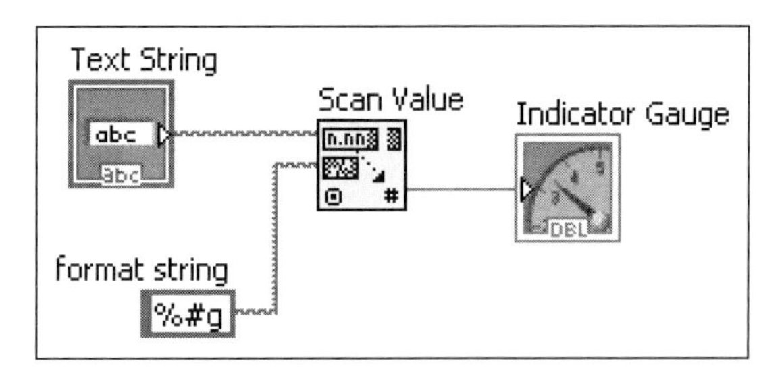

Figure 2.28 Reading a numeric value from a text string.

The front panel associated with the block diagram shown in Figure 2.28 is shown in Figure 2.29.

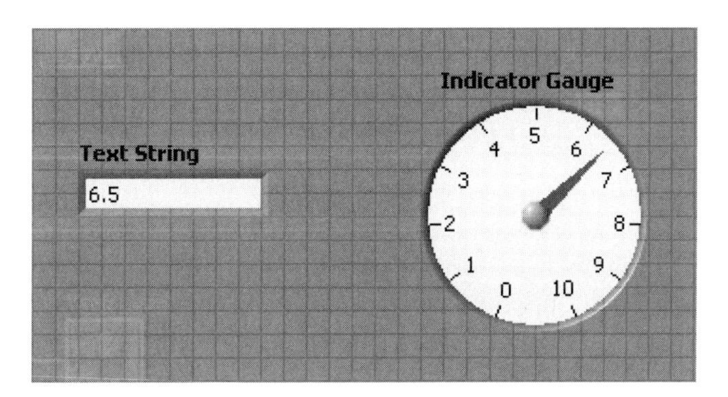

Figure 2.29 Modified front panel using a Text String control.

While getting a numeric value from a text string using the Scan Value function does work, in most cases it is a lot simpler just to use a Numeric control instead of a String control.

APPLICATION

Voltage Divider

A voltage divider (see Figure 2.30) is a very simple DC circuit that is often used to generate a desired voltage (V_{out}) when another voltage (V_{in}) is available. With the proper choice of resistors, V_{out} can be generated at any voltage between V_{in} and ground (0 V).

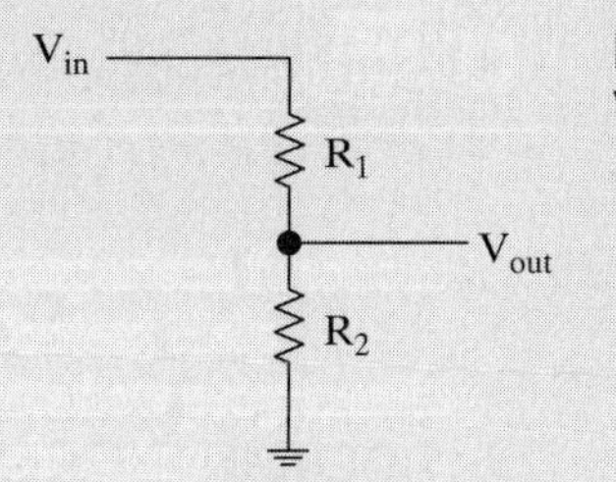

Figure 2.30 Voltage divider.

The math behind a voltage divider is simple:

$$V_{out} = V_{in}\left[\frac{R_2}{R_1 + R_2}\right]$$

Figure 2.31 shows a VI that calculates V_{out} given V_{in}, R_1, and R_2.

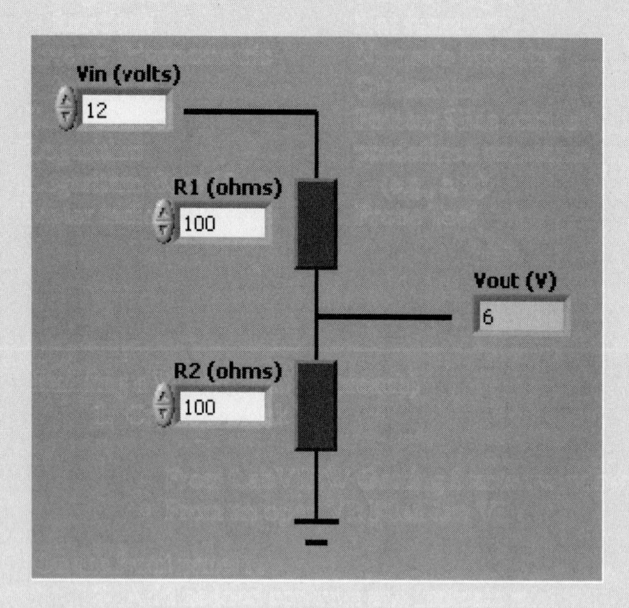

Figure 2.31
Voltage Divider VI.

This VI can be used to determine what resistances are needed to generate a V_{out} of 8 V as shown in Figure 2.32. Of course, this is "a" solution, not "the" solution since other resistor combinations could also be used.

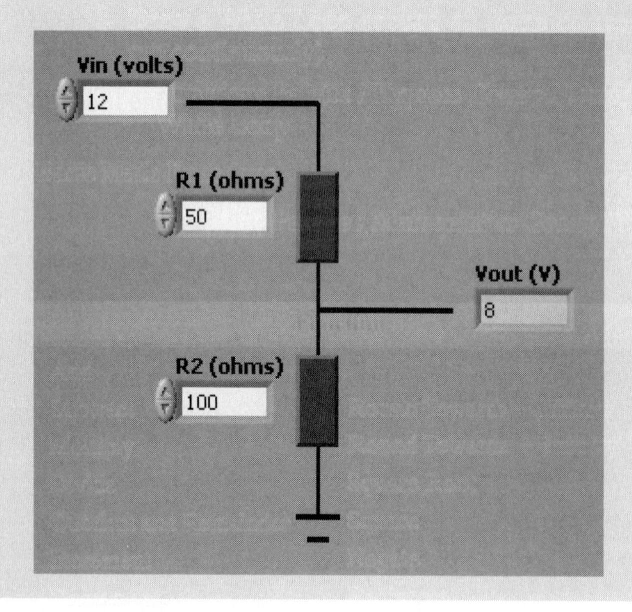

Figure 2.32
Voltage Divider VI used to determine the resistances needed to generate 8 V V_{out}.

2.6 DOCUMENTING VIS

All computer programs in any language should be well documented. Documentation is intended to indicate the purpose of the program (or program element, such as a subVI) as well as the author of the program. LabVIEW's graphical structure makes the information flow more obvious than in test-based computer languages, but a few comments on the block diagram can still go a long way toward making a complex program easier to understand.

LabVIEW provides several mechanisms that make it easy to document VIs. We begin with the simplest: labeling the block diagram and front panel.

2.6.1 Labeling VIs

As a terrible example, consider the completely undocumented VI shown in Figure 2.33. This VI performs a very common calculation, but that wouldn't be obvious to most of us.

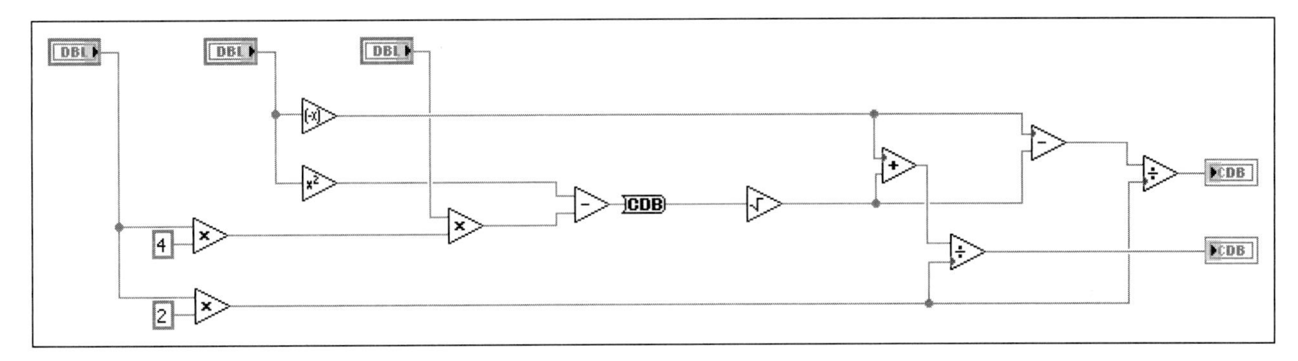

Figure 2.33
Undocumented block diagram.

Compare the undocumented VI with the documented VI in Figure 2.34. Now it's obvious what this VI is designed to do.

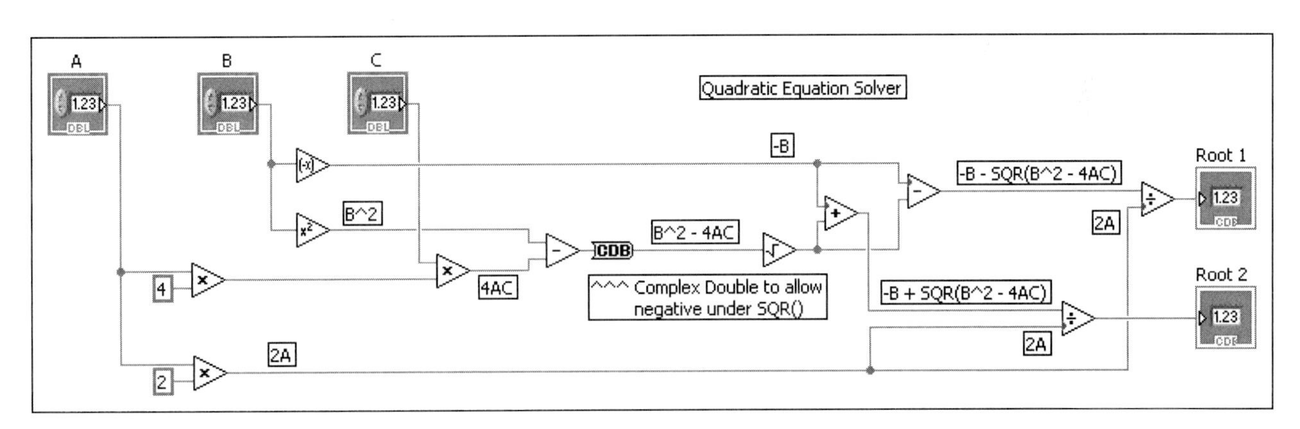

Figure 2.34
Documented block diagram.

The differences between Figures 2.33 and 2.34 are as follows:

- The block diagram has a title.
- Controls and indicators have been shown as icons with labels.
- Labels have been added to streams to make it clear what is being calculated.
- The CDB data type converter has been labeled to show why it is needed.

Labeling a block diagram doesn't take much time, but it makes the VI much easier to comprehend. To add a label to the block diagram, simply double-click in an open area and start typing. The labels can then be moved if needed.

Similarly, the front panel should be labeled to make it clear what the user needs to do. First, the unlabeled front panel is shown in Figure 2.35.

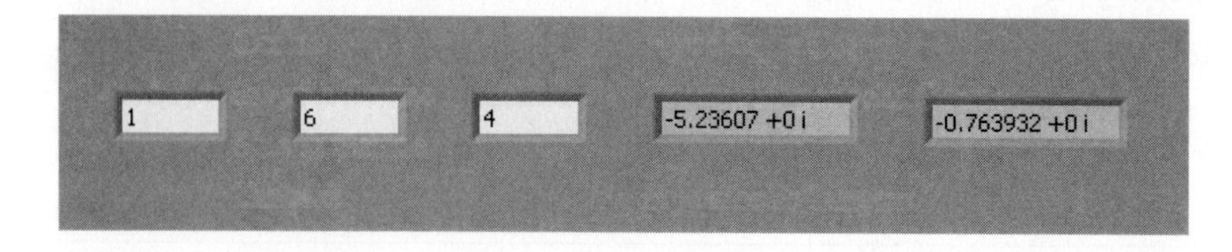

Figure 2.35
Unlabeled front panel.

A few labels help a lot, as shown in Figure 2.36.

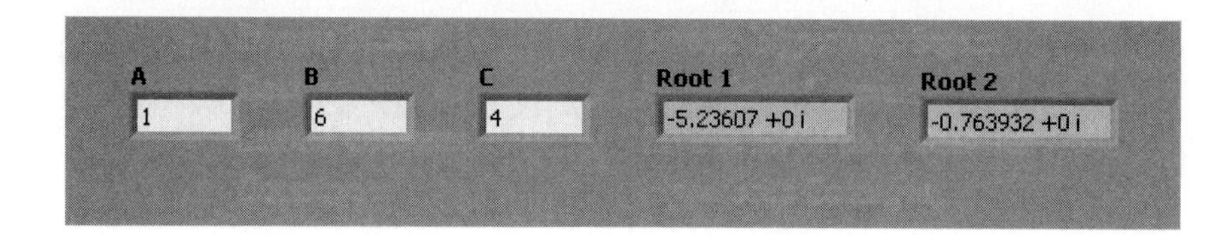

Figure 2.36
Minimally labeled front panel.

A little more information can help the user know which values are inputs and how A, B, and C relate to the coefficients of a quadratic equation. The more thoroughly labeled front panel is shown in Figure 2.37. This version is over-the-top for a

Quadratic Equation Solver
Quadratic Coefficients
Calculated Results
A 1
B 6
C 4
Root 1 -5.23607 +0 i
Root 2 -0.763932 +0 i
A x^2 + B x + C = 0

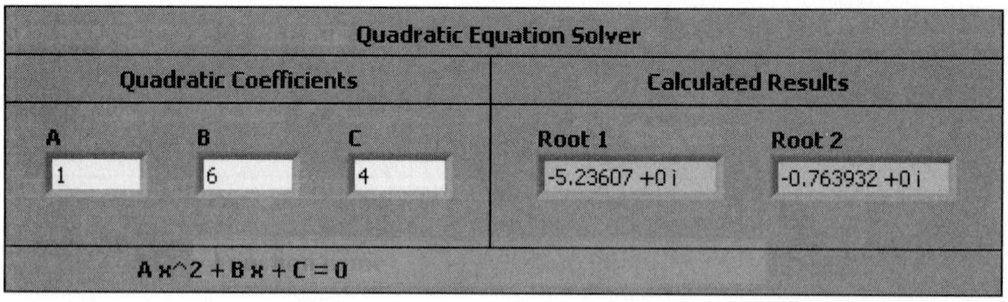

Figure 2.37
Thoroughly labeled front panel.

VI created to solve a single problem, but this degree of labeling is appropriate if the VI will be used routinely, especially if it will be used by someone other than the programmer.

2.6.2 Descriptive Information

Labeling the front panel and block diagram is an important part of documenting a VI, but another level of documentation involves providing a name and description for every VI. LabVIEW makes this easy by providing a **VI Properties** item on the File menu.

Selecting **File / VI Properties** opens the VI Properties dialog, shown in Figure 2.38. There are a dozen categories of information about every VI, but we want to use the **Documentation** category, shown in Figure 2.39.

In Figure 2.39 a description has been added that includes the following:

- VI Title
- Author and date

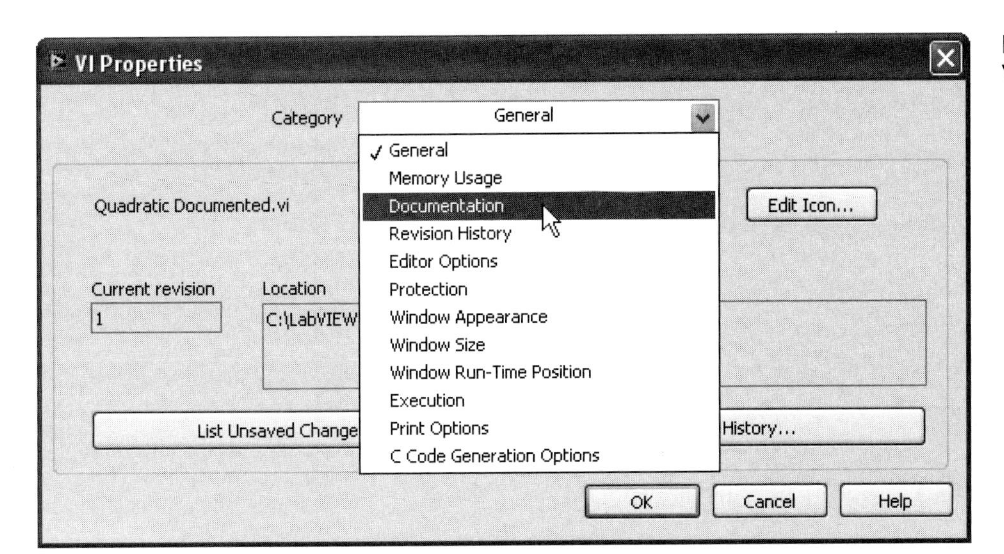

Figure 2.38
VI Properties dialog.

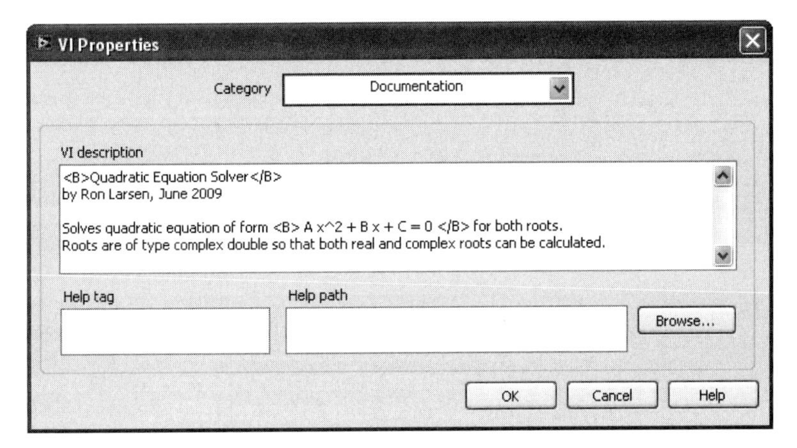

Figure 2.39
Adding a description to the VI.

- statement of what the VI is intended to do
- information about the data type of the solutions, and why complex double was used

Notice that parts of the description were surrounded by <B> and </B>; this will cause them to be shown in bold characters when the description is displayed.

Two fields were not used in this example:

- **Help tag**—used to enter an index term for a help system tied to the VI.
- **Help path**—you can enter a path to an HTML help file in this field. If you include a Help path, the link to **Detailed Help** is included in the Context Help window.

This description is saved with the VI and is displayed in LabVIEW's Context Help window if the mouse is moved over the VI's icon at the top-right corner of the front panel and block diagram (and Context Help is active). The Context Help window for this VI is shown in Figure 2.40. The file name and icon are shown first, then the VI description.

Note: Use Ctrl-H or the menu options **Help / Context Help** to open the Context Help window.

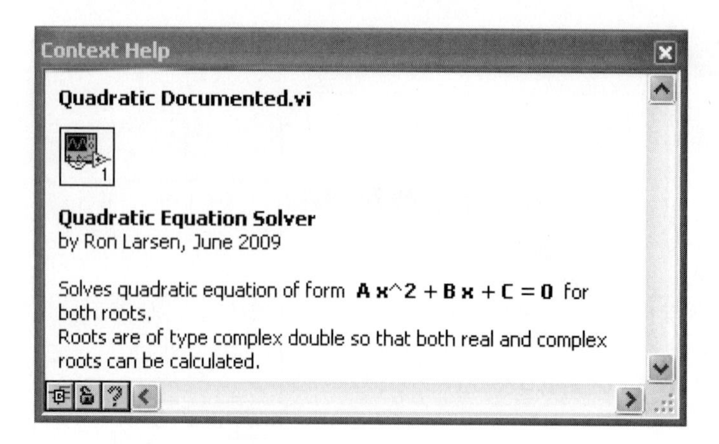

Figure 2.40
VI description shown in the Context Help window.

2.6.3 Descriptions with SubVIs

A quadratic equation solver is a fairly generic piece of computer code, and it is a good candidate for a subVI that can be used in other VIs. In Figure 2.41, the calculations required to compute the roots of the quadratic have been turned into the Quad Solve subVI (creating subVIs is presented in a later chapter).

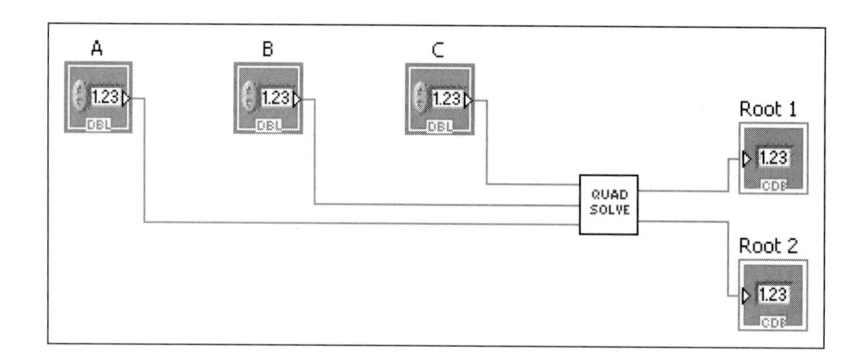

Figure 2.41
The calculations used to find the quadratic roots have been saved as a subVI called Quad Solve subVI.

The description used in Figure 2.39 was also saved with the Quad Solve subVI. Now, when the Quad Solve subVI is used on any block diagram, any time the programmer moves the mouse over the Quad Solve icon, the Context Help window shows the description of the subVI. This is shown in Figure 2.42.

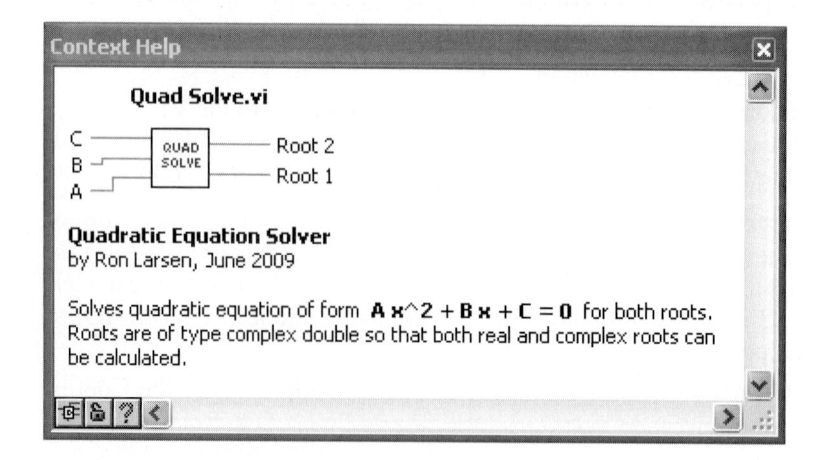

Figure 2.42
Context Help for the Quad Solve subVI.

When VIs are used within other VIs, the descriptions shown in the Context Help window are very useful.

2.7 PRINTING A VI

In a classical programming language, printing the program meant printing a listing of the programming statements. Because LabVIEW is a graphical programming environment, printing the program is a little different, and generally it involves printing some combination of the following:

- the front panel
- the block diagram
- information about the controls on the block diagram
- names of any sub-VI is (none have been used in any examples yet)

If you just want a printout of the current window (either the front panel or the block diagram), use the menu options **File / Print Window ...** This approach does not allow you to select options, but it is the quickest way to get a picture of your block diagram or front panel to a printer.

Opening the Print dialog by selecting **Print ...** from the File menu (from either the front panel or the block diagram) gives you a lot of control over:

- what is printed
- in what format
- to what destination

For most situations the **File / Print Window ...** approach is adequate. If you need more control over printing, additional details are available in the Appendix: Printing VIs.

2.8 SAVING YOUR WORK

LabVIEW's File menu provides the usual **Save** and **Save As ...** options, but also provides **Save All** and **Save for Previous Version ...** options.

- **Save** Saves the current VI (the VI from which the **File / Save** action was initiated).
- **Save As ...** Opens the **Save <VI name> As** dialog (described below).
- **Save All** Saves all open VIs
- **Save for Previous Version ...** Saves the current VI for a previous version of LabVIEW. For example, from LabVIEW 8.5 you can save a VI that will run under LabVIEW 8.0 or LabVIEW 8.2.

Note: It does not matter whether the File menu is accessed from the front panel or the block diagram. When a VI is saved, both the front panel and the block diagram are saved.

2.8.1 Using the Save <VI name> As dialog

LabVIEW's Save As ... dialog is a little different from the usual Windows Save As ... dialog because it not only allows you to save the current VI with a new name, but also allows you to choose which version (original name or new name) will remain open for further editing.

The Save As ... dialog is opened using the following menu options:

File / Save As ...

The Save As . . . dialog includes the name of the current VI in the title bar, as illustrated in Figure 2.43.

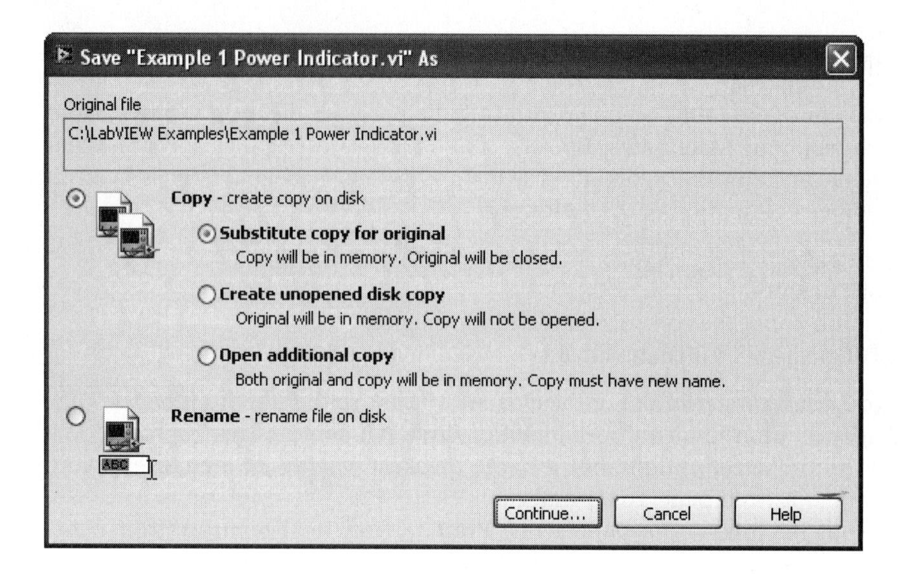

Figure 2.43
LabVIEW's Save As... dialog.

The Save As . . . dialog allows you to either create a copy of the current VI (with several options) or simply rename the existing VI file. If you choose to make a copy, you have three options for how to handle the original VI and the new (copy) VI:

Option 1: Substitute copy for original (see Figure 2.44)

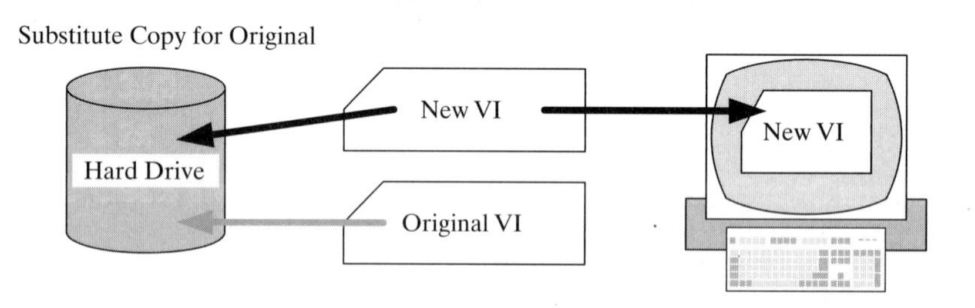

Figure 2.44
Save As: Substitute Copy for Original.

- A new VI (copy of original) is created, stored on the hard drive, and made available for editing.
- The original file is no longer available for editing, but is still available on the hard drive.

Option 1 is useful when you want to start with an existing VI and modify it for a new application.

Option 2: Create unopened disk copy (see Figure 2.45)

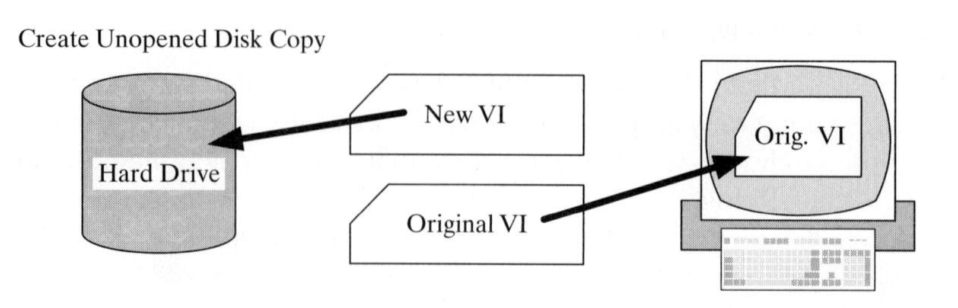

Figure 2.45
Save As: Create Unopened Disk Copy.

- A new VI (copy of original) is created and stored on the hard drive; copy is not opened for editing.
- The original file is still available for editing.

Option 2 is useful when you want to create periodic backups as you develop a complex VI.

Option 3: Open additional copy (see Figure 2.46)

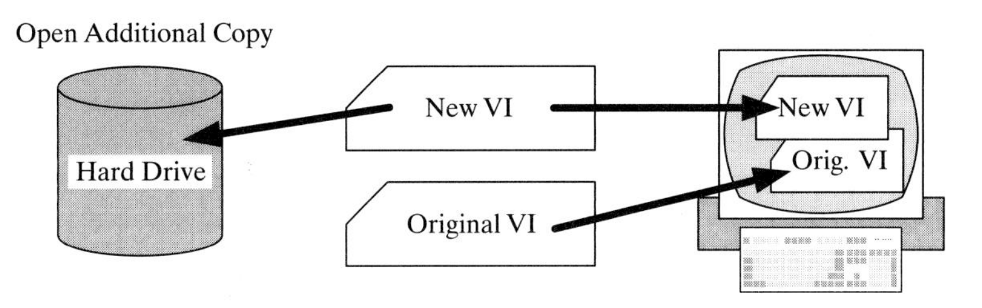

Figure 2.46
Save As: Create Additional Copy.

- A new VI (copy of original) is created, stored on the hard drive, and made available for editing.
- The original file is still available for editing.

Option 3 is useful when you need to create a new VI that is similar to the original VI, and will need to continue to edit both.

2.9 CLOSING A VI

You can close the window containing either the front panel or the block diagram by using the Close button at the top-right corner of the window, or the menu options **File / Close**.

- Closing the front panel automatically closes the block diagram as well.
- Closing the block diagram does not cause the front panel to close.

You can close all open LabVIEW files with **File / Close All**.

LabVIEW will display a warning (Figure 2.47) if you attempt to close a file that has not been saved, and give you a chance to save the changes before closing.

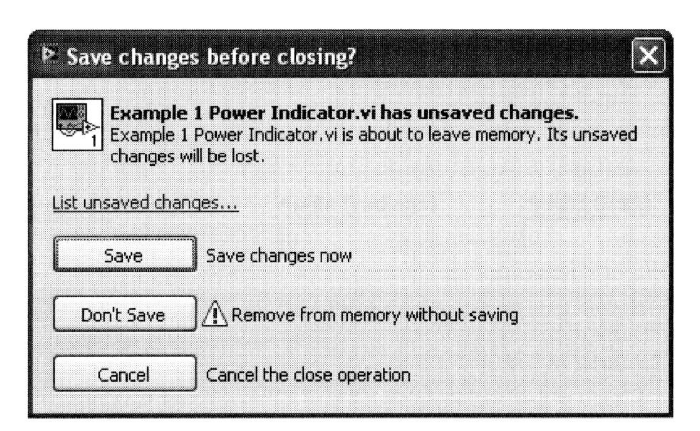

Figure 2.47
LabVIEW warning to save changes before closing a file.

This chapter introduced the basics of working with LabVIEW; there will be many opportunities to practice in the following chapters.

KEY TERMS

boolean (TF) data type
context help [ctrl H]
control
Controls Palette
data types
dataflow programming
digital indicator
double-precision (DBL) data type
Express Group
front panel
function
Functions Palette
Getting Started window
indicator
integer (I32) data type
LabVIEW
mechanical action (of a switch)
Run button
Run Continuously button
string (abc) data type
terminal
tunnel
While Loop
wire

SUMMARY

Starting LabVIEW

- Start Menu/All Programs/National Instruments LabVIEW

Getting Started window

- provides quick access to information about LabVIEW
- allows you to create a new VI or project
- allows you to open a recently used VI or project

LabVIEW Nomenclature

- VI stands for virtual instrument, a LabVIEW program
- front panel—graphical user interface
- block diagram—contains the programming elements
- a project is a collection of related program elements

Creating a VI—Basic Approach

1. create a blank VI
2. add controls and indicators to the front panel
3. wire the nodes on the block diagram
4. save the VI with a descriptive name
5. run and test the VI.

LabVIEW Palettes

- Controls Palette—controls and indicators for the front panel
- Functions Palette—functions and programming structures for the block diagram
- Tools Palette—rarely used, allows you to select tools for operating controls, wiring, etc. Automatic tool selection eliminates the need to select tools in most cases

Wiring Terminals
The mouse icon changes to a spool of wire when positioned over a terminal (or near an existing wire) to indicate that wiring is possible.

- Drag the mouse from the output terminal to the input terminal (or vice versa).

Run Button Icons
The Run Button icon shows the status of the VI. The three possible status options are shown in Figure 2.48.

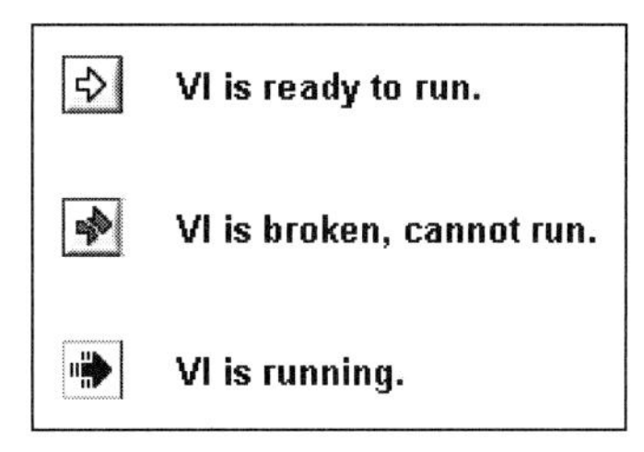

Figure 2.48
The Run button icon changes to indicate the VI status.

Running a VI Continuously
A While Loop from the Express Group on the Functions Palette is often drawn around all controls on the block diagram. When the VI is run, the While Loop will keep the VI running until the STOP button is clicked.

Alternatively,you can also use the **Run Continuously** button, just to the right of the **Run** button to keep the VI running.

Saving a VI
Use menu options **File / Save**, or **File / Save As . . .**
Save As . . . Options

- Substitute copy for original—when you want to start with an existing VI and modify it for a new application.
- Create unopened disk copy—when you want to create periodic backups.
- Open additional copy—when you need to create a new VI that is similar to the original VI, and will need to continue to edit both.

Switch Actions
To set a switch action, right-click on the switch and select Mechanical Action from the pop-up menu.

- **Switch When Pressed**—toggle switches
- **Switch When Released**—mouse buttons, usually
- **Switch Until Released**—doorbell buzzer
- **Latch When Pressed**—like a starting pistol
- **Latch When Released**
- **Latch Until Released**—behaves like the doorbell buzzer

Printing a VI
Use menu options **File / Print . . .**

Closing a VI

- Closing the front panel automatically closes the block diagram as well.
- Closing the block diagram does not cause the front panel to close.

Opening a VI

- Double-click on a VI file in a file browser.
- Start LabVIEW (**Start / All Programs > / National Instruments LabVIEW**) and open the VI file from the LabVIEW environment.

The VI block diagram does not open by default, but can be opened from the front panel using the menu options **Window / Show Block Diagram**.

Lists of recently edited VIs are available

- On the Getting Started menu
- From the File menu: **File / Recent Files**

LabVIEW Math Functions (partial sample)

- Basic Math Functions
  - Add, Subtract, Multiply, Divide
  - Increment, Decrement
  - Absolute Value
  - Square, Square Root
  - Reciprocal
- Trig and Hyperbolic Trig Functions
- Log and Exponential Functions
- Matrix Functions
- Optimization Functions
- Differential Equations Functions

Dataflow Programming
A node (or block) on a block diagram executes as soon as all of the inputs have values

Data Types

Symbol	Data type	Color	Range	Default value	Comment
DBL	Double-precision floating point numeric	Orange	4.94e–324 to 1.79e1308	0.0	Default data type for floating point numeric values.
I32	32-bit signed integer numeric	Blue	–2,147,483,648 to 2,147,483,647	0	Default data type for integer numeric values.
TF	Boolean	Green	True or False	False	
[DBL]	Matrix of double-precision numbers				Brackets indicate array or matrix. Color indicates data type of matrix elements. Wires carrying matrices are displayed with thick lines.
abc	String	Magenta		Empty string	
	Path	Gray		Empty path	Holds a file path.
	128-bit (64.64) Time stamp	Brown	01/01/1600.00:00:00 to 01/01/3001.00:00:00	12:00:00.000 AM 1/1/1904	Format is Date.Time.
	Cluster				Clusters are collections of multiple data types.
	Waveform	Brown			Holds the start time, time step, and data of a waveform. Wires carrying waveforms are displayed with thick lines.

SELF-ASSESSMENT

1. The Getting Started window is designed to provide quick access to useful things. What "things" can you access from the Getting Started window?
   ANS: A new (blank) VI, a new project, access to available VI templates, access to recently edited VIs, and access to LabVIEW resource information and examples.
2. If you wanted to place a toggle switch on the front panel, would you look in the Controls Palette or the Functions Palette?
   ANS: A toggle switch is a control, so look in the Controls Palette.
3. An LED indicator is available on the Controls Palette. Which group contains the LED indicator?
   ANS: Modern / Boolean, or **Classic / Classic Boolean**, or **Express / LEDs**
4. The mouse icon looks like a spool while wiring terminals together on the block diagram. How do you get the mouse icon to look like a spool? (i.e., how do you get LabVIEW into "wiring" mode?)
   ANS: If Automatic Tool Selection is active, just move the mouse over a terminal or near (but not over) a wire. Otherwise, select the Connect Wire tool (the spool) from the Tools Palette.
5. In some of the examples in this chapter, a While Loop was placed around all of the controls on the block diagram. What was the purpose of the While Loop? How can you accomplish the same result without the While Loop?
   ANS: The While Loop keeps the VI running until the STOP button is pressed. This allows the controls to be changed (e.g., switches clicked multiple times) and allows the VI's response to changing controls to be observed.

   ANS: The **Run Continuously** button on the front panel can also be used to keep a VI running.
6. What are the two ways that LabVIEW provides to see a list of recently edited VIs?
   ANS: (1) On the Getting Started window, and (2) on the File menu: **File / Recent Files**.
7. When numeric controls are placed on the front panel, the increment/decrement buttons are displayed by default. How do you hide the increment/decrement buttons?
   ANS: Right-click on the control, and select **Visible Items / Increment / Decrement**.
8. The Subtract function must be wired carefully to ensure that the subtraction takes place correctly. How can you learn how the Subtract function terminals are laid out?
   ANS: When you hover the mouse over the terminals, the labels are displayed. This helps, but using the context help is a better way. (**Help / Show Context Help**, or press [Ctrl H].)
9. With dataflow programming, when is a function evaluated and the function's result calculated?
   ANS: When all of the inputs to the function have data.
10. What data type is associated with toggle switches and LEDs?
    ANS: Boolean data type.

PROBLEMS

1. Use LabVIEW's Square and Square Root functions to create a VI (similar to the VI shown in Figure 2.49) that will accept a value, compute the square of the

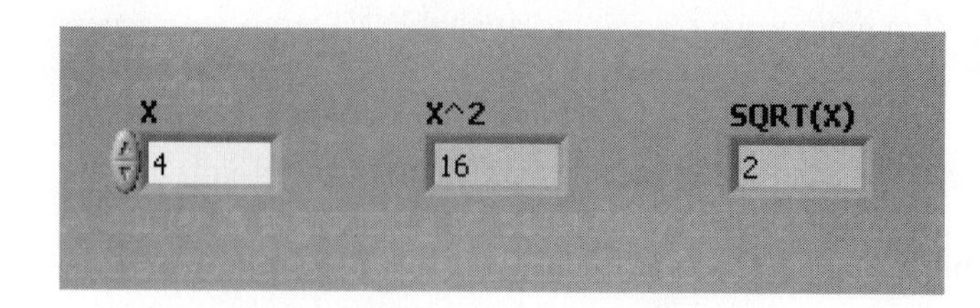

Figure 2.49
Calculating square and square root of a value.

value and the square root of the value, and display the results. What happens when $X = 0$ and $X < 0$?

Functions Palette / Mathematics Group / Numerics Group / Square Function

Functions Palette / Mathematics Group / Numerics Group / Square Root Function

2. Use LabVIEW's Natural Log and Base-10Log functions to create a VI (similar to the VI shown in Figure 2.50) that will accept a value, compute the logarithms, and display the results. What happens when $X = 0$ and $X < 0$?

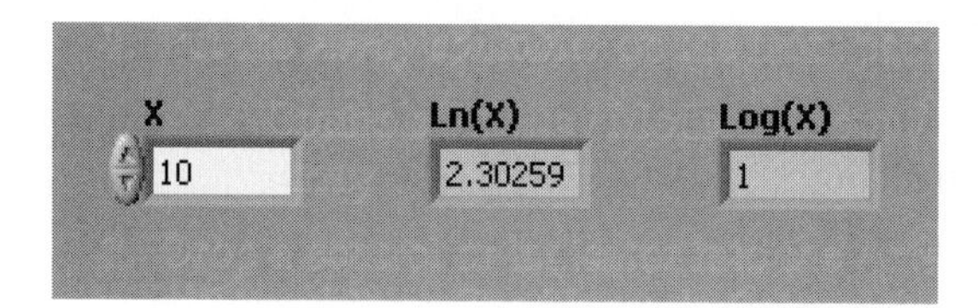

Figure 2.50
Calculating natural and base-10 logarithms.

Functions Palette / Mathematics Group / Elementary and Special Functions Group / Exponential Functions Group / Natural Logarithm

Functions Palette / Mathematics Group / Elementary and Special Functions Group / Exponential Functions Group / Logarithm Base 10

3. Create a VI that has four numeric controls and displays the sum of the four values. The front panel should look something like Figure 2.51.

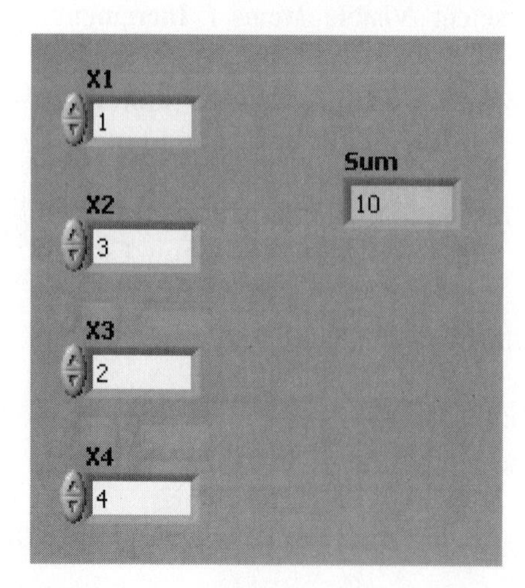

Figure 2.51
Adding four values.

a. Use several Add functions to compute the sum.

Functions Palette / Mathematics Group / Numeric Group / Add

b. Use LabVIEW's Compound Arithmetic function (only one is needed) to compute the sum. The Compound Arithmetic function icon expands (drag the bottom border) to accept any number of input values.

Functions Palette / Mathematics Group / Numeric Group / Compound Arithmetic

4. Write a quadratic equation solver that will accept values for A, B, and C, defined by

$$Ax^2 + Bx + C = 0$$

and then compute both quadratic solutions (one solution using the plus symbol, the other using the minus symbol in the following equation).

$$x = \frac{-B \pm \sqrt{B^2 - 4AC}}{2A}$$

Test your VI with the coefficients shown in Figure 2.52. When it is working, solve the following quadratic equations:

a. $2x^2 - 2x - 4 = 0$

b. $x^2 - 1.7x - 4.8 = 0$

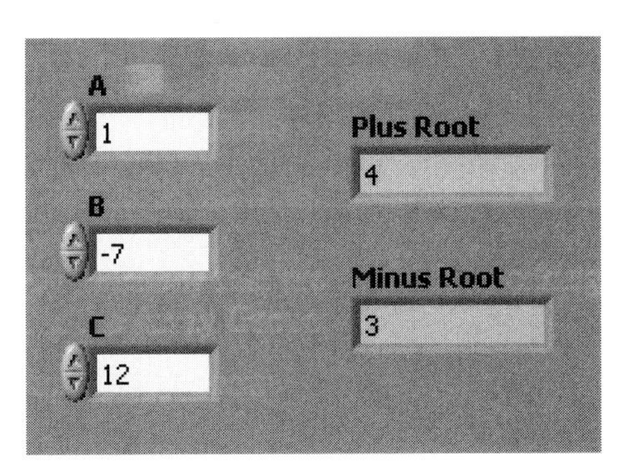

Figure 2.52 Solving quadratic equations.

c. When $4AC > B^2$, there is a negative number inside the square root operator. This is the case for equations such as

$$2 + x + 2x^2 = 0$$

What does LabVIEW show as the solutions to this equation?

5. LabVIEW provides a function that converts a Boolean (True, False) value into a 1 or 0. The function is available in the Mathematics Group:

Functions Palette / Mathematics Group / Numeric Group / Conversion Group / Boolean to (0,1)

That function can be used to convert a switch position to a zero or one, which makes it possible to calculate the digital value of a set of three switches (see Figure 2.53) used to set a three-bit binary value (101 in Figure 2.53).

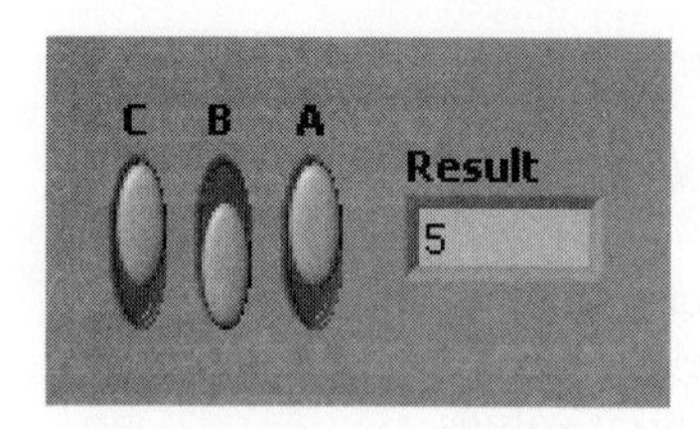

Figure 2.53
Converting binary switch settings to a decimal value.

The math is simpler than the explanation:

$$\text{Result} = C \times 2^2 + B \times 2^1 + A \times 2^0$$

Where A, B, and C each have values of 0 or 1 depending on whether the switch is open or closed.

a. Create a VI similar to the front panel shown in Figure 2.53, and use it to determine the decimal value equivalent to the following binary numbers:

  a. 001 (C is off, B is off, A is on)

  b. 010

  c. 101

b. Modify your VI to handle four-bit binary numbers by adding another switch.

6. The hypotenuse of a right triangle (C in Figure 2.55) can be calculated from the lengths of the other sides as

$$C^2 = A^2 + B^2$$

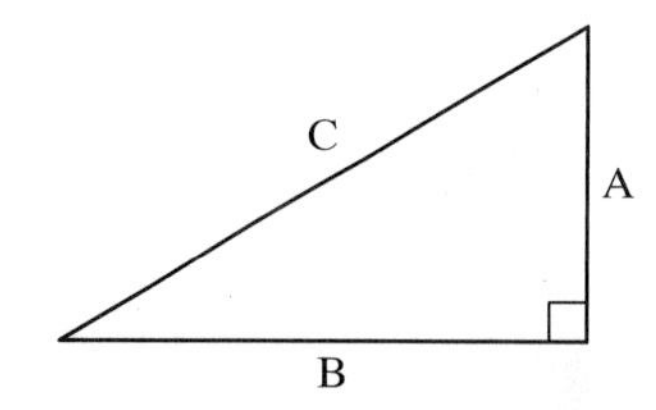

Figure 2.55
Right triangle.

Create a VI that will accept values for A and B as inputs and then calculate and display the C value.

Test your VI with these values: A = 3, B = 4, C = 5. Then, solve the following problems:

a. A = 3.3, B = 4.1
b. A = 7, B = 2

7. Given the lengths of each of the sides of the right triangle shown in Figure 2.56, create a VI that calculates the sine, cosine, and tangent of angle A.

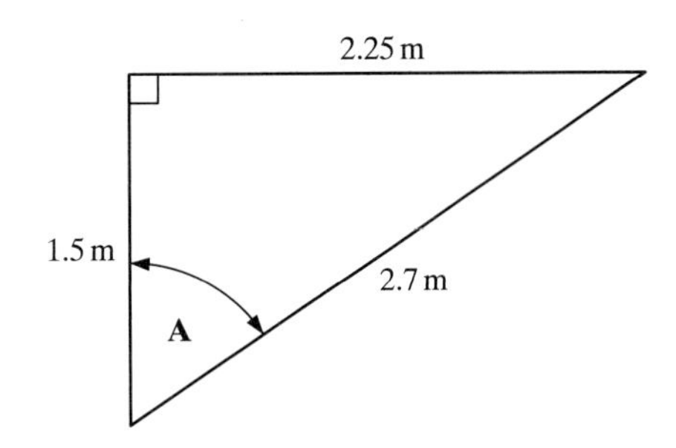

Figure 2.56
Right triangle with specified side lengths.

The following trigonometric definitions can be used:

$$\text{Sin}(A) = \frac{\text{Opposite}}{\text{Hypotenuse}}$$

$$\text{Cos}(A) = \frac{\text{Adjacent}}{\text{Hypotenuse}}$$

$$\tan(A) = \frac{\text{Opposite}}{\text{Adjacent}}$$

CHAPTER 3

LabVIEW Math Functions

Objectives

After reading this chapter, you will know:

- how to use LabVIEW functions for
  - basic mathematics
  - trigonometry
  - exponentials and logarithms
- how to work with LabVIEW's Boolean and comparison functions
- several techniques for debugging LabVIEW VIs

3.1 INTRODUCTION

In this chapter we will present LabVIEW's math functions and show how LabVIEW can be used to solve basic math problems. Most of the problems in this chapter could be more easily solved with a calculator, but this chapter forms a basis for more substantial calculations in later chapters.

Note: The Functions Palette contains functions, VIs, and Express VIs. In this text we use the term *function* loosely, applying the term to most of the programming elements on the Functions Palette.

We will build the controls and math functions into a While Loop on the block diagram in order to create VIs that allow the user to change the input values and see the calculated results. Alternatively, the **Run Continuously** button can be used to keep the VI running until the **STOP** button is pressed.

The basic structure of several of the VIs in this chapter is shown in Figure 3.1, and will include the following elements inside of an overall While Loop:

- one or more controls to set the math input values
- one or more math functions (the *Add function* is used as an example in Figure 3.1)
- an indicator to display the calculated result

The front panel for the Basic Math VI is shown in Figure 3.2.

Notes:

1. The dial controls have been used to illustrate one of the styles of controls available in LabVIEW. We will vary the style of control throughout this chapter just to show the variety available. Simple numeric controls are probably more functional than the dials in most cases. You can easily change the appearance of the controls on the front panel by right-clicking on the control to be changed and selecting **Replace** from the pop-up menu. Then select the style of control that you want to use.

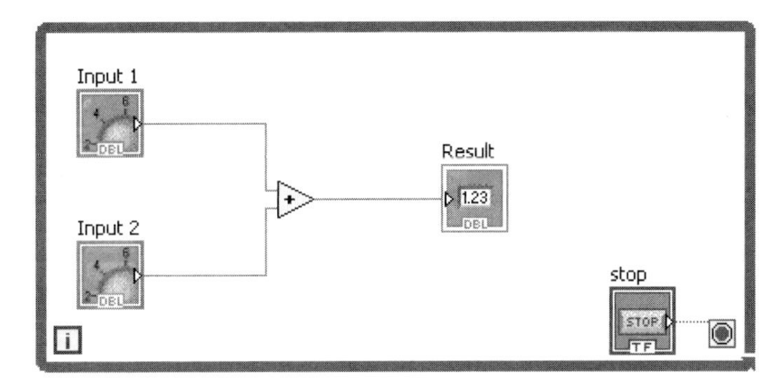

Figure 3.1
Basic block diagram for simple math VIs.

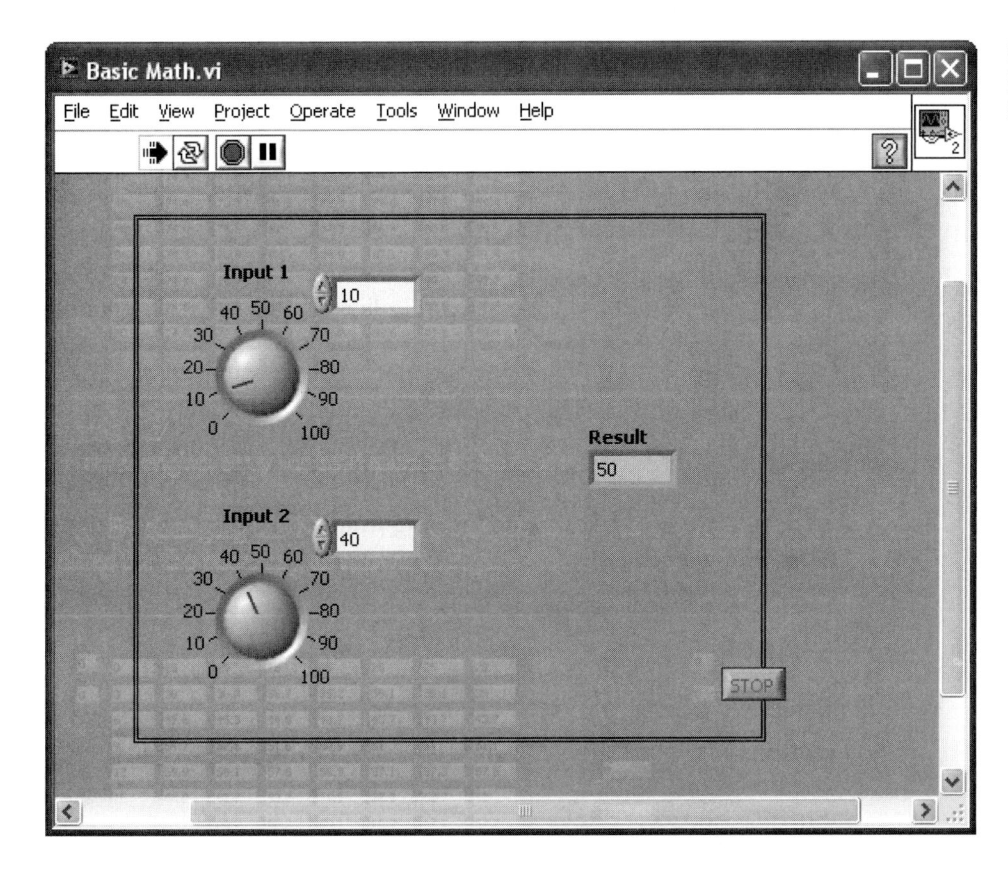

Figure 3.2
Front panel for Basic Math VI.

2. The box (called a Flat Frame in LabVIEW) is used to provide a visual clue to the user that the Inputs and Result are inside of a loop that will continue until the STOP button is pressed.

The Basic Math VI can be created with the following steps:

1. Create a blank VI. There are two ways to do this:
   - From LabVIEW Getting Started window: **New Panel / Blank VI**
   - From an open VI: **File / New VI**

On the front panel . . .

2. Add a *Dial Numeric Control* for the first input:

   Controls Palette / Express Group / Numeric Controls Group / Dial Control

   a. Change the label from "Numeric" to "Input 1".
   b. Double-click on the dial's maximum value and change it to 100.
   c. Show a digital display next to the dial.
      i. Right-click on the dial to open the pop-up menu.
      ii. Select **Visible Items / Digital Display.**
3. Add a Dial Numeric Control for the second input:

   Controls Palette / Express Group / Numeric Controls Group / Dial Control

   a. Change the label from "Numeric" to "Input 2".
   b. Double-click on the dial's maximum value and change it to 100.
   c. Show a digital display next to the dial.
      i. Right-click on the dial to open the pop-up menu.
      ii. Select **Visible Items / Digital Display.**
4. Add a Numeric Indicator for the Result:

   Controls Palette / Express Group / Numeric Indicators Group / Num Ind

   a. Change the label from "Numeric" to "Result".
5. Add a Flat Frame around the controls and indicator (indicates the controls that will be inside the While Loop).

   Controls Palette / Modern Group / Decorations Group / Flat Frame

On the block diagram . . .

6. Add the math function (the Add function has been used as an example):

   Functions Palette / Mathematics Group / Numeric Group / Add Function

7. Wire the Dial control outputs (Input 1 and Input 2) to the Add function inputs.
8. Wire the Add function output to the Result indicator input.
9. Draw a While Loop around all of the components on the block diagram:

   Functions Palette / Express Group / Execution Control Group / While Loop

Back on the front panel . . .

10. Move the While Loop STOP button (added by LabVIEW when the While Loop was added to the block diagram) near the bottom-right corner of the Flat Frame.

3.2 BASIC MATH FUNCTIONS

LabVIEW mathematics functions are accessed via the block diagram and the Functions Palette.

Notes:

1. When you create a blank VI both the front panel and block diagram are displayed. But when you open an existing VI only the front panel is displayed. To open the block diagram, use the menu options **Window / Show Block Diagram.**

2. When the block diagram is displayed, the Functions Palette is normally displayed as well—but the palette can be closed. If the palette has been closed, you can open it with the following menu options (from the block diagram): **View / Functions Palette**.

The mathematics functions that we will use in this chapter are all in the Function Palette's Mathematics Group, shown in Figure 3.3.

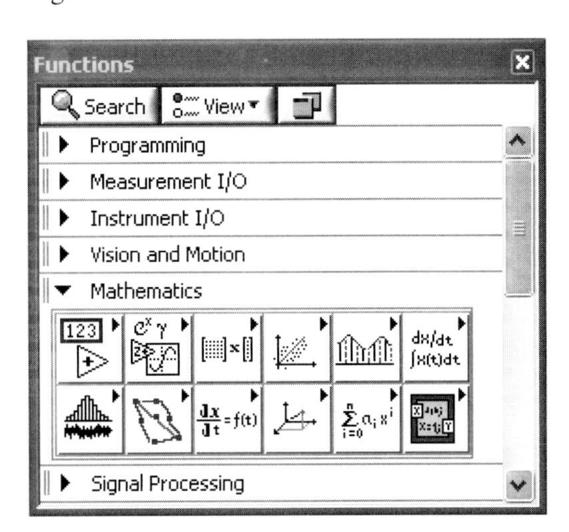

Figure 3.3
The Functions Palette with the Mathematics Group expanded.

We begin with the very basic math functions, which are in the Numeric Group within the Mathematics Group, as shown in Figure 3.4.

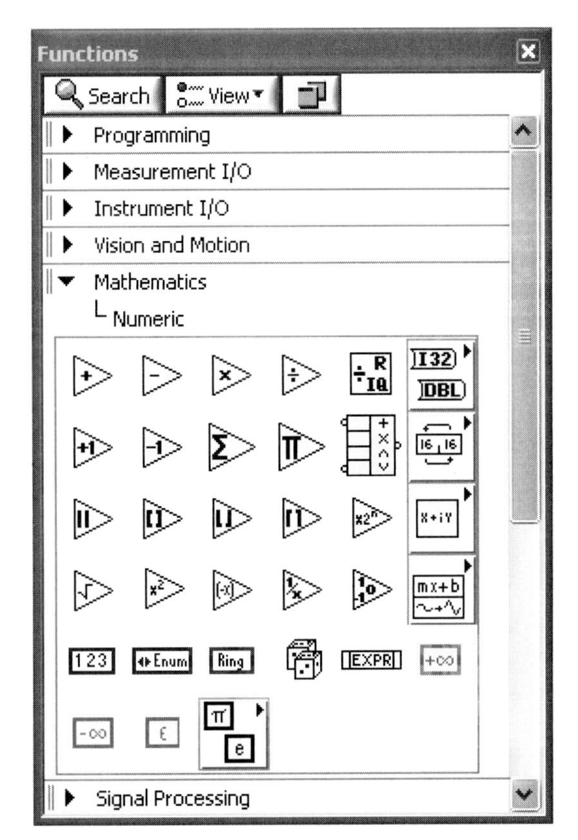

Figure 3.4
The Numeric Group (inside the Mathematics Group on the Functions Palette).

Table 3.1 Basic math functions

Function
Add
Subtract
Multiply
Divide
Quotient and Remainder
Increment
Decrement
Absolute Value
Round to Nearest
Round Towards + Infinity
Round Towards − Infinity
Square Root
Square
Negate
Reciprocal
Numeric Constant
Random Number

3.2.1 Basic Math Functions

LabVIEW provides the basic math functions in the Numeric Group:

Functions Palette / Mathematics Group / Numeric Group

The commonly used basic math functions are listed in Table 3.1.

Additionally, LabVIEW provides a number of *predefined constants*, including

- π
- e (base of natural logarithm)
- h (Planck's constant)
- c (speed of light in a vacuum)
- G (gravitational constant)
- R (molar gas constant)

We will demonstrate the use of some of these in the examples in this chapter.

EXAMPLE 3.1

Adding Two Numbers

We'll start simple and determine the results of the addition problem: 6.13 + 4.78

The Basic Math VI was set up using LabVIEW's Add function, so these problems can be solved with the Basic Math VI without changes. The result is shown in Figure 3.5.

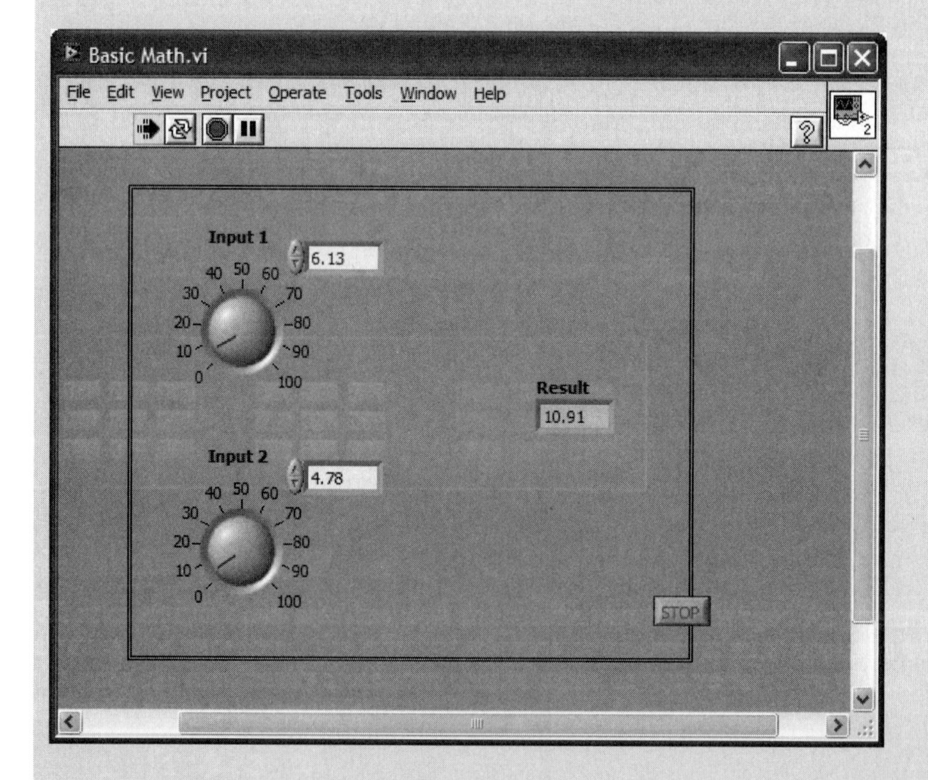

Figure 3.5 Using the Basic Math VI to add two numbers.

It is nearly impossible to enter exactly 6.13 and 4.78 using the dial controls (but they are great for setting approximate values); typing directly into the input fields is a much more efficient way to set the input values.

EXAMPLE 3.2

Dividing Two Numbers

Find the result of dividing 144 by 12.

The result (12) is probably obvious, but solving a problem with a known answer is a great way to test a VI.

To modify the Basic Math VI to divide two numbers, we need to replace the Add function with the *Divide function*. LabVIEW makes this easy. Start with the block diagram and right-click on the Add function to open the pop-up menu (see Figure 3.6). Select **Replace / Numeric Palette / Divide function** as illustrated in Figure 3.6. The result is shown in Figure 3.7.

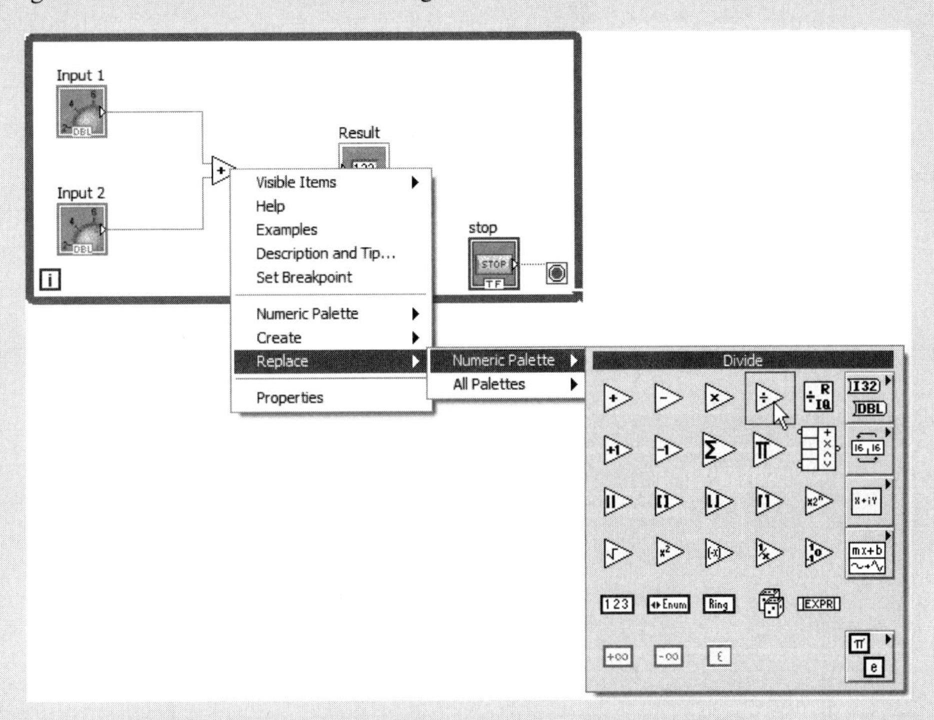

Figure 3.6
Replacing the Add function with the Divide function.

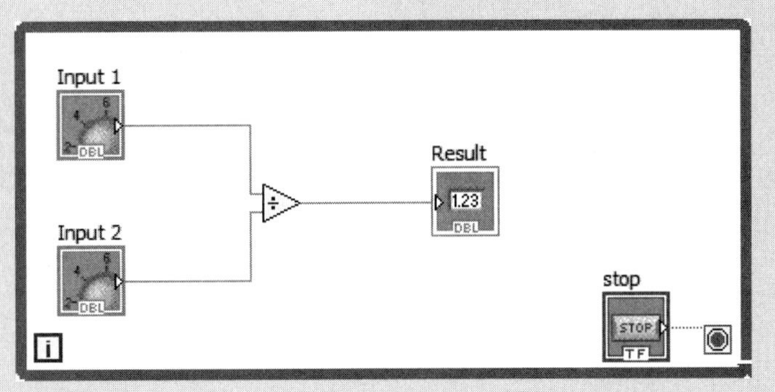

Figure 3.7
The block diagram modified to handle division.

Since order matters in division, we can use the *Context Help* system to see how the Divide function should be wired.

1. Show context help with block diagram menu options: **Help / Show ContextHelp**.
2. Click on the Divide function to see context help for that function (shown in Figure 3.8).

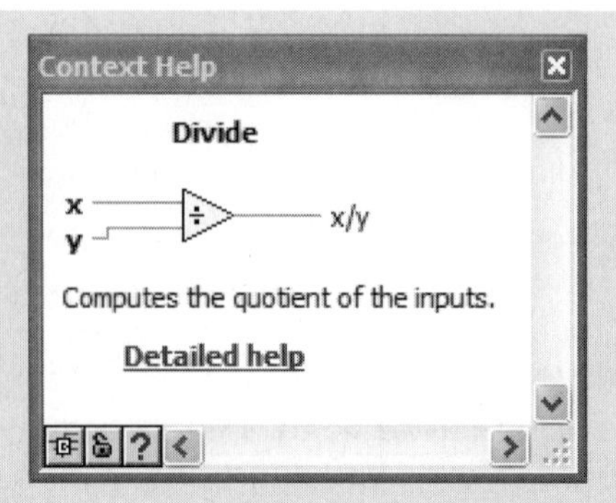

Figure 3.8
Context help for the Divide function.

From Figure 3.8 we see that the top input on the left side of the Divide function is the numerator, and the bottom input is the denominator. To make this clear to the user, we have renamed "Input 1" to "Numerator" and "Input 2" is now "Denominator".

The result of the division is shown in Figure 3.9. The dial controls have been replaced by *Pointer Slide controls*.

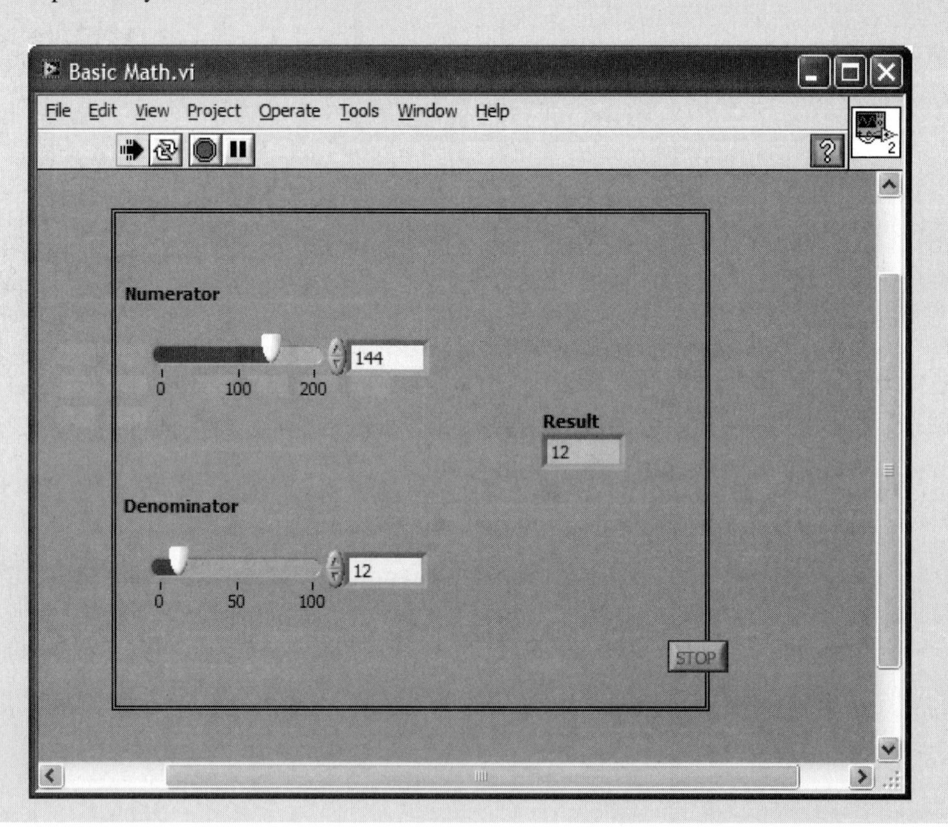

Figure 3.9
Solving the division problem.

APPLICATION

Body Mass Index Calculator

Adding and dividing a couple of numbers is a little boring, so let's create a VI that uses simple math to compute a number that people actually care about: their *Body Mass Index*, or *BMI*. With obesity on the rise, the BMI is an increasingly useful number.

A BMI is used to tell people if they are:

- Underweight (BMI < 18.5)
- Normal weight (BMI = 18.5–25)
- Overweight (BMI = 25–30)
- Obese (BMI > 30)

The formulas for BMI are very simple:

$$\text{Metric Units: BMI} = \frac{\text{weight (kg)}}{(\text{height (m)})^2}$$

$$\text{English Units: BMI} = \frac{\text{weight (lb)}}{(\text{height (ft)})^2} \times 4.88$$

The "4.88" is effectively a unit converter.

We'll create a VI that will allow people to enter their height in feet and inches and their weight in pounds, and then calculate their BMI and report the results. The front panel is shown in Figure 3.10 and the block diagram in Figure 3.11.

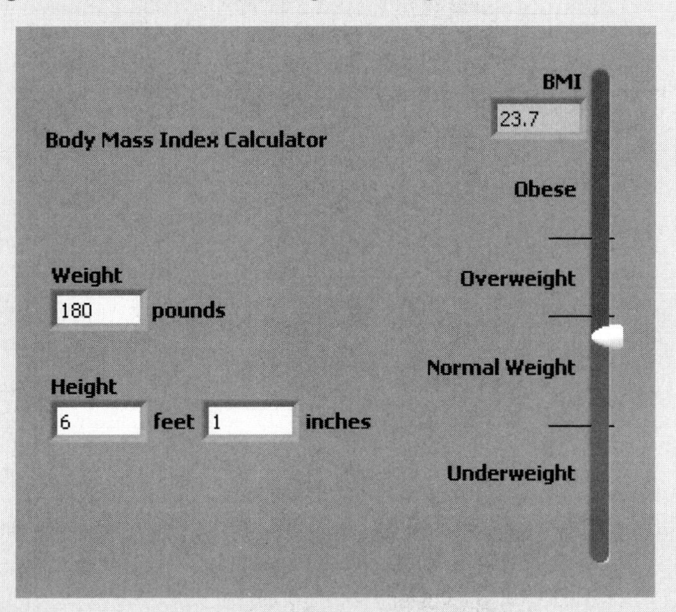

Figure 3.10 Body Mass Index VI, front panel.

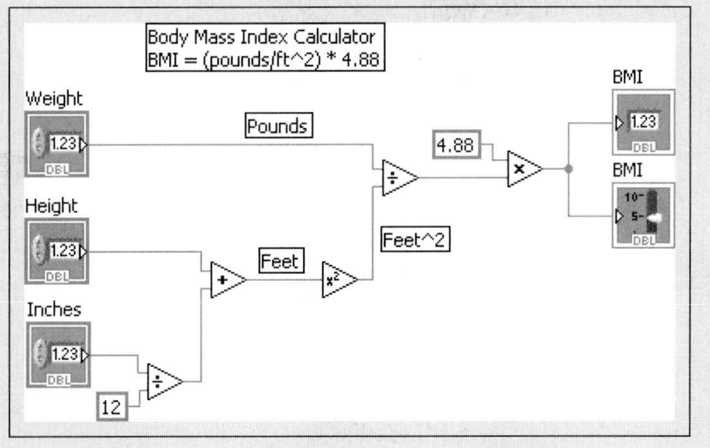

Figure 3.11 Body Mass Index VI, block diagram.

The vertical pointer scale has the scale set to range from 10 to 40. The scale markers and text labels are lines and labels placed in the correct spots on the front panel. When you are editing the front panel, you can drag the scale pointer with the mouse and the value is reported on the indicator, so it is easy to locate the lines at BMI values of 18.5, 25, and 30.

EXAMPLE 3.3

Calculating Integer Multiples of π

Find the values of π, 2π, 3π, and 4π.

The intent of this example is to learn to work with integer values and LabVIEW's predefined constant for π.

For this example, we will use a control to set the multiplier, but we will use a LabVIEW constant to set the value of π.

1. Create a blank VI.

On the front panel (see Figure 3.12) . . .

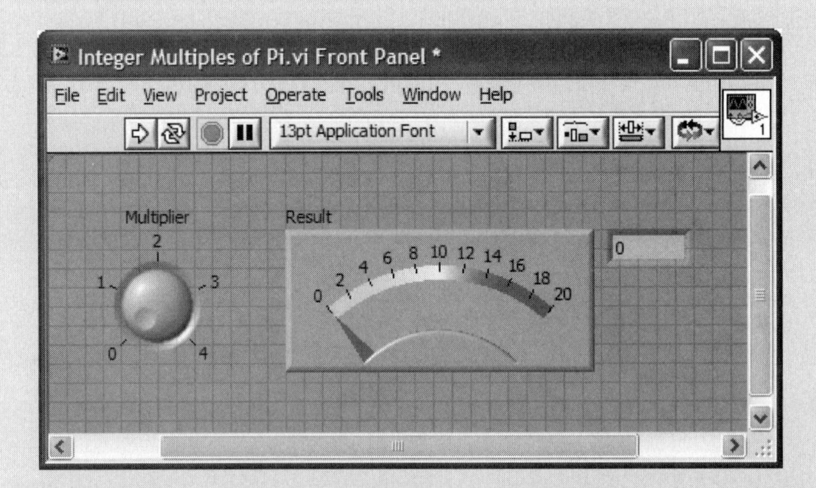

Figure 3.12
Front panel for the Integer Multiples of Pi VI.

2. Add a *Knob Numeric Control*:

   Controls Palette / Express Group / Numeric Controls Group / Knob Control

   a. Change the label to "Multiplier". (Double-click on the "Knob" label and change to "Multiplier".)
   b. Change the maximum value to "4". (Double-click on the maximum value, then enter the new value.)
   c. Change the data type to 32-bit Integer:
      i. Right-click on the Knob control (opens the pop-up menu).
      ii. Select: **Representation / I32**.

   Note: Changing the control's data type to integer will restrict the values that can be selected (by turning the knob) to integer values.

3. Add a *Meter Numeric Indicator*:

   Controls Palette / Express Group / Numeric Indicators Group / Meter Indicator

a. Change the label to "Result". (Double-click on the "Meter" label and change to "Result".)
b. Change the maximum value to "20". (Double-click on the maximum value, then enter the new value.)
c. Make the Digital Display visible. (Right-click on the meter indicator, select **Visible Items / Digital Display** from the pop-up menu.)

On the block diagram (see Figure 3.13) . . .

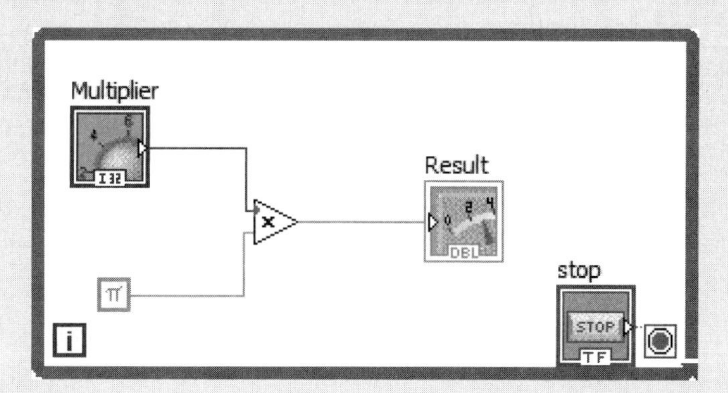

Figure 3.13
Wired block diagram for Integer Multiples of Pi VI.

4. Add the π constant using the following commands:

   Functions Palette / Mathematics Group / Numeric Group / Math & Sci. Constants / π

5. Add the *Multiply function* using the following commands:

   Functions Palette / Mathematics Group / Numeric Group / Multiply Function

6. Wire the block diagram as shown in Figure 3.13.
7. Add a While Loop around the control, constant, and indicator:

   Functions Palette / Express Group / Execution Control Group / While Loop

Now, when the VI is running (see Figure 3.14), turning the Multiplier knob causes the control to output values 0, 1, 2, 3, and 4. These integer values are multiplied by the value of π (from LabVIEW), and the result is displayed by the Result indicator.

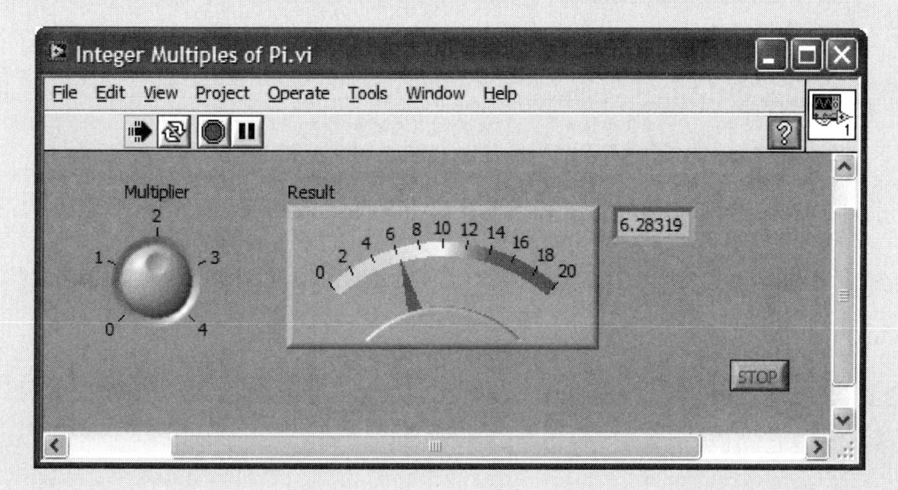

Figure 3.14
Solving for 2π.

3.2.2 Less Commonly Used Numeric Functions

In this section we will take a look at two less commonly used functions:

- Quotient and Remainder function
- Random Number function
- Formula function

We will build the Quotient and Remainder function into a VI to investigate how this function works, and to present an efficient way to create a VI for math functions. Then we'll look at an example that uses the Random Number function to simulate rolling dice. Finally, we will use LabVIEW's Formula function, which is a very flexible function that operates much like a scientific calculator.

Quotient and Remainder Function

The Quotient and Remainder function is an interesting function that receives two inputs (numerator and denominator) and returns two results:

- the integer number of times that denominator can divide into numerator (integer quotient)
- the remainder of that division

To test this function the block diagram shown in Figure 3.15 was created.

Note: The "floor" used as floor(x/y) in Figure 3.15 is a mathematical operation that returns the integer smaller than or equal to the result of (x/y). For example, 26/5 = 5.2 so floor(26/5) = 5. This is also called the *integer quotient.*

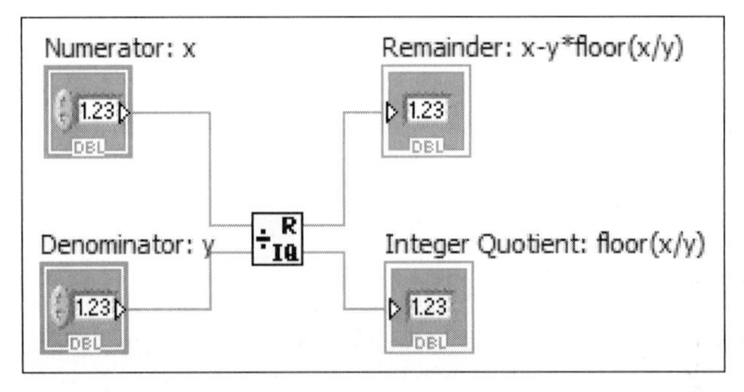

Figure 3.15
Block diagram of Quotient and Remainder VI.

This VI was created almost entirely from the block diagram, which can save a lot of steps for VIs designed to solve math problems. To create the Quotient and Remainder VI, use the following steps:

1. Create a blank VI.
2. Add the Quotient and Remainder function to the block diagram:

   Functions Palette / Mathematics Group / Numeric Group / Quotient and Remainder Function

3. Add needed controls and indicators; label and wire them.

The Quotient and Remainder function has two inputs and two outputs. If you right-click on an input or an output, a pop-up menu will be displayed, and one of the menu options is **Create / Control** (for an input) or **Create / Indicator** (for an output). By right-clicking on each input and output on the Quotient and Remainder function block, you can quickly add the needed controls and indicators to the block diagram—and they will automatically be:

- the correct data type
- labeled as shown in Figure 3.16
- wired to the function

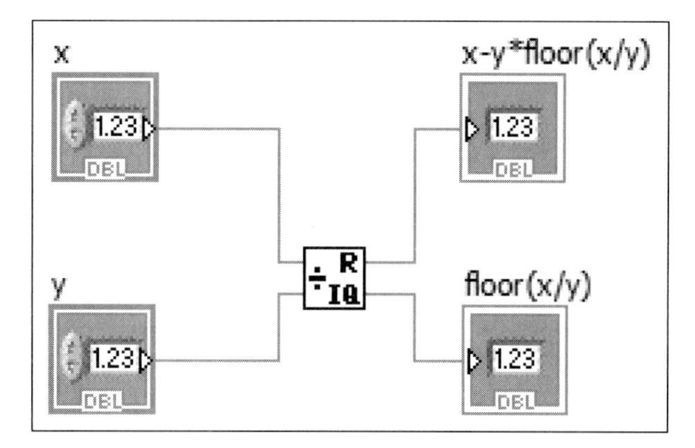

Figure 3.16
Block diagram showing default labeling.

The labels on the outputs seemed a little vague, so all of the labels were modified to add more descriptive names:

- "x" became "Numerator: x"
- "y" became "Denominator: y"
- "x-y*floor(x/y)" became "Remainder: x-y*floor(x/y)"
- "floor(x/y)" became "Integer Quotient: floor(x/y)"

At this point the VI is complete and will run. For appearance, the controls on the front panel may need to be rearranged a little.

Note: The VI will solve for the integer quotient and remainder only once, because the controls were not built into a While Loop to keep the VI running. The While Loop has been omitted here to demonstrate that you can solve problems in LabVIEW without the While Loop—you just have to remember to run the VI after entering all of the input values (or use the **Run Continuously** button).

In Figure 3.17, the value 26 was divided by 5. We expect that the integer quotient will be 5 with a remainder of 1, and that is the LabVIEW result.

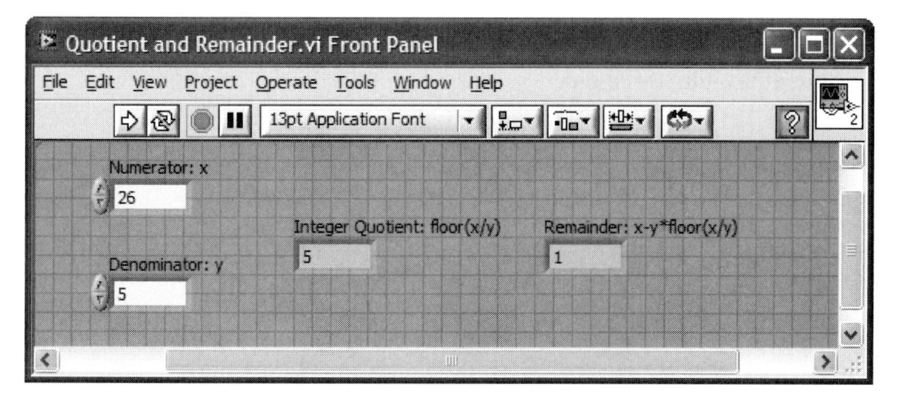

Figure 3.17
Using the Quotient and Remainder VI.

Random Number Function

The icon for LabVIEW's *Random Number function* looks like a pair of dice (see Figure 3.18), but it does not return an integer between 1 and 6. Instead it returns a

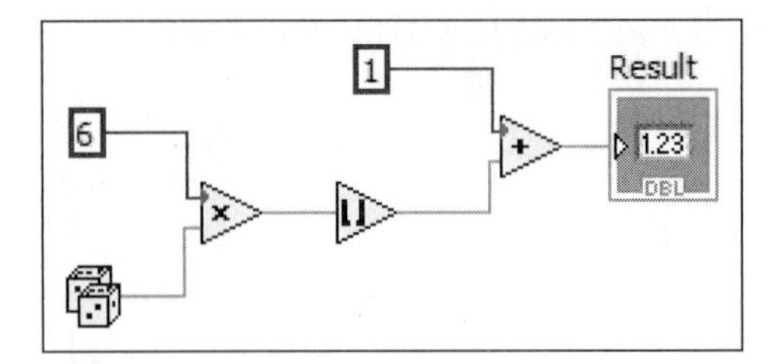

Figure 3.18
Block diagram for simulated roll of one die.

double-precision floating point value greater than or equal to 0 and less than 1. The distribution of returned values is uniform (all values in the range are equally likely to occur).

To simulate rolling a die (one of a pair of dice), we need to

- multiply the random number by 6 (values will range from 0.00 to 5.99)
- round down to the next lower integer (values will range from 0 to 5)—this requires the Round Towards – Infinity function:

  Functions Palette / Mathematics Group / Numeric Group / Round Towards – Infinity function

- add 1 (values will range from 1 to 6)

A block diagram that accomplishes this task is shown in Figure 3.18.
The front panel is shown in Figure 3.19.

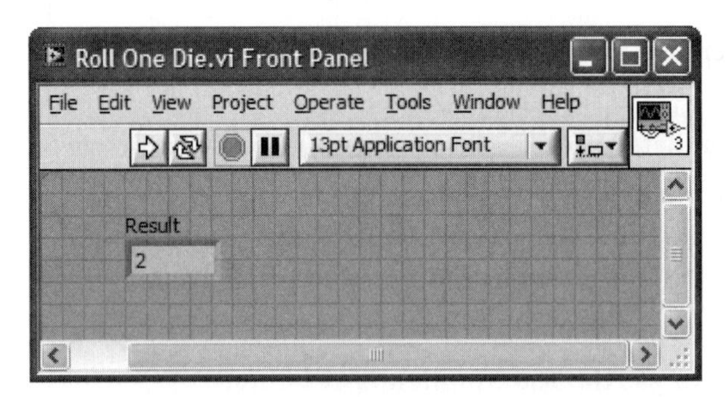

Figure 3.19
Simulated Die Roll VI.

Each time the VI is run, a value between 1 and 6 is displayed in the Result indicator.

Once you have written a VI, you can use that VI on the block diagram of another VI. For example, the VI in Figure 3.20 calls the Roll One Die.vi three times to simulate rolling three dice.

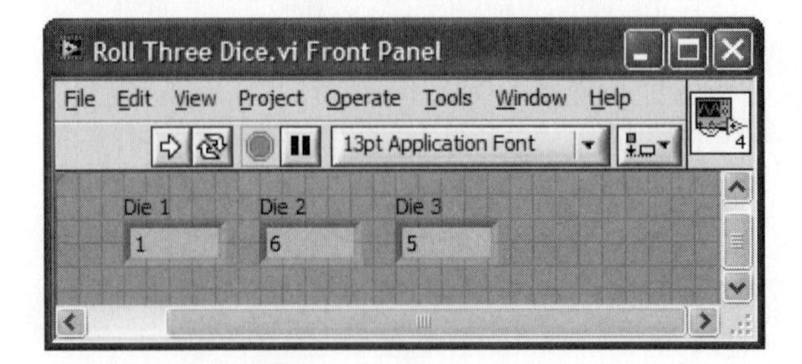

Figure 3.20
A VI simulating rolling three dice.

Formula Express VI

LabVIEW provides an interesting Express VI, called the *Formula Express VI*, that provides a lot of problem-solving power.

Express VI—an *Express VI* is a more sophisticated VI that can be configured using a dialog box. The dialog box automatically opens when the Express VI is placed on the block diagram. Double-click the VI's icon to re-open the dialog when needed. Express VIs appear on the Functions Palette with a blue strip across the top of the icon (or blue border when expanded).

The Formula Express VI functions, in many ways, like a scientific calculator. It is located in the Script and Formulas Group:

Functions Palette / Mathematics Group / Script and Formulas Group / Formula

When you move the Formula Express VI to a block diagram, the icon for the Formula (see Figure 3.21) is displayed on the block diagram, and the Configure Formula dialog (Figure 3.22) is opened.

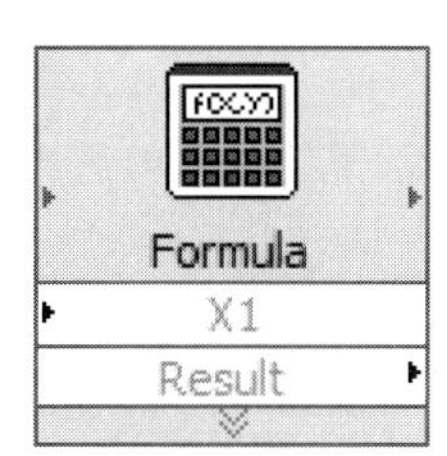

Figure 3.21
Icon for Formula Express VI.

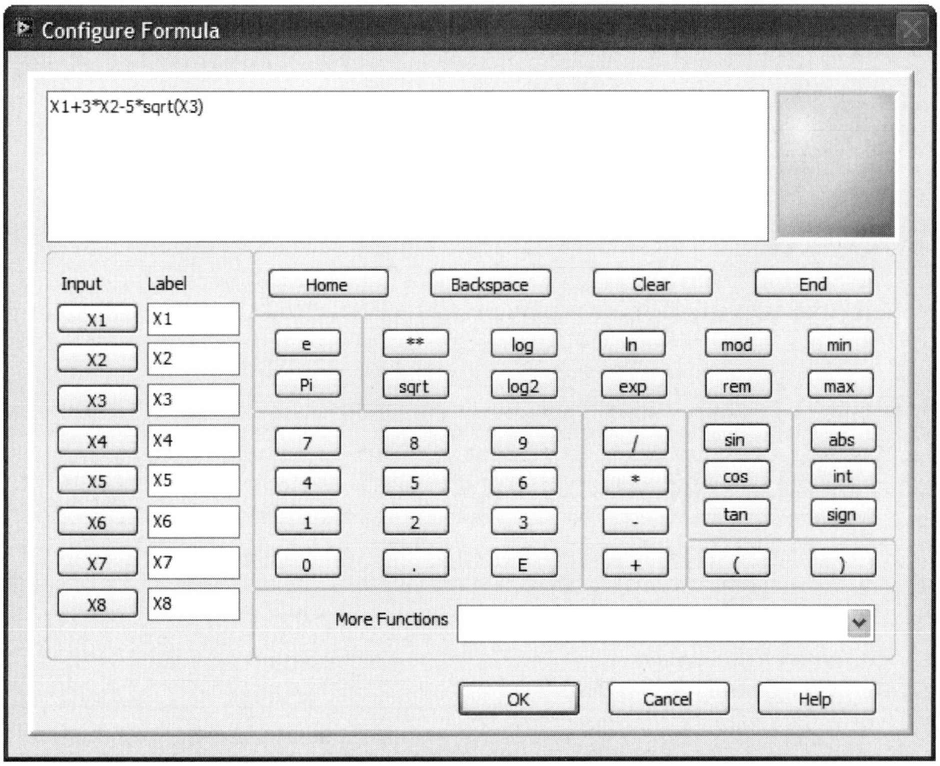

Figure 3.22
Configure Formula dialog.

Your formula can accept up to eight inputs, and you can build those input values into a formula that can include a wide range of mathematical functions. In this example, three inputs ($X1$, $X2$, and $X3$) have been used to build the formula

$$X1 + 3 * X2 - 5 * \mathrm{sqrt}(X3)$$

Clicking OK closes the Configure Formula dialog, and the Formula icon on the block diagram is modified to reflect the actual number of inputs as shown in Figure 3.23. The necessary controls and indicator have been created in the block diagram to allow values for $X1$, $X2$, and $X3$ to be set, and to display the computed result.

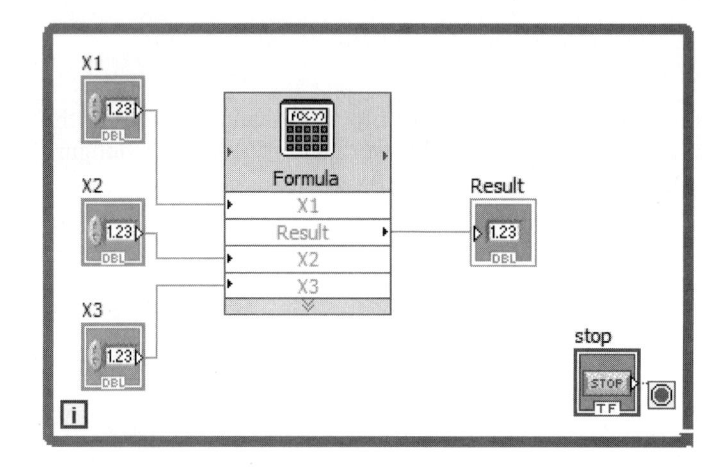

Figure 3.23
Block diagram for solving a formula (incomplete).

When the VI is run, the solution to 2 + 3 * 4 – 5 * sqrt(6) is found to be 1.753, as shown in Figure 3.24.

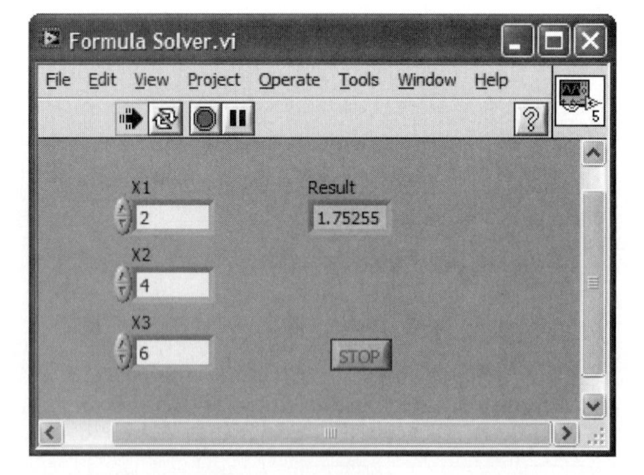

Figure 3.24
Using the Formula Express VI in Formula Solver VI.

3.3 TRIGONOMETRIC AND HYPERBOLIC TRIGONOMETRIC FUNCTIONS

LabVIEW provides a wide range of trigonometric functions:

- *Trigonometric functions*
- *Inverse Trigonometric functions*
- *Hyperbolic Trigonometric functions*
- *Inverse Hyperbolic Trigonometric functions*

The functions are available in the following groups on the Functions Palette:

- Functions Palette/Mathematics Group/Elementary & Special Functions Group/ Trigonometric Functions Group
- Functions Palette/Mathematics Group/Elementary & Special Functions Group/ Hyperbolic Functions Group

These functions are also available in the Express Group:

- Functions Palette/Express Group/Arithmetic & Comparison Group/Express Math Group/Express Trigonometric Functions
- Functions Palette/Express Group/Arithmetic & Comparison Group/Express Math Group/Express Hyperbolic Functions

The available functions are listed in Table 3.2.

All angles in these functions are in *radians*. The use of the functions is demonstrated in the following examples.

Table 3.2 LabVIEW Trigonometric Functions

Trigonometric Functions	Hyperbolic Trigonometric Functions
Sine	Hyperbolic Sine
Inverse Sine	Inverse Hyperbolic Sine
Cosine	Hyperbolic Cosine
Inverse Cosine	Inverse Hyperbolic Cosine
Sine & Cosine	
Tangent	Hyperbolic Tangent
Inverse Tangent	Inverse Hyperbolic Tangent
Inverse Tangent (2 Input)	
Secant	Hyperbolic Secant
Inverse Secant	Inverse Hyperbolic Secant
Cosecant	Hyperbolic Cosecant
Inverse Cosecant	Inverse Hyperbolic Cosecant
Cotangent	Hyperbolic Cotangent
Inverse Cotangent	Inverse Hyperbolic Cotangent
Sinc	

EXAMPLE 3.4

Build a VI that Converts Degrees to Radians

Because LabVIEW angles are always in radians, having a VI that converts degrees to radians might be handy. The relationship between degrees and radians is

$$360 \text{ degrees} = 2\pi \text{ radians}$$

so

$$\text{Angle in radians} = \text{Angle in degrees} \cdot \frac{2\pi \text{ radians}}{360 \text{ degrees}}$$

A block diagram that performs this conversion is shown in Figure 3.25, and the corresponding front panel is shown in Figure 3.26.

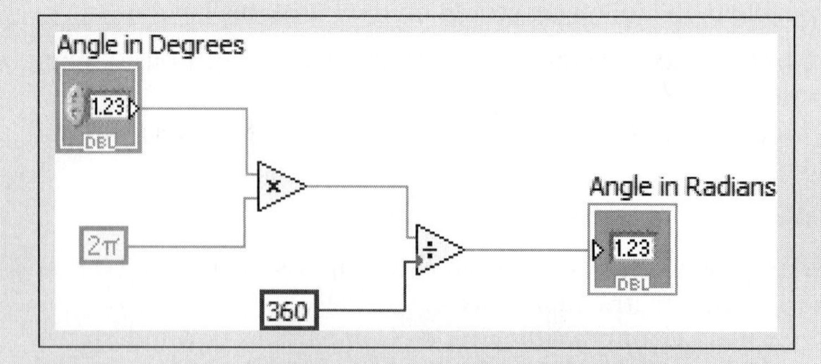

Figure 3.25
Block diagram for Deg2Rad VI.

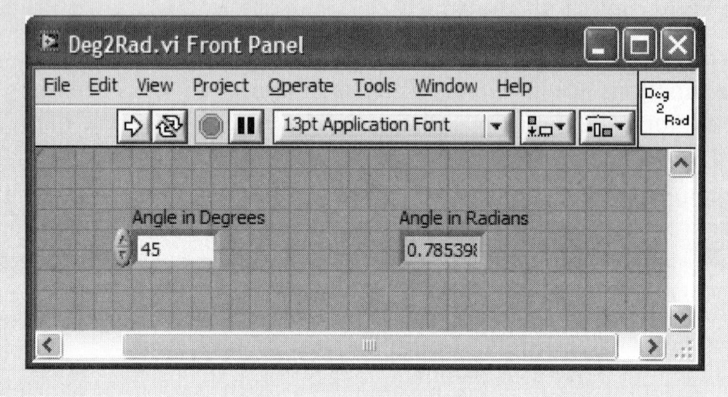

Figure 3.26
Front panel for Deg2Rad VI.

EXAMPLE 3.5

Find the Cosine of 45°.

The cosine of 45° should be 0.7071; we will see if we get that value using LabVIEW's Cosine function.

LabVIEW's Cosine function requires that the angle be specified in radians, but from the previous example we know that 45° = 0.7854 radians (see Figure 3.26).

The block diagram using LabVIEW's cosine function is shown in Figure 3.27. The calculated result is shown in Figure 3.28, and is indeed the expected value of 0.7071.

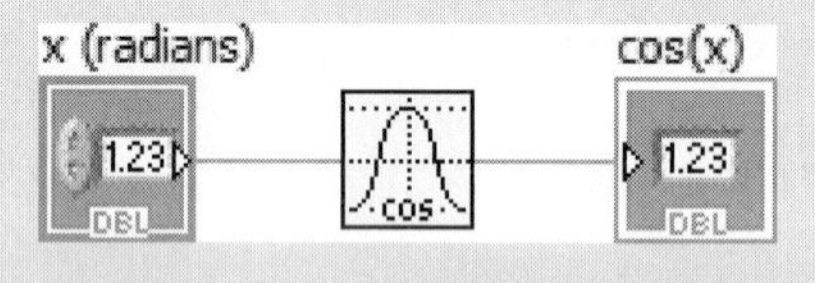

Figure 3.27
Block diagram for taking a cosine.

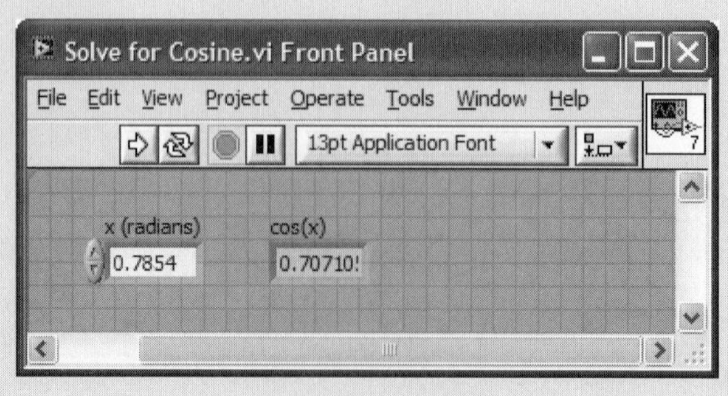

Figure 3.28
Finding the cosine of 45°.

We can combine the block diagram elements from examples 4 and 5 to create a VI that accepts an angle in degrees and returns the cosine. The block diagram is shown in Figure 3.29 and front panel in Figure 3.30.

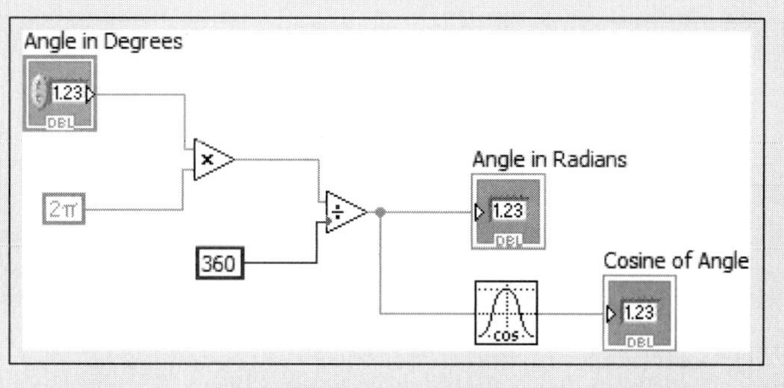

Figure 3.29
Converting angle in degrees to radians before taking the cosine.

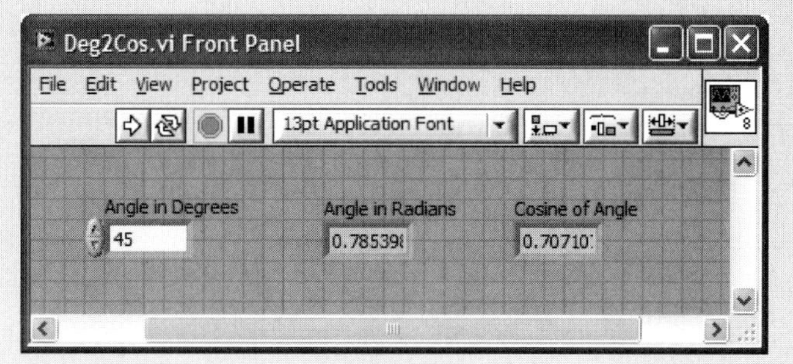

Figure 3.30
Front panel: Converting degrees to radians before taking cosine.

3.4 EXPONENTIAL AND LOGARITHM FUNCTIONS

LabVIEW provides the *exponential functions* and *logarithm functions* listed in Table 3.3.

Table 3.3 Exponential and Logarithm Functions

Function name	
Exponential	$\exp(x)$
Exponential (Arg) − 1	$\exp(x) - 1$
Natural Logarithm	$\ln(x)$
Natural Logarithm (Arg +1)	$\ln(x + 1)$
Logarithm Base 2	$\log_2(x)$
Logarithm Base 10	$\log_{10}(x)$
Logarithm Base *X*	$\log_X(x)$
Power Of 2	2^x
Power Of 10	10^x
Power Of *X*	X^x
Y-th Root of *X*	$\sqrt[Y]{X}$

EXAMPLE 3.6

First-Order Response

A *first-order response* is described by the equation

$$y(t) = (y_{\text{ult}} - y_{\text{orig}})[1 - e^{\frac{-t}{\tau}}]$$

where

y_{orig} is the *initial (original) value of y* at time $t = 0$
y_{ult} is the *ultimate value of y* at time $t = \infty$
τ is the *time constant*

Create a VI that will solve for the value of y for a specified time, given

$$y_{\text{orig}} = 0$$
$$y_{\text{ult}} = 100$$
$$\tau = 10 \text{ minutes}$$

Our equation becomes

$$y = (100 - 0)[1 - e^{\frac{-t}{10}}]$$

The block diagram and front panel of the First Order VI are shown in Figure 3.31 and Figure 3.32, respectively.

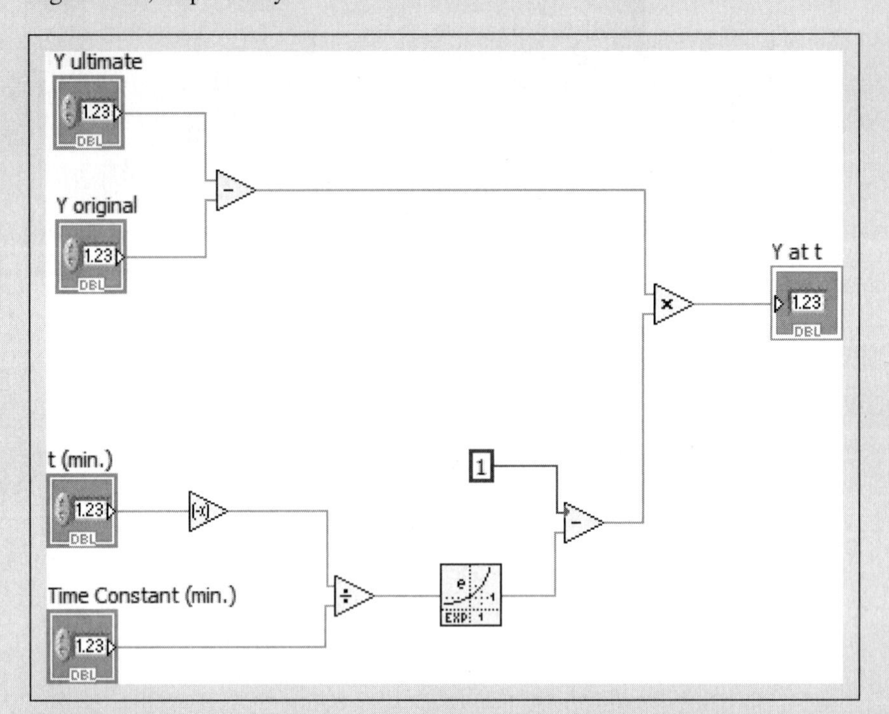

Figure 3.31 Block diagram for First Order VI.

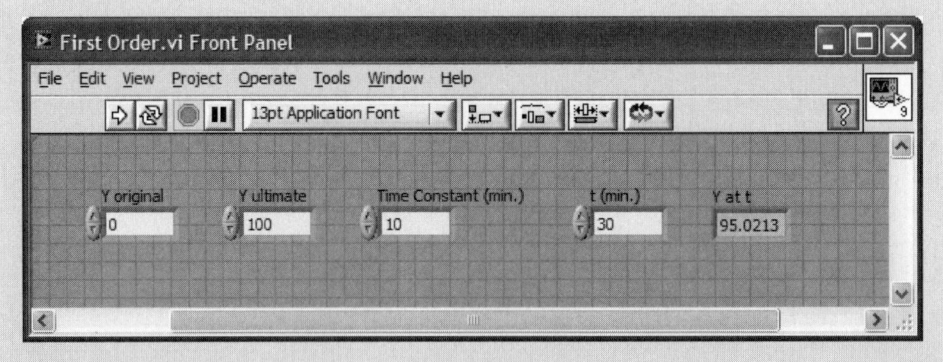

Figure 3.32 Front panel for First Order VI.

The result indicates that after 30 minutes, the value of y will be 95. This result will not surprise a lot of engineers because it is fairly common knowledge that a first-order process will be 95% of the way to the ultimate value after a time period equal to three time constants. Three time constants equals 30 minutes in this example, and the value of y is changing from 0 to 100. In this example, after a time period of 30 minutes we would expect y to equal 95.

The block diagram is fairly complicated, so it might help to look at it piece by piece:

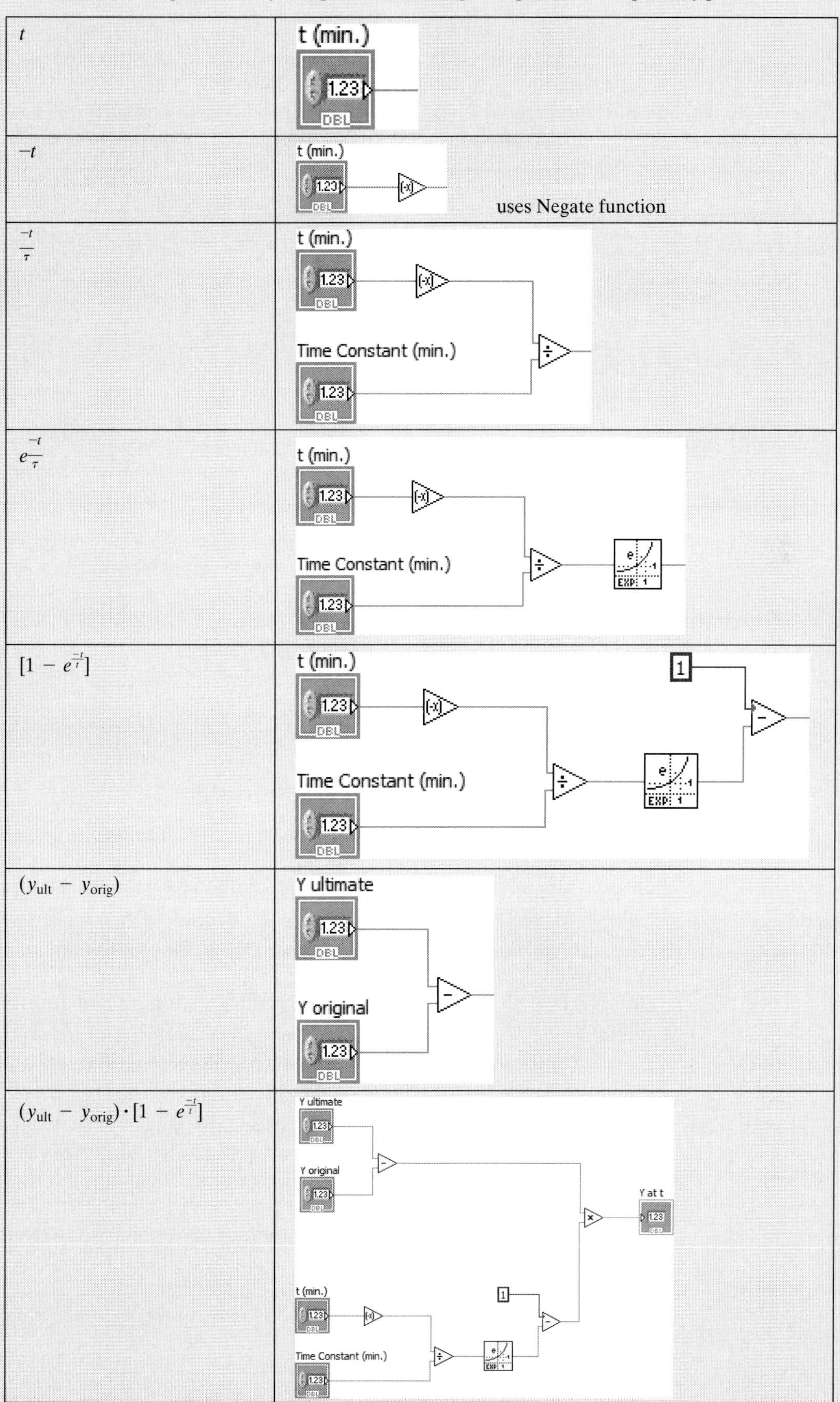

You should also add *comments* to the block diagram to help people understand what is being calculated, as illustrated in Figure 3.33. To add a comment to the block diagram, double-click on the background area. LabVIEW will insert a text field and leave it in edit mode so that you can enter the desired text. Click outside the comment area to finish entering text.

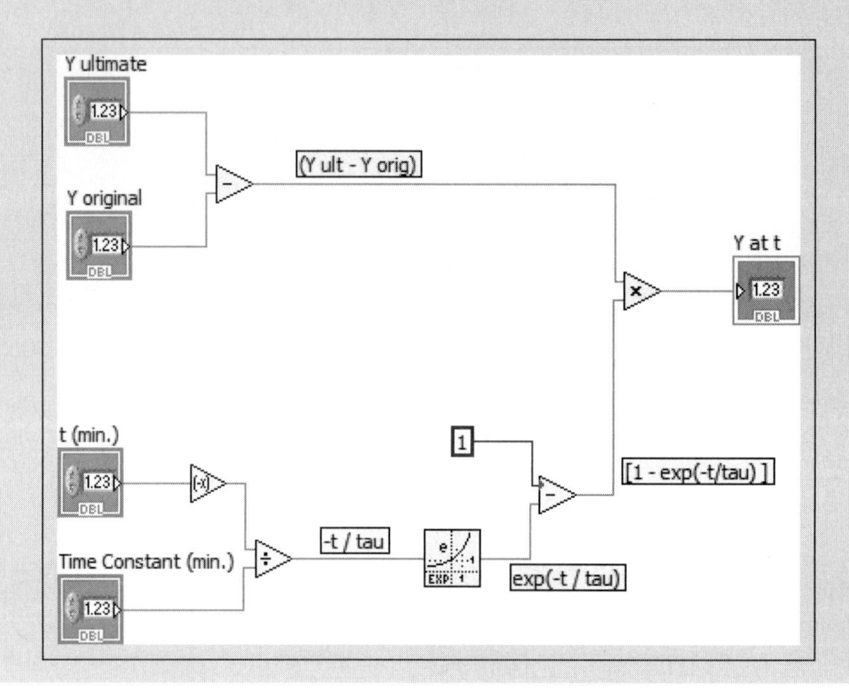

Figure 3.33 Block diagram for First Order VI with added comments.

3.5 BOOLEAN AND COMPARISON FUNCTIONS

Some fields within engineering rarely use the Boolean functions and comparison functions, but others use them frequently. Boolean and comparison functions are both used for making decisions, but there is a difference in the expected data type of the inputs to each type of function:

- *Boolean functions* take Boolean values (TRUE, FALSE) as inputs and return Boolean outputs.
- *Comparison functions* take numeric values as inputs and return Boolean outputs.

LabVIEW provides access to the following Boolean and comparison functions in two locations: in both the Programming and Express groups:

- Functions Palette/Programming Group/Boolean Group
- Functions Palette/Programming Group/Comparison Group
- Functions Palette/Express Group/Arithmetic & Comparison Group/Express Boolean Group
- Functions Palette/Express Group/Arithmetic & Comparison Group/Express Comparison Group

LabVIEW's Boolean functions are listed in Table 3.4, and the comparison functions are listed in Table 3.5.

Table 3.4 Boolean functions

Function	Comment
And	AND
Boolean To (0,1)	Converts FALSE, TRUE to 0, 1 and TF data type to 116
Compound Arithmetic	Performs certain math operations (Add, Multiply, AND, OR, or XOR) on more than two values
Exclusive Or	XOR
False Constant	Returns FALSE
Not And	NAND
Not Exclusive Or	NOT XOR
Not Or	NOR
Not	NOT
Or	OR
True Constant	Returns TRUE

Table 3.5 Comparison functions

Function		Comment (from LabVIEW help system)
Equal To 0?	$= 0$	Returns TRUE if x is equal to 0. Otherwise, this function returns FALSE.
Equal?	$=$	Returns TRUE if x is equal to y. Otherwise, this function returns FALSE.
Greater Or Equal To 0?	≥ 0	Returns TRUE if x is greater than or equal to 0. Otherwise, this function returns FALSE.
Greater Or Equal?	$\geq$	Returns TRUE if x is greater than or equal to y. Otherwise, this function returns FALSE.
Greater Than 0?	> 0	Returns TRUE if x is greater than 0. Otherwise, this function returns FALSE.
Greater?	$>$	Returns TRUE if x is greater than y. Otherwise, this function returns FALSE.
Less Or Equal To 0?	≤ 0	Returns TRUE if x is less than or equal to 0. Otherwise, this function returns FALSE.
Less Or Equal?	$\leq$	Returns TRUE if x is less than or equal to y. Otherwise, this function returns FALSE.
Less Than 0?	< 0	Returns TRUE if x is less than 0. Otherwise, this function returns FALSE.
Less?	$<$	Returns TRUE if x is less than y. Otherwise, this function returns FALSE.
Max & Min		Compares x and y and returns the larger value at the top output terminal and the smaller value at the bottom output terminal. This function accepts time stamp values if all inputs are time stamp values. If the inputs are time stamp values, the function returns the later time at the top and the earlier time at the bottom. The wire is broken if the inputs are not the same data type.
Not Equal To 0?	$\neq 0$	Returns TRUE if x is not equal to 0. Otherwise, this function returns FALSE.
Not Equal?	$\neq$	Returns TRUE if x is not equal to y. Otherwise, this function returns FALSE.
Select		Returns the value wired to the t input or f input, depending on the value of s. If s is TRUE, this function returns the value wired to t. If s is FALSE, this function returns the value wired to f.

EXAMPLE 3.7

Check Status of Safety Interlock Switches before Activating Machinery

One example of the use of Boolean functions is checking the status of two safety switches that both must be in the ON position before a piece of machinery can be started. In this example, if either switch A or switch B is in the OFF position (as in Figure 3.34), then the equipment cannot be started. But, if switch A and switch B are both ON (as in Figure 3.35), an LED indicator will be illuminated indicating that it is OK to start the equipment.

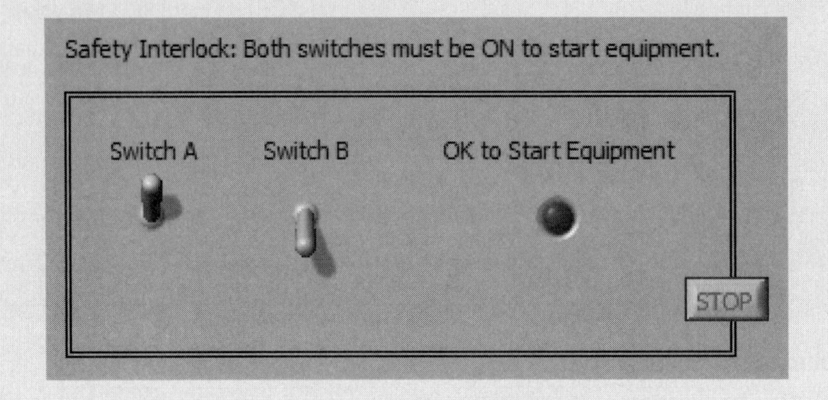

Figure 3.34 Safety Interlock: Equipment will not start unless both switches are ON.

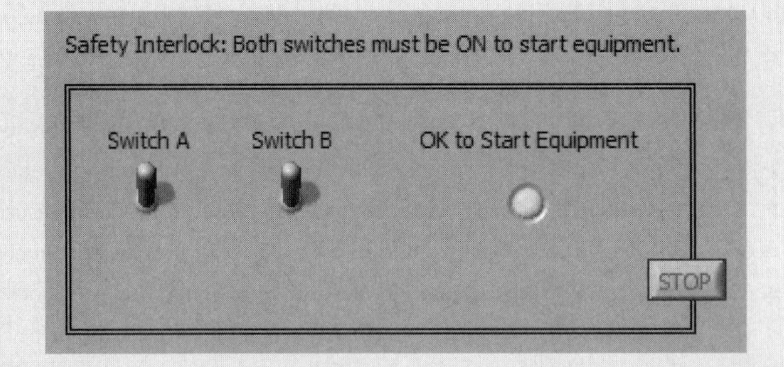

Figure 3.35 When both switches are ON, it is OK to start equipment.

The block diagram for this VI is shown in Figure 3.36.

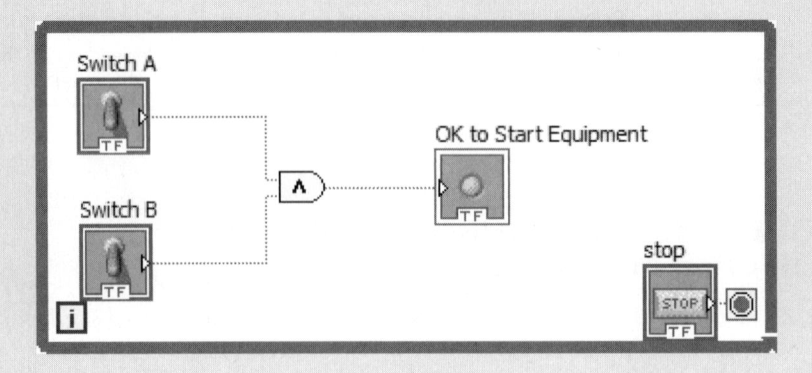

Figure 3.36 Block diagram for the safety interlock (uses an AND function).

PRACTICE

Using Boolean values

LabVIEW provides a *Select function* that will perform one of two actions depending on Boolean value. The function is available in either of two groups on the Functions Palette:

> **Functions Palette / Express Group / Arithmetic & Comparison Group / Express Comparison Group / Select**
>
> **Functions Palette / Programming Group / Comparison Group / Select**

For practice using the Select function, create a VI that will display either the natural logarithm of a value, or the base 10 logarithm depending on the position of a toggle switch. The front panel is shown in Figure 3.37.

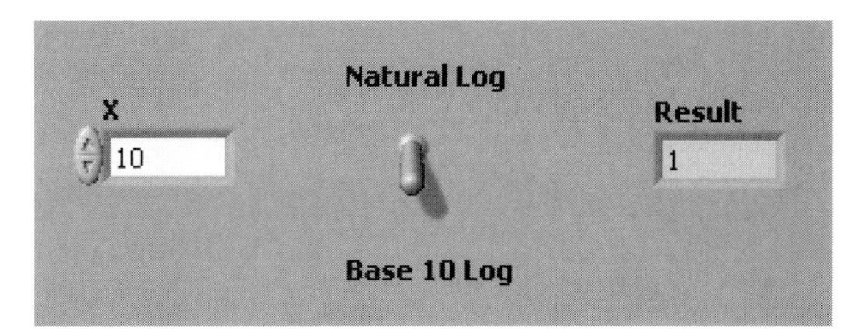

Figure 3.37
Front panel of a VI that allows the user to select type of logarithm.

Solution

The block diagram for this problem is shown in Figure 3.38. Notice that both logarithms are calculated, the Select function is used to determine which value is displayed. The toggle switch sends a Boolean value into the Select function.

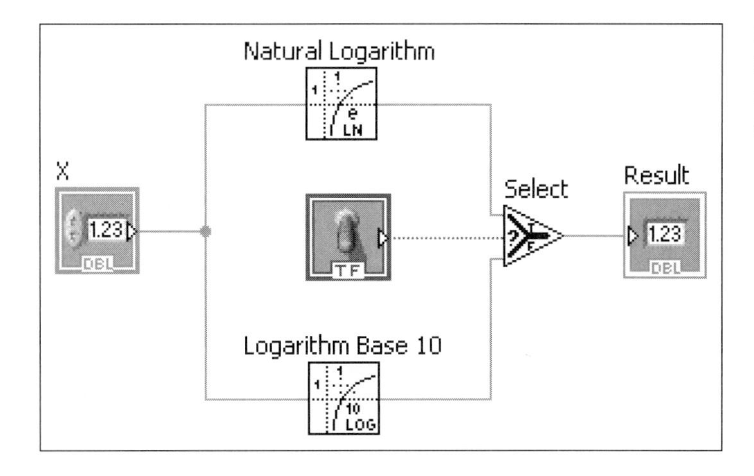

Figure 3.38
Block diagram of a VI that allows the user to select type of logarithm.

- When the switch sends a True, the upper path of the Select function is used and the natural logarithm is passed to the Result indicator.
- When the switch sends a False, the lower path of the Select function is used and the base 10 logarithm is passed to the Result indicator.

APPLICATION

Volume in a cylindrical tank

Gasoline is often stored in underground cylindrical tanks (see Figure 3.39).

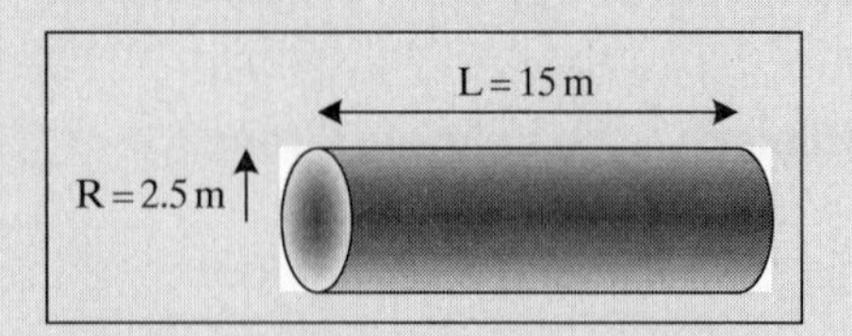

Figure 3.39
Underground storage tank.

To determine how much gasoline is left in the tank, someone will lower a measuring rod into the tank and see how much of the rod comes out wet; this is the liquid level, LL. Measuring the liquid level is easy, but determining the liquid volume is not quite so simple. If this is a task that needs to be done routinely, you can write a LabVIEW VI to easily compute liquid volume from liquid level. The front panel is shown in Figure 3.40.

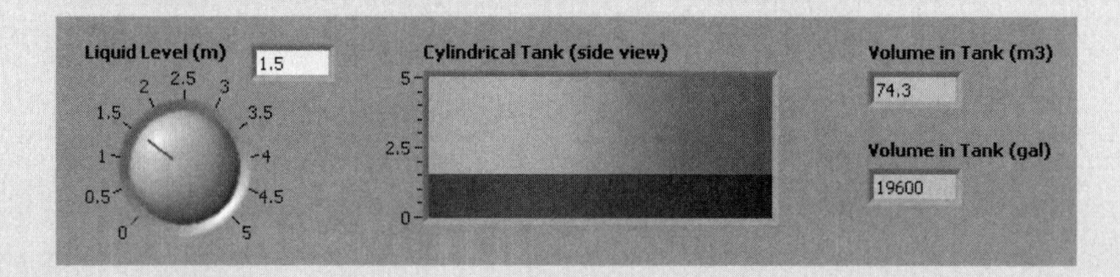

Figure 3.40
Front panel of a VI to compute liquid volume in a cylindrical tank.

The math involved is reasonably straightforward. With variables defined as shown in Figure 3.41, the equation for liquid volume is

$$V = \frac{1}{2} LR^2[\theta - sin(\theta)]$$

where

$$\theta = 2\,\mathrm{Acos}\left(\frac{R - LL}{R}\right)$$

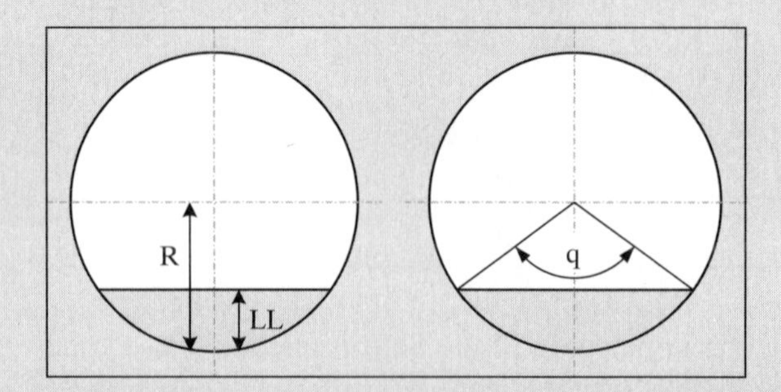

Figure 3.41
Variables used to determine liquid volume.

The block diagram is shown without annotations in Figure 3.42 for those who like to figure things out on their own, and with annotations in Figure 3.43.

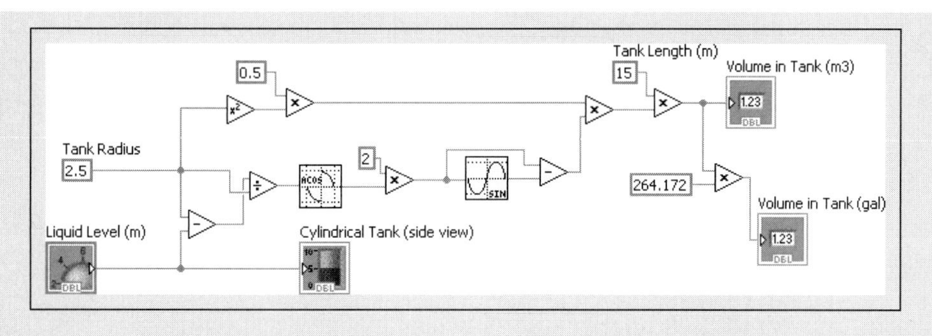

Figure 3.42
Cylindrical tank volume block diagram, without annotations.

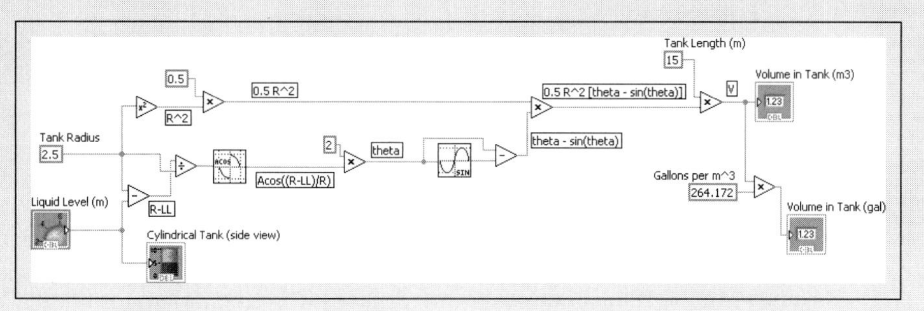

Figure 3.43
Cylindrical tank volume block diagram, annotated.

APPLICATION

Using comparison functions to check tank status

An overflowing tank can be an environmental disaster. In this example comparison functions and Boolean functions are used to deactivate a tank fill pump when the liquid level in a tank exceeds a specified Alarm Level. Figures 3.44 and 3.45 illustrate how the front panel looks before and after the High Level Warning.

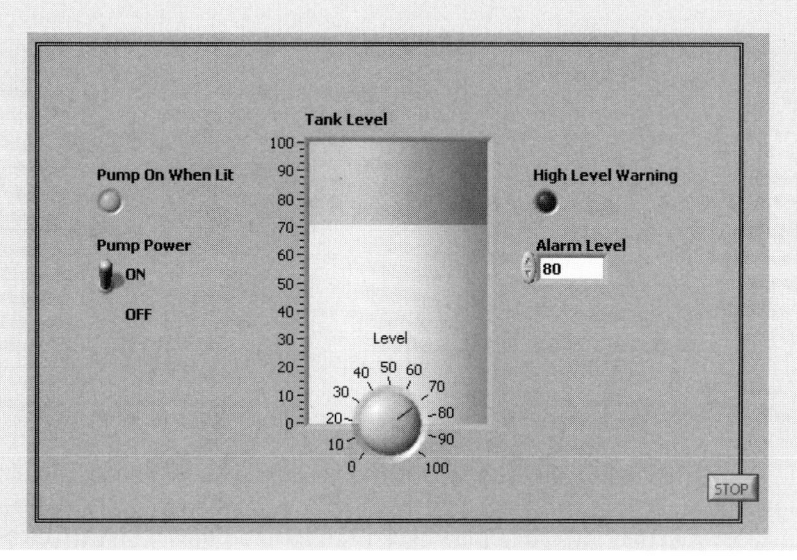

Figure 3.44
When level is below 80% the pump stays on.

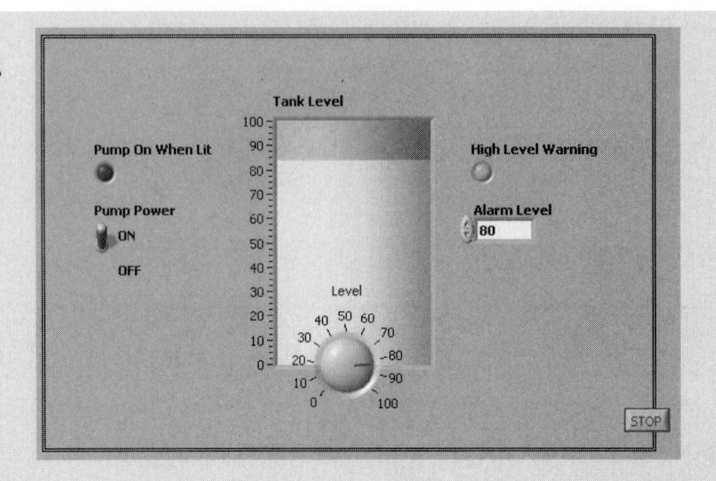

Figure 3.45
When level rises above 80% the high level warning shuts down the pump.

The block diagram for this VI is shown in Figure 3.46.

In Figure 3.46 the Greater Than comparison function tests to see if the actual tank level is greater than the alarm level. If it is, a TRUE is sent to the High Level Warning LED, illuminating that LED.

The NOT in the Boolean wire downstream of the Greater Than turns any TRUE generated by the Greater Than into a FALSE. The AND will send a TRUE to the Pump On When Lit LED (and, presumably, power to the pump) only when the pump power switch is on AND the actual tank level is NOT greater than the alarm level.

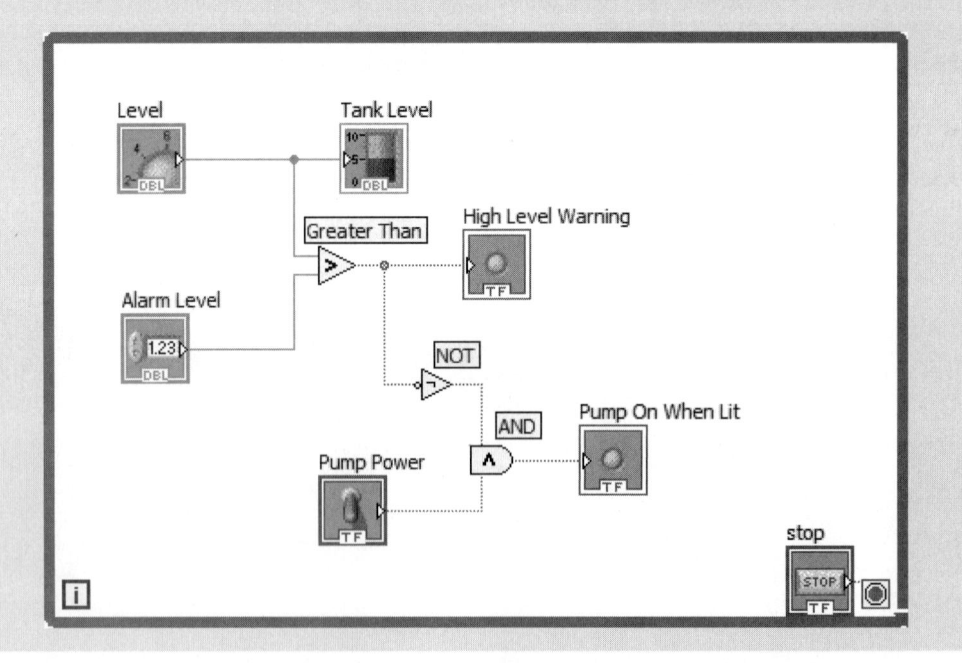

Figure 3.46
The block diagram for the high level warning system.

The goal of this chapter was to introduce some of the commonly used LabVIEW functions and show how they can be combined into VIs. There are many, many more functions available than have been covered here. LabVIEW's help system can assist you in becoming familiar with the functions that you might need in the future.

3.6 PROGRAMMING PREVIEW: DEBUGGING

LabVIEW's graphical programs mean that the process of *debugging* will be highly visual as well. While that makes things easier, LabVIEW is such a different approach to programming for most new users, it is not difficult to generate errors that can occasionally be hard to find and fix

LabVIEW provides several tools to help get your VIs running, and running correctly.

- **Broken wires**—an early warning that something is not working correctly
- **Broken Run Button**—indicates that the VI will not run, but if you click the broken Run button you will see a list of error messages
- **Execution Highlighting**—you can watch the flow of information through your block diagram
- **Single-Step Execution**—you can run the VI step by step to see where something goes wrong
- **Probe Tool**—you can test to see what any wire contains
- **Breakpoints**—you can set a breakpoint to stop the program at a point of interest

3.6.1 Fixing Broken Wires

The first error checking is done at the point of wiring the nodes on the block diagram. If a connection is not allowed, LabVIEW will show a *broken wire*. Broken wires must be fixed before the VI will run.

If you move your mouse over the broken wire, LabVIEW will display a message indicating the problem with the attempted connection. This is illustrated in Figure 3.47.

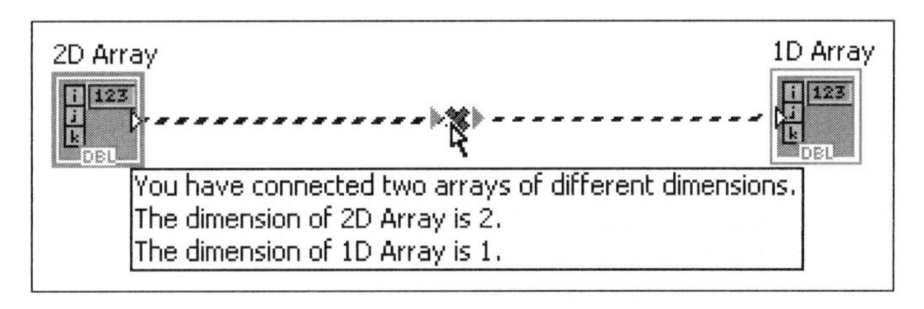

Figure 3.47
Error messages about broken wires appear when you move the mouse over the break.

The most common problem with broken wires is mismatched data types. There are two solutions:

- Determine why the data type is mismatched and change the incorrect item.
- Convert from one data type to another.

The latter approach often seems reasonable to people new to LabVIEW; it usually leads to trouble eventually. Unless you know why you need to use a data type conversion function, it's probably a good idea to try to fix the problem instead.

3.6.2 Using the Broken Run Button

The *broken Run* button, shown in Figure 3.48 is an indicator that the VI will not run in its current form. You commonly see the broken Run button when you are in the process of wiring the VI. When that last connection is made and the broken Run button goes away, you have a lot of hope that the VI will run successfully.

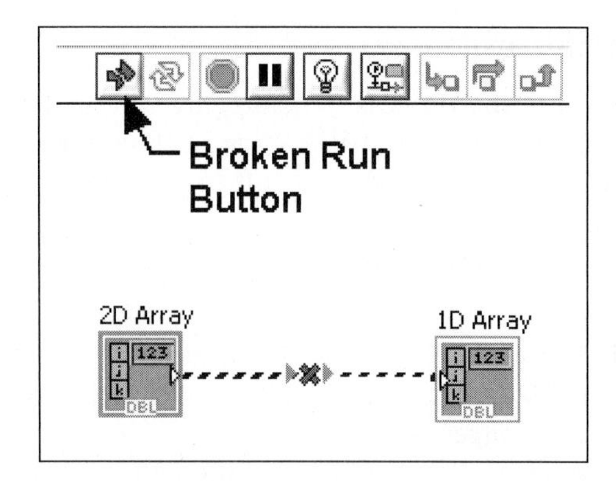

Figure 3.48
The Broken Run Button indicates a problem with the VI.

But the broken Run button does even more than tell you there's a problem. If you click the broken Run button LabVIEW will display the *Error List*, as shown in Figure 3.49.

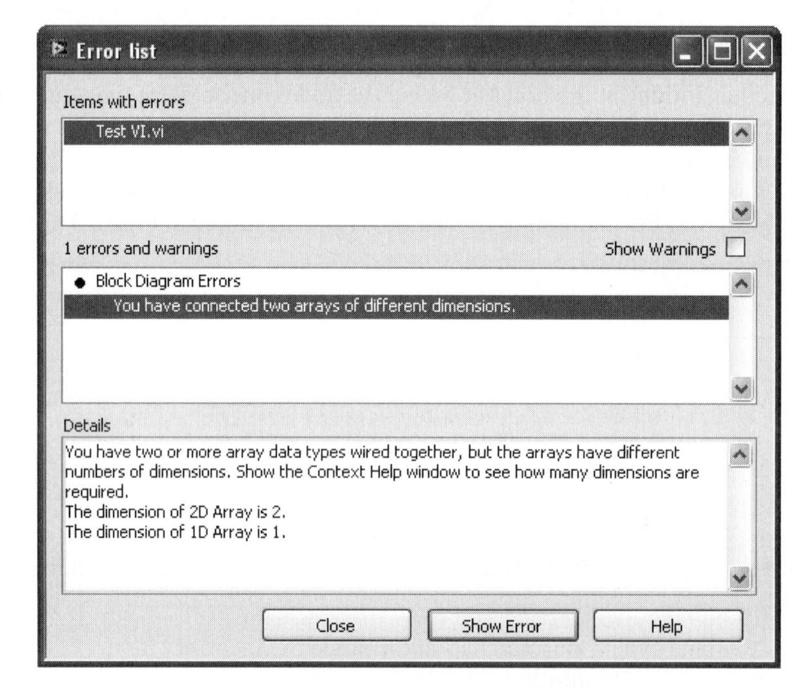

Figure 3.49
The Error List is displayed when the broken Run button is clicked.

The Error List describes each error and, if you click the **Show Error** button, will highlight the problem on the VI itself.

3.6.3 Execution Highlighting

LabVIEW is a *dataflow programming language*. Dataflow programming means that a node (or block) on a block diagram executes as soon as all of the inputs have values. Sometimes seemingly odd results happen because LabVIEW is performing calculations in a different order than you expected. You can use *execution highlighting* to determine the order in which tasks are carried out in your LabVIEW programs.

To highlight execution (from the block diagram):

1. Click the Highlight Execution button (looks like a lightbulb).
2. Click the Run button.

The VI will slowly execute, and LabVIEW will display the progress on the block diagram by showing the wires containing calculated data values in full color instead of being grayed out. In Figure 3.50 the Tank High Warning VI is about halfway through. Notice that the Level and Alarm controls show values. That indicates that they have already been evaluated. The Boolean controls (e.g., Pump Power switch) show "F" when they evaluate False.

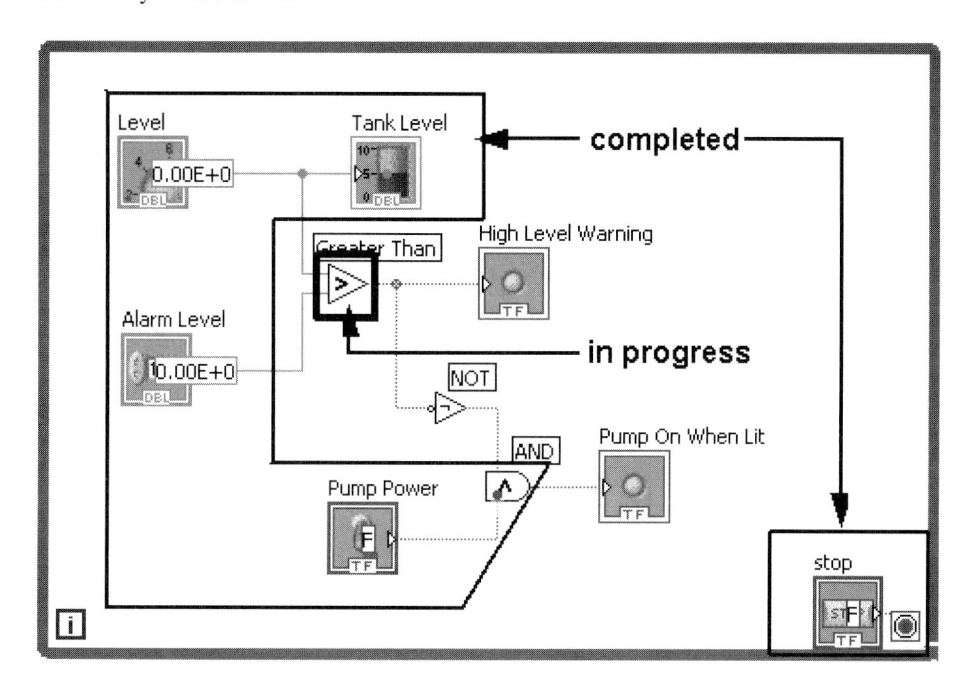

Figure 3.50
The Tank High Warning VI, running with highlighted execution.

Execution highlighting is kind of fun to watch once in a while, but it can also be very useful in determining how the VI is actually calculating.

3.6.4 Single-Step Execution

Sometimes you need to slow down the execution even further, and you can execute the VI step by step using the **Step Into** and **Step Over** buttons on the block diagram's toolbar.

- **Step Into**—Follow the progress into Loop structures.
- **Step Over**—Follow the progress but skip over Loop structures.

With *single-step execution* you can pinpoint where a problem is occurring.

3.6.5 Probes

You can use the *Probe Tool* to determine the contents of any wire while the VI is running (either at full speed, or with execution highlighting on to slow the speed of execution) or before running the VI.

To set up a probe on a wire before the VI is running, display the Tools Palette (**View/Tools Palette**) and select the Probe Tool. Then, click on the wire that you

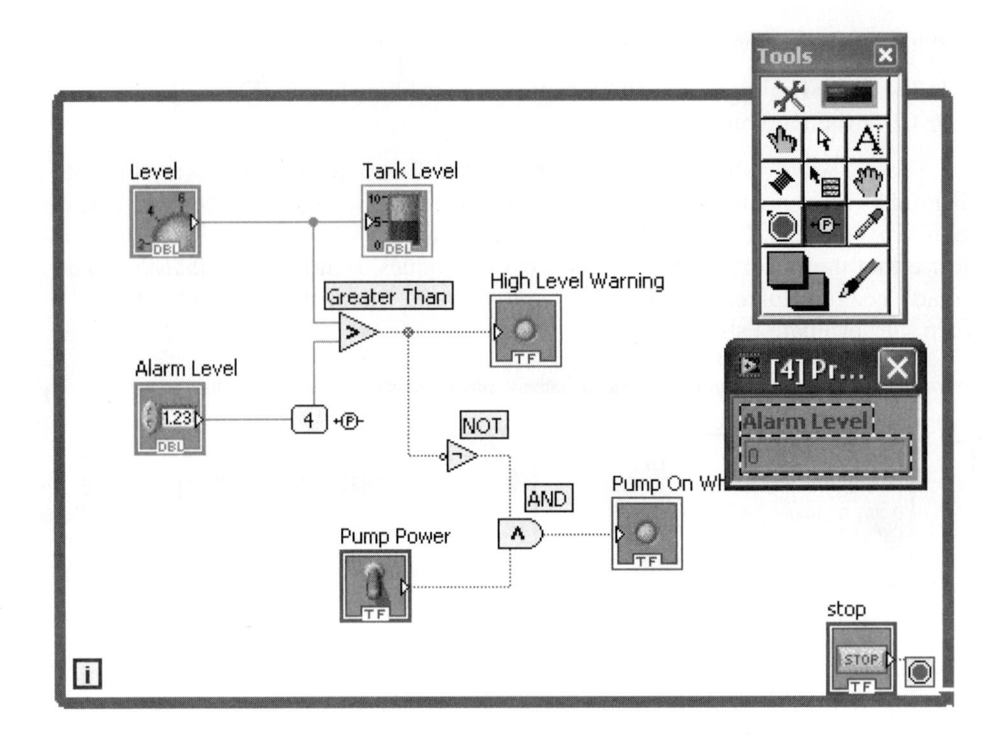

Figure 3.51
Placing a probe on the wire leading to the A + B indicator.

want to probe. In Figure 3.51 the probe is being placed on the wire leaving the Alarm Level control.

Once the probe is in place, when the VI is running, the value in the probed wire is displayed. This is illustrated in Figure 3.52.

3.6.6 Breakpoints

A *breakpoint* is a location in a program where the program execution is temporarily halted. This gives the programmer a chance to probe around to see what's happening.

- If a program is crashing, breakpoints can be used to see what is happening just before the VI crashes.
- If a program is calculating strange values, breakpoints can be used to follow the progress of the program to see where things go awry.

To set a breakpoint, right-click on a node or wire in the block diagram, and select **Set Breakpoint** from the pop-up menu. In Figure 3.53 a breakpoint was set just downstream of the Greater Than function.

When the VI executes, execution will stop at the breakpoint, but the wires will still have values so you can use the probe tool to see what's happening.

To clear the breakpoint, right-click on a breakpoint, and select **Clear Breakpoint** from the pop-up menu.

Debugging any program is a chore, but LabVIEW provides the tools to let you know what's going on inside the program, and that is a big help.

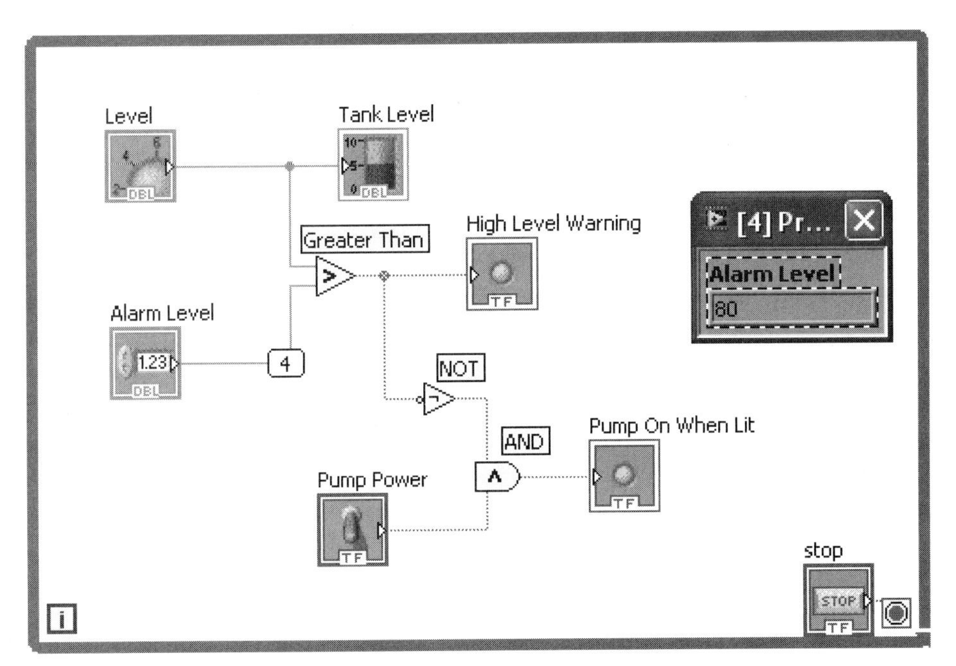

Figure 3.52
Probe 4 indicates that the value in the wire is 80.

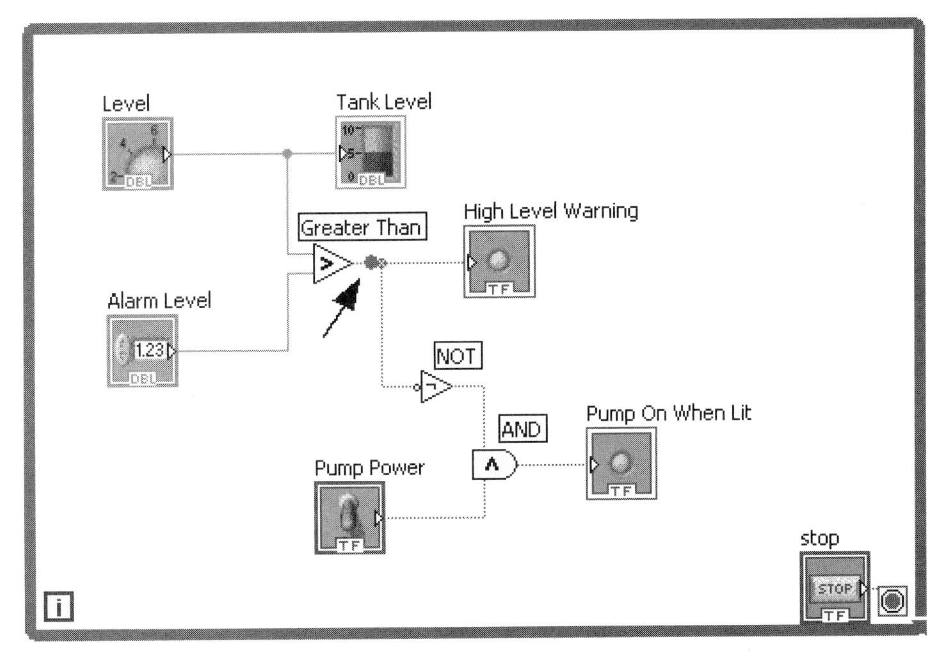

Figure 3.53
A breakpoint has been set just after the Greater Than function.

KEY TERMS

Add function
And function
Body Mass Index (BMI)
Boolean functions
breakpoint
Broken Run Button
broken wires
comments
comparison functions
Context Help
dataflow programming
debugging
Dial Numeric Control
Divide function
Error List
Exclusive Or function
Execution Highlighting
exponential
Express VI
first-order response
Formula Express VI
Formula function
Function
Hyperbolic Trigonometric functions
integer quotient

Inverse Hyperbolic Trigonometric functions
Inverse Trigonometric functions
Knob control
Logarithm
Logarithm Base 10
Meter indicator
Multiply function
Natural Logarithm
Not function
Numeric indicator
Or function
Pointer Slide controls
predefined constants (e.g., e, π)
Probe Tool
radians
Random Number function
remainder
Run Continuously button
Select function
Single-Step Execution
trigonometric functions

SUMMARY

Building a VI (General Approach)

1. Create a blank VI:
   - From LabVIEW Getting Started window: **New Panel/Blank VI**
   - From an open VI: **File/New VI**

On the front panel . . .

2. Add needed controls and indicators

On the block diagram . . .

3. Add needed functions
4. Wire the controls, functions, and indicators
5. Draw a While Loop around all items on the block diagram (if VI is to be run continuously)

   Functions Palette / Express Group / Execution Control Group / While Loop

Basic Math Functions

Functions Palette / Mathematics Group / Numeric Group

Function	
Add	
Subtract	Round Towards − Infinity
Multiply	Round Towards + Infinity
Divide	Square Root
Quotient and Remainder	Square
Increment	Negate
Decrement	Reciprocal
Absolute Value	Numeric Constant
Round to Nearest	Random Number

Predefined Constants

- π
- e (base of natural logarithm)
- h (Planck's constant)
- c (speed of light in a vacuum)
- G (gravitational constant)
- R (molar gas constant)

Context Help
Provides information on how to wire math functions.

Help / Show Context Help

Changing the Data Type Associated with a Control
Example: Change the data type to 32-bit Integer:

1. Right-click on the control (opens the pop-up menu)
2. Select: **Representation/I32**

Automatically Creating Controls and Indicators

Input: Right-click on an input, select **Create/Control**
Output: Right-click on an output, select **Create/Indicator**

Benefits:

- correct data type
- labeled
- wired to the function

Quotient and Remainder function
Returns two results:

- the integer number of times that denominator can divide into numerator (*integer quotient*)
- the *remainder* of that division

Random Number function
Returns a double-precision floating point value greater than or equal to 0 and less than 1.

Express VI

- configured using a dialog box
- dialog box opens when the Express VI is placed on the block diagram.
- double-click to re-open the dialog when needed.

Formula Express VI

- functions like a scientific calculator
- accepts up to eight inputs
- enter formula using a dialog box

Trigonometric Functions (angles in radians)

Trigonometric functions	Hyperbolic Trigonometric functions
Sine	Hyperbolic Sine
Inverse Sine	Inverse Hyperbolic Sine
Cosine	Hyperbolic Cosine
Inverse Cosine	Inverse Hyperbolic Cosine
Sine & Cosine	
Tangent	Hyperbolic Tangent
Inverse Tangent	Inverse Hyperbolic Tangent
Inverse Tangent (2 Input)	
Secant	Hyperbolic Secant
Inverse Secant	Inverse Hyperbolic Secant
Cosecant	Hyperbolic Cosecant
Inverse Cosecant	Inverse Hyperbolic Cosecant
Cotangent	Hyperbolic Cotangent
Inverse Cotangent	Inverse Hyperbolic Cotangent
Sinc	

Convert degrees to radians

$$\text{Angle in radians} = \text{Angle in degrees} \cdot \frac{2\pi \text{ radians}}{360 \text{ degrees}}$$

Exponential and Logarithm Functions

Function name	
Exponential	$\exp(x)$
Exponential (Arg) − 1	$\exp(x) - 1$
Natural Logarithm	$\ln(x)$
Natural Logarithm (Arg + 1)	$\ln(x + 1)$
Logarithm Base 2	$\log_2(x)$
Logarithm Base 10	$\log_{10}(x)$
Logarithm Base X	$\log_X(x)$
Power Of 2	2^x
Power Of 10	10^x
Power Of X	X^x
Y-th Root of X	$\sqrt[y]{X}$

First-order response

A *first-order response* is described by the equation

$$y(t) = (y_{\text{ult}} - y_{\text{orig}})[1 - e^{\frac{-t}{\tau}}]$$

where

y_{orig} is the *initial (original) value of* y at time $t = 0$
y_{ult} is the *ultimate value of* y at time $t = \infty$
τ is the *time constant*

Boolean and Comparison Functions

- Boolean functions take Boolean values (TRUE, FALSE) as inputs and return Boolean outputs.
- Comparison functions take numeric values as inputs and return Boolean outputs.

Locations

- Functions Palette/Programming Group/Boolean Group
- Functions Palette/Programming Group/Comparison Group
- Functions Palette/Express Group/Arithmetic & Comparison Group/Express Boolean Group
- Functions Palette/Express Group/Arithmetic & Comparison Group/Express Comparison Group

Boolean Functions

Function	Comment
And	AND
Boolean To (0,1)	Converts FALSE, TRUE to 0, 1 and TF data type to 116
Compound Arithmetic	Performs certain math operations (Add, Multiply, AND, OR, or XOR) on more than two values
Exclusive Or	XOR
False Constant	Returns FALSE
Not And	NAND
Not Exclusive Or	NOT XOR
Not Or	NOR
Not	NOT
Or	OR
True Constant	Returns TRUE

Comparison Functions

Function		Comment (from LabVIEW help system)
Equal To 0?	$= 0$	Returns TRUE if x is equal to 0. Otherwise, this function returns FALSE.
Equal?	$=$	Returns TRUE if x is equal to y. Otherwise, this function returns FALSE.
Greater Or Equal To 0?	≥ 0	Returns TRUE if x is greater than or equal to 0. Otherwise, this function returns FALSE.
Greater Or Equal?	≥ 0	Returns TRUE if x is greater than or equal to y. Otherwise, this function returns FALSE.
Greater Than 0?	> 0	Returns TRUE if x is greater than 0. Otherwise, this function returns FALSE.
Greater?	$>$	Returns TRUE if x is greater than y. Otherwise, this function returns FALSE.

Less Or Equal To 0?	≤ 0	Returns TRUE if x is less than or equal to 0. Otherwise, this function returns FALSE.
Less Or Equal?	$\leq$	Returns TRUE if x is less than or equal to y. Otherwise, this function returns FALSE.
Less Than 0?	< 0	Returns TRUE if x is less than 0. Otherwise, this function returns FALSE.
Less?	$<$	Returns TRUE if x is less than y. Otherwise, this function returns FALSE.
Max & Min		Compares x and y and returns the larger value at the top output terminal and the smaller value at the bottom output terminal. This function accepts time stamp values if all inputs are time stamp values. If the inputs are time stamp values, the function returns the later time at the top and the earlier time at the bottom. The wire is broken if the inputs are not the same data type.
Not Equal To 0?	$\neq 0$	Returns TRUE if x is not equal to 0. Otherwise, this function returns FALSE.
Not Equal?	$\neq$	Returns TRUE if x is not equal to y. Otherwise, this function returns FALSE.
Select		Returns the value wired to the t input or f input, depending on the value of s. If s is TRUE, this function returns the value wired to t. If s is FALSE, this function returns the value wired to f.

Debugging

Debugging is the process of analyzing a program for errors, and removing them.

LabVIEW debugging tools

- **Broken wires**—an early warning that something is not working correctly
- **Broken Run Button**—indicates that the VI will not run, but if you click the broken Run button you will see a list of error messages
- **Execution Highlighting**—you can watch the flow of information through your block diagram
- **Single-Step Execution**—you can run the VI step by step to see where something goes wrong
- **Probe Tool**—you can test to see what any wire contains
- **Breakpoints**—you can set a breakpoint to stop the program at a point of interest

SELF-ASSESSMENT

1. How do you add a digital display to a dial control?
   ANS: Right-click on the control and select **Visible Items/Digital Display** from the pop-up menu.
2. The Controls Palette and the Functions Palette are both commonly used as LabVIEW VIs are created.

   a. Which palette is used when developing the front panel?
   b. Which palette is used when developing the block diagram?

   ANS: The Controls Palette is used when developing the front panel and the Functions Palette is used when developing the block diagram.
3. Where are the basic math functions located (which palette and which group)?
   ANS: Functions Palette/Mathematics Group
4. Find the following functions on the Functions Palette:

   a. Add
   b. Round to Nearest

c. π (constant)
d. Tangent
e. Exponential (e^x)

ANS: They are all found in the **Functions Palette/Mathematics Group**, then

a. Add	**Numeric Group/Add**
b. Round to Nearest	**Numeric Group/Round to Nearest**
c. π (constant)	**Numeric Group/Math & Scientific Constants/Pi**
d. Tangent	**Elementary & Special Functions/Trig. Functions/Tangent**
e. Exponential (e^x)	**Elem. & Special Functions/Exponential Functions/ Exponential**

Note: These functions are also available in the **Express Group/Arithmetic & Comparison Group**

5. By default, a numeric control includes increment and decrement buttons. What are these buttons used for?
ANS: You can click these buttons with the mouse to increase or decrease the value displayed in the control.

6. Engineers commonly use constants such as π, h (Planck's constant), and G (gravitational constant). Where can these constants be found on the Functions Palette?
ANS: Functions Palette/Mathematics Group/Numeric Group/Math & Scientific Constants

7. What steps are required to replace one function with another on the block diagram?
ANS:

1. Right-click on the function to be replaced.
2. Select **Replace** on the pop-up menu. Two options will typically be displayed:
   a. The palette (or group) that the current function came from.
   b. All palettes.
3. Locate and click on the function to be placed on the block diagram.

8. How do you set up a control to handle integer values?
ANS: Right-click on the control, select **Representation/I32** (or another integer data type) from the pop-up menu.

9. LabVIEW's random number function returns uniformly distributed values in what range?
ANS: Greater than or equal to zero, less than one.

10. LabVIEW's Formula Express VI accepts a number of inputs and allows you to easily build the input values into a formula. How many inputs will the Formula Express VI allow?
ANS: Up to eight.

11. Do LabVIEW's trigonometric functions work with angles in degrees or radians?
ANS: Only radians.

12. How do wires get broken?
ANS: Wires get broken when nodes are deleted from the block diagram, or when you attempt to connect two terminals that are incompatible.

13. How can you get a list of error messages?
ANS: Click the broken **Run** button to see a list of errors.

14. What is execution highlighting used for?
ANS: Execution highlighting causes each node on the block diagram to be highlighted when it is solved; it allows you to watch how LabVIEW solves the VI. If the VI is not functioning correctly, execution highlighting can help you find out where something is going wrong.

15. How is the Probe tool used for debugging?
ANS: When a VI stops running the values in the controls, indicators, and wires are left in their final state. The Probe tool allows you to poke around on the block diagram and check the values in wires. This can help you to find out why a VI is malfunctioning.

16. How do you set and clear breakpoints?
ANS: You set a breakpoint by right-clicking on a node or wire on a block diagram and selecting **Set Breakpoint** from the pop-up menu. To clear a break point, right-click on the breakpoint and select **Clear Breakpoint** from the pop-up menu.

PROBLEMS

1. Create a VI that converts an angle value in radians to degrees. Use the following values to test your VI: 180° is equal to π radians. Then convert the following values:

a. $\pi/2$

b. $3\pi/2$

c. 2π

2. Create a VI that converts temperature values in °F to °C and K.

$$T(°C) = \frac{T(°F) - 32}{1.8}$$
$$T(K) = T(°C) + 273.15$$

The following values can be used to test the VI: 32°F = 0°C = 273.15 K; −40°F = −40°C = −233.15 K. Then convert the following values:

a. 212°F

b. 98.6°F

c. 350°F

d. 1400°F

3. Create a VI that converts temperature values in °C to °F. The following values can be used to test the VI: 32°F = 0°C; −40°F = −40°C. Then convert the following values:

a. 100°C

b. 37°C

c. 85°C

d. 1200°C

4. Maintenance needs to get a ramp up to a height of 2 meters, and company regulations do not allow a ramp to be used with an angle greater than 30°. Create a VI to determine the required length of the ramp (pictured in Figure 3.54).

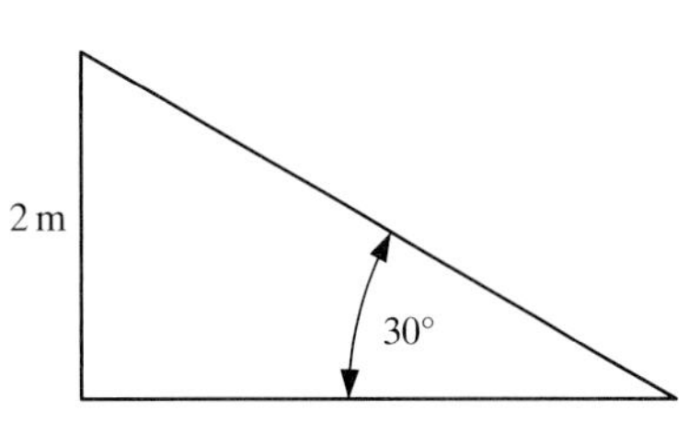

Figure 3.54 Determining required ramp length.

5. Some neighbors just added a particularly ugly addition to their home. You have a suspicion that the addition is not in compliance with the homeowners' association limit of 32 feet to the highest point on the home. To surreptitiously check it out, you wait until mid afternoon and take a photo from the street. From the photo you determine that the home's shadow is making an angle of 40° from the ground, as indicated in Figure 3.55. Then you step off the length of the shadow along the ground (32 steps).

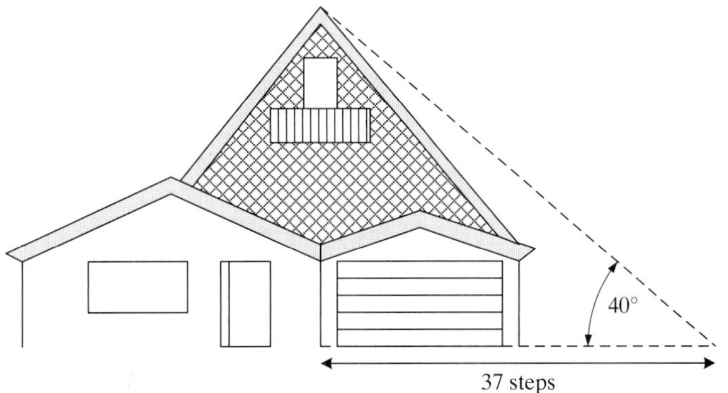

Figure 3.55 Determining the height of an addition.

Create a LabVIEW VI to determine if the neighbors are in compliance with the neighborhood height requirement. A possible VI front panel is shown in Figure 3.56.

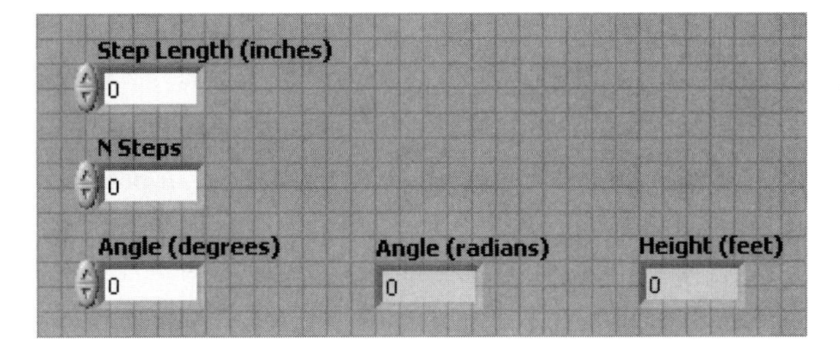

Figure 3.56 Front panel of a possible VI configuration.

a. Assume your feet are exactly one foot in length and determine the height of the addition.

b. How far off can the "length of foot" assumption be before the conclusion (in compliance, or not in compliance) would change?

6. Pneumatically controlled valves are operated by changing the air pressure applied to a diaphragm, as illustrated in Figure 3.57. The pressure, acting on the area of the diaphragm, generates a force that attempts to move the valve stem.

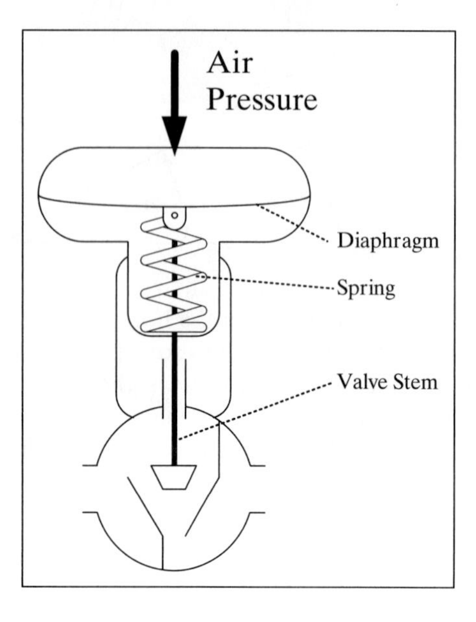

Figure 3.57
Pneumatic control valve.

$$F_{air} = PA$$

That force is countered by a spring that pushes on the diaphragm in the other direction.

$$F_{spring} = k_{spring}x$$

where

k_{spring} is the spring constant
x is the extension of the spring

Control valves typically use air pressures ranging between 3 psi (lb_f/in^2) and 15 psi. The change in applied force because of the pressure change causes the valve stem to move approximately 1 inch.

Determine:

a. The force applied to a 10-inch diameter diaphragm by 3 psi air.
b. The force applied to a 10-inch diameter diaphragm by 15 psi air.
c. The change in applied force (ΔF) as the air pressure increases from 3 to 15 psi.
d. The spring constant required to cause the valve stem to move 1 inch in response to the calculated change in force.

Note: The difference version of the spring equation may be helpful:

$$\Delta F_{spring} = k_{spring}\Delta x$$

7. When fluids have to be transported long distances, pipelines can be an efficient, safe, and economical alternative to trucking. But, because of fluid friction, the pressure of the fluid falls over distance, and pumping stations are required at intervals to re-pressurize the fluid.

The pressure drop between pumping stations can be predicted (if it is due to fluid friction only) with the following equation:

$$\Delta P = \frac{1}{2} f \rho \frac{L}{D} V_{avg}^2$$

where

f is the Moody (or Darcy) friction factor (no units)
ρ is the fluid density (kg/m^3)
L is the distance between pumping stations (m)
D is the pipe diameter (m)
V_{avg} is the average fluid velocity (m/s)

The friction factor depends in a complicated fashion on fluid, flow rate, and pipe surface conditions. For this problem, assume that $f = 0.008$.

Create a VI similar to Figure 3.58 that will allow you to enter the required values, and then calculate and display the pressure drop in Pa (Pa = 1 N/m^2)) and bars (1 bar = 10^5 Pa). Test your VI with these values:

$f = 0.008$
$\rho = 800$ kg/m^3
$L = 15{,}000$ m
$D = 0.2$ m
$V_{avg} = 1.5$ m/s
$\Delta P = 5.4$ bar

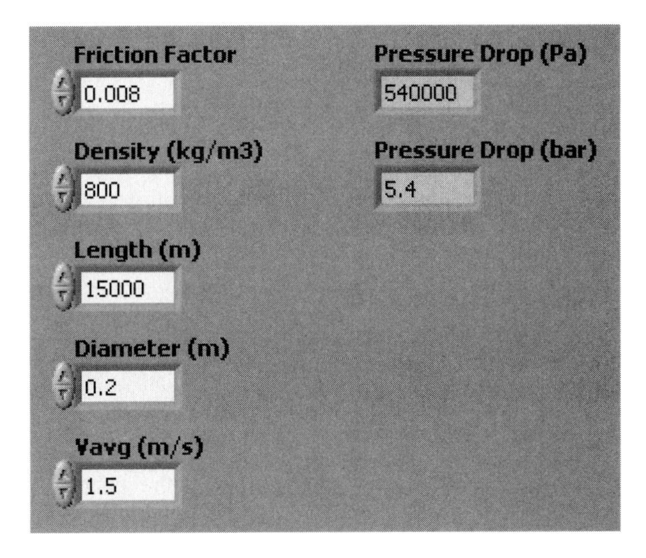

Figure 3.58
VI for estimating pipeline pressure drop.

Then, answer the following questions:

a. What will the pressure drop increase to if the distance between piping stations is increased to 20,000 m?

b. What is the maximum distance between pumping stations that will keep the pressure drop from exceeding 5 bars?

c. To increase throughput, it has been proposed to increase the average fluid velocity to 2.2 m/s. What is the expected pressure drop at the higher flow rate? Assume $L = 15{,}000$ m.

d. What will happen to the expected pressure drop if a different fluid with a density of 1100 kg/m³ is transported in the pipeline? Assume $V_{avg} =$ 1.5 m/s.

8. Three resistances in parallel (shown in Figure 3.59) are subjected to the same voltage drop. Create a VI similar to the one shown in Figure 3.60 that will calculate the current through each resistor. Ohm's law will be used for each resistor:

$$V = IR$$

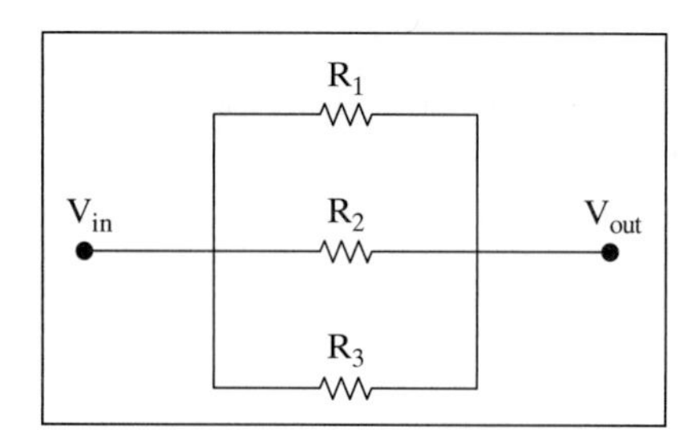

Figure 3.59
Resistors in parallel.

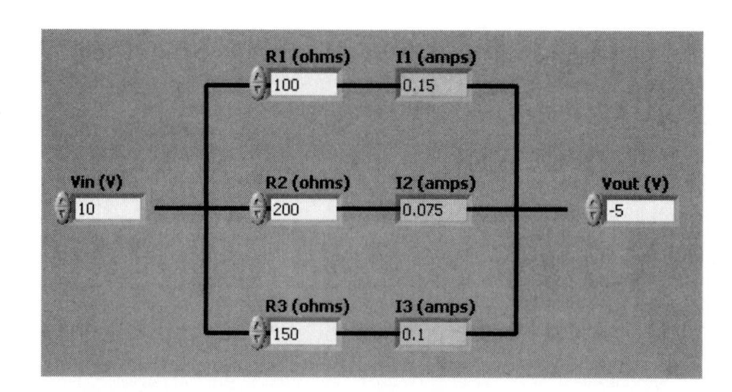

Figure 3.60
VI for finding current through resistors in parallel.

Test your VI with these values:

- $V_{in} = +10$ V
- $V_{out} = -5$ V
- $R_1 = 100$ ohms $I_1 = 0.15$ amp
- $R_2 = 200$ ohms $I_2 = 0.075$ amp
- $R_3 = 400$ ohms $I_3 = 0.1$ amp

Then, use your VI to solve for the currents through each resistor in this problem:

- $V_{in} = +12$ V
- $V_{out} = 0$ V
- $R_1 = 2500$ ohms
- $R_2 = 1000$ ohms
- $R_3 = 1200$ ohms

CHAPTER 4

Matrix Math Using LabVIEW

Objectives

After reading this chapter, you will know:

- how arrays and matrices are used in LabVIEW
- how to extract a subarray from a larger array or matrix
- how to use LabVIEW functions for matrix mathematics
    - adding arrays
    - transposing arrays
    - multiplying an array by a scalar
    - matrix multiplication
    - finding the condition number for a matrix
    - calculating the determinant of a matrix
    - matrix inversion
- how to solve simultaneous linear equations using LabVIEW
- how to use For Loops to create and process array data

4.1 WORKING WITH MATRICES AND ARRAYS IN LABVIEW

Matrices and arrays provide an efficient way to store and manipulate data sets. LabVIEW provides an extensive set of basic matrix math functions including

- Adding Matrices or Arrays
- Transpose of a Matrix or Array
- Multiplying Matrices or Arrays
- Condition Number
- Matrix Determinant
- Inverse Matrix

LabVIEW also provides more complex functions that can be used directly to solve matrix problems. For example, LabVIEW provides a function for solving simultaneous linear equations.

4.1.1 Should You Use Arrays or Matrices?

The term *array* has a long history in the field of computer programming, and there are a number of standard *array operations*, like sorting and finding minimum and maximum values, that are common knowledge to programmers. The term *matrix* comes from the field of mathematics, and there are a number of standard *matrix math operations*, like transposing and inverting, that are known to anyone who works with matrices.

LabVIEW supports both arrays and matrices; there are a lot of similarities, but a couple of subtle distinctions. In LabVIEW, both arrays and matrices are collections of related values, but a matrix is always 2D (two-dimensional), whereas an array can have any number of dimensions. Some LabVIEW functions (e.g., graphing functions) required 1D arrays, so arrays must be used with those LabVIEW functions. Since arrays are more flexible in LabVIEW, we will primarily use arrays in this chapter.

In LabVIEW both array operations and matrix math operations can be used on either arrays or matrices. We will use the term *matrix math* when describing matrix operations that are common in the field mathematics, but we will apply the matrix math operations to LabVIEW arrays.

Actually, LabVIEW provides three ways to collect values:

- Clusters
- Arrays
- Matrices

Clusters

A *cluster* is a grouping of potentially different data types. One possible use of a cluster would be to store information about a person, such as

- Birthdate (date.time data)
- Age (integer numeric data)
- Height (floating point numeric data)
- Photograph (image data)

Clusters are used to collect related data (all about the same person, for example) that require a variety of data types. Clusters are not used for matrix math, so we won't mention them again in this chapter.

Arrays

In LabVIEW, an *array* is a collection of controls or indicators that all hold values with the same data type. Because of this, building an array is a multi-step process:

1. Put the *Array container* on the front panel:

   Controls Palette / Modern Group / Array, Matrix & Cluster Group / Array

2. Drop a control or indicator into the Array container (only one array element will be visible). The Numeric Control (used in this example) can be found in either of these groups:

   Controls Palette / Modern Group / Numeric Group /Numeric Control

   Controls Palette / Express Group / Numeric Controls Group / Num Ctrl

3. Resize the array to show the required number of elements.

This is illustrated in Figure 4.1

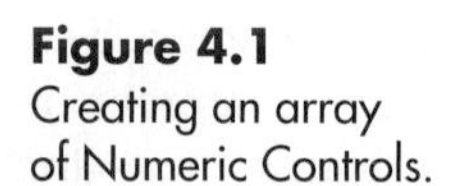

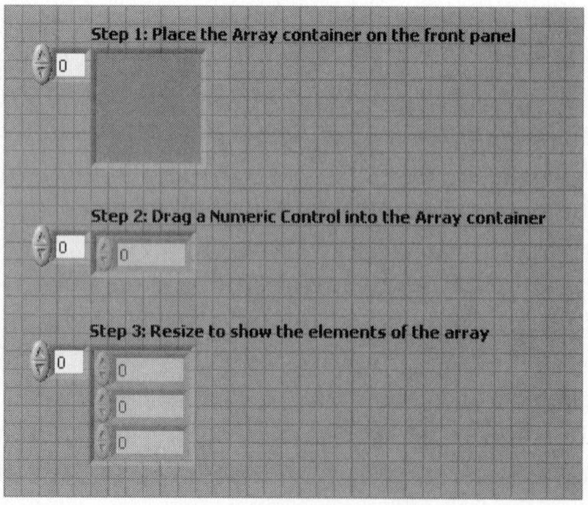

Figure 4.1
Creating an array of Numeric Controls.

By default, arrays have a single column of elements, called a 1D array. If you need a 2D array, right-click on the *Index Display* (on the left side of the Array container) and select **Add Dimension** from the pop-up menu. This is illustrated in Figure 4.2. An array can have many dimensions, if needed.

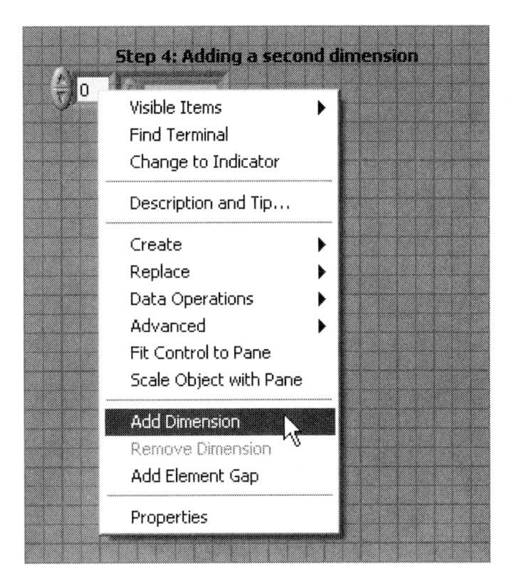

Figure 4.2
Adding a second dimension to create a 2D array.

The resulting array is shown in Figure 4.3

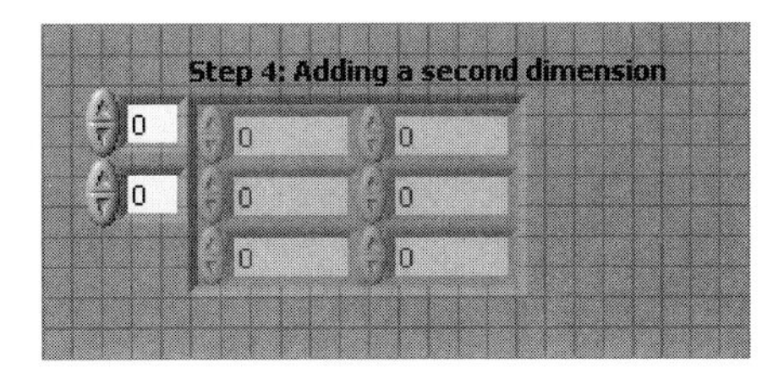

Figure 4.3
The 2D array of Numeric Controls.

LabVIEW provides a collection of *array functions* that are useful for working with arrays. These array functions are found in the Array Group:

Functions Palette / Programming Group / Array Group

The array functions include the following:

- Array Max & Min
- Array Size
- Array Subset
- Search 1D Array
- Sort 1D Array
- Split 1D Array
- Reverse 1D Array
- Rotate 1D Array
- Delete From Array
- Insert Into Array
- Replace Array Subset
- Reshape Array
- Transpose 2D Array
- Cluster To Array

- Array To Cluster
- Matrix to Array
- Array to Matrix

Matrices

A *matrix* is a single control that holds a collection of values of the same data type. A LabVIEW matrix is always 2D, although you can use a single *column* or *row*. A LabVIEW matrix cannot have more than two dimensions. The first element of a matrix is element 0, not 1.

A matrix can be placed on the front panel in a single step; simply move the Real Matrix control from the Array, Matrix & Cluster Group onto the front panel.

Controls Palette / Modern Group / Array, Matrix & Cluster Group / Real Matrix

This is illustrated in Figure 4.4.

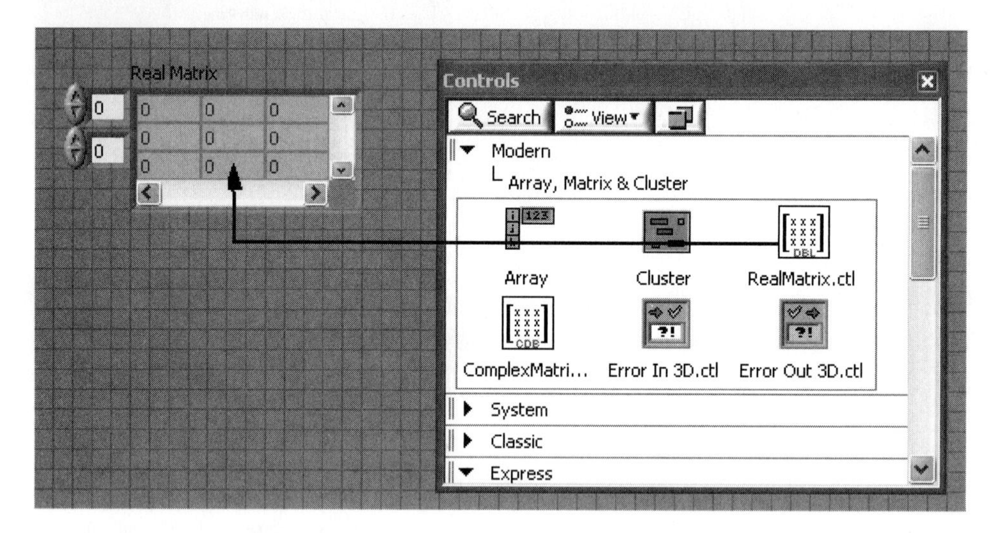

Figure 4.4
Placing a Matrix Control on the front panel.

Matrices can be placed on the front panel with fewer steps, and a collection of values (matrix) is a bit simpler to deal with than a collection of controls containing values (array).

LabVIEW is very flexible when it comes to matrix operations.

- If you need a subset of a matrix, you can use LabVIEW's array functions on matrices (or arrays).
- Most of the numeric math functions can be used on matrices as well.

LabVIEW will perform matrix math on both matrices and arrays, but LabVIEW was originally developed around arrays, and using matrices can cause problems (admittedly small problems) making downstream operations work correctly. For example, LabVIEW's graphing controls generally require 1D arrays as inputs. If you have used matrices to manipulate data sets, an added conversion step from matrix to array may be needed prior to graphing.

4.2 EXTRACTING A SUBARRAY FROM A LARGER ARRAY OR MATRIX

When you are working with large arrays or matrices, you will occasionally need to work with a portion of the entire array or matrix. LabVIEW's array functions make this possible.

One of the most common matrix tasks is extracting one column or one row from a larger matrix. For example, we might have collected a data set containing temperature values from seven thermocouples over a period of time, such as the data set shown in Table 4.1. It would be convenient to have the time and temperature values in separate arrays, and perhaps we only need the data from thermocouple 1 (TC1). LabVIEW provides the Array Subset function to select a portion of an array:

Table 4.1 Time and Temperature Data

Time (min.)	TC1	TC2	TC3	TC4	TC5	TC6	TC7
0	25.0	25.0	25.0	25.0	25.0	25.0	25.0
3	38.0	36.3	35.7	35.2	35.1	35.1	35.0
6	47.6	45.3	44.5	43.7	43.7	43.7	43.7
9	54.7	52.5	51.6	50.9	51.0	51.0	51.1
12	59.9	58.1	57.5	56.9	57.1	57.3	57.5
15	63.8	62.7	62.3	61.9	62.3	62.7	63.0
18	66.7	66.2	66.2	66.1	66.7	67.2	67.7
21	68.9	69.1	69.5	69.7	70.4	71.2	71.8
24	70.5	71.3	72.1	72.6	73.6	74.5	75.3
27	71.6	73.1	74.2	75.1	76.3	77.3	78.3
30	72.5	74.5	76.0	77.2	78.5	79.8	80.9
33	73.2	75.7	77.5	79.0	80.4	81.9	83.2
36	73.6	76.6	78.6	80.4	82.1	83.6	85.1
39	74.0	77.3	79.6	81.6	83.4	85.1	86.8
42	74.3	77.8	80.4	82.7	84.6	86.4	88.2
45	74.4	78.3	81.1	83.5	85.6	87.5	89.4
48	74.6	78.6	81.6	84.3	86.4	88.5	90.5
51	74.7	78.9	82.0	84.9	87.1	89.3	91.4
54	74.8	79.1	82.4	85.4	87.7	90.0	92.2
57	74.8	79.3	82.7	85.8	88.2	90.6	92.8
60	74.9	79.5	82.9	86.2	88.6	91.1	93.4

Note: The large data sets used in this text are available as .txt files at the text's website: www.chbe.montana.edu/LabVIEW

Functions Palette / Programming Group / Array Group / Array Subset

Note: When only a single column or row is needed, the Index Array function should be used. The Index Array function returns a 1D array.

Functions Palette / Programming Group / Array Group / Index Array

The LabVIEW connection pane used for the Array Subset function is shown in Figure 4.5. LabVIEW's nomenclature is shown on the left, and a more descriptive nomenclature, which only applies to 2D arrays, is shown on the right. The two terminals near the top are the input for the original array and the output for the selected *subarray*. The icon expands depending on the number of dimensions in the original array.

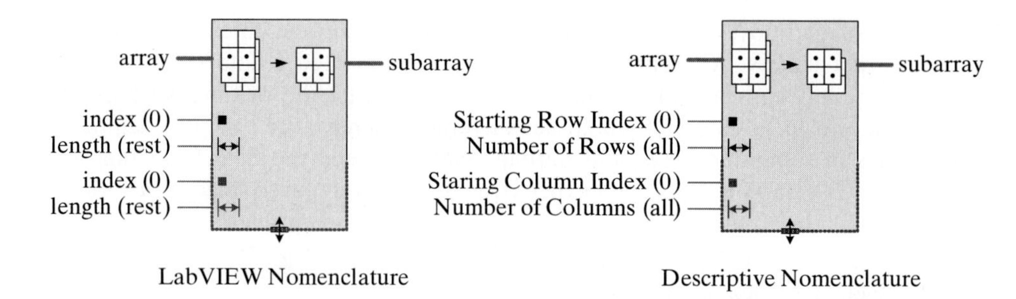

Figure 4.5 Array Subset function description.

The four terminals on the lower left side of the icon (see Figure 4.5) are used to inform LabVIEW which portion of the original array to use for the subarray.

- **Starting Row Index** (0)—the index of the first row in the original array that should be included in the subarray. The default value is (0), which is the top row in the original matrix.
- **Number of Rows** (all)—the number of rows from the original array to include in the subarray. The default is (all) rows (i.e., all rows after the starting row).
- **Starting Column Index** (0)—the index of the first column in the original array that should be included in the subarray. The default value is (0), which is the left column in the original matrix.
- **Number of Columns** (all)—the number of columns from the original array to include in the subarray. The default is (all) columns (i.e., all columns after the starting column).

Notice that array indexing starts at 0 (not 1) in LabVIEW. The top element of array X is called X[0].

Figure 4.6 shows the block diagram of a VI designed to allow any subset of the data set shown in Table 4.1 to be selected.

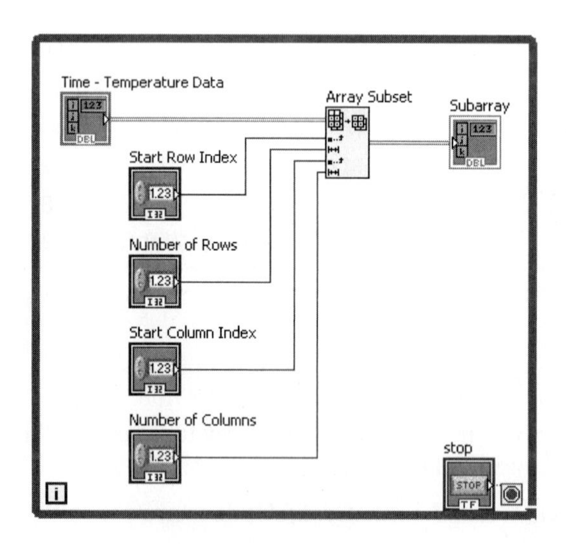

Figure 4.6 Block diagram of a VI allowing a subarray to be selected at run time.

Functions Palette / Programming Group / Array Group / Array Subset

Note: When only a single column or row is needed, the Index Array function should be used. The Index Array function returns a 1D array.

Functions Palette / Programming Group / Array Group / Index Array

The front panel is shown in Figure 4.7. Initially the entire original array has been selected.

Time - Temperature Data
0
0
0 25 25 25 25 25 25 25 0
3 38 36.3 35.7 35.2 35.1 35.1 35 0
6 47.6 45.3 44.5 43.7 43.7 43.7 43.7 0
9 54.7 52.5 51.6 50.9 51 51 51.1 0
12 59.9 58.1 57.5 56.9 57.1 57.3 57.5 0
15 63.8 62.7 62.3 61.9 62.3 62.7 63 0
18 66.7 66.2 66.2 66.1 66.7 67.2 67.7 0
21 68.9 69.1 69.5 69.7 70.4 71.2 71.8 0
24 70.5 71.3 72.1 72.6 73.6 74.5 75.3 0
27 71.6 73.1 74.2 75.1 76.3 77.3 78.3 0
30 72.5 74.5 76 77.2 78.5 79.8 80.9 0
33 73.2 75.7 77.5 79 80.4 81.9 83.2 0
36 73.6 76.6 78.6 80.4 82.1 83.6 85.1 0
39 74 77.3 79.6 81.6 83.4 85.1 86.8 0
42 74.3 77.8 80.4 82.7 84.6 86.4 88.2 0
45 74.4 78.3 81.1 83.5 85.6 87.5 89.4 0
48 74.6 78.6 81.6 84.3 86.4 88.5 90.5 0
51 74.7 78.9 82 84.9 87.1 89.3 91.4 0
54 74.8 79.1 82.4 85.4 87.7 90 92.2 0
57 74.8 79.3 82.7 85.8 88.2 90.6 92.8 0
60 74.9 79.5 82.9 86.2 88.6 91.1 93.4 0
0 0 0 0 0 0 0 0 0
Start Row Index
0
Number of Rows
21
Start Column Index
0
Number of Columns
8
Subarray
0
0
0 25 25 25 25 25 25 25 0
3 38 36.3 35.7 35.2 35.1 35.1 35 0
6 47.6 45.3 44.5 43.7 43.7 43.7 43.7 0
9 54.7 52.5 51.6 50.9 51 51 51.1 0
12 59.9 58.1 57.5 56.9 57.1 57.3 57.5 0
15 63.8 62.7 62.3 61.9 62.3 62.7 63 0
18 66.7 66.2 66.2 66.1 66.7 67.2 67.7 0
21 68.9 69.1 69.5 69.7 70.4 71.2 71.8 0
24 70.5 71.3 72.1 72.6 73.6 74.5 75.3 0
27 71.6 73.1 74.2 75.1 76.3 77.3 78.3 0
30 72.5 74.5 76 77.2 78.5 79.8 80.9 0
33 73.2 75.7 77.5 79 80.4 81.9 83.2 0
36 73.6 76.6 78.6 80.4 82.1 83.6 85.1 0
39 74 77.3 79.6 81.6 83.4 85.1 86.8 0
42 74.3 77.8 80.4 82.7 84.6 86.4 88.2 0
45 74.4 78.3 81.1 83.5 85.6 87.5 89.4 0
48 74.6 78.6 81.6 84.3 86.4 88.5 90.5 0
51 74.7 78.9 82 84.9 87.1 89.3 91.4 0
54 74.8 79.1 82.4 85.4 87.7 90 92.2 0
57 74.8 79.3 82.7 85.8 88.2 90.6 92.8 0
60 74.9 79.5 82.9 86.2 88.6 91.1 93.4 0
0 0 0 0 0 0 0 0 0
STOP

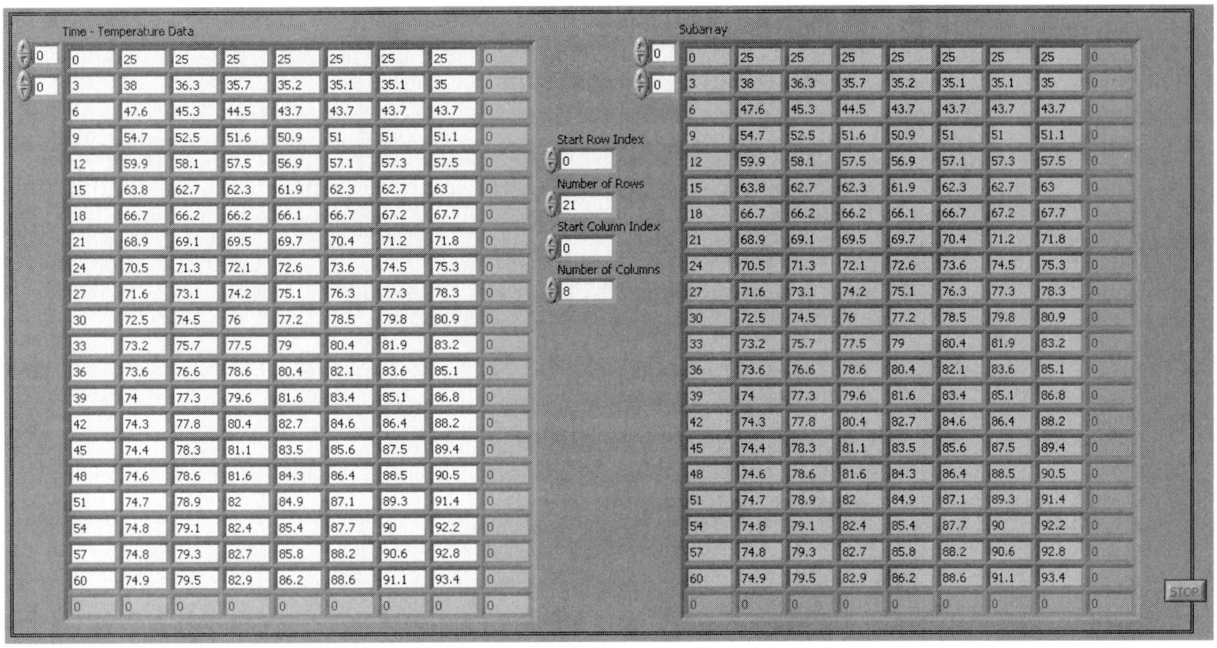

Figure 4.7
The entire original array has been selected as the subarray.

By changing the **Number of Columns** from 8 to 1, we can select only the time values for the subarray, as shown in Figure 4.8.

Time - Temperature Data
0
0
0 25 25 25 25 25 25 25 0
3 38 36.3 35.7 35.2 35.1 35.1 35 0
6 47.6 45.3 44.5 43.7 43.7 43.7 43.7 0
9 54.7 52.5 51.6 50.9 51 51 51.1 0
12 59.9 58.1 57.5 56.9 57.1 57.3 57.5 0
15 63.8 62.7 62.3 61.9 62.3 62.7 63 0
18 66.7 66.2 66.2 66.1 66.7 67.2 67.7 0
21 68.9 69.1 69.5 69.7 70.4 71.2 71.8 0
24 70.5 71.3 72.1 72.6 73.6 74.5 75.3 0
27 71.6 73.1 74.2 75.1 76.3 77.3 78.3 0
30 72.5 74.5 76 77.2 78.5 79.8 80.9 0
33 73.2 75.7 77.5 79 80.4 81.9 83.2 0
36 73.6 76.6 78.6 80.4 82.1 83.6 85.1 0
39 74 77.3 79.6 81.6 83.4 85.1 86.8 0
42 74.3 77.8 80.4 82.7 84.6 86.4 88.2 0
45 74.4 78.3 81.1 83.5 85.6 87.5 89.4 0
48 74.6 78.6 81.6 84.3 86.4 88.5 90.5 0
51 74.7 78.9 82 84.9 87.1 89.3 91.4 0
54 74.8 79.1 82.4 85.4 87.7 90 92.2 0
57 74.8 79.3 82.7 85.8 88.2 90.6 92.8 0
60 74.9 79.5 82.9 86.2 88.6 91.1 93.4 0
0 0 0 0 0 0 0 0 0
Start Row Index
0
Number of Rows
21
Start Column Index
0
Number of Columns
1
Subarray
0
0
0 0 0 0 0 0 0 0 0
3 0 0 0 0 0 0 0 0
6 0 0 0 0 0 0 0 0
9 0 0 0 0 0 0 0 0
12 0 0 0 0 0 0 0 0
15 0 0 0 0 0 0 0 0
18 0 0 0 0 0 0 0 0
21 0 0 0 0 0 0 0 0
24 0 0 0 0 0 0 0 0
27 0 0 0 0 0 0 0 0
30 0 0 0 0 0 0 0 0
33 0 0 0 0 0 0 0 0
36 0 0 0 0 0 0 0 0
39 0 0 0 0 0 0 0 0
42 0 0 0 0 0 0 0 0
45 0 0 0 0 0 0 0 0
48 0 0 0 0 0 0 0 0
51 0 0 0 0 0 0 0 0
54 0 0 0 0 0 0 0 0
57 0 0 0 0 0 0 0 0
60 0 0 0 0 0 0 0 0
0 0 0 0 0 0 0 0 0
STOP

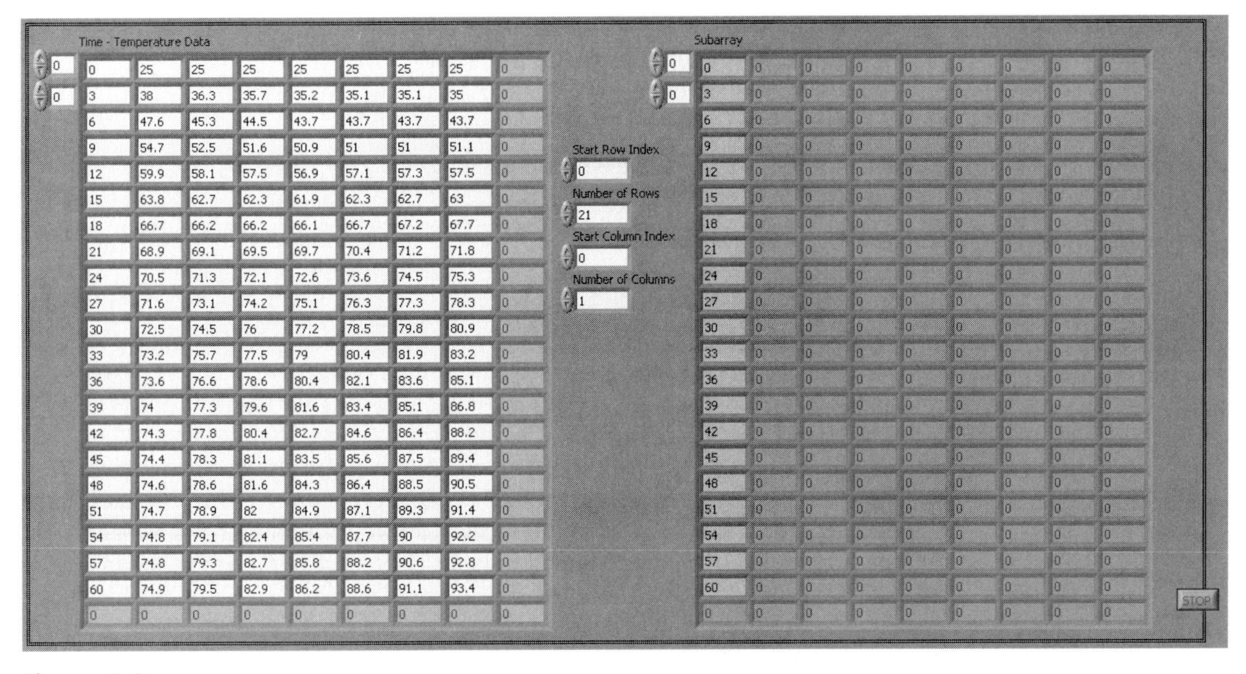

Figure 4.8
Only the time values (left column) have been selected for the subarray.

By setting the **Start Column Index** to 1, and the **Number of Columns** to 7, we can select all of the temperature values, as shown in Figure 4.9.

Time - Temperature Data

0	25	25	25	25	25	25	25	0
3	38	36.3	35.7	35.2	35.1	35.1	35	0
6	47.6	45.3	44.5	43.7	43.7	43.7	43.7	0
9	54.7	52.5	51.6	50.9	51	51	51.1	0
12	59.9	58.1	57.5	56.9	57.1	57.3	57.5	0
15	63.8	62.7	62.3	61.9	62.3	62.7	63	0
18	66.7	66.2	66.2	66.1	66.7	67.2	67.7	0
21	68.9	69.1	69.5	69.7	70.4	71.2	71.8	0
24	70.5	71.3	72.1	72.6	73.6	74.5	75.3	0
27	71.6	73.1	74.2	75.1	76.3	77.3	78.3	0
30	72.5	74.5	76	77.2	78.5	79.8	80.9	0
33	73.2	75.7	77.5	79	80.4	81.9	83.2	0
36	73.6	76.6	78.6	80.4	82.1	83.6	85.1	0
39	74	77.3	79.6	81.6	83.4	85.1	86.8	0
42	74.3	77.8	80.4	82.7	84.6	86.4	88.2	0
45	74.4	78.3	81.1	83.5	85.6	87.5	89.4	0
48	74.6	78.6	81.6	84.3	86.4	88.5	90.5	0
51	74.7	78.9	82	84.9	87.1	89.3	91.4	0
54	74.8	79.1	82.4	85.4	87.7	90	92.2	0
57	74.8	79.3	82.7	85.8	88.2	90.6	92.8	0
60	74.9	79.5	82.9	86.2	88.6	91.1	93.4	0
0	0	0	0	0	0	0	0	0

Start Row Index: 0
Number of Rows: 21
Start Column Index: 1
Number of Columns: 7

Subarray

25	25	25	25	25	25	25	0	0
38	36.3	35.7	35.2	35.1	35.1	35	0	0
47.6	45.3	44.5	43.7	43.7	43.7	43.7	0	0
54.7	52.5	51.6	50.9	51	51	51.1	0	0
59.9	58.1	57.5	56.9	57.1	57.3	57.5	0	0
63.8	62.7	62.3	61.9	62.3	62.7	63	0	0
66.7	66.2	66.2	66.1	66.7	67.2	67.7	0	0
68.9	69.1	69.5	69.7	70.4	71.2	71.8	0	0
70.5	71.3	72.1	72.6	73.6	74.5	75.3	0	0
71.6	73.1	74.2	75.1	76.3	77.3	78.3	0	0
72.5	74.5	76	77.2	78.5	79.8	80.9	0	0
73.2	75.7	77.5	79	80.4	81.9	83.2	0	0
73.6	76.6	78.6	80.4	82.1	83.6	85.1	0	0
74	77.3	79.6	81.6	83.4	85.1	86.8	0	0
74.3	77.8	80.4	82.7	84.6	86.4	88.2	0	0
74.4	78.3	81.1	83.5	85.6	87.5	89.4	0	0
74.6	78.6	81.6	84.3	86.4	88.5	90.5	0	0
74.7	78.9	82	84.9	87.1	89.3	91.4	0	0
74.8	79.1	82.4	85.4	87.7	90	92.2	0	0
74.8	79.3	82.7	85.8	88.2	90.6	92.8	0	0
74.9	79.5	82.9	86.2	88.6	91.1	93.4	0	0
0	0	0	0	0	0	0	0	0

STOP

Figure 4.9
Selecting only the temperature values for the subarray.

In practice you would rarely use controls to select the subarray. The next VI uses constants on the block diagram to create two subarrays: one containing the time values and one containing the temperature values. The block diagram is shown in Figure 4.10 and the front panel in Figure 4.11.

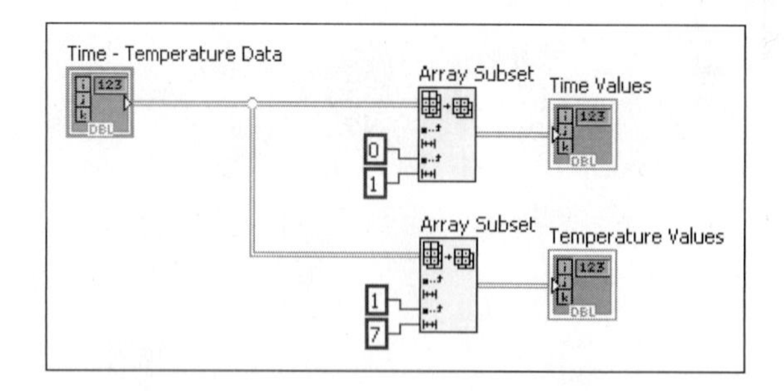

Figure 4.10
Block diagram for selecting Time and Temperature subarrays from original array.

Notice in Figure 4.10 that constants (0, 1 and 1, 7) were used to indicate the desired columns for each subarray, but no constants were sent into the Array Subset functions for the row terminals. Because we wanted all rows, we accepted the default values for Starting Row Index and Number of Rows by leaving them unwired.

Time - Temperature Data

0	25	25	25	25	25	25	25	0
3	38	36.3	35.7	35.2	35.1	35.1	35	0
6	47.6	45.3	44.5	43.7	43.7	43.7	43.7	0
9	54.7	52.5	51.6	50.9	51	51	51.1	0
12	59.9	58.1	57.5	56.9	57.1	57.3	57.5	0
15	63.8	62.7	62.3	61.9	62.3	62.7	63	0
18	66.7	66.2	66.2	66.1	66.7	67.2	67.7	0
21	68.9	69.1	69.5	69.7	70.4	71.2	71.8	0
24	70.5	71.3	72.1	72.6	73.6	74.5	75.3	0
27	71.6	73.1	74.2	75.1	76.3	77.3	78.3	0
30	72.5	74.5	76	77.2	78.5	79.8	80.9	0
33	73.2	75.7	77.5	79	80.4	81.9	83.2	0
36	73.6	76.6	78.6	80.4	82.1	83.6	85.1	0
39	74	77.3	79.6	81.6	83.4	85.1	86.8	0
42	74.3	77.8	80.4	82.7	84.6	86.4	88.2	0
45	74.4	78.3	81.1	83.5	85.6	87.5	89.4	0
48	74.6	78.6	81.6	84.3	86.4	88.5	90.5	0
51	74.7	78.9	82	84.9	87.1	89.3	91.4	0
54	74.8	79.1	82.4	85.4	87.7	90	92.2	0
57	74.8	79.3	82.7	85.8	88.2	90.6	92.8	0
60	74.9	79.5	82.9	86.2	88.6	91.1	93.4	0
0	0	0	0	0	0	0	0	0

Time Values

0	0
3	0
6	0
9	0
12	0
15	0
18	0
21	0
24	0
27	0
30	0
33	0
36	0
39	0
42	0
45	0
48	0
51	0
54	0
57	0
60	0
0	0

Temperature Values

25	25	25	25	25	25	25	0
38	36.3	35.7	35.2	35.1	35.1	35	0
47.6	45.3	44.5	43.7	43.7	43.7	43.7	0
54.7	52.5	51.6	50.9	51	51	51.1	0
59.9	58.1	57.5	56.9	57.1	57.3	57.5	0
63.8	62.7	62.3	61.9	62.3	62.7	63	0
66.7	66.2	66.2	66.1	66.7	67.2	67.7	0
68.9	69.1	69.5	69.7	70.4	71.2	71.8	0
70.5	71.3	72.1	72.6	73.6	74.5	75.3	0
71.6	73.1	74.2	75.1	76.3	77.3	78.3	0
72.5	74.5	76	77.2	78.5	79.8	80.9	0
73.2	75.7	77.5	79	80.4	81.9	83.2	0
73.6	76.6	78.6	80.4	82.1	83.6	85.1	0
74	77.3	79.6	81.6	83.4	85.1	86.8	0
74.3	77.8	80.4	82.7	84.6	86.4	88.2	0
74.4	78.3	81.1	83.5	85.6	87.5	89.4	0
74.6	78.6	81.6	84.3	86.4	88.5	90.5	0
74.7	78.9	82	84.9	87.1	89.3	91.4	0
74.8	79.1	82.4	85.4	87.7	90	92.2	0
74.8	79.3	82.7	85.8	88.2	90.6	92.8	0
74.9	79.5	82.9	86.2	88.6	91.1	93.4	0
0	0	0	0	0	0	0	0

Figure 4.11
Result of selecting Time and Temperature subarrays.

4.3 ADDING ARRAYS

One of the most fundamental matrix operations is adding two arrays, such as the $[A]$ and $[B]$ arrays shown here:

$$A = \begin{bmatrix} 1 & 2 & 3 \\ 2 & 1 & 4 \\ 3 & 4 & 7 \end{bmatrix} \qquad B = \begin{bmatrix} 3 & 5 & 7 \\ 2 & 4 & 8 \\ 1 & 3 & 6 \end{bmatrix}$$

In order to add two arrays, they must be the same size (same number of rows and columns). The process used to add two matrices is to add corresponding elements from each matrix. For example, when $[A]$ and $[B]$ are added together, the top-left element of the resulting array will be $1 + 3 = 4$. All of the corresponding elements are similarly added together. LabVIEW's Add function is used to add arrays and matrices.

Functions Palette / Mathematics Group / Numeric Group / Add function

The following steps are used to add two arrays in LabVIEW:
On the front panel (see Figure 4.12) . . .

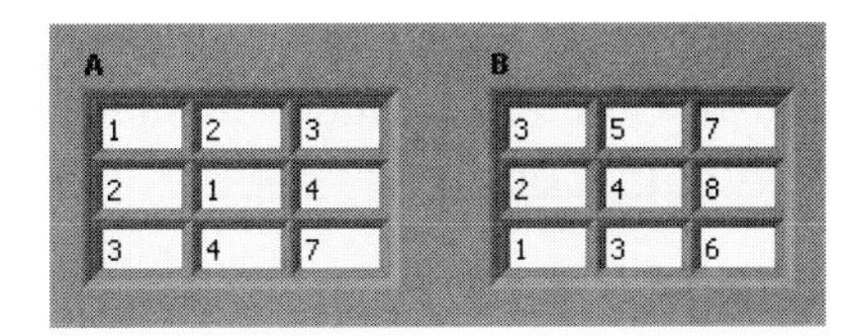

Figure 4.12
The *A* and *B* arrays on the front panel.

1. Create two array controls on the front panel. The steps required to create an array are as follows:
   - Place the Array container on the front panel.

     Controls Palette / Modern Group / Array, Matrix & Cluster Group / Array

   - Place one numeric control (from either location) inside the Array container.

     Controls Palette / Modern Group / Numeric Group / Numeric Control

     Controls Palette / Express Group / Numeric Controls Group / Num Ctrl

   - Add a dimension to create a 2D array. Right-click on the Index Display and select **Add Dimension** from the pop-up menu.
   - Expand the size of the array. Drag the handles at the sides of the array to change the number of displayed array elements.
2. Enter values into the arrays. Once the right number of rows and columns are displayed, double-click in each array element to enter the value for the element.

   Note: The *Index Display* for each matrix has been hidden in Figure 4.12. The Index Display is used to scroll through a large matrix. When the entire matrix can be seen in the control, the Index Display is not needed. Hiding the Index Display simplifies the front panel display.

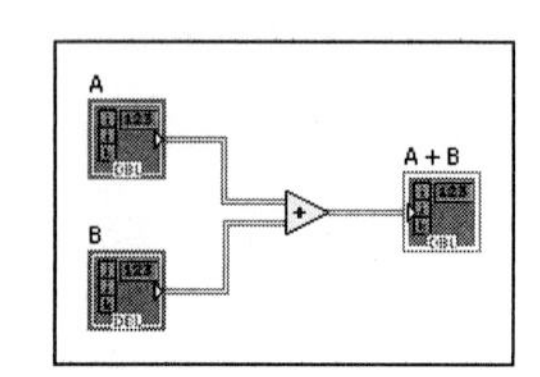

Figure 4.13
Block diagram for adding arrays *A* and *B*.

On the block diagram (see Figure 4.13) . . .

3. Place an Add function on the block diagram.

   Functions Palette / Mathematics Group / Numeric Group / Add function

4. Wire the matrix output terminals to the Add function input terminals.
5. Right-click on the Add function output terminal and select **Create / Indicator** from the pop-up menu. (The created array indicator will be 2D, but it will need to be resized on the front panel to show three rows and three columns.)

Run the VI to add the arrays. The solution is shown in Figure 4.14.

Figure 4.14
The added arrays, *A* + *B*.

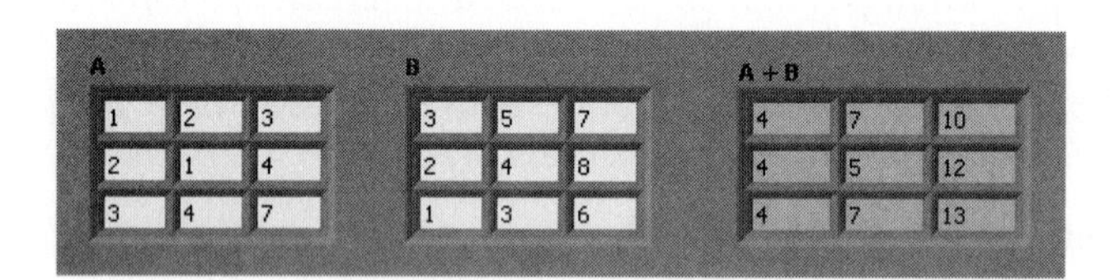

4.4 TRANSPOSE ARRAY

When an array is *transposed*, the rows and columns are interchanged. Any matrix or array can be transposed. The effect of transposing an array is most apparent when the array has significantly more rows than columns, or vice versa, so we will use the following array as an example:

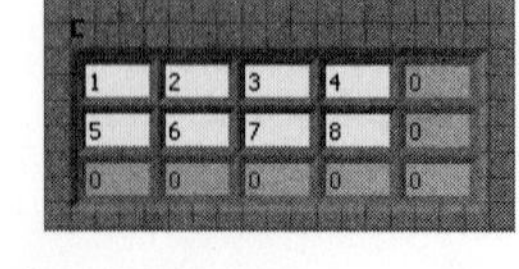

Figure 4.15
The C array defined on the front panel.

$$C = \begin{bmatrix} 1 & 2 & 3 & 4 \\ 5 & 6 & 7 & 8 \end{bmatrix}$$

The first step is to create the array on the front panel, as shown in Figure 4.15.

In Figure 4.15 the visible portion of the array has been expanded to illustrate how LabVIEW handles variable array sizes. An array can be as large as needed, but unused elements are indicated in gray. As values are entered into the array, the active elements are shown with a white background. With this approach, it is easy to create arrays of whatever size is needed.

LabVIEW provides two functions for transposing an array:

- Matrix function: **Transpose Matrix**

  Function Palette / Mathematics Group / Linear Algebra Group / Transpose Matrix

- Array function: **Transpose 2D Array**

  Function Palette / Programming Group / Array Group / Transpose 2D Array

To demonstrate that both functions generate the same result, we used both functions in the Matrix Transpose VI shown in Figure 4.16 (front panel) and Figure 4.17 (block diagram).

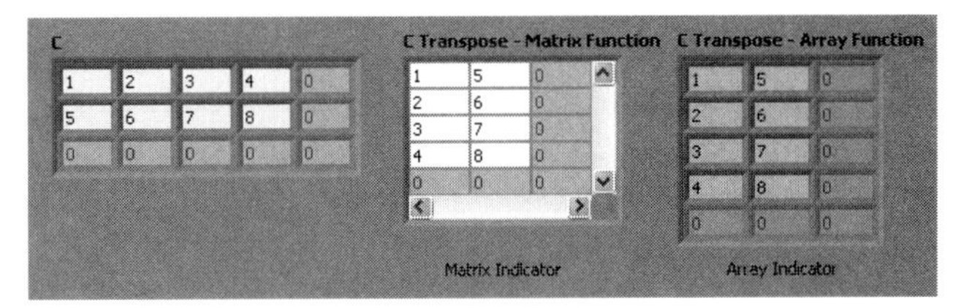

Figure 4.16 The results of transposing using the Transpose Matrix function and the Transpose 2D Array function.

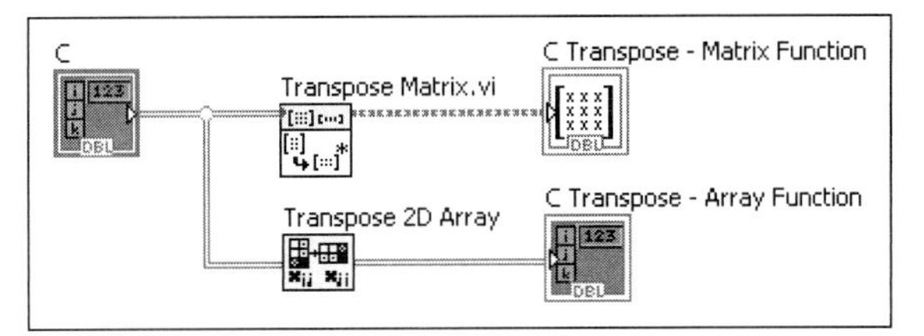

Figure 4.17 Transposing the C array using two methods.

Notice that the Transpose Matrix function created a matrix result, whereas the Transpose 2D Array created an array result. The results are numerically equivalent, but the values are stored in variables of differing data types.

4.5 MULTIPLYING AN ARRAY BY A SCALAR

The process of multiplying an array or matrix by a *scalar* (a single value) is termed *scalar multiplication*. Each element of the array is multiplied by the scalar value. For example, if the C array is multiplied by the scalar 10, each element of the C array is multiplied by 10.

$$C = \begin{bmatrix} 1 & 2 & 3 & 4 \\ 5 & 6 & 7 & 8 \end{bmatrix}$$

Functions Palette / Mathematics Group / Numeric Group / Multiply Function

A VI for multiplying an array by a scalar is illustrated in Figure 4.18 (front panel) and Figure 4.19 (block diagram).

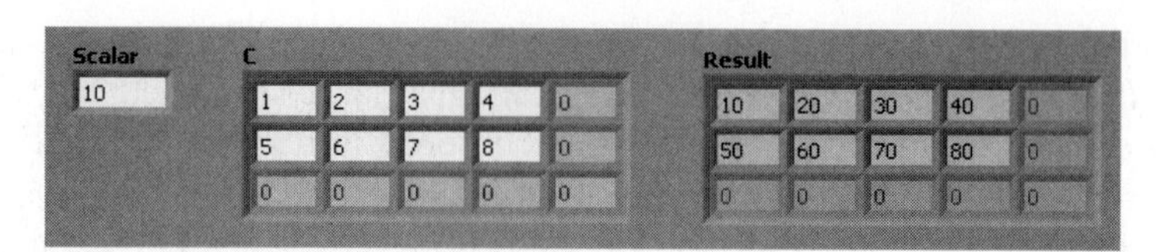

Figure 4.18
Multiplying an array by a scalar (front panel).

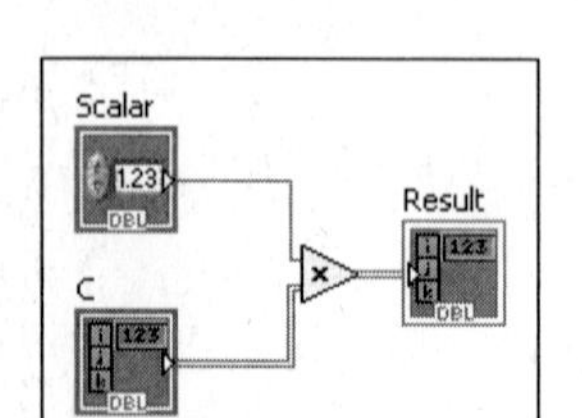

Figure 4.19
Multiplying an array by a scalar (block diagram).

4.6 MATRIX MULTIPLICATION

LabVIEW provides a *matrix multiplication* function called **A × B** for matrix multiplication (works with matrices or arrays).

Matrix multiplication can be performed on two arrays if the number of columns in the first array equals the number of rows in the second array. For example, $A \times B$ is allowed because A has two columns and B has two rows.

$$A_{3\times 2} = \begin{bmatrix} 2 & 4 \\ 3 & 2 \\ 1 & 5 \end{bmatrix} \quad B_{2\times 4} = \begin{bmatrix} 2 & 1 & 3 & 4 \\ 4 & 6 & 1 & 3 \end{bmatrix}$$

Notice that $B \times A$ is not allowed, because B has four columns and A has three rows.

The resulting array has as many rows as the first array, and as many columns as the second array, so $A_{3\times 2} \times B_{2\times 4}$ should generate an array with 3 rows and 4 columns.

As usual, the first step is to define the input arrays on the front panel, as shown in Figure 4.20.

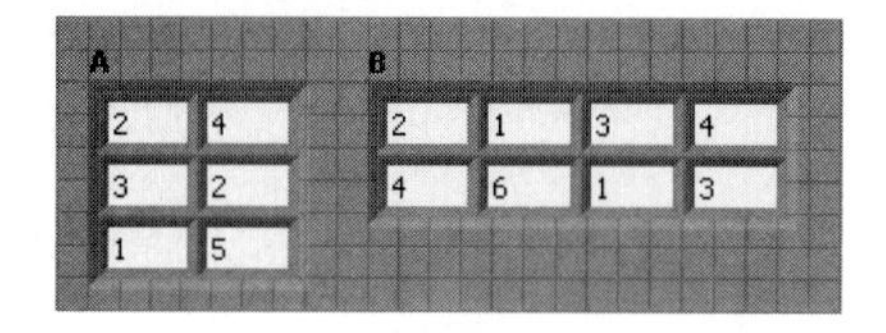

Figure 4.20
Defining the arrays to be multiplied.

LabVIEW provides the $A \times B$ function to perform matrix multiplication:

Function Palette / Mathematics Group / Linear Algebra Group / A × B Function

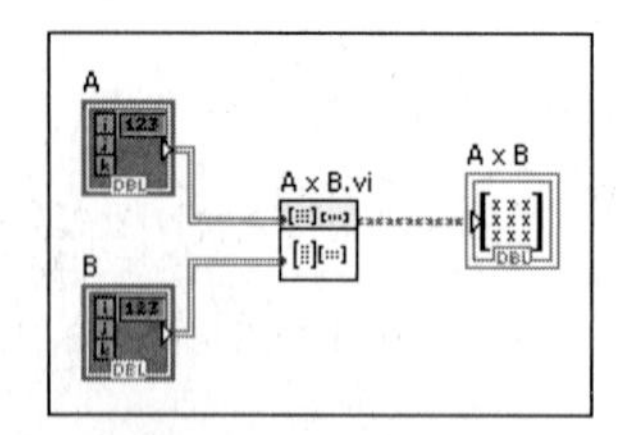

Figure 4.21
Block diagram for multiplying two matrices.

The block diagram is shown in Figure 4.21, and the result (of size 3 × 4 (3 rows by 4 columns), as expected) is shown in Figure 4.22.

Notice that the $A \times B$ function generated a matrix result, not an array. (The scrollbars at the right and bottom edges of the matrix indicator in Figure 4.22 are a visual clue that a matrix result was obtained.) The matrix result is the default for the $A \times B$ function. However, you can send the results to an array if you explicitly create the array on the front panel, and then wire the $A \times B$ output to the array. This result

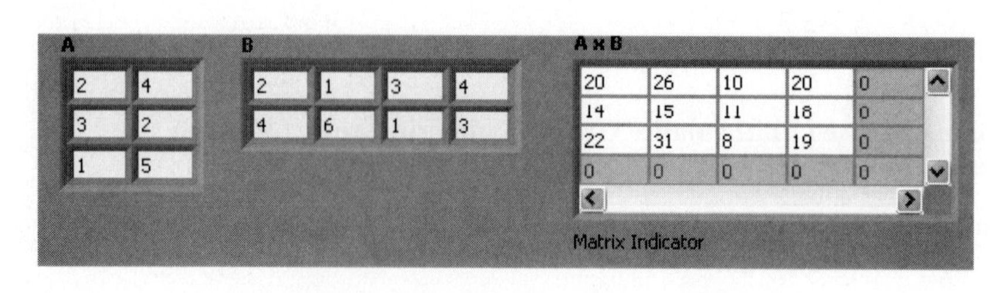

Figure 4.22
Result of multiplying $A \times B$, matrix result.

is shown in Figure 4.23. The computed values are the same whether the result is displayed as a matrix or as an array.

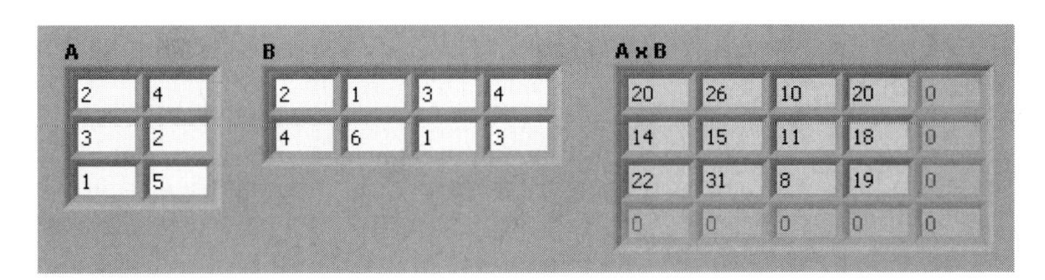

Figure 4.23
Result of multiplying $A \times B$, array result.

A word of caution: LabVIEW does not automatically display the entire array. After running the VI, you will need to resize the result array to see all the array elements.

So, how does matrix multiplication work? Officially, the formula for matrix multiplication is

$$[AB]_{i,j} = \sum_{k=1}^{n} A_{i,k}\, B_{k,j}$$

where

i is the row number
j is the column number
k is a counter
n is the number of columns in A, or the number of rows in B (they're equal).

In practice, people think of multiplying across the first matrix and down the second matrix, adding terms as they go.

Starting with the first row of $[A]$:

- Element $[A \times B]$ 1,1 (top, left) $\quad 2 \times 2 + 4 \times 4 = 20$
- Element $[A \times B]$ 1,2 $\quad 2 \times 1 + 4 \times 6 = 26$
- Element $[A \times B]$ 1,3 $\quad 2 \times 3 + 4 \times 1 = 10$
- Element $[A \times B]$ 1,4 $\quad 2 \times 4 + 4 \times 3 = 20$

Then, using the second row of $[A]$:

- Element $[A \times B]$ 2,1 (middle, left) $\quad 3 \times 2 + 2 \times 4 = 14$
- Element $[A \times B]$ 2,2 $\quad 3 \times 1 + 2 \times 6 = 15$
- Element $[A \times B]$ 2,3 $\quad 3 \times 3 + 2 \times 1 = 11$
- Element $[A \times B]$ 2,4 $\quad 3 \times 4 + 2 \times 3 = 18$

Finally, using the third row of $[A]$:

- Element $[A \times B]$ 2,1 (bottom, left) $\quad 1 \times 2 + 5 \times 4 = 22$
- Element $[A \times B]$ 2,2 $\quad 1 \times 1 + 5 \times 6 = 31$
- Element $[A \times B]$ 2,3 $\quad 1 \times 3 + 5 \times 1 = 8$
- Element $[A \times B]$ 2,4 $\quad 1 \times 4 + 5 \times 3 = 19$

For large arrays, matrix multiplication is straightforward, but tedious. Nowadays, all spreadsheets and math software packages will perform matrix multiplication.

4.7 ELEMENT BY ELEMENT MULTIPLICATION

You can also multiply arrays and matrices using the standard **Multiply** function, but the results may or may not be what you want.

> **Functions Palette / Mathematics Group / Numeric Group / Multiply Function**

1. If you multiply matrices (not arrays) using the Multiply function, the Multiply function will perform matrix multiplication (see Figure 4.24, top panel).
2. If you multiply arrays using the Multiply function, the Multiply function will perform *element-by-element multiplication* and ignore extraneous elements (see Figure 4.24, bottom panel).

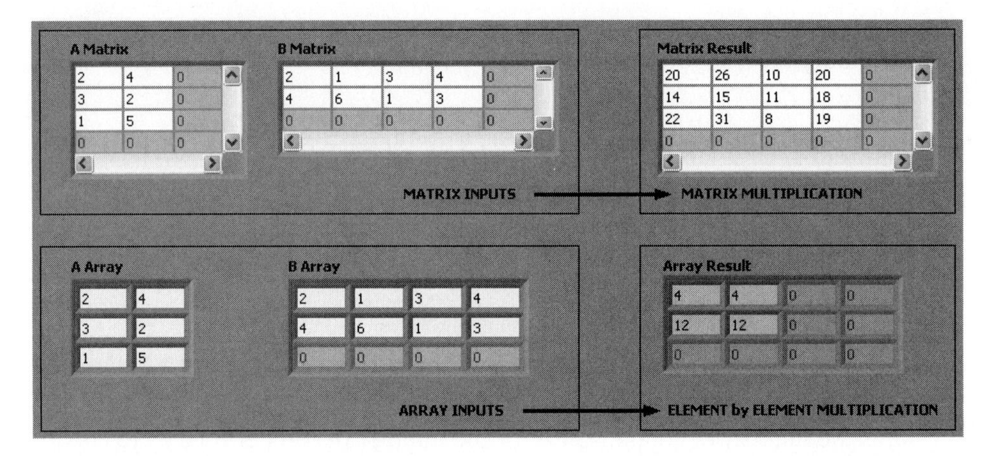

Figure 4.24
The Multiply function gives different results when used with matrices and arrays.

As you can see in the block diagram (Figure 4.25), the same Multiply function was used to get these very different results. This is because the Multiply function treats array inputs and matrix inputs differently.

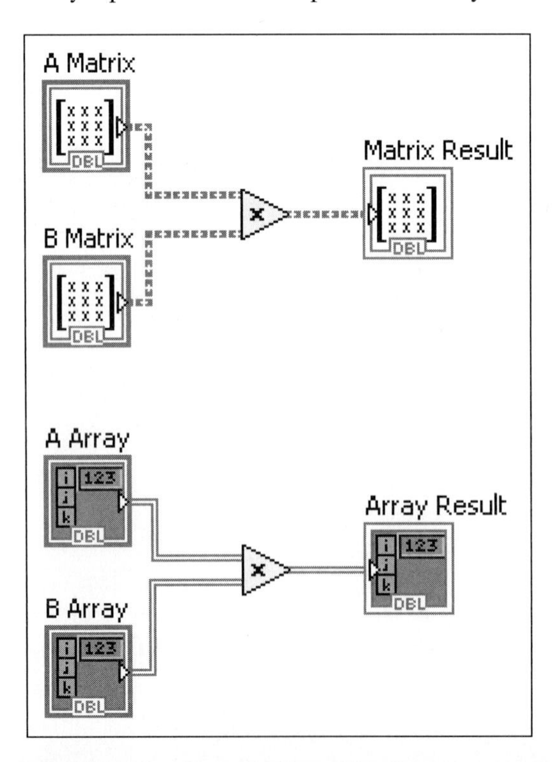

Figure 4.25
The block diagram showing how the Multiply function was used.

There are times when element-by-element multiplication is useful. When used, the arrays (not matrices) should have the same number of rows and columns, and the **Multiply** function should be used.

4.8 CONDITION NUMBER

A common use of matrix math is the solution of systems of linear algebraic equations, such as the following:

$$
\begin{aligned}
2x_1 + 3x_2 + 4x_3 &= 7 \\
2x_1 + 3x_2 + 4x_3 &= 8 \\
3x_1 + 5x_2 + 7x_3 &= 4
\end{aligned}
$$

In matrix form, this set of equations can be written as a coefficient matrix $[C]$, an unknown vector $[x]$, and a right-hand-side vector $[r]$ as

$$
C = \begin{bmatrix} 2 & 3 & 4 \\ 2 & 3 & 4 \\ 3 & 5 & 7 \end{bmatrix} \quad x = \begin{bmatrix} x_1 \\ x_2 \\ x_3 \end{bmatrix} \quad r = \begin{bmatrix} 7 \\ 8 \\ 4 \end{bmatrix}
$$

The two identical rows in the coefficient matrix $[C]$ are a sure sign that this set of equations does not have a solution.

The matrix *condition number* gives an indication of whether or not an error-free solution is likely with a given coefficient matrix. When a coefficient matrix has two identical rows, the condition number should be infinity.

A small condition number indicates that a good solution is likely.

LabVIEW provides the **Matrix Condition Number** function to calculate this quantity. It is available in the Linear Algebra group.

Function Palette / Mathematics Group / Linear Algebra Group / Matrix Condition Number

To investigate this, a Matrix Condition Number VI was written. The front panel is shown in Figure 4.26 and the corresponding block diagram is shown in Figure 4.27. We didn't get an infinite condition number, but 1.15×10^{17} is still huge.

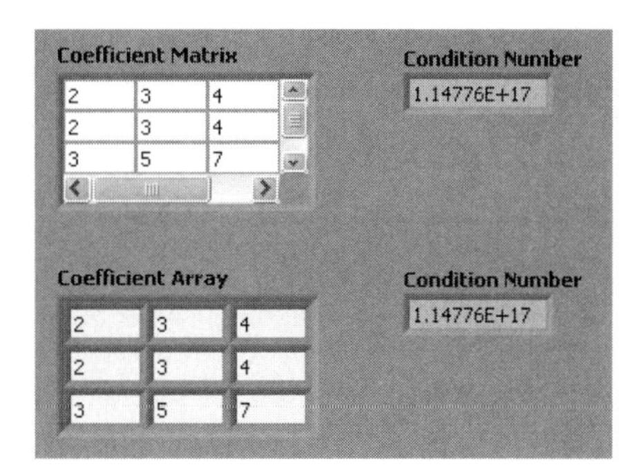

Figure 4.26 Matrix Condition Number VI front panel.

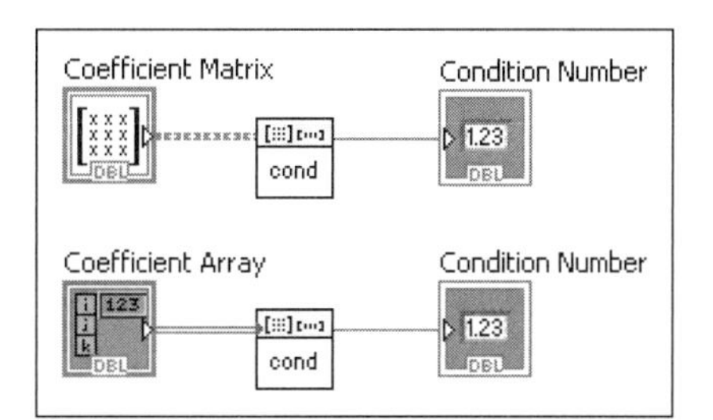

Figure 4.27 Matrix Condition Number VI block diagram.

Next, a While Loop will be added so that we can make changes in the coefficient matrix and observe the effect on the matrix condition number without having to keep hitting the **Run** button.

If we modify one of the coefficients very slightly, we would expect the condition number to get smaller, but obtaining a solution when two rows of coefficients are very close should still be difficult. What is the matrix condition number when the "3" in the second row is changed to "3.01"? We can see the answer in Figure 4.28.

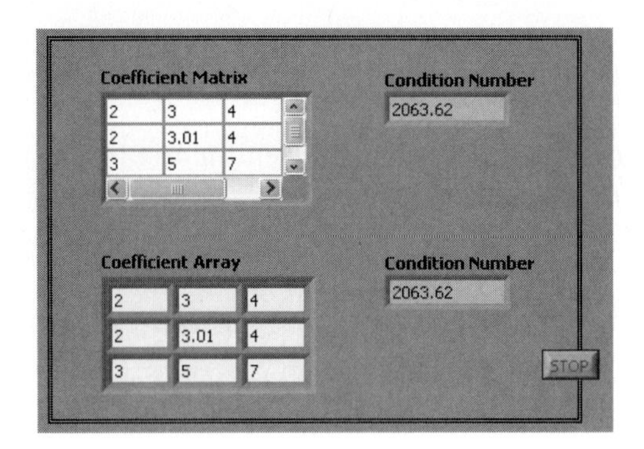

Figure 4.28 What happens to the condition number when one coefficient is changed slightly?

The matrix condition number has dropped from 10^{17} to slightly over 2000. This gives us a clue that a matrix condition number of 2000 is pretty bad and likely to yield a poor solution.

So, what is the matrix condition number when the rows are all clearly different? The solution shown in Figure 4.29 can help answer that question.

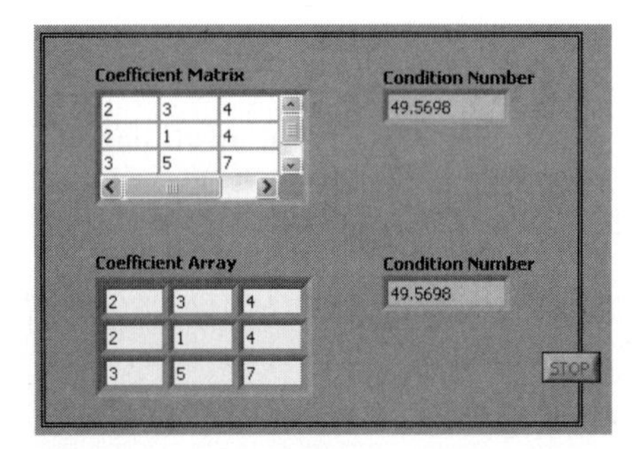

Figure 4.29 Finding the condition number when the rows are clearly distinct.

When the middle coefficient in the second row was changed to "1", making the three rows quite distinct, the matrix condition number fell to 50. An ideal matrix condition number is 1, so 50 still sounds kind of high. We'll look at the solution of this set of equations in Section 4.10.

4.9 MATRIX DETERMINANT

The *determinant* is a commonly calculated quantity when working with *square matrices* (matrices with the same number of rows and columns). For example, you might want to check the value of the determinant of the coefficient matrix before attempting to solve a set of simultaneous linear equations. If the determinant is 0, no solution is possible.

We have already seen a system of equations that has no solution, and those equations are repeated here:

$$2x_1 + 3x_2 + 4x_3 = 7$$
$$2x_1 + 3x_2 + 4x_3 = 8$$
$$3x_1 + 5x_2 + 7x_3 = 4$$

Or, in matrix form:

$$C = \begin{bmatrix} 2 & 3 & 4 \\ 2 & 3 & 4 \\ 3 & 5 & 7 \end{bmatrix} \quad x = \begin{bmatrix} x_1 \\ x_2 \\ x_3 \end{bmatrix} \quad r = \begin{bmatrix} 7 \\ 8 \\ 4 \end{bmatrix}$$

Because there are two identical rows in the C matrix, we know that this set of equations has no solution. We would expect that the determinant of C, which is written as $|C|$, has a value of 0. We can use LabVIEW's **Determinant** function (from the Linear Algebra Group) to verify this.

Function Palette / Mathematics Group / Linear Algebra Group / Determinant

Figure 4.30 shows a block diagram for calculating the determinant of an array and displaying the result. The front panel is shown in Figure 4.31.

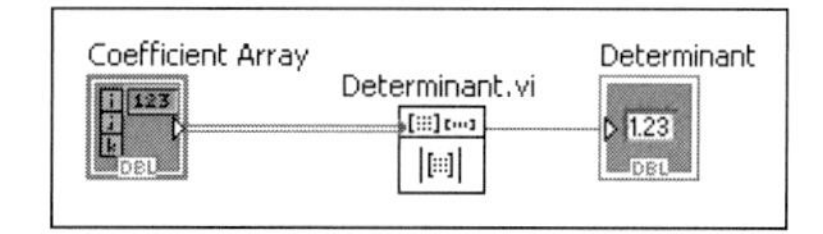

Figure 4.30
Block diagram of a VI to calculate the determinant of an array.

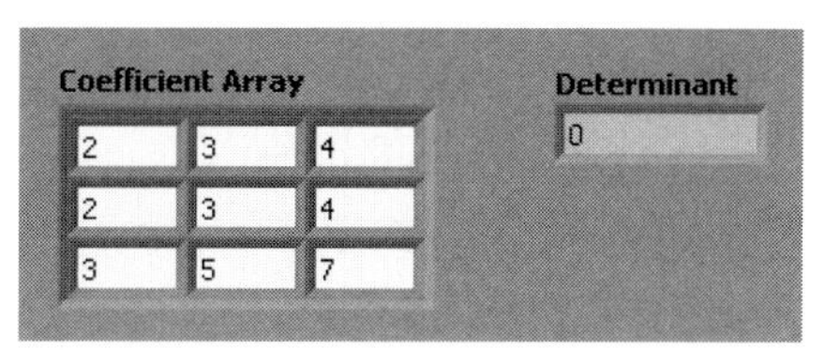

Figure 4.31
Front panel of Determinant VI.

As expected, the determinant of the coefficient array C was 0, indicating that the array cannot be inverted and there will be no solution to the system of equations.

A modified system of equations (changed second coefficient in second equation) removes the problem of two identical rows:

$$2x_1 + 3x_2 + 4x_3 = 7$$
$$2x_1 + 1x_2 + 4x_3 = 8$$
$$3x_1 + 5x_2 + 7x_3 = 4$$

Or, in matrix form:

$$C = \begin{bmatrix} 2 & 3 & 4 \\ 2 & 1 & 4 \\ 3 & 5 & 7 \end{bmatrix} \quad x = \begin{bmatrix} x_1 \\ x_2 \\ x_3 \end{bmatrix} \quad r = \begin{bmatrix} 7 \\ 8 \\ 4 \end{bmatrix}$$

This system of equations should have a solution. Let's check the determinant of the new coefficient array to be sure. After modifying the C array in the Determinant VI,

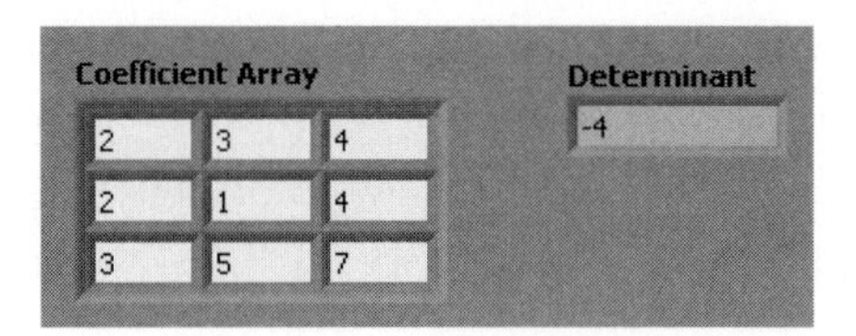

Figure 4.32
Calculating the determinant of the modified coefficient array.

and running the program again, the result can be seen in Figure 4.32. The determinant is indeed non-zero, so there should be a solution to this set of equations. We will solve this system of equations in Section 4.9.

4.10 INVERSE MATRIX

One method for solving systems of *simultaneous linear equations* uses an inverse of the coefficient matrix $[C]^{-1}$, as illustrated in the following derivation.

$$
\begin{aligned}
[C][x] &= [r] \\
[C]^{-1}[C][x] &= [C]^{-1}[r] \\
[I][x] &= [C]^{-1}[r] \\
[x] &= [C]^{-1}[r]
\end{aligned}
$$

The original system of equations can be written in matrix form as

$$[C][x] = [r]$$

To "divide out" the $[C]$ matrix (apologies to math teachers), we actually multiply through by the inverse $[C]$ matrix, which is labeled $[C]^{-1}$. The matrix product of $[C]^{-1}$ $[C]$ is the identity matrix [I]. Multiplying the identity matrix [I] by vector $[x]$ is like multiplying $[x]$ by 1; it leaves $[x]$ unchanged. Since $[x]$ is the vector of unknowns in the system of equations, we can solve for the unknowns if we can invert the $[C]$ matrix.

LabVIEW provides the *Inverse Matrix* function to invert a matrix. It can be found in the Linear Algebra Group:

Function Palette / Mathematics Group / Linear Algebra Group / Inverse Matrix

Note: Only *square matrices* (number of rows = number of columns) can be inverted, and they must be *non-singular matrices* (i.e., the determinant cannot be zero).

A VI that calculates an *inverse matrix* is shown in Figure 4.33 (block diagram) and Figure 4.34 (front panel).

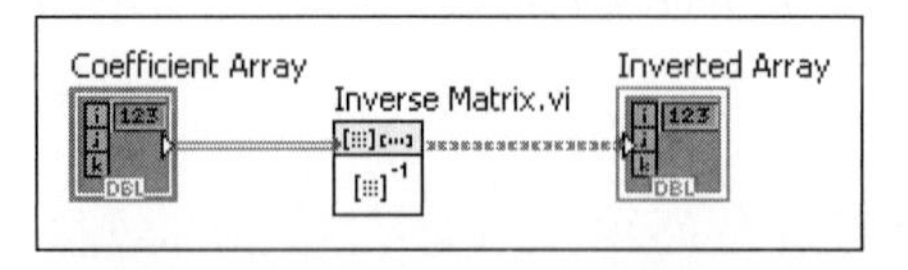

Figure 4.33
Matrix Inverse VI, block diagram.

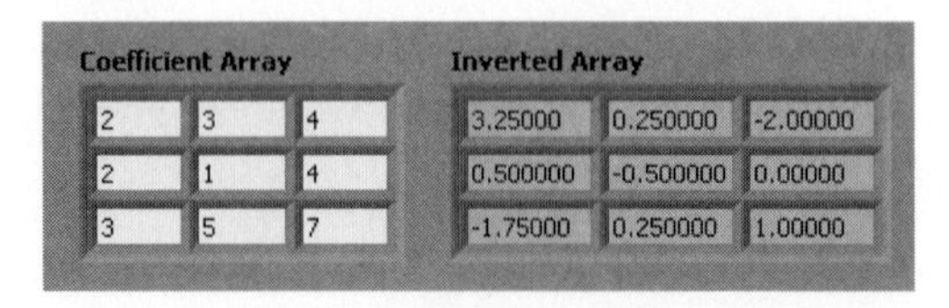

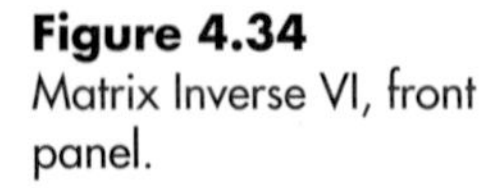
Figure 4.34
Matrix Inverse VI, front panel.

The matrix condition number and determinant are both good checks to perform on the coefficient matrix. Another check is to multiply the $[C]^{-1}$ and $[C]$ to see if the result really is an identity matrix. The VI shown in Figure 4.35 (block diagram) and Figure 4.36 (front panel) performs these tests.

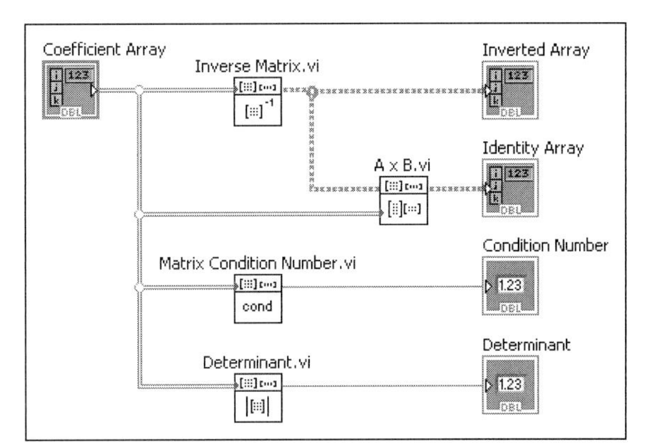

Figure 4.35
Array Inverse with Checks VI, block diagram.

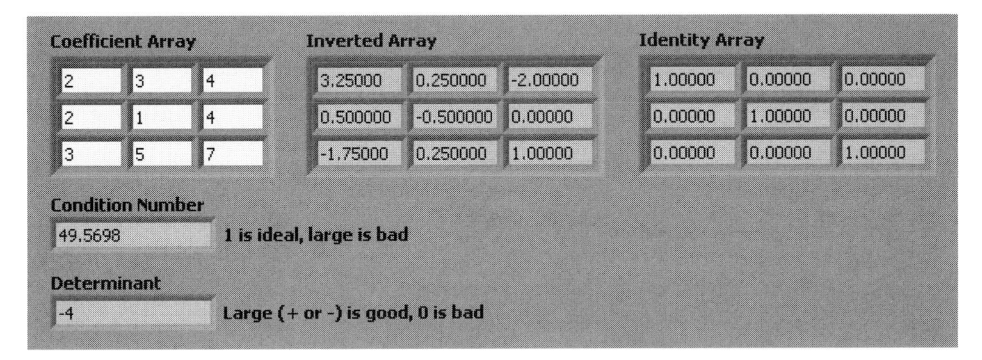

Figure 4.36
Array Inverse with Checks VI, front panel.

As a test, we can modify the coefficient array to cause the first two rows to be identical. This should make inverting the array impossible, so we can see how LabVIEW responds to a singular array. The result is shown in Figure 4.37.

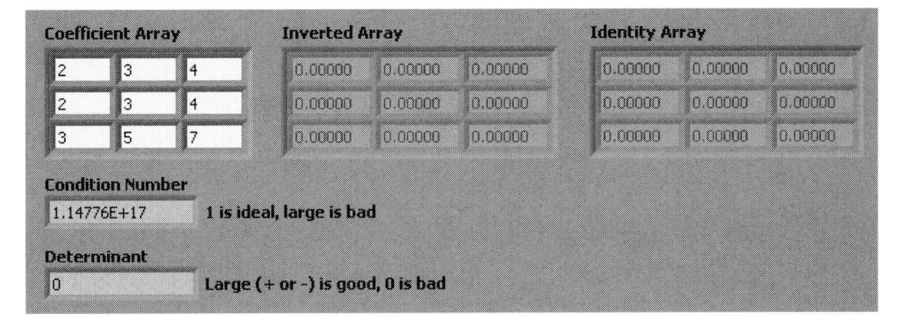

Figure 4.37
Observing LabVIEW's response to a singular matrix.

In Figure 4.37 we can see that LabVIEW shows that the matrix condition number is huge, and the determinant is 0. No matrix inversion is possible. The failed inverse array is shown with a gray background—that's what an array indicator looks like when it received no values. The array inversion failed and no information (no result values) flowed to the Inverted Array indicator, or the Identity Array indicator.

4.11 SOLVING SIMULTANEOUS LINEAR EQUATIONS

As an example of solving a system of simultaneous linear equations, we will solve the "modified" system of equations that has been presented earlier:

$$2x_1 + 3x_2 + 4x_3 = 7$$
$$2x_1 + 1x_2 + 4x_3 = 8$$
$$3x_1 + 5x_2 + 7x_3 = 4$$

Or, in matrix form

$$C = \begin{bmatrix} 2 & 3 & 4 \\ 2 & 1 & 4 \\ 3 & 5 & 7 \end{bmatrix} \quad x = \begin{bmatrix} x_1 \\ x_2 \\ x_3 \end{bmatrix} \quad r = \begin{bmatrix} 7 \\ 8 \\ 4 \end{bmatrix}$$

The short derivation in the previous section showed one way to solve for the x values (called vector $[x]$). The result was

$$[x] = [C]^{-1}[r]$$

A VI using this approach is shown in Figure 4.38 (block diagram) and Figure 4.39 (front panel).

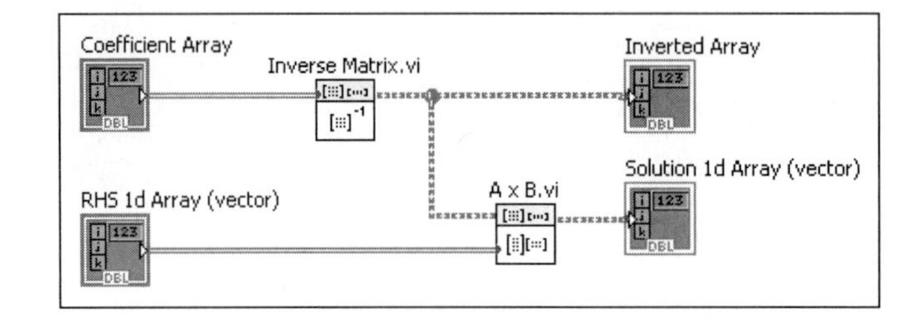

Figure 4.38
Block diagram for solving simultaneous linear equations by matrix inversion.

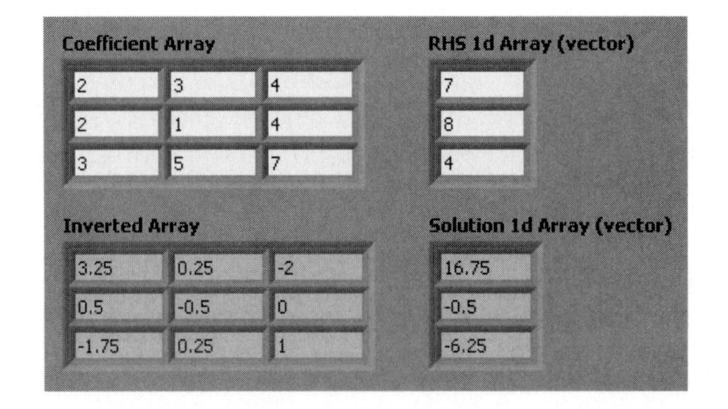

Figure 4.39
Front panel for solving simultaneous linear equations by matrix inversion.

While the method used in Figure 4.38 works, there are methods that don't require a complete matrix inversion that takes less computer time and creates less round-off error. In addition, LabVIEW provides the **Solve Linear Equations** function (in the Linear Algebra Group), which allows you to get a solution in one step.

Function Palette / Mathematics Group / Linear Algebra Group / Solve Linear Equations

A VI using the **Solve Linear Equations** function to solve the same problem is shown in Figure 4.40 (block diagram) and Figure 4.41 (front panel).

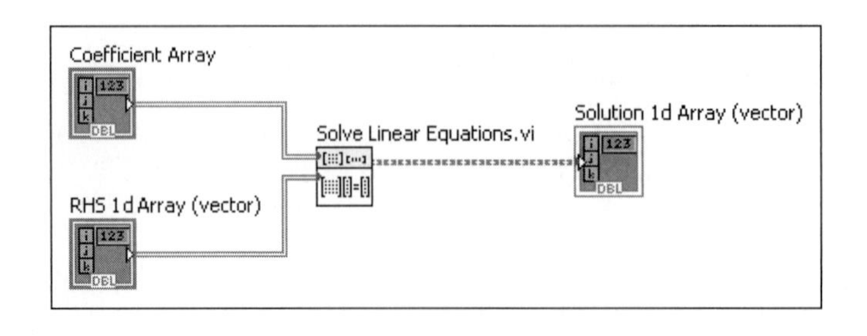

Figure 4.40
Block diagram for solving simultaneous linear equations using Solve Linear Equations function.

Coefficient Array			RHS 1d Array (vector)	Solution 1d Array (vector)
2	3	4	7	16.75
2	1	4	8	-0.5
3	5	7	4	-6.25

Figure 4.41
Front panel for solving simultaneous linear equations using Solve Linear Equations function.

APPLICATION

Circuit Analysis

When you see a complex resistor network such as the one shown in Figure 4.42, the immediate response is to wonder how you can ever solve for all of the currents in all of the circuit segments. But, applying a few basic laws allows the circuit to be analyzed.

The laws include

Ohm's Law—applies to any resistor in the network

$$V = IR$$

Resistors in Series—applies whenever multiple resistors are connected in series

$$R_{\text{total}} = \sum_{i=1}^{N_{\text{resistors}}} R_i$$

Kirchhoff's Voltage Law—applies to any loop through the circuit

For a closed loop, the algebraic sum of all changes in voltage must be 0.

Kirchhoff's Current Law—applies at any junction in the circuit

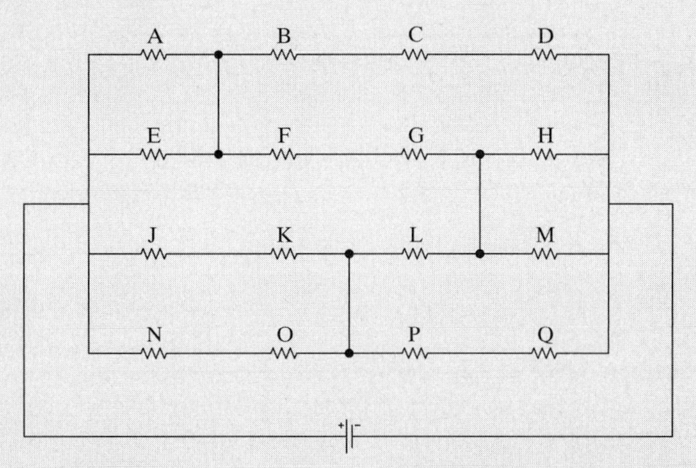

Figure 4.42
Resistor network.

At any junction, the sum of the input currents must equal the sum of the output currents.

In this problem, all of the resistance values (R_A, R_B, etc.) are known, as well as the battery voltage (12 volts). We are trying to find the currents in each segment of the circuit.

First, we look for resistors in series and combine them. The result is shown in Figure 4.43.

Next, we look for the unique current flow paths through the network; there are 13 of them as indicated by heavy arrows in Figure 4.44. The arrows for the crossover connections (labeled Top, Mid, and Bot) might be drawn in the wrong direction, but if they are, the calculated currents in those segments will have a negative sign.

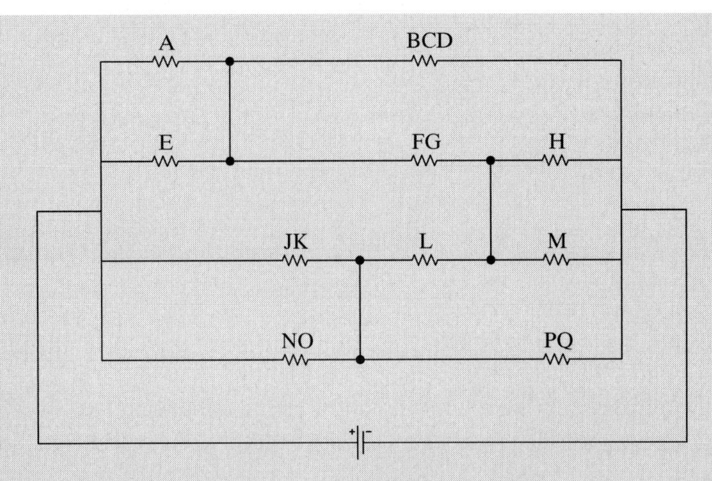

Figure 4.43
Resistor network, resistors in series combined.

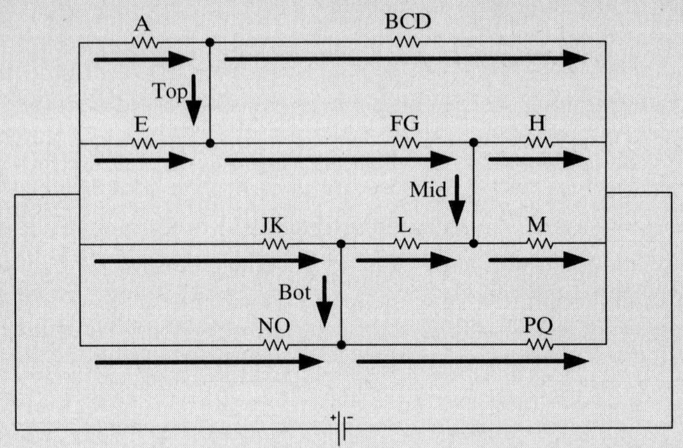

Figure 4.44
Resistor network, current paths.

We can apply Kirchhoff's voltage law to any path through the network. That generates the following equations (V_{Bat} represents the voltage increase from the battery):

$$V_{\text{Bat}} - V_{\text{A}} - V_{\text{BCD}} = 0$$
$$V_{\text{Bat}} - V_{\text{A}} - V_{\text{FG}} - V_{\text{H}} = 0$$
$$V_{\text{Bat}} - V_{\text{A}} - V_{\text{FG}} - V_{\text{M}} = 0$$
$$V_{\text{Bat}} - V_{\text{E}} - V_{\text{FG}} - V_{\text{M}} = 0$$
$$V_{\text{Bat}} - V_{\text{JK}} - V_{\text{L}} - V_{\text{M}} = 0$$
$$V_{\text{Bat}} - V_{\text{JK}} - V_{\text{PQ}} = 0$$
$$V_{\text{Bat}} - V_{\text{NO}} - V_{\text{PQ}} = 0$$

Next, we use Ohm's Law to relate the voltages to resistances and currents:

$$V_{\text{Bat}} - I_{\text{A}}R_{\text{A}} - I_{\text{BCD}}R_{\text{BCD}} = 0$$
$$V_{\text{Bat}} - I_{\text{A}}R_{\text{A}} - I_{\text{FG}}R_{\text{FG}} - I_{\text{H}}R_{\text{H}} = 0$$
$$V_{\text{Bat}} - I_{\text{A}}R_{\text{A}} - I_{\text{FG}}R_{\text{FG}} - I_{\text{M}}R_{\text{M}} = 0$$
$$V_{\text{Bat}} - I_{\text{E}}R_{\text{E}} - I_{\text{FG}}R_{\text{FG}} - I_{\text{M}}R_{\text{M}} = 0$$
$$V_{\text{Bat}} - I_{\text{JK}}R_{\text{JK}} - I_{\text{L}}R_{\text{L}} - I_{\text{M}}R_{\text{M}} = 0$$
$$V_{\text{Bat}} - I_{\text{JK}}R_{\text{JK}} - I_{\text{PQ}}R_{\text{PQ}} = 0$$
$$V_{\text{Bat}} - I_{\text{NO}}R_{\text{NO}} - I_{\text{PQ}}R_{\text{PQ}} = 0$$

Finally, we apply Kirchhoff's current law to each of the junctions (black dots) in the circuit, yielding the following equations:

$$I_A = I_{Top} + I_{BCD}$$
$$I_{FG} = I_{Mid} + I_H$$
$$I_{JK} = I_{Bot} + I_L$$
$$I_E + I_{Top} = +I_{FG}$$
$$I_L + I_{Mid} = +I_M$$
$$I_{NO} + I_{Bot} = +I_{PQ}$$

We now have 13 equations and 13 unknown currents. The known resistance values are as follows:

Resistor	**Resistance**
R_A	100
R_{BCD}	350
R_E	50
R_{FG}	125
R_H	75
R_{JK}	150
R_L	100
R_M	200
R_{NO}	250
R_{PQ}	175

In matrix form, leaving out zeros for clarity, we can write the equations in terms of a coefficient matrix and a right-hand-side (RHS) vector.

Coefficient Matrix (all empty cells must contain zeros before any matrix math)

100	350											
100			125	75								
100			125				200					
		50	125				200					
					150	100	200					
					150				175			
								250	175			
1	−1									−1		
			1	−1							−1	
					1	−1						−1
		1	−1							1		
						1	−1				1	
								1	−1			1

Right-Hand-Side Vector

12
12
12
12
12
12
12
0
0
0
0
0
0

A VI to solve this set matrix problem is shown in Figure 4.45. The condition number for the coefficient matrix is far from ideal, but the determinant is at least non-zero. The solution indicates the current values in amperes.

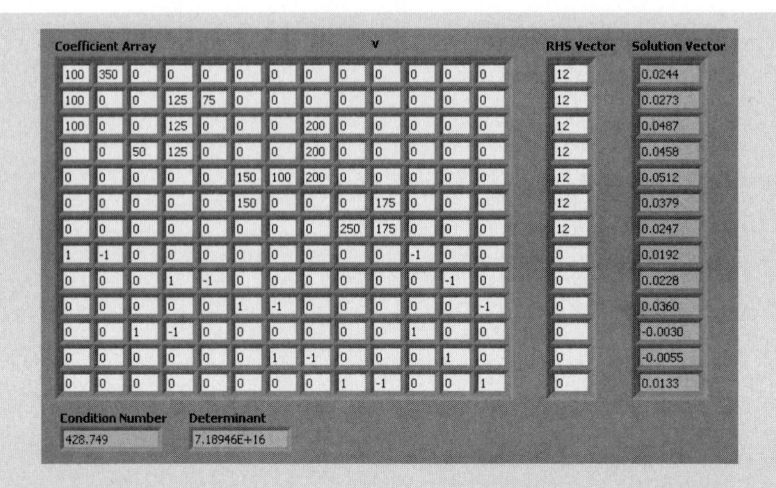

Figure 4.45
VI for solving the resistor network for current values.

The block diagram for this problem is shown in Figure 4.46.

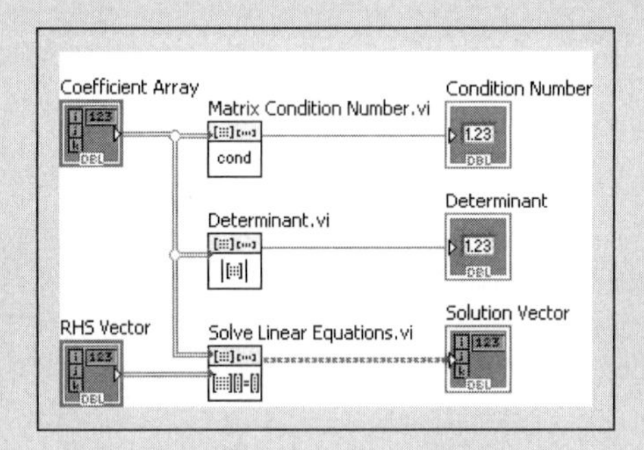

Figure 4.46
Block diagram for solving the resistor network for current values.

The solution, with labels, is shown below. Note that the units are in milliamps in the following table.

Variable	Current (mA)
I_A	24.37
I_{BCD}	27.32
I_E	48.73
I_{FG}	45.77
I_H	51.23
I_{JK}	37.94
I_L	24.66
I_M	19.21
I_{NO}	22.77
I_{PQ}	36.05
I_{TOP}	−2.96
I_{MID}	−5.45
I_{BOT}	13.28

4.12 PROGRAMMING PREVIEW: FOR LOOPS

Most of the programming features of LabVIEW will be covered in a later chapter, but there is one programming structure that is commonly used to build arrays: the *For Loop*.

A For Loop is a programming structure that is designed to loop through a set of programming instructions a specified number of times. A very simple VI that demonstrates some features of a For Loop is shown in Figure 4.47.

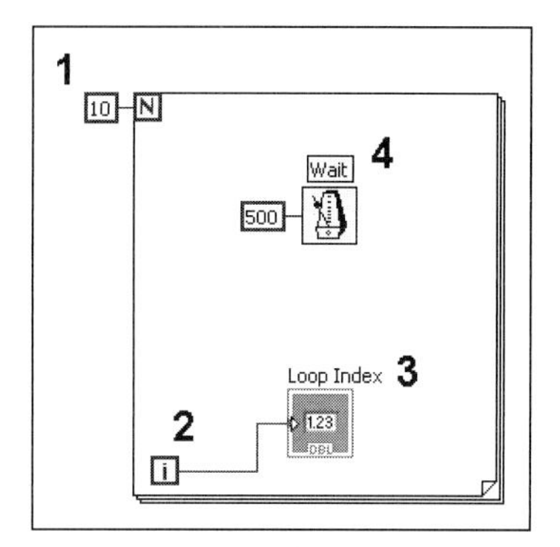

Figure 4.47
A simple For Loop VI.

1. The constant wired to the ***N*** input (called the *count terminal*) tells the For Loop to cycle 10 times.
2. The ***i*** is the loop index (called the *iteration terminal*). The loop index is incremented each time the For Loop cycles. Indexing starts at 0, so the index will take on values of 0–9 in this VI.
3. The index value is displayed in an indicator on the front panel. When the VI is running, the values 0–9 are displayed one after another (see Figure 4.48).
4. The Wait function wired to a constant of 500 makes the For Loop wait half a second (500 ms) between cycles. This gives the user time to see the numbers change on the front panel.

Figure 4.48
The front panel shows the current loop index.

Whereas While Loops are used when you want the loop to cycle until some condition is met (like the STOP button being clicked), For Loops are typically used when you know how many times the loop should cycle.

Some useful features of For Loops in LabVIEW:

- If you wire an array control to an edge of a For Loop, the loop will cycle once for every element in the array (this is termed *auto-indexing*). For example, in Figure 4.49, Array contains five elements. When Array is connected to the For Loop, the loop will cycle five times and the index will take on values from 0 to 4.
- The connection to the left edge of the For Loop is called a *tunnel* (indicated in Figure 4.50). Tunnels allow values to pass through, into the loop. When you wire an Array control to an *input tunnel*, the values of the array are passed into the loop and can be used inside the loop. In Figure 4.51 the array values are now being displayed one after another as the loop cycles. (The block diagram is shown in Figure 4.52.)

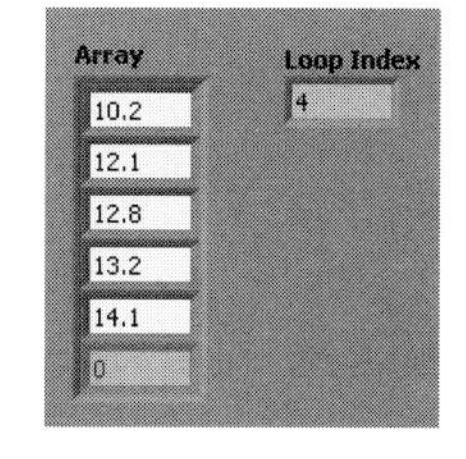

Figure 4.49
Using the number of elements in an array to set the number of For Loop iterations.

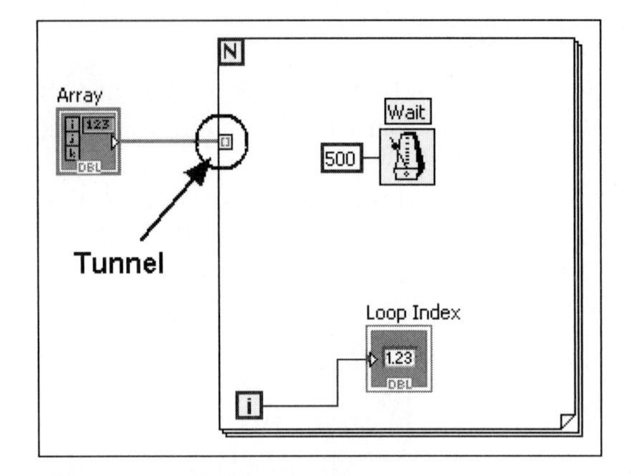

Figure 4.50
The block diagram showing how Array is wired to the left edge of the For Loop.

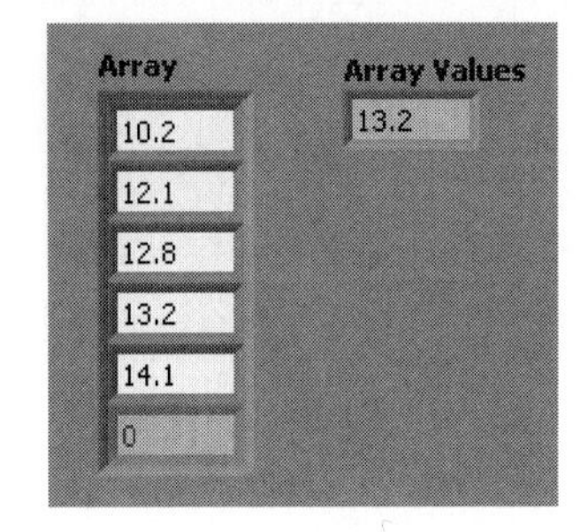

Figure 4.51
Displaying Array values one after another using a For Loop.

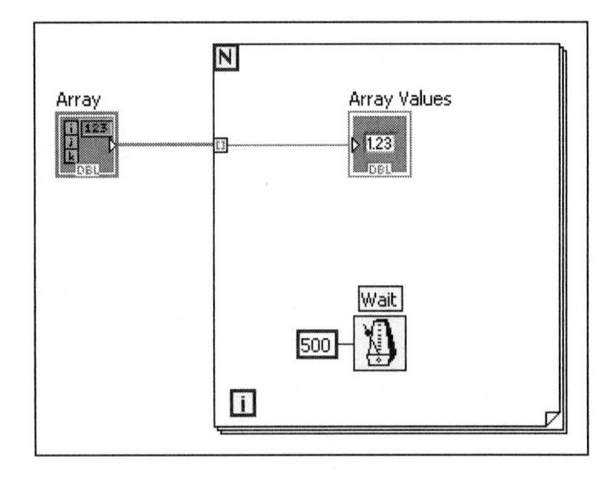

Figure 4.52
Using Array values inside the For Loop.

Note: Input tunnels are typically placed on the left side of the loop for readability, but actually an input tunnel can be on any For Loop boundary. LabVIEW knows values are coming into the loop because the values are being sent from a control output.

- Tunnels can pass information into the loop in two ways: with *indexing enabled* (as used here) and with *indexing disabled.* The "indexing" being referred to is array indexing. When indexing is enabled, one element of the array is passed into the loop with each iteration. When indexing is disabled, the entire array is passed into the loop when the loop initializes. Right-click on the tunnel to enable or disable indexing.
- When values are passed out of a loop through a tunnel (with indexing enabled), an array of values is created. The For Loop shown in Figure 4.53 cycles six times, and the index takes on values of 0–5. Those values are passed outside the loop

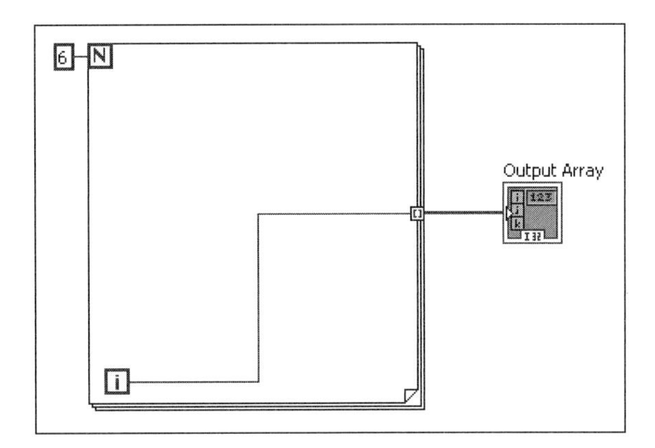

Figure 4.53
Creating an array using a For Loop.

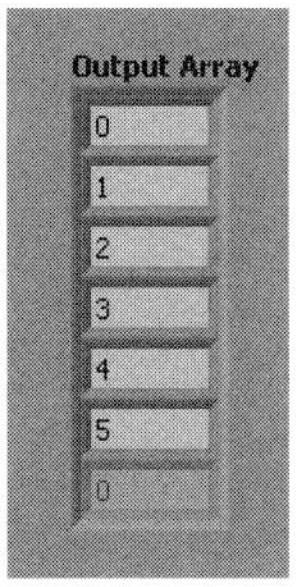

as an array when the loop terminates. The resulting output array is shown in Figure 4.54.

Note: Again, an output tunnel can be placed on any loop boundary. It is common to place them on the right boundary because information flow in LabVIEW programs tends to be from left to write.

Note: If indexing is disabled on an output tunnel, only the final element of the array is passed through the tunnel.

- A For Loop is an easy way to create an array of calculated values. In the VI shown in Figure 4.55, the loop index is used to create two arrays: x and $\sin(x)$. These are then plotted using an XY Graph as shown in Figure 4.56.

Figure 4.54
The output array created using the For Loop.

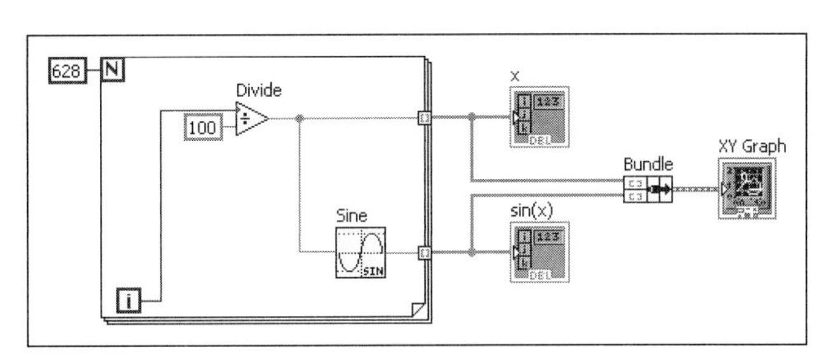

Figure 4.55
For Loop used to create two calculated arrays for graphing.

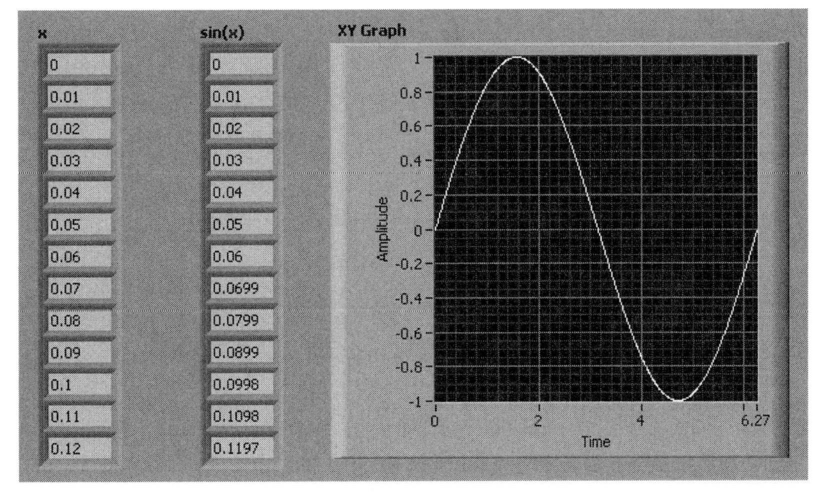

Figure 4.56
The calculated arrays and XY graph.

PRACTICE

Using Probes

The creator of the VI shown in Figure 4.57 was attempting to graph a sine wave (one cycle). It didn't work out that way.

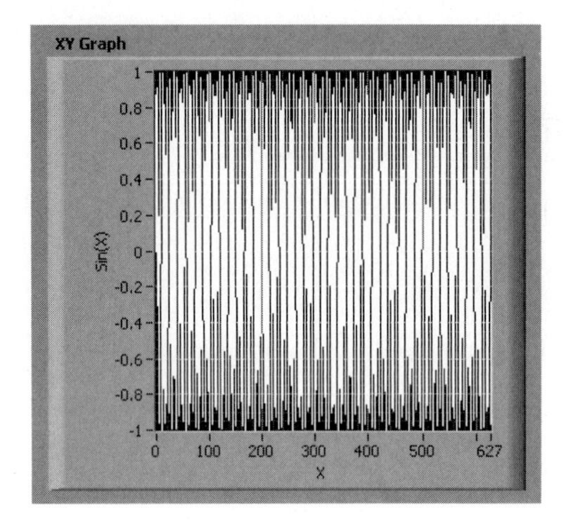

Figure 4.57
Failed attempt to plot one sine wave.

When you look at their block diagram, you notice that the For Loop runs from 0 to 628. The creator of the VI intended to divide "*i*" by 100 to create *x* values ranging from 0 to 2π, but he or she forgot to include the Divide function.

Use the Probe tool to position a probe on the wire leaving the "*i*". Then run the VI to demonstrate that the value being sent out of the For Loop as *X* is ranging from 0 to 627 instead of 0 to 6.27.

The value in the wire under the "20" probe (627) is shown in the probe window in Figure 4.58.

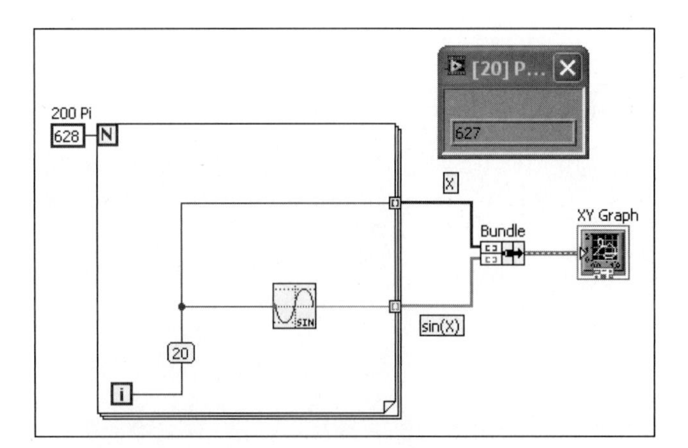

Figure 4.58
The VI with the probe in place (marked "20").

PROGRAMMING APPLICATION

Automatic Array Maker

A common programming task is creating an array of **Nvalues** values spread over a specified range (**Xmin** to **Xmax**). We will develop a VI that is capable of creating the array, and then create a subVI that can be used whenever such an array is required.

For development purposes, assume Nvalues = 5, Xmin = –20, Xmax = 20. With these values it is easy to predict that the final array will contain the values [–20, –10, 0, 10, 20]. We can use these values to test the VI.

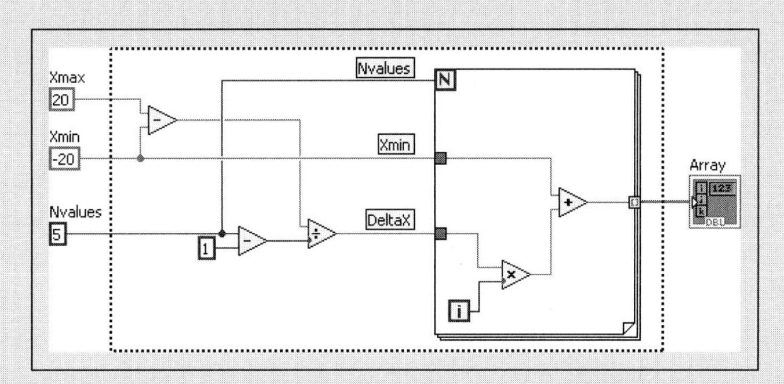

Figure 4.59
VI for creating an array (subVI elements in dashed box).

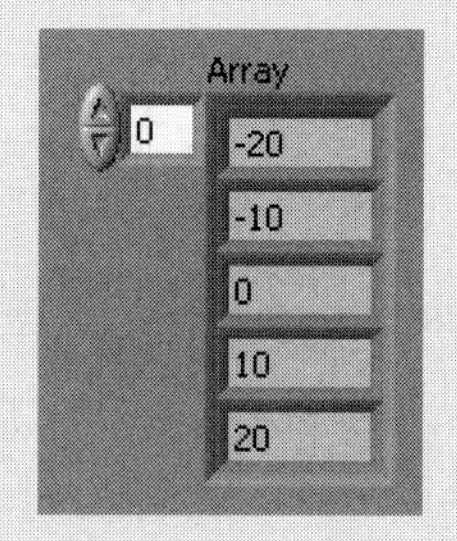

Figure 4.60
Front panel showing successful array creation.

The Nvalues output will be wired to the loop count terminal (the "*N*") of the For Loop. The step size, **DeltaX**, is calculated as

$$DeltaX = \frac{Xmax - Xmin}{Nvalues - 1}$$

The VI is shown in Figure 4.59, and we can see on the front panel in Figure 4.60 that it is generating the desired array values.

Next, we select the elements in the dashed box shown in Figure 4.59, and create a subVI using menu options: **Edit / Create subVI**. The result is shown in Figure 4.61. The subVI has been assigned a default icon and default number ("5" in this example, but it is arbitrary).

Double-click on the subVI icon to open the subVI for editing. The subVI (Figure 4.62) looks a lot like the original VI, except that the constants on the left side of Figure 4.59 have been replaced by controls in Figure 4.62.

The front panel of the subVI is shown in Figure 4.63. Double-click the icon in the upper right corner to open the *icon editor*.

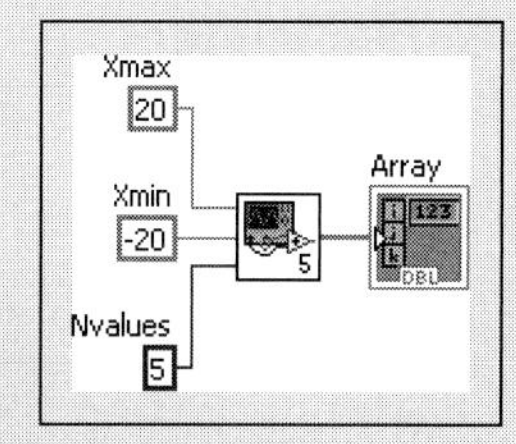

Figure 4.61
After creating the subVI.

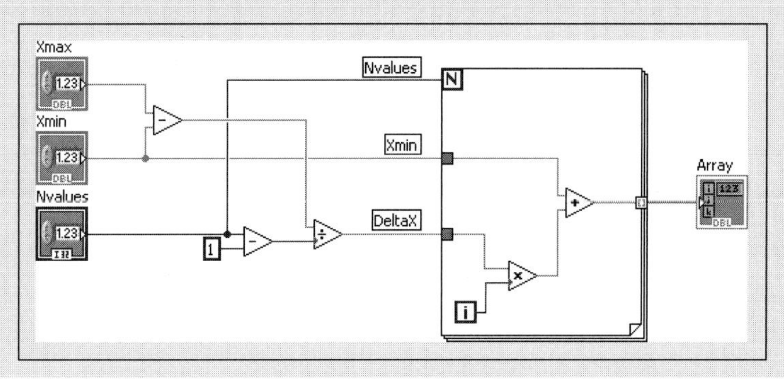

Figure 4.62
The subVI block diagram.

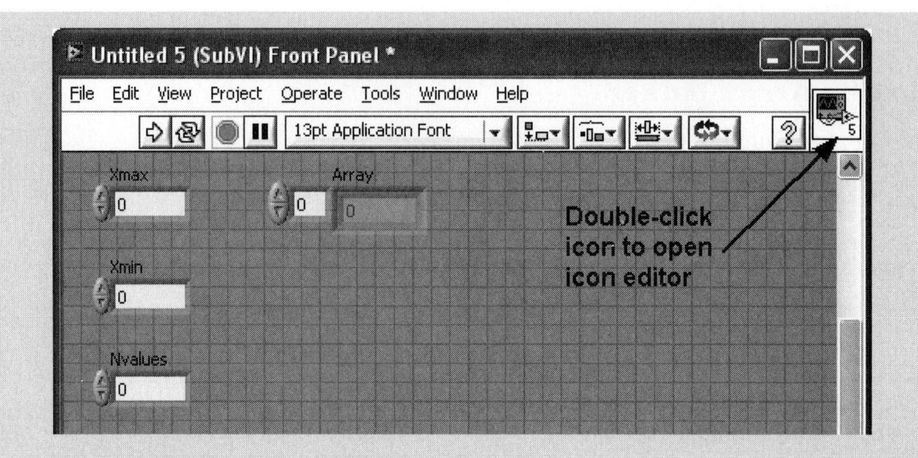

Figure 4.63
The front panel of the new subVI.

The LabVIEW 2009 Icon Editor is shown in Figure 4.64; the default icon image has been replaced with one of the many glyphs now available in the Icon Editor. Click OK to close the Icon Editor and return to editing the subVI. The new icon is now used for the subVI.

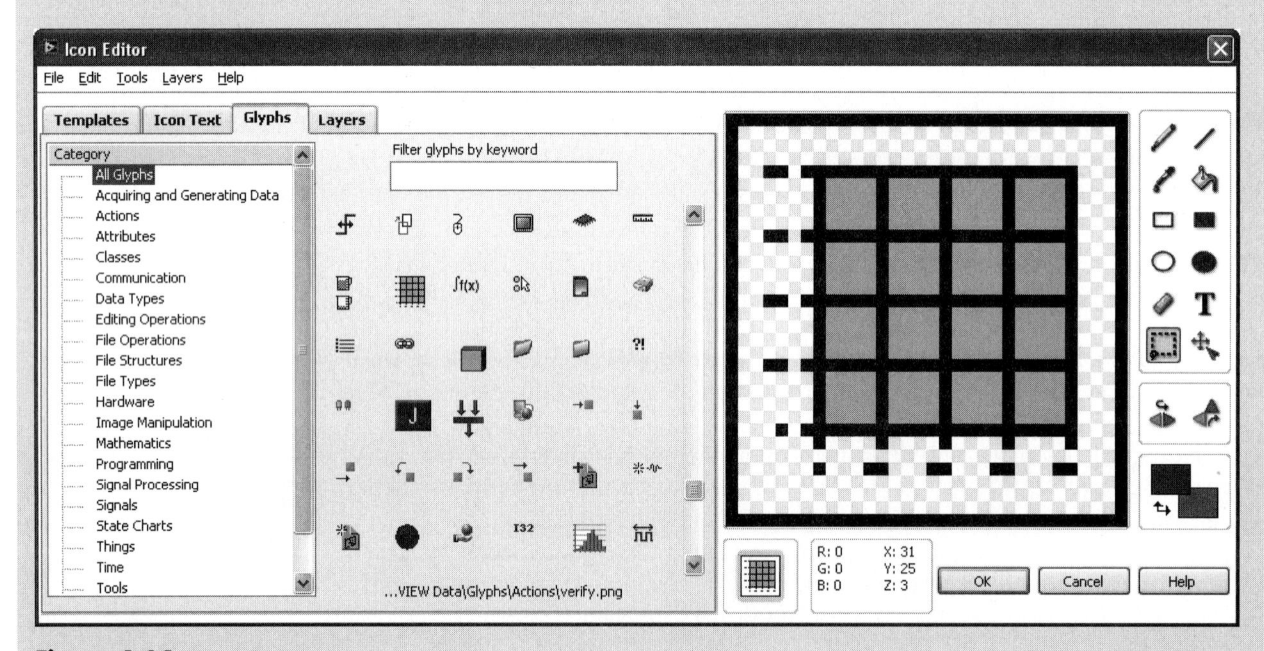

Figure 4.64
LabVIEW 2009 Icon Editor.

Be sure to save the subVI from the LabVIEW editor (**File / Save**). The name you use to save the subVI will be used to identify the subVI whenever it is used. In this example it has been named Make Array.vi.

Now, if we want to create two arrays for plotting: an *X* array containing 20 elements ranging from −80 to −40, and a *Y* array 20 with elements ranging from 240 to 360, we can call the Make Array subVI to create them as shown in Figure 4.65 (block diagram) and Figure 4.66 (front panel).

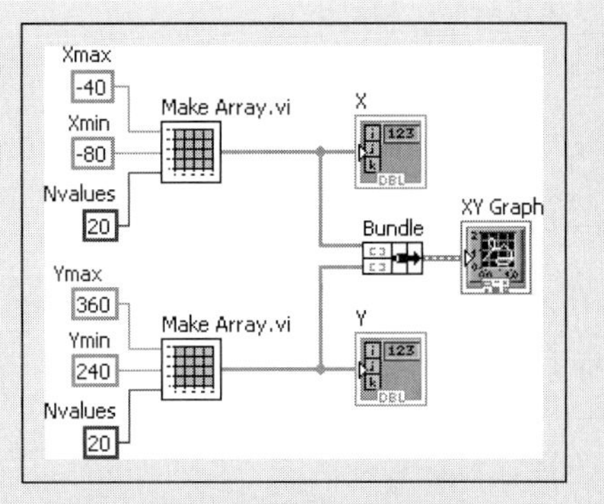

Figure 4.65
Creating two arrays for plotting using subVI calls.

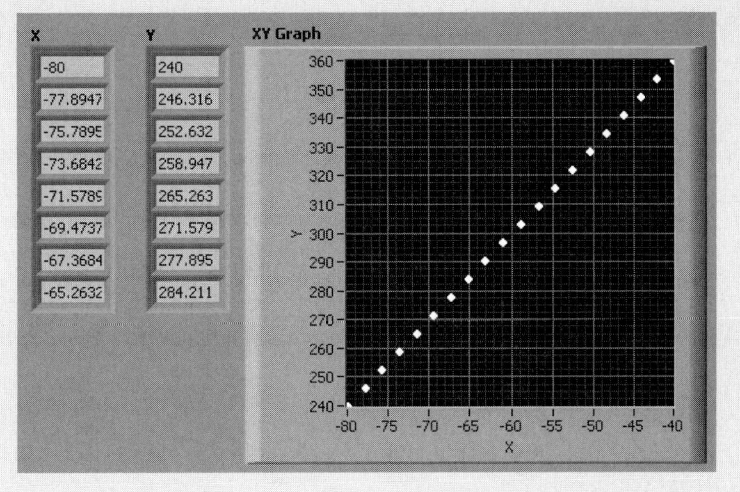

Figure 4.66
The results of using the subVI to create two arrays.

KEY TERMS

array
Array container
array functions
array operations
auto-indexing
cluster
column
condition number
count terminal (N)
determinant
element by element multiplication
For Loop
Icon Editor
Index Display
indexing disabled (tunnel)
indexing enabled (tunnel)
inverse matrix
iteration terminal (i)
matrix
matrix math operations
matrix multiplication (A×B function)
non-singular matrix
row
scalar
scalar multiplication
simultaneous linear equations
square matrix
subarray
transpose
tunnel

SUMMARY

Matrix Mathematics

- Adding Matrices or Arrays
- Transpose a Matrix or Array
- Multiplying Matrices or Arrays
- Condition Number
- Matrix Determinant
- Inverse Matrix

Arrays or Matrices?
In LabVIEW, both arrays and matrices are collections of related values, but a matrix is always 2D, whereas an array can have any number of dimensions. Some LabVIEW functions (e.g., graphing functions) required 1D arrays, so arrays must be used with those LabVIEW functions.

Array Index Origin
Array (and matrix) indexing begins at zero

Creating an Array (front panel)

1. Put the *Array container* on the front panel:

   Controls Palette / Modern Group / Array, Matrix & Cluster Group / Array

2. Drop a control or indicator into the Array container (only one array element will be visible):

   Controls Palette / Modern Group / Numeric Group / Numeric Control

   Controls Palette / Express Group / Numeric Controls Group / Num Ctrl

3. Resize the array to show the required number of elements
4. Hide the Index Display (optional): Right-click, **Visible Items / Index Display**

Placing a Matrix on Front Panel
Move the Real Matrix control from the Array, Matrix & Cluster Group onto the front panel.

Controls Palette / Modern Group / Array, Matrix & Cluster Group / Real Matrix

Array Functions

- Array Max & Min
- Array Size
- Array Subset
- Search 1D Array
- Sort 1D Array
- Split 1D Array
- Reverse 1D Array
- Rotate 1D Array
- Delete From Array
- Insert Into Array
- Replace Array Subset
- Reshape Array
- Transpose 2D Array
- Cluster To Array
- Array To Cluster
- Matrix to Array
- Array to Matrix

Extracting a Subarray

Subsets of various size: **Functions Palette / Programming Group / Array Group / Array Subset**

- **Starting Row Index** (0)—the index of the first row in the original array that should be included in the subarray. The default value is (0) which is the top row in the original matrix.
- **Number of Rows** (all)—the number of rows from the original array to include in the subarray. The default is (all) rows (i.e., all rows after the starting row).
- **Starting Column Index** (0)—the index of the first column in the original array that should be included in the subarray. The default value is (0) which is the left column in the original matrix.
- **Number of Columns** (all)—the number of columns from the original array to include in the subarray. The default is (all) columns (i.e., all columns after the starting column).

Single rows or columns: **Functions Palette / Programming Group / Array Group / Index Array**

- **Row Index** (0)—the index of the row in the original array that should be extracted.
- **Column Index** (0)—the index of the column in the original array that should be extracted.

Note: Wire only the Row Index or Column Index, not both.

Adding Arrays

Requirement: Arrays must be same size
Process: Add element by element

Functions Palette / Mathematics Group / Numeric Group / Add function

Transpose Array

Requirement: Any array can be transposed
Process: Interchange rows and columns

- Matrix function: **Transpose Matrix**

  Function Palette / Mathematics Group / Linear Algebra Group / Transpose Matrix

- Array function: **Transpose 2D Array**

  Function Palette / Programming Group / Array Group / Transpose 2D Array

Multiply an Array by a Scalar

Definition: scalar—single value (a number)
Requirement: Any array can be multiplied by a scalar
Process: multiply each element in array by scalar

Functions Palette / Mathematics Group / Numeric Group / Multiply Function

Matrix Multiplication

Requirement: Number of columns in the first array must equal the number of rows in the second array. Product array will have as many rows as first array, as many columns as second array.

Process: Multiply across the first matrix and down the second matrix, adding terms.

Function Palette / Mathematics Group / Linear Algebra Group / A × B Function

Element by Element Multiplication

Requirement: Arrays must be same size

Process: Multiply element by element

Functions Palette / Mathematics Group / Numeric Group / Multiply Function

Note: The Multiply function performs element by element multiplication on arrays, but matrix multiplication on matrices.

Condition Number

Requirement: Array must be square

Result: A small condition number indicates that a good solution is likely

Function Palette / Mathematics Group / Linear Algebra Group / Matrix Condition Number

Determinant

Requirement: Array must be square

Result: A zero determinant indicates that matrix is singular (cannot be inverted)

Function Palette / Mathematics Group / Linear Algebra Group / Determinant

Inverse Matrix

Requirement: Array must be square and non-singular

Result: matrix inverse

Function Palette / Mathematics Group / Linear Algebra Group / Inverse Matrix

Solving Simultaneous Linear Equations

Requirement: Number of columns in the coefficient array must equal the number of rows in the right-hand-side.

Process:

$$[x] = [C]^{-1}[r]$$

Function Palette / Mathematics Group / Linear Algebra Group / Solve Linear Equations

For Loops

Commonly used to create arrays of calculated values.

- The constant wired to the ***N*** input (called the *count terminal*) instructs the For Loop how many times to cycle.

- The ***i*** is the loop index (called the *iteration terminal.*) The loop index is incremented each time the For Loop cycles. Indexing starts at zero.
- If you wire an array control to an edge of a For Loop, the loop will cycle once for every element in the array (this is termed *auto-indexing*).
- A wire connection to the boundary of a For Loop is called a *tunnel.*
- Tunnels can pass information into a loop in two ways:
  - *indexing enabled*—one element of the array is passed into the loop with each iteration
  - *indexing disabled*—entire array is passed into the loop when the loop initializes
- Tunnels can pass information out of a loop in two ways:
  - *indexing enabled*—an array of values is created
  - *indexing disabled*—only the final value is sent out of the loop

Creating a subVI

1. Create a working VI containing the programming elements that will become the subVI.
2. Select the programming elements that will become the subVI.

   Note: The wires entering and leaving the selection will become the subVI inputs and outputs.
3. Use menu options: **Edit / Create subVI** to create the subVI.
4. Double-click on the subVI icon to open the subVI for editing.
5. Double-click the icon in the upper right corner of the subVI to open the icon editor.
6. Edit the icon, as desired.
7. Click OK to close the Icon Editor and return to editing the subVI.
8. Save the subVI from the LabVIEW editor (**File / Save**).

SELF-ASSESSMENT

1. What LabVIEW function is used to extract a subarray from an array?
   ANS: Array Subset function:

   Functions Palette / Programming Group / Array Group / Array Subset

2. Can the following arrays be added? Why, or why not?
   - A and C
   - B and D
   - C and E

$$A = \begin{bmatrix} 1 & 2 & 3 \\ 2 & 1 & 4 \\ 3 & 4 & 7 \end{bmatrix} \quad B = \begin{bmatrix} 3 & 5 & 7 \\ 2 & 4 & 8 \\ 1 & 3 & 6 \end{bmatrix} \quad C = \begin{bmatrix} 2 & 3 \\ 5 & 7 \end{bmatrix} \quad D = \begin{bmatrix} 1 \\ 3 \\ 4 \end{bmatrix} \quad E = \begin{bmatrix} 1 & 5 \\ 8 & 3 \end{bmatrix}$$

   ANS: Arrays must be the same size if they are to be added
   - A and C—NO
   - B and D—NO
   - C and E—YES

3. Can the following arrays be multiplied? If so, what will be the size of the product array?
   - A and C
   - B and D
   - C and B

$$A_{3\times 2} = \begin{bmatrix} 2 & 4 \\ 3 & 2 \\ 1 & 5 \end{bmatrix} \quad B_{2\times 4} = \begin{bmatrix} 2 & 1 & 3 & 4 \\ 4 & 6 & 1 & 3 \end{bmatrix} \quad C_{2\times 2} = \begin{bmatrix} 2 & 3 \\ 5 & 7 \end{bmatrix} \quad D_{3\times 1} = \begin{bmatrix} 1 \\ 3 \\ 4 \end{bmatrix}$$

   ANS: In order to multiply, the number columns in the first array must equal the number of rows in the second. The size of the result is equal to the number of rows in the first array by the number of columns in the second.
   - [A] [C]—YES, result will be 3 × 2
   - [B] [D]—NO
   - [C] [B]—YES, result will be 2 × 4

4. Which LabVIEW function should be used to multiply arrays?
   ANS: A × B function, located at

   Function Palette / Mathematics Group / Linear Algebra Group / A × B Function

5. When trying to solve simultaneous equations, is a large condition number a good thing or a bad thing?
   ANS: A bad thing—a large condition number suggests the solution may be subject to round-off and truncation errors. A condition number near 1 is ideal.

6. When trying to solve simultaneous equations, what does a determinant of zero on the coefficient matrix tell you?
   ANS: The coefficient matrix is singular; no solution is possible.

PROBLEMS

1. Create a VI that adds the following arrays, if array addition is possible.

   a. $A = \begin{bmatrix} 3.1 & 2.4 \\ 4.8 & 6.5 \end{bmatrix} \quad B = \begin{bmatrix} 2.2 & 4.9 \\ 5.3 & 8.1 \end{bmatrix}$

   b. $C = \begin{bmatrix} 2 & 7 & 1 \\ 3 & 5 & 2 \\ 1 & 4 & 9 \end{bmatrix} \quad D = \begin{bmatrix} 3 \\ 5 \\ 1 \end{bmatrix}$

   c. $E = \begin{bmatrix} 2 & 7 & 1 \\ 3 & 5 & 2 \\ 1 & 4 & 9 \end{bmatrix} \quad F = \begin{bmatrix} 6 & -2 & 4 \\ -1 & 3 & 6 \\ 0 & 2 & 7 \end{bmatrix}$

2. Create a VI that multiplies the following arrays, if multiplication is possible.
   a. [A][B]

$$A = \begin{bmatrix} 3.1 & 2.4 \\ 4.8 & 6.5 \end{bmatrix} \qquad B = \begin{bmatrix} 2.2 & 4.9 \\ 5.3 & 8.1 \end{bmatrix}$$

   b. [C][D]

$$C = \begin{bmatrix} 2 & 7 & 1 \\ 3 & 5 & 2 \\ 1 & 4 & 9 \end{bmatrix} \qquad D = \begin{bmatrix} 3 \\ 5 \\ 1 \end{bmatrix}$$

   c. [D][C]

$$C = \begin{bmatrix} 2 & 7 & 1 \\ 3 & 5 & 2 \\ 1 & 4 & 9 \end{bmatrix} \qquad D = \begin{bmatrix} 3 \\ 5 \\ 1 \end{bmatrix}$$

3. Create a VI that multiplies the following arrays, if multiplication is possible.
   a. [E][F]

$$E = \begin{bmatrix} 2 & 7 & 1 \\ 2 & 7 & 1 \\ 1 & 4 & 9 \end{bmatrix} \qquad F = \begin{bmatrix} 2 & 4 & 1 & 7 & 8 & 9 \\ 1 & 3 & 1 & 9 & 2 & 6 \\ 8 & 2 & 4 & 1 & 3 & 5 \end{bmatrix}$$

   b. [G][H]

$$G = \begin{bmatrix} 2 & 7 & 1 \\ 2 & 7 & 1 \\ 1 & 4 & 9 \end{bmatrix} \qquad H = \begin{bmatrix} 1 & 4 & 2 & 5 \\ 2 & 4 & 1 & 2 \\ 7 & 2 & 4 & 3 \end{bmatrix}$$

4. Check the determinant to see if the following arrays can be inverted. If inversion is possible, create a VI that inverts the arrays.

   a. $E = \begin{bmatrix} 2 & 7 & 1 \\ 3 & 5 & 2 \\ 1 & 4 & 9 \end{bmatrix}$

   b. $G = \begin{bmatrix} 2 & 7 & 1 \\ 2 & 7 & 1 \\ 1 & 4 & 9 \end{bmatrix}$

   c. $F = \begin{bmatrix} 2 & 4 & 1 & 7 & 8 & 9 \\ 1 & 3 & 1 & 9 & 2 & 6 \\ 8 & 2 & 4 & 1 & 3 & 5 \end{bmatrix}$

5. Check the condition number of the following arrays.

a. $E = \begin{bmatrix} 2 & 7 & 1 \\ 3 & 5 & 2 \\ 1 & 4 & 9 \end{bmatrix}$

b. $G = \begin{bmatrix} 2 & 7 & 1 \\ 2 & 7 & 1 \\ 1 & 4 & 9 \end{bmatrix}$

c. $F = \begin{bmatrix} 2 & 4 & 1 & 7 & 8 & 9 \\ 1 & 3 & 1 & 9 & 2 & 6 \\ 8 & 2 & 4 & 1 & 3 & 5 \end{bmatrix}$

6. The following matrix systems represent sets of simultaneous linear equations written in matrix form. Check the condition number and determinant of the coefficient matrix, then, if a solution is possible, create a VI to solve the equations.

a. $C = \begin{bmatrix} 2 & 7 & 1 \\ 3 & 5 & 2 \\ 1 & 4 & 9 \end{bmatrix} \quad rhs = \begin{bmatrix} 3 \\ 5 \\ 1 \end{bmatrix}$

b. $C = \begin{bmatrix} 2 & 7 & 1 \\ 2 & 7 & 1 \\ 1 & 4 & 9 \end{bmatrix} \quad rhs = \begin{bmatrix} 3 \\ 5 \\ 1 \end{bmatrix}$

c. $C = \begin{bmatrix} -2 & 4 & 1 & 3 \\ 1 & -4 & 6 & 2 \\ 8 & 3 & 3 & 1 \\ 3 & 7 & -3 & 2 \end{bmatrix} \quad rhs = \begin{bmatrix} -2 \\ -4 \\ -1 \\ -5 \end{bmatrix}$

7. Solve the following sets of simultaneous linear equations, if possible. Check the determinant of the coefficient matrix to see if a solution is possible.

a.
$$\begin{aligned} 2x_1 + 7x_2 + 1x_3 &= 3 \\ 3x_1 + 5x_2 + 2x_3 &= 5 \\ 1x_1 + 4x_2 + 9x_3 &= 1 \end{aligned}$$

b.
$$\begin{aligned} 2x_1 + 7x_2 + 1x_3 &= 3 \\ 3x_1 + 5x_2 + 2x_3 &= 5 \\ -x_1 + 2x_2 - x_3 &= 1 \end{aligned}$$

c.
$$\begin{aligned} 7a + 2b + c + d &= 4 \\ 3a + 8b - 2c + 2d &= 1 \\ a + 4b + 9c - 2d &= 3 \\ 5a - 2b - 3c + 4d &= 2 \end{aligned}$$

8. LabVIEW provides two ways to solve sets of simultaneous linear equations: inverting the coefficient matrix and multiplying by the right-hand-side vector, and using the Solve Linear Equations function (actually a VI). Use both methods to solve the following set of equations and compare the results. Do both methods yield the same result?

$$2x_1 + 7x_2 + 1x_3 = 3$$
$$3x_1 + 5x_2 + 2x_3 = 5$$
$$1x_1 + 4x_2 + 9x_3 = 1$$

9. Use Kirchhoff's Laws to develop three equations to solve for the three currents (I1, I2, I3) indicated in Figure 4.67. The known quantities are as follows:

- E = 20 V
- R1 = 120 ohms
- R2 = 150 ohms
- R3 = 30 ohms

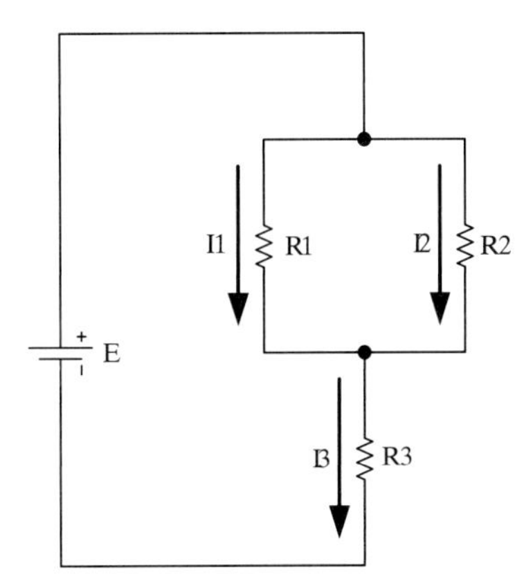

Figure 4.67
Series-parallel circuit.

CHAPTER 5

Data Acquisition with LabVIEW

Objectives

After reading this chapter, you will know:

- how LabVIEW can be used for automated data acquisition
- why signal conditioning is often needed with data acquisition systems
- how data acquisition hardware functions
- how to write two types of LabVIEW VIs for data acquisition
    - acquiring data sets
    - acquiring data point-by-point

5.1 OVERVIEW OF DATA ACQUISITION

The ability to pull in data from an outside *source*, process the data, and send signals back out to control devices is what sets LabVIEW apart from numerous other software products that can be used to analyze data. In this chapter we take a look at acquiring data with LabVIEW.

From the viewpoint of a researcher in a laboratory, the goal of *data acquisition* is to capture data from one or more laboratory instruments on a computer so that it can be analyzed and stored. If you design data acquisition systems for a living, the term *data acquisition* refers to the process of automatically importing data from one or more *sensors* or *transducers* directly into a computer system. Some nomenclature will help:

- A *sensor* is a device that responds to a physical change and outputs an electrical signal.
- A *transducer* is a device that converts energy from one form to another.

For example, a thermocouple is a sensor that generates an *electromotive force (emf)* due to two dissimilar metals joined at the thermocouple junction. The emf generated is a low-level voltage (typically millivolts), which, when sent down a wire, becomes a voltage *signal*. A transducer can be used to convert the low-level voltage to a higher voltage. Modern measurement devices routinely bundle sensors and transducers together to generate and transform the output signal to a useful form.

A simple system (see Figure 5.1) requires a transducer that outputs a signal, *data acquisition hardware (DAQ)*, and a computer. It sounds simple, and sometimes it can be—but there are lots of things that can make the process non-trivial.

First, not all laboratory instruments generate a signal that can be transmitted over a wire. A thermometer, for example, is a simple device for measuring temperature, but the temperature is measured visually by looking at the level of fluid in the thermometer capillary. Reading the level with your eye is a great way to get the temperature value into your brain, but visual readings are hard to get into computer systems.

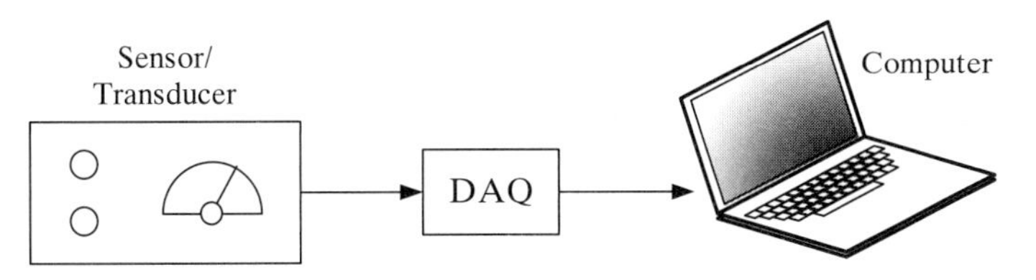

Figure 5.1
Simple data acquisition system.

Fortunately, there are commonly available temperature sensors:

- *Thermocouples* output a low-level voltage signal that is related to the temperature of the thermocouple junction.
- *Resistance temperature devices (RTDs)* have a varying resistance that depends on the temperature of the device's sensor. Running a known current through the varying resistance generates a varying voltage that can be related to the temperature of the RTD sensor.

Once the sensor is outputting a signal that can be sent through a wire to a computer, there still may be a mismatch between the type or range of the signal, and the type and range of signal the data acquisition system has been designed to handle. For example,

- Some industrial transducers have been designed to output current signals in the 4–20 mA range. If the data acquisition system has been designed to accept voltage signals, there is a signal type mismatch between the current signal from the transducer and the voltage signal required by the data acquisition system.
- If a transducer's output is in the range 50–100 mV and the data acquisition system has been designed to accept voltage signals in the range 0–10 V, there is a mismatch between the transducer output signal range and the allowable input signal range of the data acquisition system.

Signal conditioning may be required to adjust the signal type and range of the output signal to align with the requirements of the data acquisition system (see Figure 5.2).

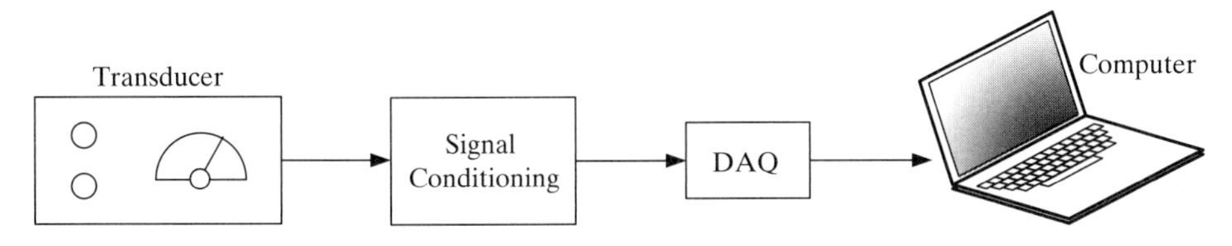

Figure 5.2
Data acquisition system with external signal conditioning.

Once the output signal from the instrument has been adjusted to align with the requirements of the data acquisition system, data can be collected. The computer driving the data acquisition system will need to know

- what signal(s) to measure
- how often to take readings
- how many readings to collect, or how long to continue reading the signal(s)

You configure the data acquisition process by providing this required information prior to collecting data. The LabVIEW approach is to create a data acquisition *task* that contains this information.

The collected data (termed a *waveform* in LabVIEW) is available to the computer system and can be displayed, modified (e.g., digitally filtered), analyzed, or simply stored.

The sections in this chapter provide additional detail on the steps commonly involved in using a data acquisition system.

5.2 SENSORS, SIGNALS, AND SIGNAL CONDITIONING

The bottom line when it comes to measuring a variable of interest is identifying some physical phenomenon that changes when the variable of interest changes. For example, the volume of a fluid expands as its temperature increases, and that physical phenomenon has been utilized to create thermometers. While lots of physical phenomena have been used for measurements, phenomena that can be related to an electrical property are more useful for automated data acquisition systems.

A *sensor* is a device that converts a physical property change into an electrical signal. Sensors are available for the following common measurements (and many others):

- Temperature (thermocouple, RTD)
- Force or Pressure (strain gauge, load cell)
- Position (potentiometer)
- Light (phototransistor)
- Sound (microphone)

The basic output of a sensor is rarely ready to be connected directly to a data acquisition system. Data acquisition systems are typically designed to measure voltage, and have predefined input ranges. The ranges 0–10 V and −5 to +5 V are common.

Note: We are assuming here that the signal from the sensor is an *analog signal*. Analog simply implies that the signal level (e.g., voltage level) can vary continuously (smoothly), as opposed to a *digital signal* which can take only finite values. Many data acquisition systems have separate channels for analog and digital signals. The majority of this chapter focuses on analog signals, although digital signals are mentioned briefly.

5.2.1 Signal Conditioning

The process of modifying the output of a sensor is called *signal conditioning*. Signal conditioning is often needed to make the output from a sensor compatible with data acquisition systems. In the past, signal conditioning was routinely needed and often a source of problems when collecting data using computers. Today, many sensors output a signal that is compatible with common data acquisition systems. Effectively, the engineers designing the sensors have built the signal conditioning right into the sensors so that the users of the sensors don't have to do it themselves.

Common reasons to condition signals include the following:

- dealing with noisy signals (Filtering)
- aligning a sensor output to a data acquisition system input constraint (amplification and offset)

Dealing with Noisy Signals

While taking a measurement, a gauge needle bounces around if the signal is noisy. Signal noise is very common, sometimes it comes from the system being measured, and sometimes it comes from the electronics in the sensor itself. There are several ways to try to deal with a noisy signal:

- Ignore it.
- Try to modify the system.
- Get higher quality sensors.
- Use multiple measurements so that the measured results can be averaged.
- Filter the signal before it gets to the data acquisition system.
- Filter the signal after it has gone through the data acquisition system.

EXAMPLE 5.1

Determining Level in a Tank

One way to determine the height of fluid in a tank is to use sonar. A sonar system (see Figure 5.3) bounces a sound wave off the surface and measures the time between generating the sound pulse and detecting the echo. Using the speed of sound, you can calculate the distance between the sonar unit and the liquid surface. If the fluid in the tank is moving or bubbling, the signal from the sonar detector will be very noisy.

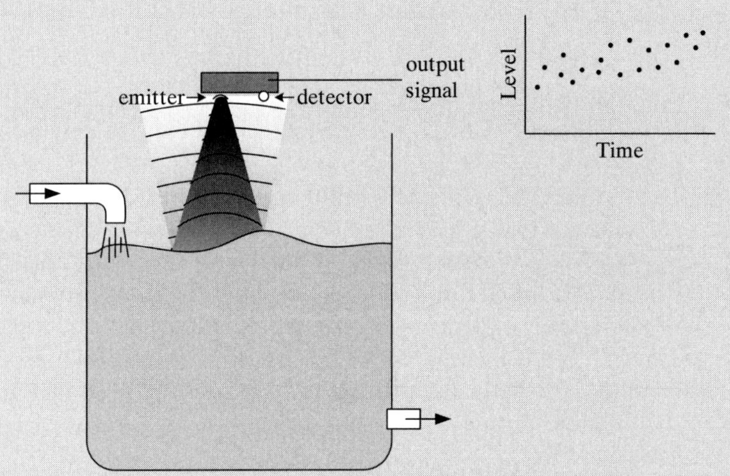

Figure 5.3
Sonar detector for level measurement.

If the purpose for measuring the liquid level is simply to prevent the level from rising higher than 80% of the tank height, a noisy signal may not matter. However, if the goal is to report the level in the tank to the exact millimeter, the moving fluid that is generating a noisy signal is a serious problem.

One approach to fixing the problem would be to modify the process to try to reduce the fluid movement. Here are a couple of options:

- If the fluid entering the tank is being dropped onto the surface, the inlet pipe might be extended so that the fluid enters below the liquid level.
- A baffle might be added to separate the level measurement area from the inlet flow.

Trying to get better sensors won't help because the noise is coming from the physical process. (Actually, there could be a noisy sensor too, but you'd never know it because of the bouncing fluid level. You can't tell if you need to replace the sensor until the moving fluid problem has been fixed.)

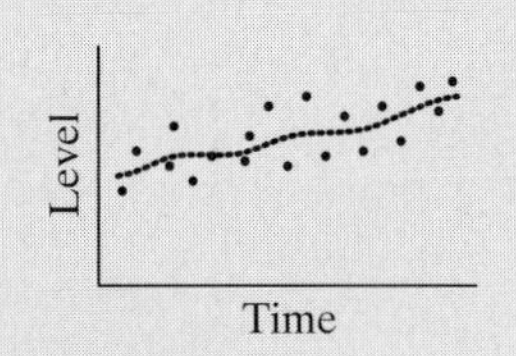

Figure 5.4
Averaging measurements to deal with noisy signal.

Signal Filtering

Taking multiple measurements and averaging is a common way to try to reduce signal noise. Averaging is a type of *signal filtering*. In this example averaging multiple readings of liquid level can work as long as the average liquid level changes slowly compared to the time required to take the multiple measurements; that is, the rate at which the level fluctuates because of bouncing fluid must be fast compared with the rate at which the tank level rises and falls. Other ways to say this include

- The frequency of the noise component of the signal must be significantly higher than the frequency of the desired signal component (the average tank level).
- The *signal-to-noise ratio* must be significantly greater than one.

Other Types of Filters

Filtering the signal can be done before or after data acquisition. Modern *filters* are designed to eliminate certain portions of a frequency spectrum. Figure 5.5 is an example of a sine wave with a varying frequency from low frequency (left) to high frequency (right).

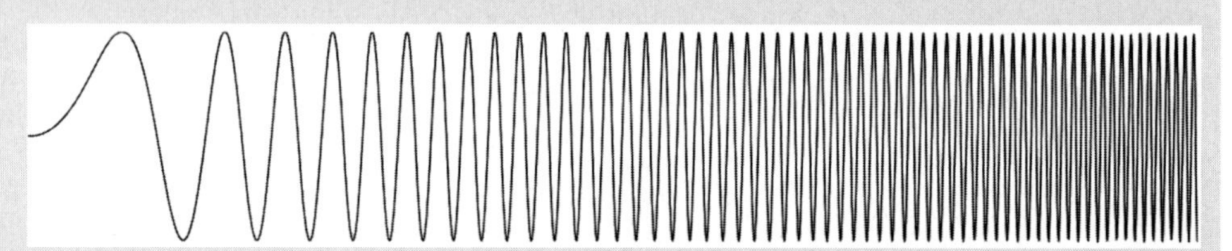

Figure 5.5
Low frequency (left) and high frequency (right) oscillations.

Low-pass filters are commonly used to filter data signals. (This assumes that the noise signal is at a higher frequency than the signal of interest.) A low-pass filter allows low-frequency signals to pass through, but stops high-frequency signals. A low-pass filter applied to the wave shown in Figure 5.5 might produce the results shown in Figure 5.6. The cut-off frequency is an adjustable filter parameter. In practice, you want the cut-off frequency set so that the filter leaves all of the desired signal and removes all of the high-frequency noise. (Real filters generate a less perfect cut off than the result shown in Figure 5.6.)

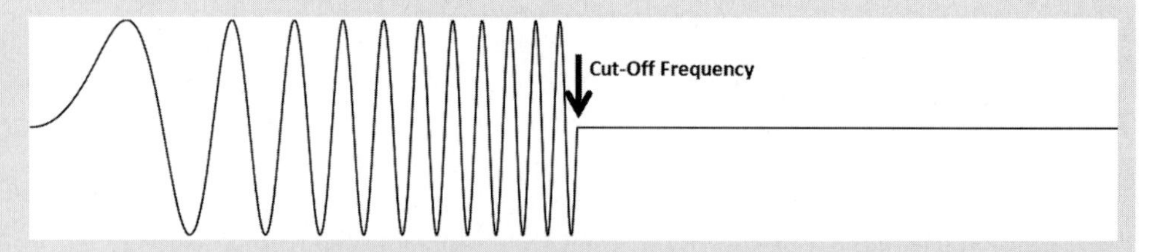

Figure 5.6
Waveform after ideal low-pass filtering.

If noise in the sonar detector's signal is at a much higher frequency than the level signal, a low-pass filter could be used to remove the noise.

High-pass filters allow the high-frequency signal components to pass, and filter out the low-frequency components, as illustrated in Figure 5.7. High-pass filters are

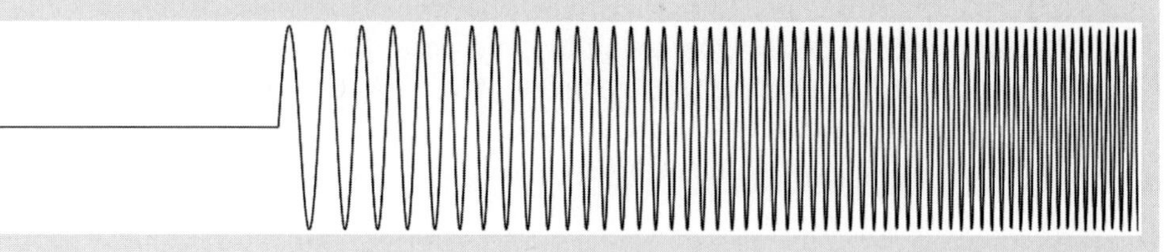

Figure 5.7
Waveform after ideal high-pass filtering.

less commonly used for signal conditioning than low-pass filters, but they are sometimes used to deal with low-frequency baseline drift on a signal.

Band-pass filters allow a specified frequency range to pass. Low and high cut-off frequencies are set with band-pass filters.

Signal filtering can be done before or after the signal passes through the data acquisition system. There are pros and cons for each approach:

Filtering after acquiring the signal (in the computer)

- PRO: No external filter is needed.
- PRO: Digital filters can be used (they are available in LabVIEW, and easy to use).
- CON: Aliasing the signal is a possibility (aliasing is described below).

Filtering before data acquisition

- CON: An external filter is required (another expense, cut-off frequency may be harder to adjust).
- PRO: Aliasing the signal is not a concern.

If you do not filter out high-frequency noise before sampling your data with a data acquisition system, you could have a problem with *aliasing*. Aliasing is best described by example.

EXAMPLE 5.2

Aliasing

Ideally, you want to sample your signal fast enough that a smooth curve through the sampled values faithfully reproduces the original signal. Another way to say this is that you should always sample at a rate that is at least twice the frequency of the highest frequency component in the signal. This ensures that there are at least two data points on every bump in the original signal. If you sample at too low a frequency, the sampled data can show artifacts that can mislead. This is called *aliasing*.

To demonstrate aliasing, we will create a noisy signal by combining a slow sine wave (amplitude = 3, frequency = $1/2\ \pi\ \text{sec}^{-1}$) and a fast cosine wave (amplitude = 1, frequency of $^{1}/_{0.2}\ \pi\ \text{sec}^{-1}$). The individual components are shown in Figure 5.8 and the combined noisy signal is shown in Figure 5.9.

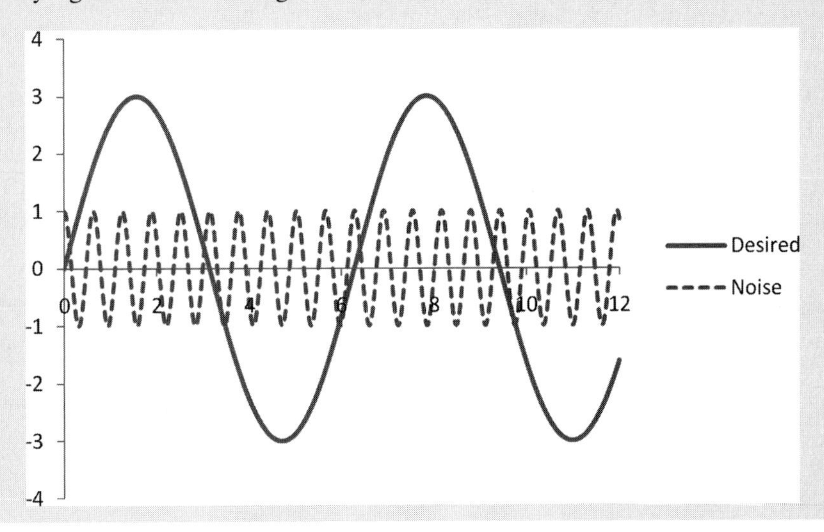

Figure 5.8
The components of the noisy signal.

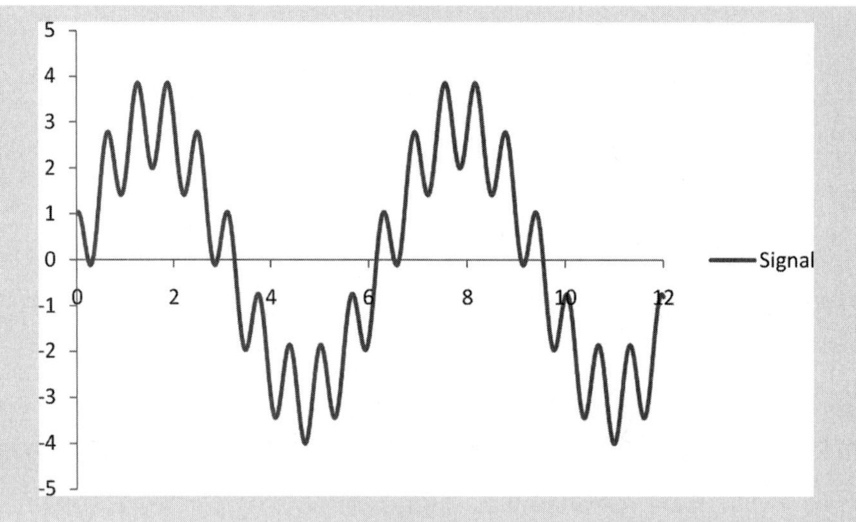

Figure 5.9
The noisy signal sent to the data acquisition system.

Note: In this example we are creating an artificial noisy signal with a signal-to-noise ratio of 3:1.

To avoid aliasing, the data acquisition system should sample at a rate at least twice the frequency of the fast cosine, or at least $^1/_{0.1}\,\pi\ \mathrm{sec}^{-1}$ (3.18 sec^{-1}) in this example. To demonstrate the effect of aliasing, we will sample at a lower rate of 1.59 sec^{-1}, which is equal to the frequency of the fast cosine curve. The result is shown in Figure 5.10. The dots represent the values recorded by the data acquisition system. At this sampling rate, the recorded data look like a sine wave, but the range is wrong (−2 to 4 rather than −3 to 3). This offset is due to aliasing.

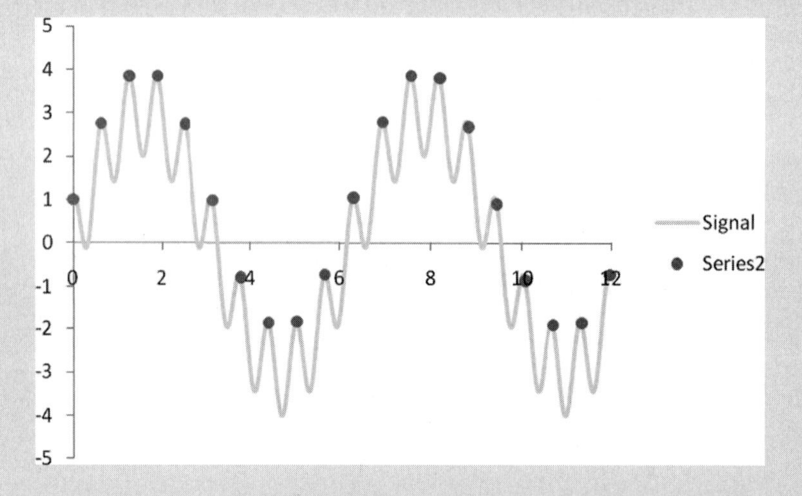

Figure 5.10
The sampled data values taken every 0.63 seconds superimposed on the noisy signal.

But it gets worse. What happens if you sample at a slightly quicker rate, say a frequency of 1.67 sec^{-1}? Now, you start taking samples from different points on the high-frequency cosine wave and the result is that the sampled data no longer even look like a sine wave (see Figure 5.11).

Aliasing can ruin a data set. There are a couple of ways to avoid aliasing:

- Sample at a rate at least twice the frequency of the highest frequency component in your signal.
- Filter out the high-frequency noise <u>before</u> sampling the data.
- *Oversample* and use digital filters <u>after</u> sampling the data.

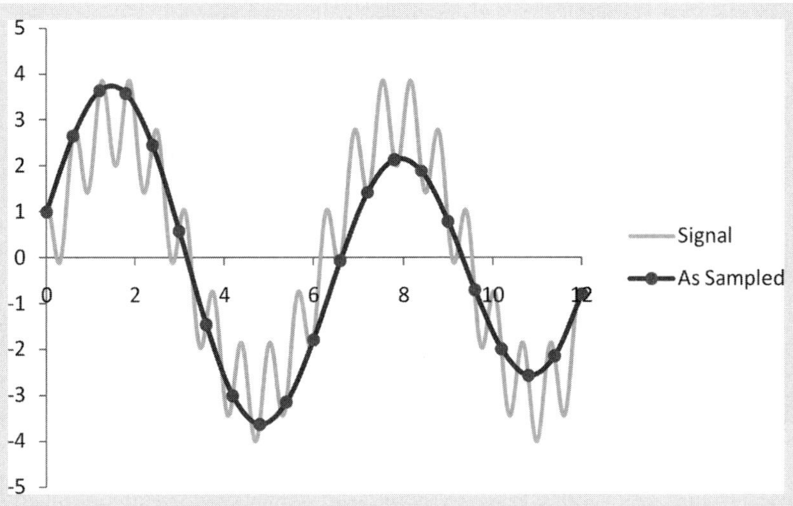

Figure 5.11
The sampled data values taken every 0.60 seconds superimposed on the noisy signal.

With the fast data acquisition systems available today, oversampling is often an option. With oversampling, you sample at a rate significantly faster than the highest frequency component in your signal so that the acquired data still faithfully represent the signal, noise and all. Then you can use digital filters on the sampled data to reduce the noise.

APPLICATION

High-Pass Filtering to Remove Baseline Drift

Measurements on human patients are often subject to baseline drift (DC offset). One common measurement is the electrocardiogram, or ECG. The ECG is a measurement of heart electrical activity taken at the surface of the chest wall. There are several inches of tissue between the signal source and the measurement site, and any movement of those tissues can show up in the signal.

The ECG waveform shown in Figure 5.12 is a simulated waveform created using the ECGSYN program developed by Patrick McSharry from the Department of Engineering Science, University of Oxford, and by Gari Clifford of the Laboratory for Computational Physiology at MIT. The program is available at www.physionet.org/physiotools/ecgsyn/. Baseline drift is so common with ECG waveforms that it is included in the simulation. The dashed line in Figure 5.12 is illustrating the extent of baseline drift in this data set.

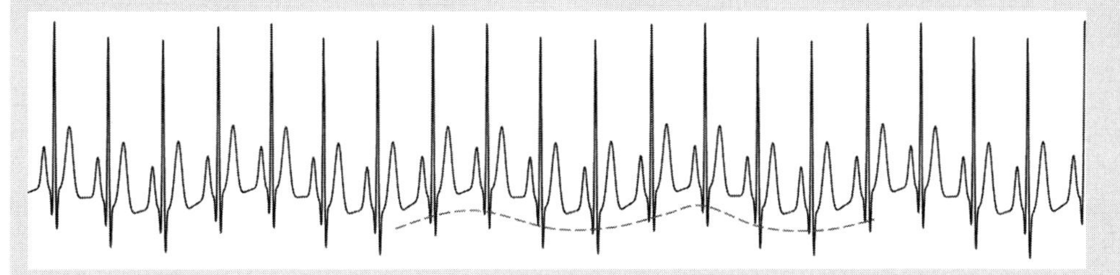

Figure 5.12
Simulated ECG waveform.

In this application we will apply a high-pass filter to the ECG data to demonstrate how such a filter can be used to reduce baseline drift. The effect of the filter is shown in Figure 5.13, in which the baseline drift has been dramatically reduced after filtering.

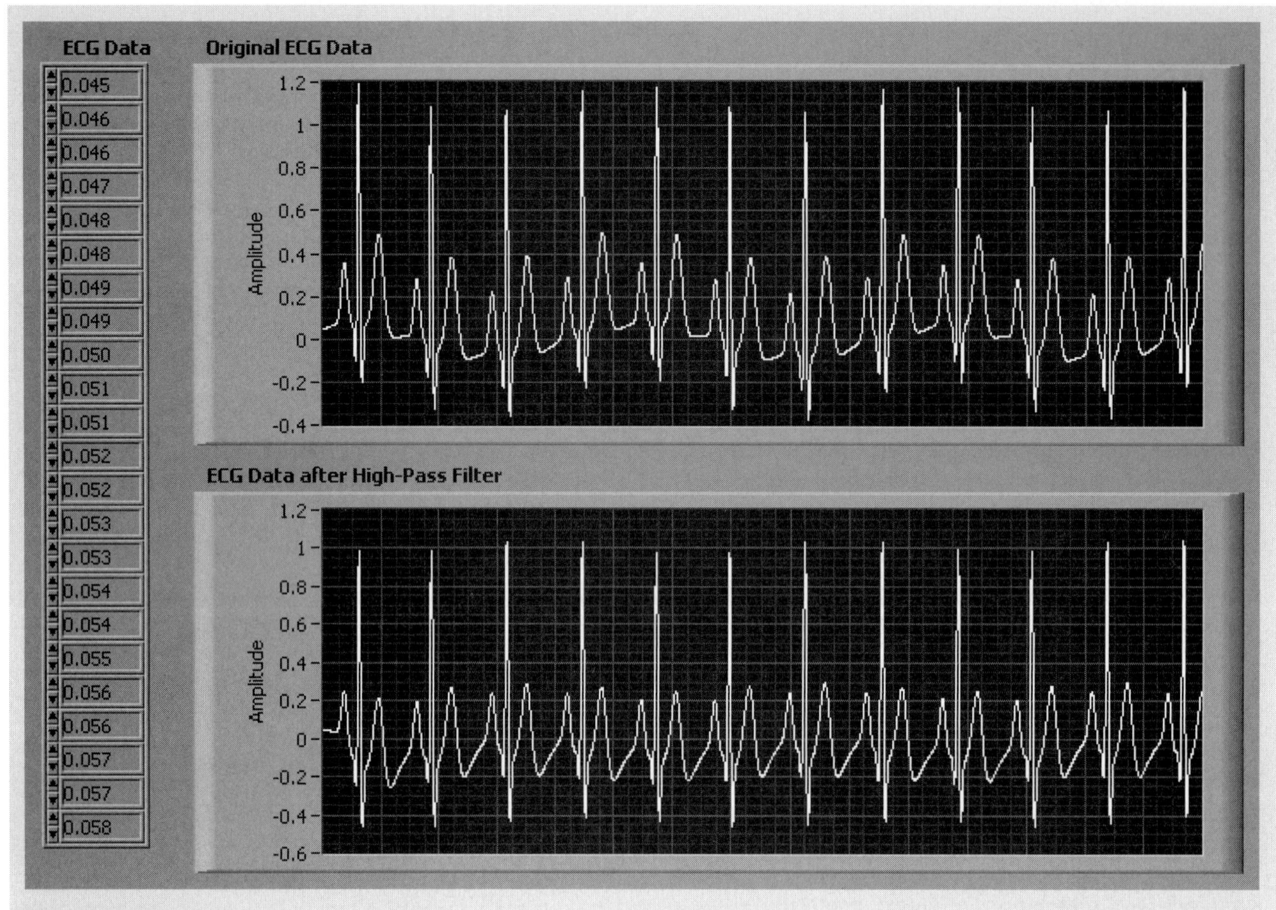

Figure 5.13
ECG Filtering VI, front panel.

The block diagram for the ECG Filtering VI is shown in Figure 5.14.

Figure 5.14
ECG Filtering VI, block diagram.

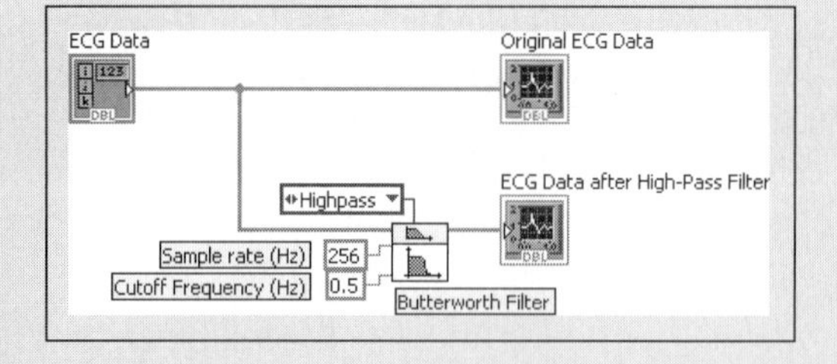

A Butterworth filter was used. It was configured as a high-pass filter with the cut-off frequency set at 0.5 Hz. The sample rate of 256 Hz was set based on the sample rate used to generate the simulated waveform.

Functions Palette / Signal Process Group / Filters Group / Butterworth Filter.vi

You have to be careful when applying filters because they can radically change the characteristics of the signal. But, when used judiciously, filtering can help make the most significant characteristics of a signal easier to identify.

Note: Low-pass filtering is not the preferred way to eliminate baseline drift in ECG signals. Instead, peak detection with spline fitting is used to define a curve through the baseline, and this curve is subtracted from the entire ECG waveform. LabVIEW can be used to perform this task as well.

Aligning Sensor Output to a Data Acquisition System Input

When the range of a sensor's signal is different than the input range of the data acquisition system, signal conditioning can be used to rescale the signal from the sensor. Common changes include *amplification* and *offset*. Amplification causes the span to increase whereas adjusting the offset causes all values in the span to be increased or decreased by the same amount.

Consider the sensor signal range illustrated in Figure 5.15. The sensor outputs voltages between 2 and 6 V.

- The *range* of the sensor output is 2–6 V
- The *span* is 4 V.

We will assume that the data acquisition system (DAQ) is designed to accept values in the range 0–10 V.

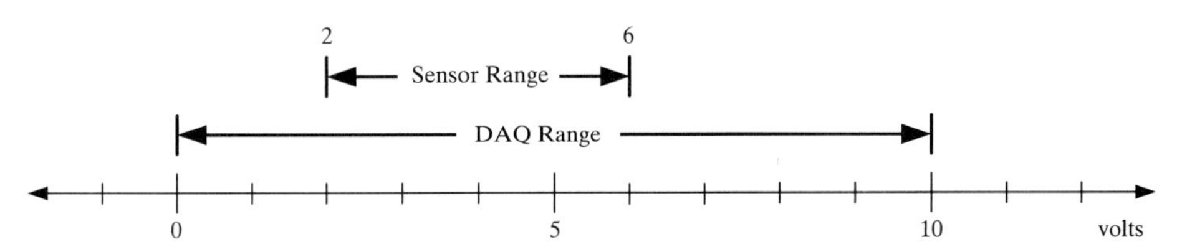

Figure 5.15
Original sensor range, before signal conditioning.

If we amplify the sensor signal by a factor of 2, the span will increase from 4 to 8 V, as shown in Figure 5.16.

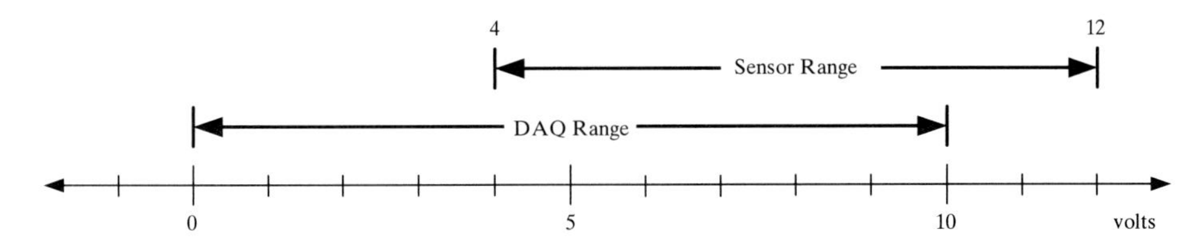

Figure 5.16
After amplification of the sensor output.

Now, the span (8 V) is closer to the span that the DAQ accepts (10 V), but the ranges don't line up. Any sensor values greater than 10 V would be misread by the data acquisition system. We can align the signals by including a −3 V offset to the amplified sensor signal. The result is shown in Figure 5.17.

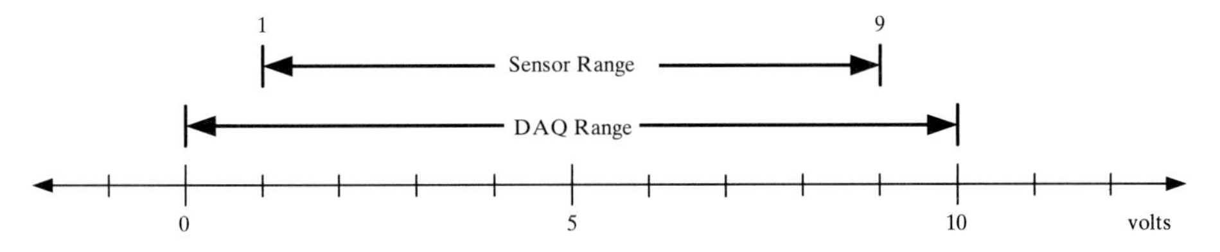

Figure 5.17
Sensor output after amplification and offset.

Note: In this example the sensor signal was amplified first, and then offset to demonstrate that amplifying a signal multiplies output values in the entire range by the amplification factor. In practice, electronic systems may saturate at 12 V (this is common, but not universal) so you must keep the amplified values within the working range of the electronic equipment. We could have avoided sensor output values approaching 12 V by first applying an offset (of −1.5 V) and then amplifying by a factor of 2. The resulting sensor range would still be 1–9 V.

Electronic equipment that has built in signal conditioning typically has controls labeled *zero* and *span*. The zero control is used to adjust the offset, and the span control is used to adjust the amplification.

EXAMPLE 5.3

Thermocouple Signal Conditioning

A very common type of signal that often requires signal conditioning is the voltage output from a thermocouple. A thermocouple is made by connecting two dissimilar metal wires by means of a welded junction. When two different metals are connected they generate an emf and that emf signal changes as the temperature of the junction changes. For a common Type K (Ni–Cr/Ni–Al) thermocouple, a temperature change from 300 to 500°C causes the output voltage (referenced to 0°C) to change from 12.2 to 20.6 mV.[1] The span of 8.4 mV is much smaller than the 10 V span common on many data acquisition systems, so signal conditioning is commonly used for thermocouples.

If the signal conditioning system has a variable gain, you could increase the span clear to 10 V by amplifying the thermocouple output signal by a factor of 1190. More likely, the signal conditioning equipment will have present scale factors, such as a factor of 1000. If a scale factor of 1000 is applied to the 8.4 mV signal, the conditioned signal will have a span of 8.4 V, which would work nicely with a 0–10 V data acquisition system, except that when the 12.2–20.6 mV signal is amplified by a factor of 1000, the result is a signal ranging from 12.2 to 20.6 V. The span is appropriate, but the actual voltage values are outside the working range of the data acquisition system. We need to not only amplify the signal (change the span) but apply an offset to the values as well (change the zero).

The signal conditioning for this thermocouple application includes the following:

1. Adjusting the zero to slide the output signal from 12.2 to 20.6 mV, to change the signal range to 0–8.4 mV.
2. Amplifying the signal by a factor of 1000 to generate a signal ranging from 0 to 8.4 V.
3. (Optional) Adjust the zero by 0.8 V to center the signal in the 0–10 V range.

The last step is optional and may not even be desirable, depending on your application.

- If you want to maximize resolution, use the entire span available on the data acquisition system (0–10 V in these examples).
- When the sensor output range is slightly smaller than the data acquisition system range, values that are outside the expected sensor range can be detected. These values might indicate, for example, that the sensor calibration has changed and should be checked. If the sensor output range is set to be exactly the same as the data acquisition system input range, there is no way to detect sensor values outside the expected range.

[1] G. W. Burns, M. G. (Scroger) Kaeser, G. F. Strouse, M. C. Croarkin, and W. F. Guthrie, *Temperature-Electromotive Force Reference Functions and Tables for the Letter-Designated Thermocouple Types Based on the ITS-90.* National Institute of Standards and Technology Monograph 175; 1993.

Sensor Calibration

Sometimes there is an equation relating a sensor's output to the measured variable. For example, the temperature at the junction of a thermocouple can be calculated if the emf voltage (and thermocouple type) is known. If you use signal conditioning to adjust the range of the sensor output signal, you must account for the offset and amplification values to determine the measured value from the sensor output. This is extremely important when the sensor output signal must be related to an external standard, such as an NIST (National Institutes of Standards and Technology) reference. Signal conditioning that breaks the connection between an NIST reference and the sensor's output can turn a $20,000 instrument into a $200 instrument and leave your data suspect.

In many cases, the offset and amplification values used in signal conditioning are not precisely known. In other cases, there may be no equation linking the sensor output to the desired measurement at all. The solution in either case is the same: you must *calibrate* the sensor to determine the relationship between sensor output and measured variable.

Calibration involves determining the sensor output for several known values of the quantity to be measured. For example, ice water and boiling water are commonly used to calibrate temperature sensors, and they work well when the desired measurement range is close to 0–100°C. Ice water and boiling water would be totally inadequate for calibrating a temperature sensor intended to measure values between 250 and 3000°C. And, using two points to create a calibration curve is only appropriate when you know that the equation relating the measured value to the sensor output is linear—this is rare. In general, calibration requires finding the sensor output for a set of known values that cover the range of the desired measurements. Additionally, since sensor performance can change over time, the sensor's calibration should be checked periodically by determining the sensor output while measuring a known quantity.

Note: A sensor must be recalibrated any time the zero or span settings are changed.

5.3 DATA ACQUISITION HARDWARE

Data acquisition devices are designed to provide a communication bridge between a laboratory instrument or sensor and a computer system.

When considering which data acquisition system to use, there are several things to consider:

1. What types of signals will the data acquisition system need to handle?
   - How many *AI channels (analog inputs)* are required
   - How many *AO channels (analog outputs)* are required
   - How many *DI channels (digital inputs)* are required
   - How many *DO channels (digital outputs)* are required
2. Will your AI channels be wired as differential inputs or single-ended?
3. What level of precision is required in the analog-to-digital converter?
4. How fast will you need to take samples?

5.3.1 Types of Signals

Analog Input

The heart of most data acquisition systems is an *analog-to-digital converter (ADC)* that can receive an analog signal and convert it to a digital form that can be used and stored on a computer system. The sensor's analog signals are connected to an analog input (AI). Many data acquisition systems provide several AIs (called *channels*) that share a single ADC by means of a *multiplexer*.

Analog Output

Common data acquisition systems also provide *analog output (AO)* channels which use a *digital-to-analog converter (DAC)* to convert a digital value specified in the computer system (entered by the user or set programmatically) into a voltage at a terminal on the data acquisition system's connector panel. That voltage can be sent to an instrument to cause it to take an action. For example, a 7 V output might be used to instruct an automatic valve to move the valve stem to 70% open. Or, the output voltage might set the rotation rate on a robotic vehicle's driveshaft.

Digital Input and Output

A *digital input (DI)* receives a signal that can have only two values, one representing "high" and the other "low". A *digital output (DO)* on the data acquisition system's connector panel will have either a high- or low-voltage value. The actual voltage values are established by conventions such as *TTL* (*transistor–transistor logic*) and *CMOS* (*complementary metal-oxide-semiconductor*). For TTL devices, a voltage between 0.0 and 0.8 V is "low", and a voltage between 2.0 and 5.0 V is "high".

Note: The gap between 0.8 and 2.0 V offers protection against a noisy signal. If the signal sent to a DI is approximately 0.7 V, but there is noise of ±0.15 V, the signal will still read as "low" even though the signal level sometimes exceeds the 0.8 V threshold. It reads as "low" because the voltage never crosses the 2.0 V threshold for a "high" level at the DI.

Digital signals are used for Boolean applications, and "low" and "high" voltage levels typically have meanings like 0 or 1, true or false, open or closed, or start or stop.

5.3.2 Differential or Single-Ended Inputs

AIs can be wired in several ways:

- Differential Inputs
- Single-Ended, Ground Referenced Inputs
- Single-Ended, Non-referenced Inputs

Differential Inputs

When AIs are wired as *differential inputs*, the two signal wires are connected to two AI channels that are configured to work together to measure the *voltage difference* between the two wires. Because two AI channels are needed for each measurement, using differential inputs cuts the number of potential measurement channels in half. There is a good reason for using them when possible: differential inputs reject *ground loop* errors. Ground loops occur when the signal source and the measurement system are both referenced to ground, but the grounds are at different potentials. Grounds at differing potentials are actually pretty common. When two instruments, each referenced to a different ground potential, are connected, current will flow because of the voltage difference between the grounds. This is called a ground loop, and it interferes with the signals you are trying to measure. Differential inputs reject ground loop errors and are they way to go whenever possible.

Single-Ended Inputs

Single-ended inputs only require one AI channel per measurement and can work under the right conditions. The requirements listed here are from National Instruments' document entitled *Ground Loops and Returns.*[2]

Requirements for Single-Ended Inputs

- Input signals are high level (greater than 1 V as a rule of thumb).
- Signal cabling is short and travels through a noise-free environment or is properly shielded.
- All input signals can share a common reference signal at the source.

High input signal level means that even if there is some signal degradation due to ground loops, there will still be a signal strong enough to measure. Keeping the signal source and measurement system close together helps reduce the difference between the ground potentials and reduces signal transmission noise. Having all input signals share a common reference at the signal source keeps from creating additional ground loops between the various input signals.

Single-ended inputs still have two signal wires connected to the data acquisition system, but only one connects to an AI port. The other wire connects to AIGND (measurement system ground) or AISENSE, which is not automatically connected to measurement system ground. These wiring configurations are termed *Single-Ended, Ground Referenced* (connected to AIGND) and *Single-Ended, Non-referenced* (connected to AISENSE).

- When your signal source is not referenced to building ground (termed *floating*), which is common with battery powered devices, you should connect to AIGND to reference the AI signal to the measurement system ground.
- When your signal source is connected to building ground (*grounded*), which is common with electrical devices using three-prong plugs, you should connect to AISENSE to reference the AI signal to the source ground.

Finding and fixing ground loop problems can be a challenge. Using differential inputs can help avoid the problem.

5.3.3 Analog to Digital Converters

Once the signal passes through the AI port, it must be converted to a digital value corresponding to the signal voltage. An ADC performs this action.

One of the characteristics of an ADC is the number of *bits* used to describe the voltage level. The higher the bit count, the greater the resolution of the ADC. Common ADCs are 12–22 bit devices, but to describe how ADCs work, we will begin with a much simpler ADC.

Consider a two-bit ADC. Each bit can be either on or off, so there are four possible combinations (00, 01, 10, 11). These four combinations correspond to digital values 0 through 3. The ADC will determine where the signal voltage falls in the AI channel's allowable signal range and assign a digital value based on the input voltage.

If the AI channels are designed with an allowable signal range of 0–10 V, spreading 10 V over the four possible digital values means the signal voltages

[2]Ground Loops and Returns, National instruments NI Developer Zone document, http://zone.ni.com/devzone/cda/tut/p/id/3394,retrieved 6/4/2009.

between 0 and 2.5 V would be assigned the digital value 0, signal voltages between 2.5 and 5 V would be assigned the digital value 1, and so on.

For example, if an input signal was 6.5 V as shown in Figure 5.18, that value falls between 5 and 7.5 V, and so would be assigned a digital value of 2 by a two-bit ADC, but the voltage would be reported as the average voltage assigned to that bit, 6.25 V as shown in Figure 5.18.

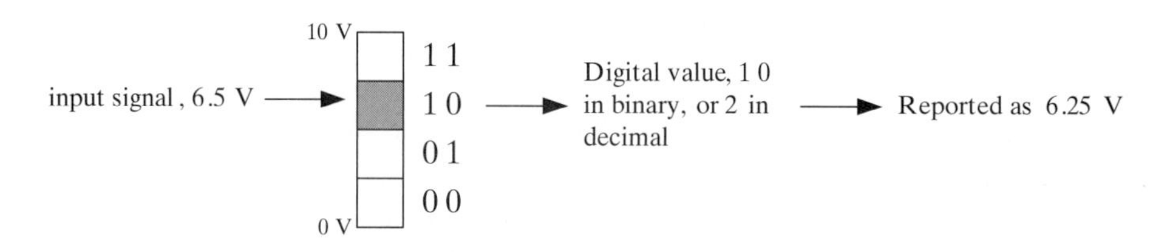

Figure 5.18 Analog to digital conversion.

Similarly, an input voltage of 2.6 V (Figure 5.19, upper panel) would be assigned a digital value of 1 by a two-bit ADC and reported as 3.75 V. But so would an input voltage of 4.5 V (Figure 5.19, lower panel). A two-bit ADC cannot tell the difference between 2.6 and 4.5 V, and would record both as 3.75 V. To get better resolution, we need more bits.

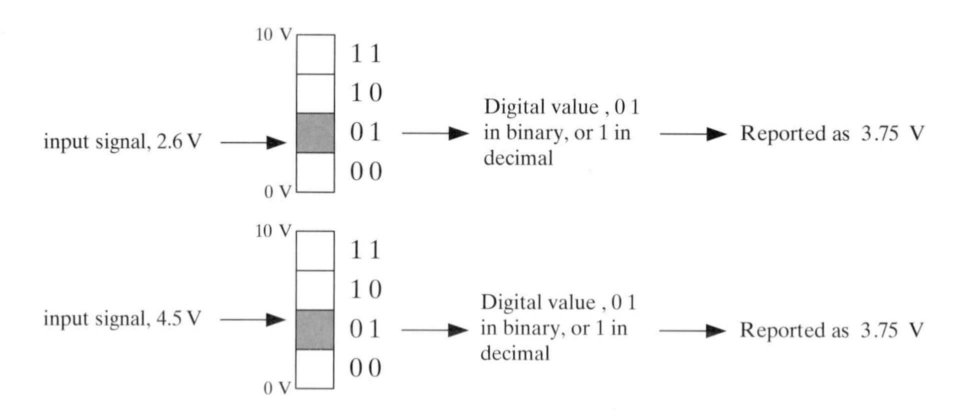

Figure 5.19 Analog to digital conversion, resolution.

More bits allow an ADC to more precisely describe the signal value. For example, 7 bits can take on 128 values (0–127). The 10 V allowable signal span can now be divided into 128 parts, so a 7-bit ADC can provide a resolution of better than 0.1 V (actually 0.078 V = 10 V / 128).

An ADC's resolution is related to the number of bits and the allowable input voltage range (expressed in terms of the DAQ's allowable span) as follows:

$$R_{\mathrm{ADC}} = \frac{Span_{\mathrm{DAQ}}}{2^{N_{\mathrm{bits}}}}$$

For the example used here, the $Span_{\mathrm{DAQ}} = 10$ V, and $N_{\mathrm{bits}} = 7$, so $R_{\mathrm{ADC}} = 0.078$ V.

A 12-bit ADC can provide a resolution of 2.4 mV (assuming a 10 V allowable signal span). With 14 bits, the resolution becomes 0.61 mV. Fourteen bits are considered the minimum number of bits that will allow a thermocouple to be sampled without amplification. Twenty-two bits are available and can resolve 0.0024 mV differences, but you must have a very clean signal to make the extra resolution useful.

ADCs with more bits carry higher prices; how do you know what you need? Consider the characteristics of your input signal. For example, a signal of 4.523 ± 0.042 V is uncertain at the 10 mV level. A 12-bit ADC that can tell the difference between a 4.522 V signal and a 4.525 V signal is adequate.

5.3.4 Sample Rate

Another consideration when selecting a data acquisition system is the required *sampling rate*. The analog-to-digital conversion process takes a finite amount of time. If you want to take one sample a second, any data acquisition system will work fine. If you need to sample each of 32 AI channels at 1000 samples a second (total of 32,000 samples/second), that is a tougher challenge but there are data acquisition systems that can handle it.

Most data acquisition systems use a multiplexer (MUX) between the AI ports and the ADC so that one ADC is used to handle all of the AI channels. This is illustrated in Figure 5.20.

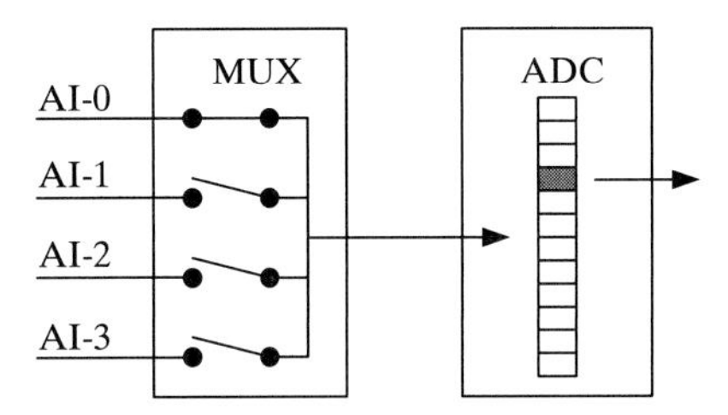

Figure 5.20
A multiplexer (MUX) allows one DAC to convert multiple input signals.

Whenever one ADC is used for all AIs (a very common situation), the total number of samples on all AI channels per second must be considered, not just the sample rate on one channel.

5.3.5 Installing a Data Acquisition System

A data acquisition system is a hardware extension of a computer system and must be installed like other hardware components like printers and hard drives. The process involves physical installation of the device and the installation of appropriate drivers. There are a variety of ways to connect a computer to a DAQ system:

- PCI
- PCI Express
- USB
- PCMCIA
- Ethernet
- Wireless Ethernet
- PXI, PXI Express
- Firewire

Some of the methods are geared toward ease of use (e.g., USB), some are for speed (e.g., PCI Express), and some are designed to meet the rugged demands of industrial electronic systems (e.g., PXI, PXI Express).

Multiple data acquisition devices can be installed on a single computer system. During the installation and configuration process, each data acquisition device is assigned a unique device number.

5.4 USING LABVIEW TO COLLECT DATA

Operating a data acquisition system requires software, and that's where LabVIEW comes into play; LabVIEW was designed to work with National Instruments data acquisition systems.

Before you can collect data you need to decide the following:

- Which data acquisition device will be used?
- Which AI channel(s) will be used?
- How often will each channel be sampled?
- How many samples should be collected?

In LabVIEW, a request for a data acquisition device to collect one or more data values is called a *task*. When you define a data acquisition task, you provide the answers to the questions listed above.

A task can be defined from inside LabVIEW using the *DAQ Assistant VI* or from outside of LabVIEW using National Instruments Measurement and Automation Explorer. Either way, you can test the tasks as you define them, which helps get the tasks configured correctly.

5.4.1 Configuring Tasks Using the Measurement and Automation Explorer

The National Instruments *Measurement and Automation Explorer* can be opened in several ways:

- From the Windows Start menu, use

  Start / All Programs / National Instruments / Measurement and Automation

- From the LabVIEW Getting Started dialog, use

  Tools / Measurement & Automation Explorer . . .

- While editing a LabVIEW VI, use

  Tools / Measurement & Automation Explorer . . .

However the Measurement and Automation Explorer is started, it opens and looks something like Figure 5.21. The items in the Configuration list may vary, depending on what is installed on your computer system.

In the Configuration list,

- The **Data Neighborhood** node can be expanded to show all previously defined data acquisition tasks.
- The **Devices and Interfaces** node can be expanded to show all data acquisition devices installed on the computer.

In Figure 5.22, the Devices and Interfaces node has been expanded to show that there is one data acquisition device installed, an NI USB-6009 data acquisition system installed as "Dev 1" (provided by National Instruments for testing associated with this text). This device has 8 single-ended (4 differential) AI ports, 2 AO ports, and 12 digital IO ports and connects to a computer via a USB port

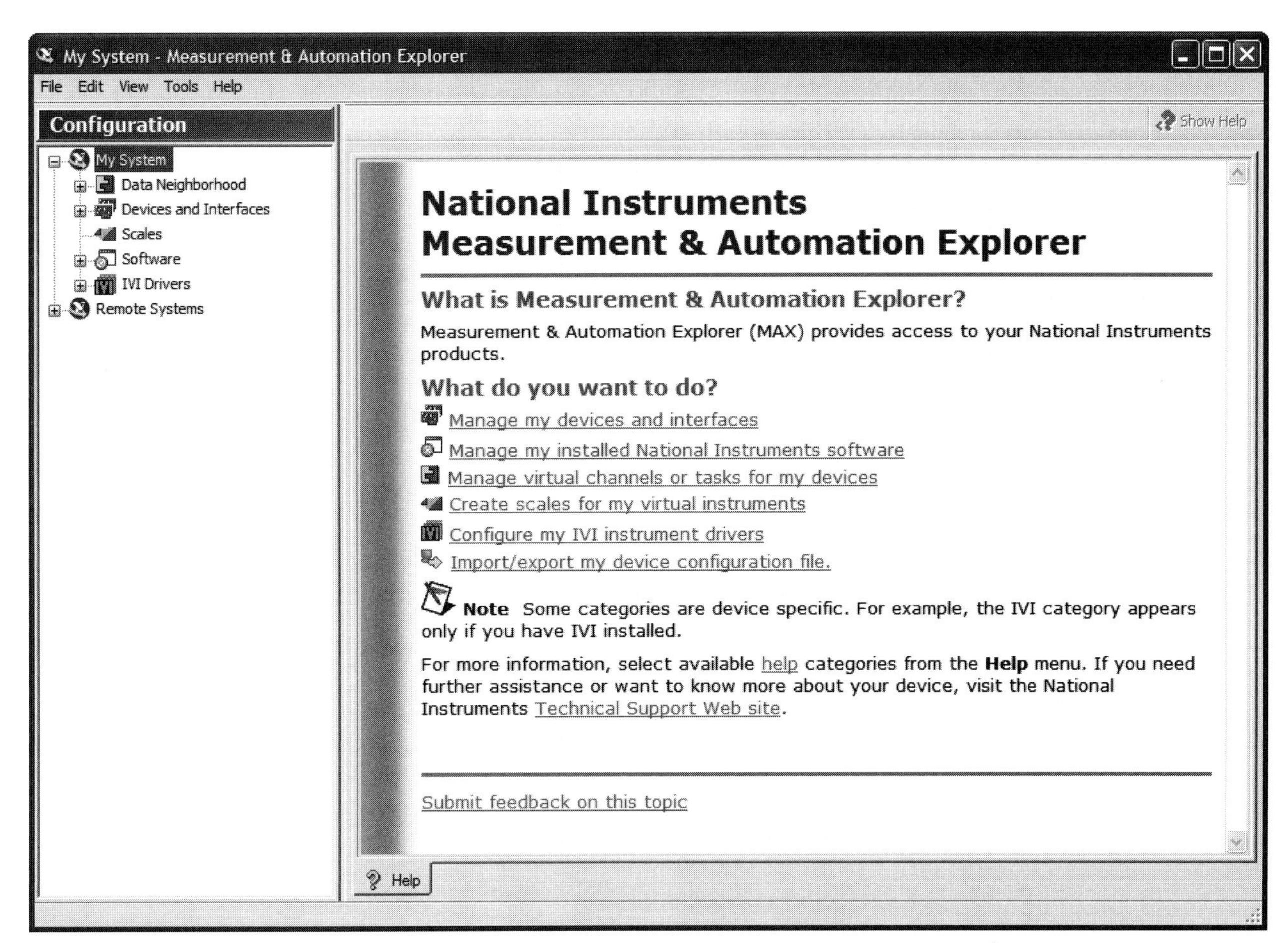

Figure 5.21
NI Measurement and Automation Explorer.

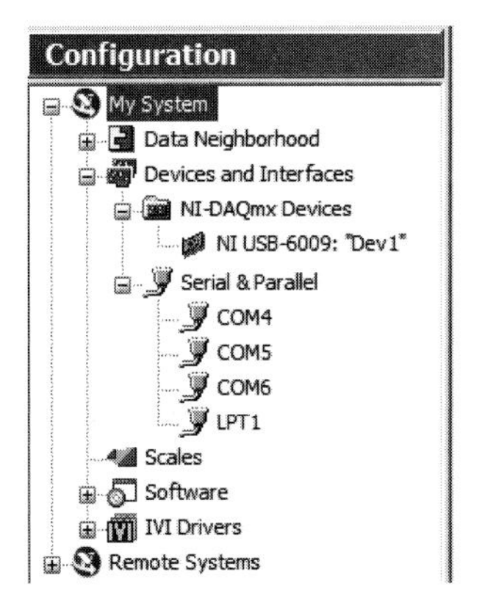

Figure 5.22
Expanding the Devices and Interfaces node.

(great for laptops). It utilizes a 14-bit ADC and can handle up to 48 K samples per second.

If you right-click on the device name in the list, a pop-up menu appears (see Figure 5.23) that offers some useful options.

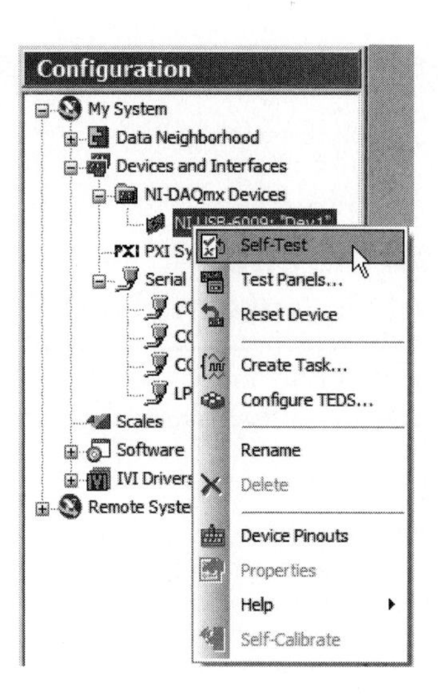

Figure 5.23
Pop-up menu of device options.

- **Self-Test**—Allows you to quickly see if the data acquisition system is connected and working.
- **Test Panels . . .**—Opens a dialog to allow you to quickly test the inputs and outputs on a device.
- **Reset Device**—Resets the device back to factory default values. (Sets AO values back to 0 V, for example.)
- **Create Task . . .**—Opens the dialog that is used to configure a data acquisition task.
- **Device Pinouts**—Shows how the connections to the data acquisition device are organized.

Testing the Data Acquisition Device

The Self-Test is quick and easy and should be used at least once right after installing the device just to ensure that it is communicating correctly with the computer.

The Test Panels . . . dialog allows you to see what is coming in through your AI channels, and what is being sent out through your AO channels. Together these can be used to test the AI channels. When you instruct an AO channel to output a voltage, it will continue to do so until you instruct it to output a different value, reset the device, or power down the device. You can wire the AO output to any of the AI ports to make sure the AI ports are receiving and reporting the correct values.

To use the Analog Output test panel (see Figure 5.24),

1. Select the **Analog Output** panel on the Test Panels dialog.
2. Select the Device and **Channel Name** that you want to use.
3. Verify the **Mode** of the output signal. On the USB-6009 device, **DC Value** is the only option.
4. Set the desired **Output Value** either by typing the value in the text box or by using the slider control.
5. Verify that the **Output Limits** are correct for your device. If you attempt to output a value outside the allowable range for the device, you will get an error message.
6. Click the **Update** button to have the selected AO begin outputting the indicated voltage.

You should be able to check the output voltage with a voltmeter, or you can send the AO signal into an AI to check the value. To see what is entering

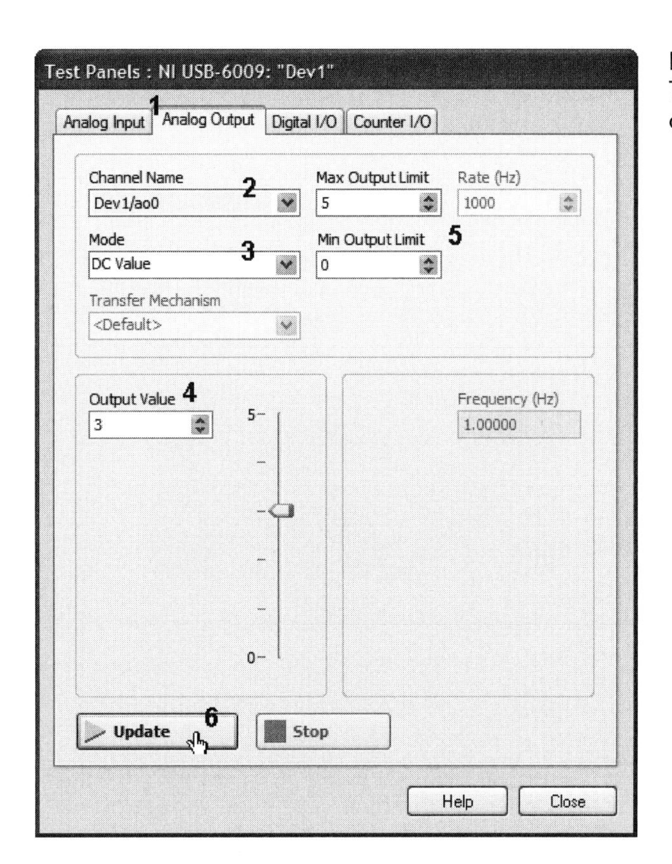

Figure 5.24
Test Panels—setting an AO value.

your AI ports, use the Test Panels dialog and the Analog Input panel (shown in Figure 5.25).

The output from AO-0 (analog output zero) was connected to AI-1 (analog input one), which was wired as a differential input. As shown in Figure 5.24, AO-0 was instructed to output 3 V. AI-1 was tested to verify that both the AO and AI are working correctly. The result is shown in Figure 5.25.

To use the Analog Input panel, do the following:

1. Choose the Device and **Channel Name** that will be tested. In Figure 5.25, channel AI-1 was used rather than the default, AI-0.
2. Select the **Mode** for data collection. In this example, **On Demand** was used meaning the data collection will begin when the **Start** button is clicked and continue until the **Stop** button is clicked. The values are graphed as collected. Other options include the following:
   - **Finite**—collects the number of samples specified in the **Samples to Read** field at the sample rate specified in the **Rate (Hz)** field. Collection begins when the **Start** button is clicked, and stops when the specified number of samples has been collected. The values are graphed once the entire data set has been collected.
   - **Continuous**—collects the number of samples specified in the **Samples to Read** field at the sample rate specified in the **Rate (Hz)** field. Collection begins when the **Start** button is clicked. However, for continuous sampling, once one batch of samples has been collected and displayed, data acquisition continues and another batch is collected. Data collection continues until the **Stop** button is clicked.

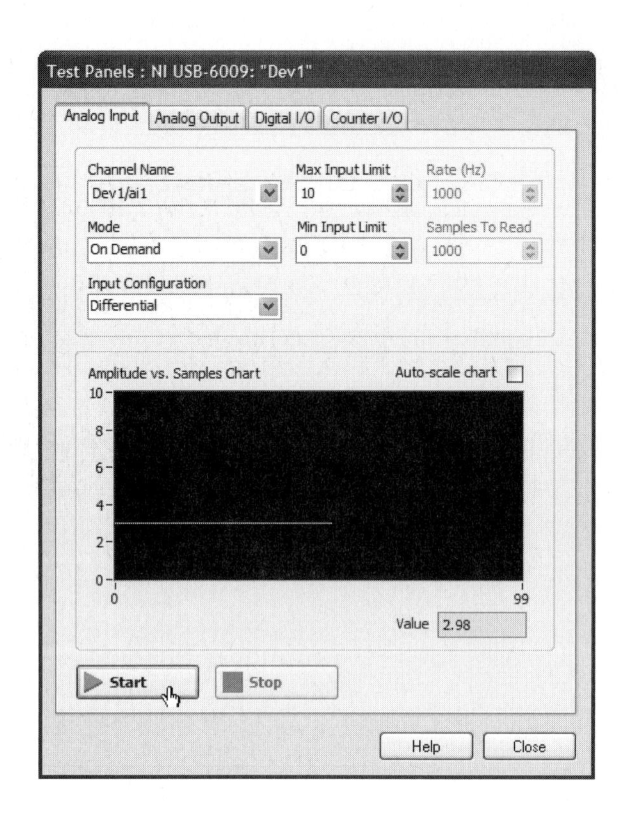

Figure 5.25
Testing an Analog Input (Channel 1).

3. Select the **Input Configuration** used with the analog input. Since AI-1 was wired as a differential input, that option was selected in the Input Configuration field. Options include the following:
    - **Differential**—Two AI ports are used and a voltage difference is measured.
    - **RSE**—Referenced single-ended. The source signal wire that is not connected to the AI input is connected to the ground on the data acquisition device.
    - **NRSE**—Non-referenced single-ended. The source signal wire that is not connected to the AI input is connected to the AISENSE connection on the data acquisition device. This connects the negative side of the AI measurement to the sensor ground. (This is not an option for the USB-6009 device used in these examples.)
    - **Pseudodifferential**—The signal wire is connected to the positive side of the AI measurement, and the negative side is connected to the sensor ground. A small resistor is used to minimize current between the sensor ground and the data acquisition device ground. (This is not an option for the USB-6009 device used in these examples.)
4. The **Input Limits** are used to scale the graphical display.
5. Click the **Start** button to begin data acquisition.

After testing the device, you typically create a data acquisition task to prepare for data collection using LabVIEW.

Creating Data Acquisition Tasks from Measurement and Automation Explorer

You can create a data acquisition task either from the Measurement and Automation Explorer or from inside LabVIEW using the DAQ Assistant Express VI. Creating a task from the Measurement and Automation Explorer is shown first.

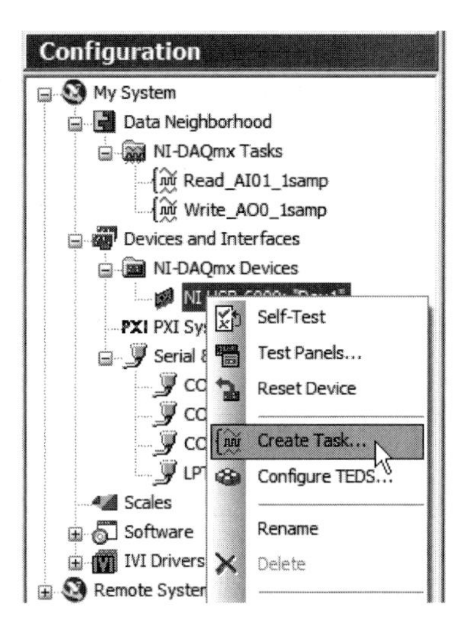

Figure 5.26
Opening the Create Task dialog from the Configuration list.

In the Configuration List, expand the Devices and Interfaces node to show the NI-DAQmx devices installed on your computer. Right-click on the device that will be used in the new task, and select **Create Task . . .** from the pop-up menu (illustrated in Figure 5.26).

The Create Task dialog collects information in several steps. The first step is shown in Figure 5.27. This part of the dialog is used to provide basic information on what the task is designed to do.

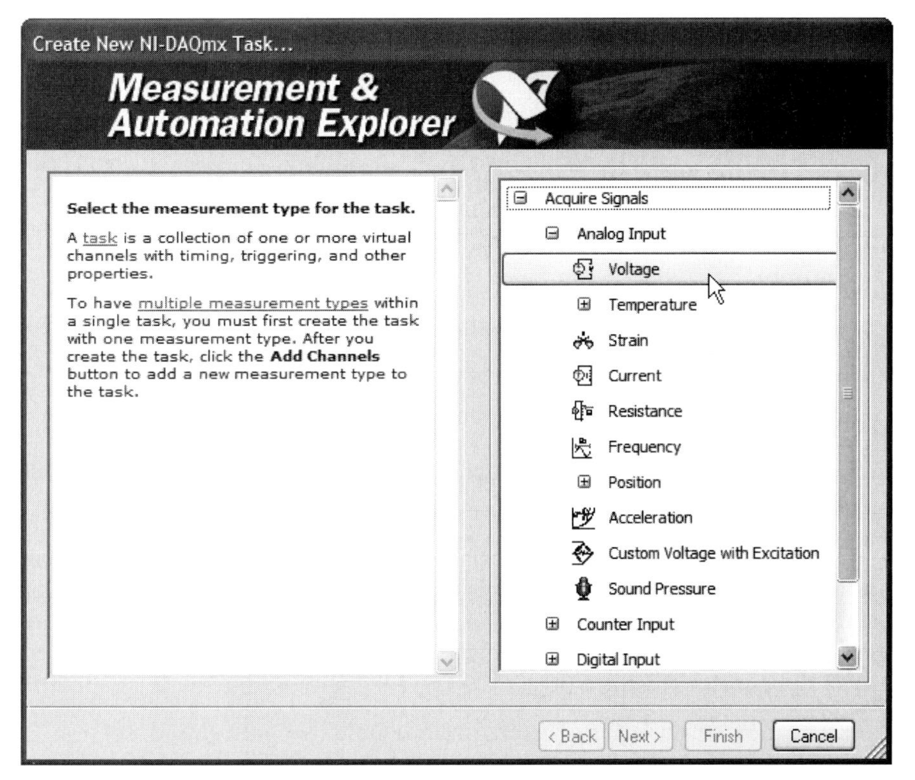

Figure 5.27
Create Task dialog, page 1—basic information.

When you click on the Voltage option (Acquire Signals/Analog Input/Voltage), the second page of the Create Task dialog (Figure 5.28) will open to ask which channel(s) will be used to read the voltage values. In Figure 5.28, AI-1 has been selected. Once all needed AI channels have been selected, click **Next >** to move to the third page of the Create Task dialog, shown in Figure 5.29.

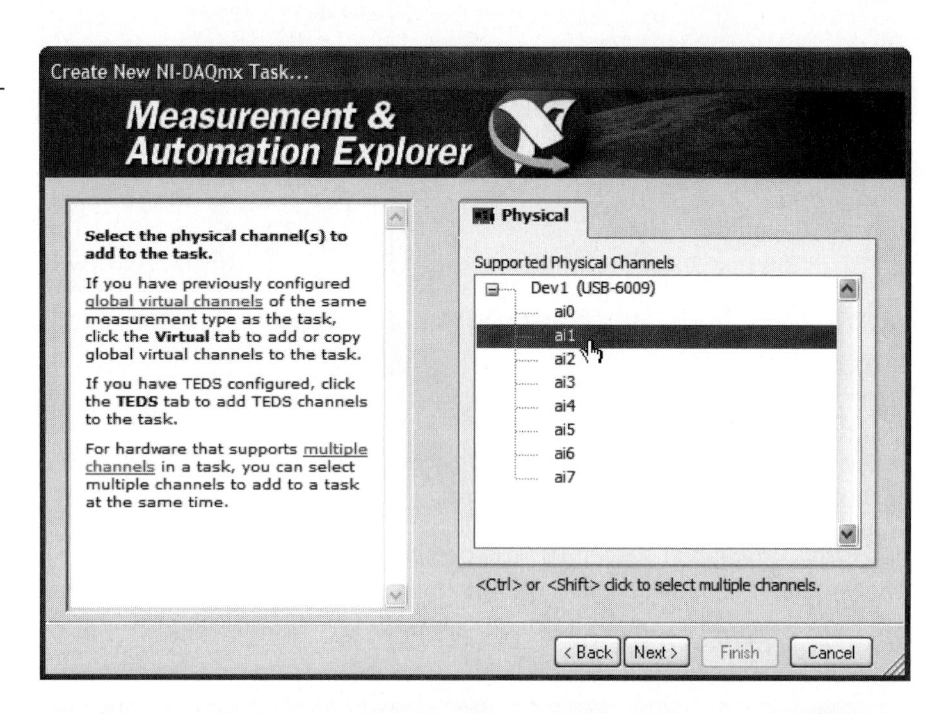

Figure 5.28 Create Task dialog, page 2—data channel.

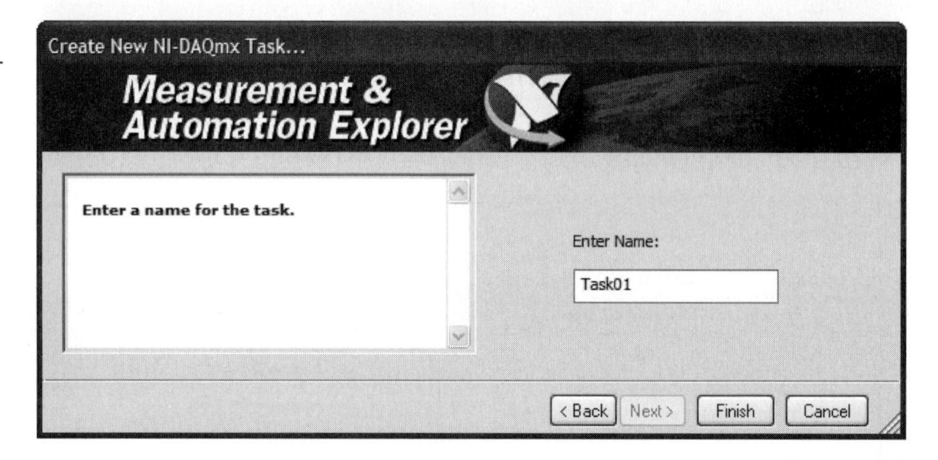

Figure 5.29 Create Task dialog, page 3—assigning a task name.

The task name allows the task to be stored. In this example the task has been named "Task01". When you click **Finish**, the new task will be shown in the Measurement and Automation Explorer, as shown in Figure 5.30. Many of the settings for the task have been assigned default values; they can be edited in the Explorer.

Comments on Figure 5.30:

1. The new task is now listed in the **Data Neighborhood** section of the Configuration list.
2. The basic data to be collected are shown in the **Channel Settings** list. In this example, only one channel of AI will be used, and we indicated in Figure 5.27 that we would collect voltage values.

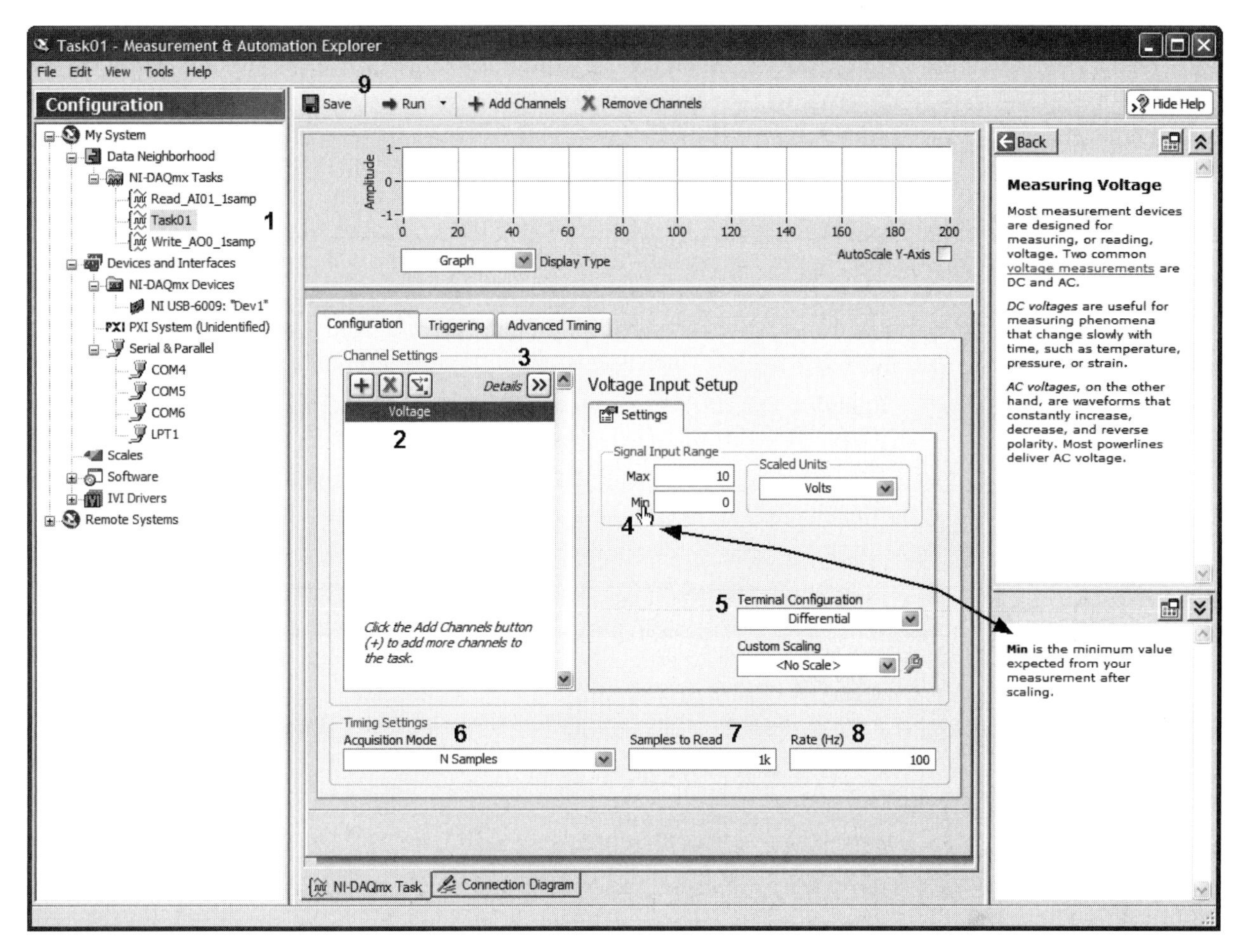

Figure 5.30
Task01 displayed in the Measurement and Automation Explorer.

3. Click the **Details** button to see which device and AI channel are associated with this task.
4. When you move the mouse pointer over the various fields, information about the fields is displayed in the Explorer.
5. You must indicate how the AI has been wired. Here, a **Differential** input has been used.
6. The Acquisition Mode indicates how the data will be collected. Options include the following:

   - **1 sample (On Demand)**—collects one sample each time the task is called.
   - **1 sample (HW Timed)**—collects one sample based on a hardware clock.
   - **N Samples** (default)—collects the number of samples specified in the **Samples to Read** field each time the task is called.
   - **Continuous Samples**—causes data collection to continue until the task is stopped.

7. **Samples to Read**—indicates how many samples should be collected each time the task is called.

8. **Rate (Hz)**—indicates how fast the samples should be collected. The default is 1 kHz, or 1000 samples per second. In Figure 5.30 the sampling rate has been reduced to 100 samples per second.
9. At the top of the task pane in the Measurement and Automation Explorer, there are buttons that allow you to **Run** the task to see if it works correctly, and to **Save** the task.

Once the task has been configured, it can be used within LabVIEW.

Creating Data Acquisition Tasks inside LabVIEW—DAQ Assistant Express VI

You can also create a data acquisition task directly from LabVIEW using the DAQ Assistant Express VI, which is located on the Function Palette:

> **Function Palette / Measurement I/O Palette / DAQmx—Data Acquisition Group / DAQ Assistant Express VI**

When the DAQ Assistant Express VI is placed on a block diagram, it automatically opens the Create Task dialog as shown in Figure 5.31.

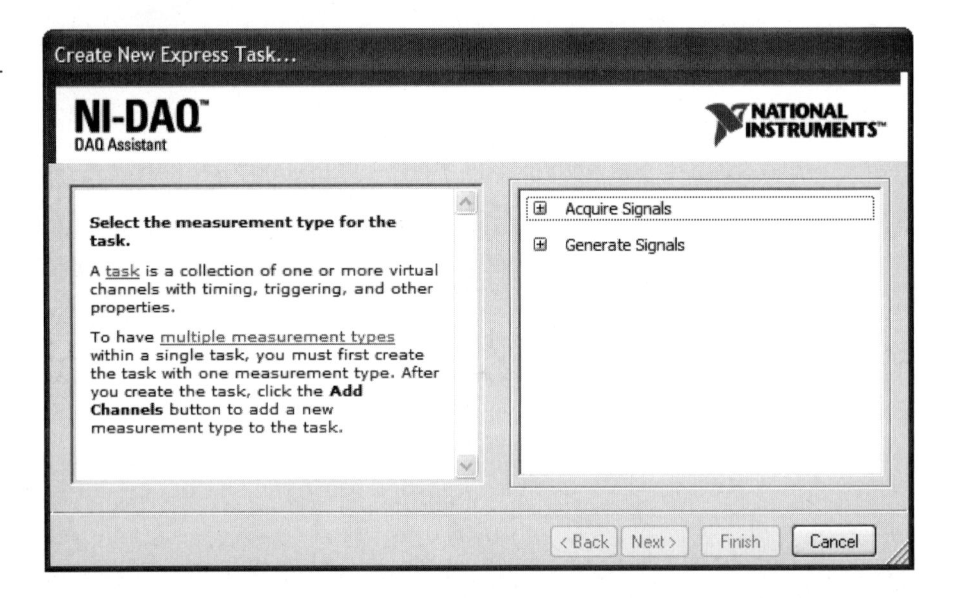

Figure 5.31
Create Task dialog, page 1—called using DAQ Assistant.

This is the same dialog as shown in Figure 5.27, except that the Title indicates that it has been called from the DAQ Assistant Express VI. The process to create a task is the same as before:

1. Provide basic information—we will acquire voltage values (see Figure 5.27).
2. Select Device and Channel(s)—we will use "Dev1" and Channel AI-1 (see Figure 5.28).

Next, the DAQ Assistant opens (Figure 5.32); it looks a lot like the task editor in the Measurement and Automation Explorer.

In the example shown in Figure 5.32, the task has been set up to collect 1000 samples at a rate of 500 samples per second when the task is called. When the OK button is clicked to close the DAQ Assistant, the DAQ Assistant Express VI appears on the block diagram, as shown in Figure 5.33. If you need to modify the task configuration, double-click on the DAQ Assistant icon to open the DAQ Assistant dialog.

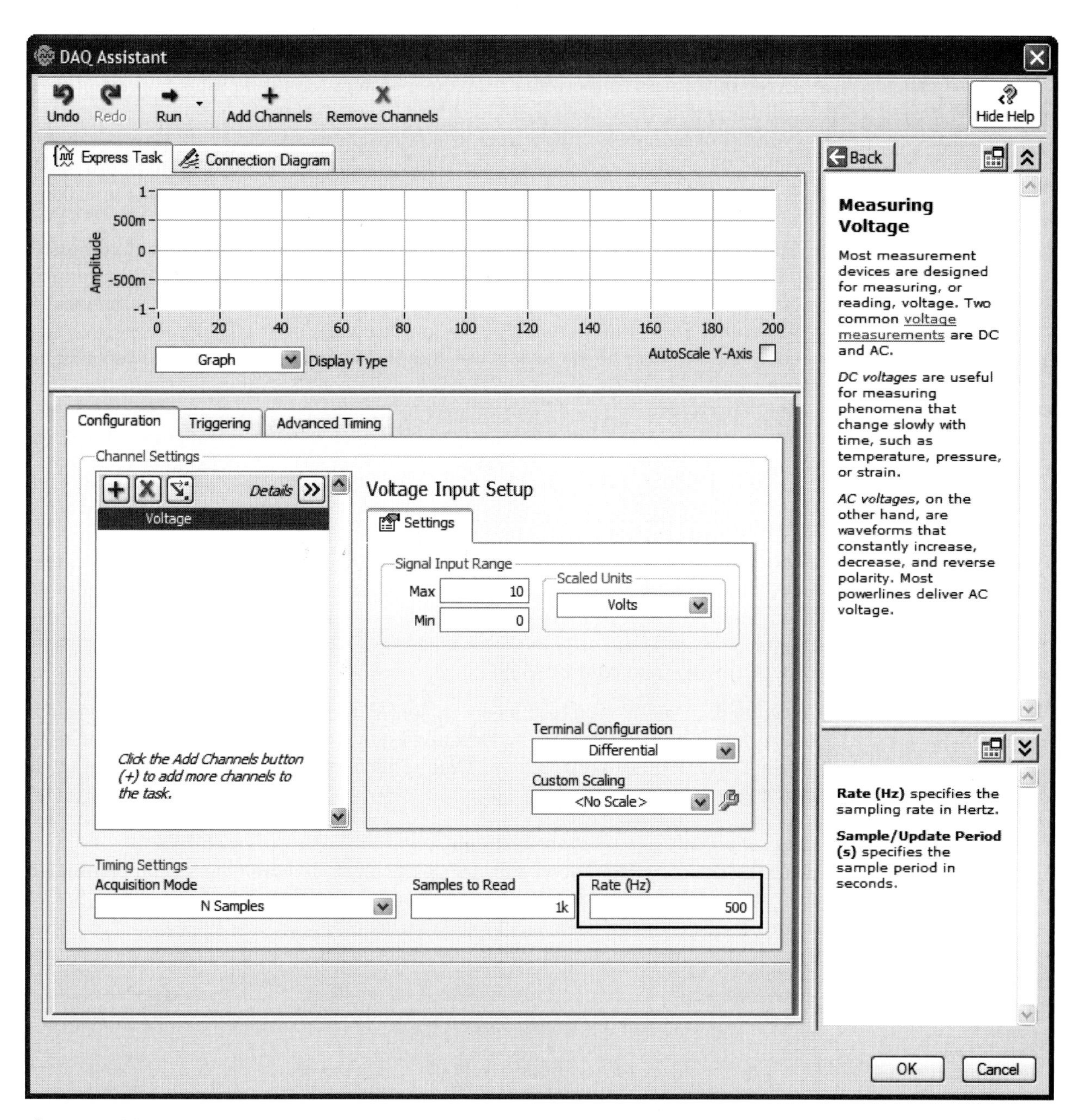

Figure 5.32
DAQ Assistant.

The DAQ Assistant icon shown in Figure 5.33 indicates numerous inputs and outputs. Expanding the icon provides more information (Figure 5.34).
The DAQ Assistant outputs include the following:

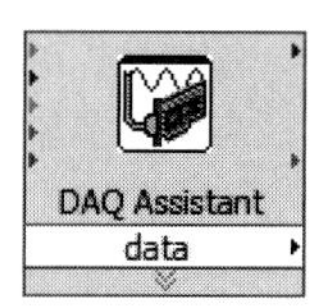

Figure 5.33
The DAQ Assistant Express VI on the block diagram after configuring the task.

- **Data**—the collected voltage values.
- **Error Out**—access to the LabVIEW error system (could be used to programmatically respond to a data collection error).
- **Task Out**—access to the data acquisition task (could be used to force the task to stop, for example).

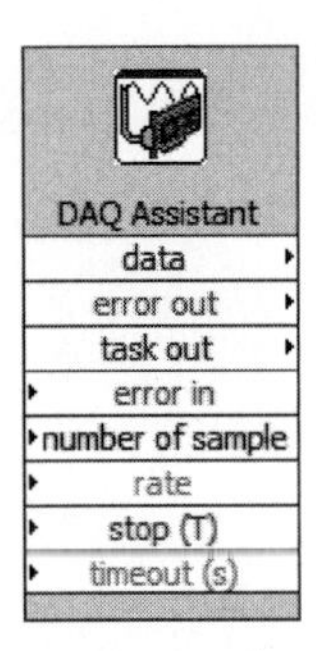

Figure 5.34
The DAQ Assistant icon, expanded.

The DAQ Assistant inputs include the following:

- **Error In**—access to the LabVIEW error system (could be used to prevent data collection if an error was detected before the task started).
- **Number of Samples**—the number of samples to be collected each time the task is called. This value was set using the DAQ Assistant dialog (1000 in Figure 5.32).
- **Rate**—the sample rate in Hz. Also set using the DAQ Assistant dialog (500 in Figure 5.32).
- **Stop**—This is a Boolean variable that indicates how the task should complete. "True" causes the task to stop and release resources when the DAQ Assistant VI terminates. "False" is used to indicate that data acquisition is to be continuous.
- **Timeout**—This value indicates how long to wait for the DAQ Assistant VI to complete the task. If the task is not completed in less than Timeout seconds, an error is generated. The Timeout value is specified on the Advanced Timing Panel on the DAQ Assistant. The default value is 10 seconds; if you plan to collect data over a long period of time, the Timeout value will have to be changed from the default.

You may have noticed that when the data acquisition task was created using the DAQ Assistant, we were never asked to assign a name to the task. Tasks created using the DAQ Assistant are stored with the VI that contains the DAQ Assistant icon; they do not appear in the list of stored tasks in the Configuration list in the Measurement and Automation Explorer's Data Neighborhood node. This means you must go through the DAQ Assistant to edit the task created using the DAQ Assistant.

5.4.2 Acquiring Data with LabVIEW

Once the data acquisition task has been configured, using either the Measurement and Automation Explorer or the DAQ Assistant Express VI, the task can be used within LabVIEW to collect data. Data Collection using the DAQ Assistant is shown first.

Data Collection Using the DAQ Assistant

Once the DAQ Assistant icon appears on the block diagram, the VI can be run and data will be collected. But you probably want to add a couple of controls to the VI to display the collected data, as shown in Figure 5.35.

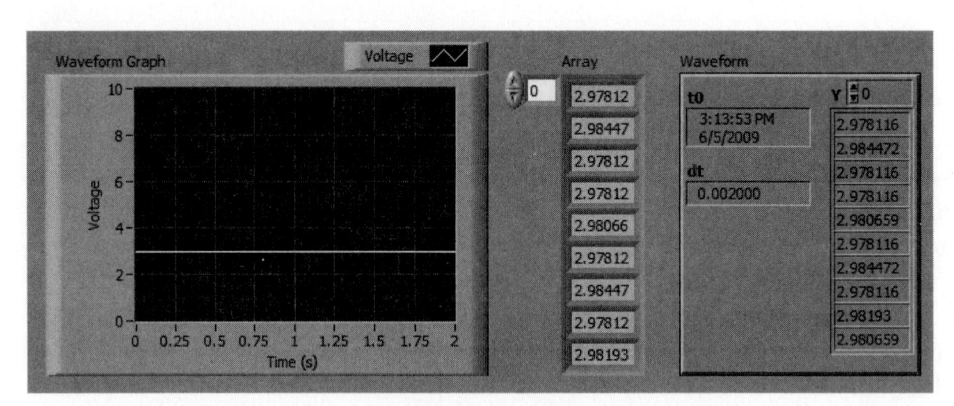

Figure 5.35
Data Acquisition VI, front panel.

It's not a very interesting acquired signal, the AI is still measuring the 3 V output being generated by AO-0 (see Figure 5.24).

The collected values have been shown both as an Array of double precision values and as a Waveform. The data collected using the DAQ Assistant are *dynamic data*, and contain time information as well as voltage information. This time information is ignored when the values are displayed as a simple array.

The block diagram for the Data Acquisition VI is shown in Figure 5.36. Notice that when the Data output was wired to the Array input, a converter was automatically added to change the data type from dynamic to array. This Convert from Dynamic Data converter essentially strips out the time information and leaves only the voltages.

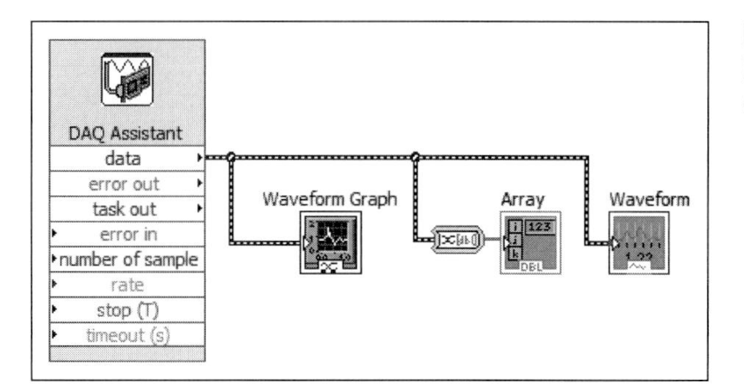

Figure 5.36
Data Acquisition VI, block diagram.

While the DAQ Assistant input values were specified using the DAQ Assistant dialog, you can still send other values to the DAQ Assistant inputs. For example, if you decide to collect 2000 samples at a rate of 200 samples per second, you could double-click the DAQ Assistant icon to open the DAQ Assistant dialog to change the parameters, or you could simply send the new values into the DAQ Assistant on the block diagram. This latter approach is illustrated in Figures 5.37 and 5.38. The AI was also wired to an external source to acquire a slightly more interesting signal. Notice that the time span shown in the front panel waveform graph is now 10 seconds.

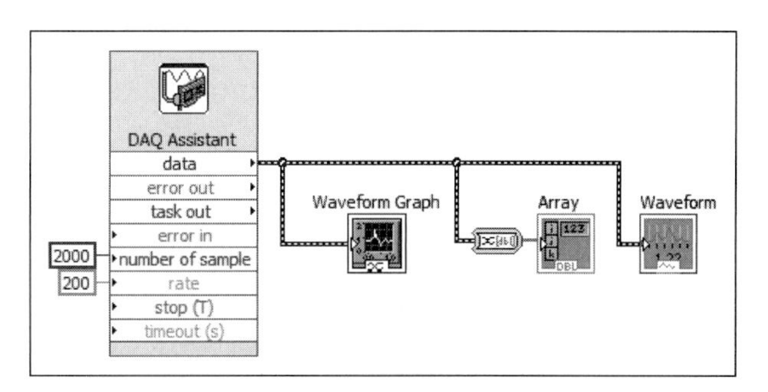

Figure 5.37
Data Acquisition block diagram modified for 2000 samples at 200 per second.

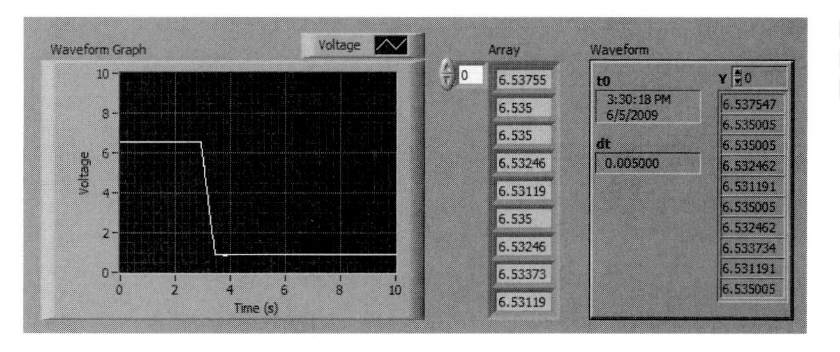

Figure 5.38
Front panel of modified Data Acquisition VI.

Data Collection Using Saved Tasks

When the Measurement and Automation Explorer is used to create and save a task, the saved task can be used within LabVIEW. Earlier in the chapter we created a task called "Task01" that reads 1000 values through AI-1 at a rate of 100 samples per second. The block diagram shown in Figure 5.39 uses that task.

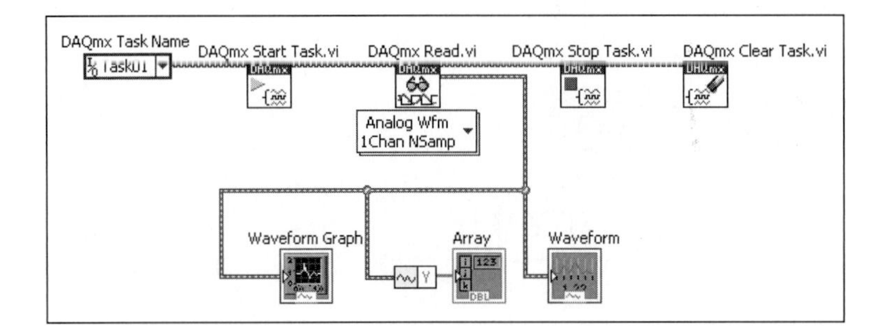

Figure 5.39
Block diagram of Data Acquisition with Task VI.

When you are using a stored task, the procedure is to

1. Start the task using DAQmx Start Task.vi
2. Use the task, in this case using DAQmx Read.vi
3. Stop the task using DAQmx Stop Task.vi
4. Clear the task using DAQmx Clear Task.vi

These data acquisition VIs are available in the Measurement I/O Group on the Functions Palette:

Functions Palette / Measurement I/O Group / DAQmx—Data Acquisition Group

Stopping the task makes the task unavailable unless it is restarted, while clearing the task releases the resources reserved for the task for reuse.

- If the DAQmx Start Task.vi and DAQmx Stop Task.vi are not used, the task will be started and stopped automatically when the DAQmx Read.vi runs. If DAQmx Read.vi is used inside a loop, starting and stopping the task each time the DAQmx Read.vi runs will severely degrade performance.
- If the DAQmx Clear Task.vi is used the task will be stopped before being cleared. Therefore, it is not necessary to explicitly stop the task before clearing it.

The DAQmx Read.vi is a polymorphic VI; it can be used for the following:

- Analog or digital inputs
- Single or multiple channels
- Single or multiple samples
- Results can be returned as array values (voltages only) or waveform (time information and voltages)

In Figure 5.39, the DAQmx Read.vi was used to sample:

- AIs
- Single channel
- N samples
- Results returned as waveform

By returning the results as a waveform, the time displays correctly on the Waveform Graph (shown in Figure 5.40).

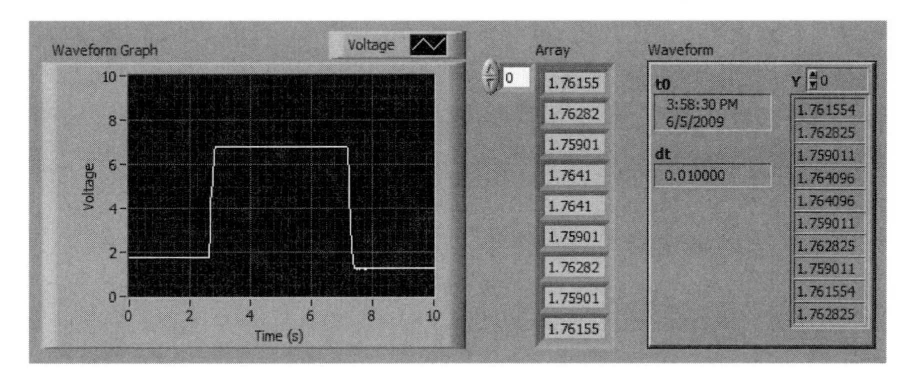

Figure 5.40 Data Acquisition with Task VI, front panel.

Using the Data as It Is Collected

The examples shown so far collect a batch of data and then provide access to the results. If your goal is to collect data for later analysis, that approach works well. However, if you need to use the data as it is collected in your LabVIEW program, you will need to acquire the data values point by point and accumulate the results as they are used.

For the next example, we will create two saved tasks:

- **Read_AI01_1samp**—reads one sample through AI-1, on demand.
- **Write_AO0_1samp**—writes one value to AO-0, on demand.

With these two tasks we can read or write single values, and, using a loop in LabVIEW, we can collect or write as many values as we need.

The basic process required to read one voltage value is shown in the block diagram in Figure 5.41. The task name is sent to the Start Task VI to instruct LabVIEW to prepare to use the data acquisition system. The Read VI reads the analog input and outputs the acquired voltage value. The Clear Task VI stops the task and releases the resources used to handle data acquisition.

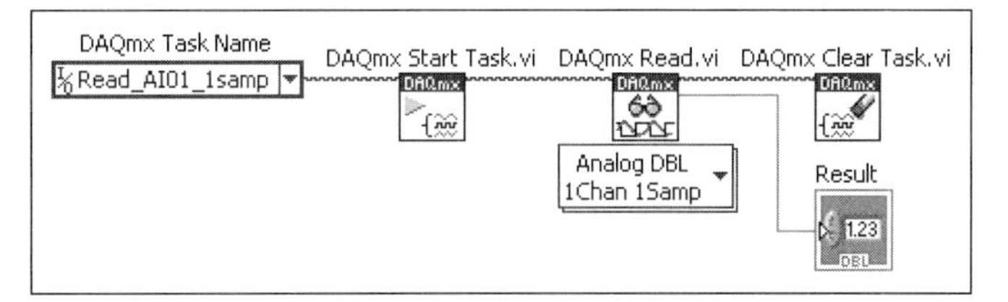

Figure 5.41 Reading one voltage value using a stored task.

Every time the VI shown in Figure 5.41 is run, one value is collected and displayed. To collect a preset number of values, we can build the read step into a For Loop as shown in Figure 5.42. Notice that the start task and clear task actions are outside the For Loop. It is very inefficient to start and stop the task each time you read a value. Instead, start the task once, read the values as often as needed, and then stop and clear the task when it is no longer required.

The Result indicator displays the most recently read voltage value, while the Array indicator holds the entire array of read values, and is displayed only after the For Loop terminates. The front panel for this VI is shown in Figure 5.43.

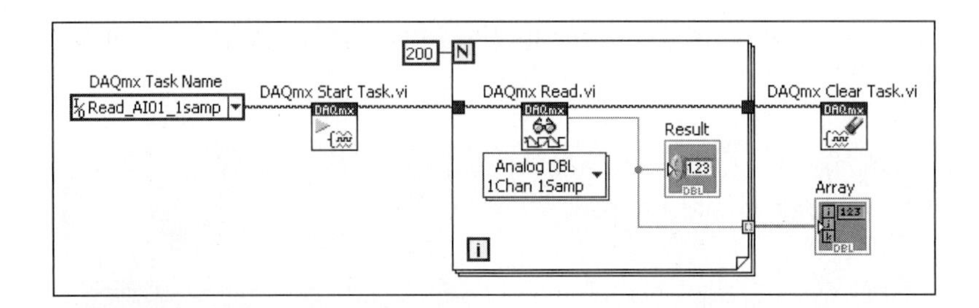

Figure 5.42
Reading 200 values with a For Loop.

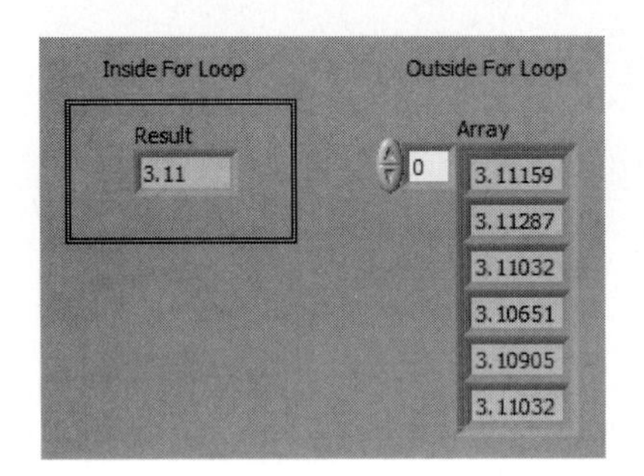

Figure 5.43
Front panel for data acquisition using For Loop VI.

There is no timing associated with the data acquisition using the VI shown in Figure 5.42; LabVIEW will collect the 200 values as fast as it can. To slow down the data acquisition process, you can include a Wait function in the For Loop, as shown in Figure 5.44.

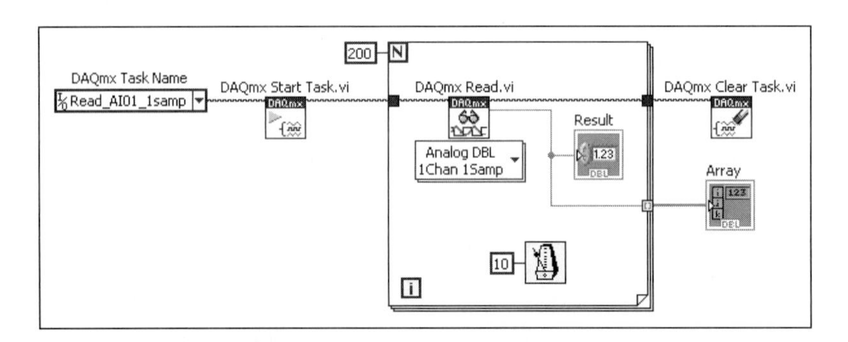

Figure 5.44
Adding a 10 ms wait to the For Loop.

With the added wait, the VI shown in Figure 5.44 will take one sample approximately every 10 ms for a period of 2 seconds. The "approximately" is there because the 10 ms wait does not guarantee that the samples will be collected exactly once every 10 ms, but if the data acquisition process is fast the sample rate should be very close to one every 10 ms.

To check the timing, we can build the For Loop into a series of sequence frames and check the clock tick value before and after the For Loop, as shown in Figure 5.45. If the timing is perfect, we would expect the elapsed time to be 2000 ms.

Using the VI shown in Figure 5.45 with the USB-6009 data acquisition device, we found that the time elapsed ranged from 1992 to 2026 ms. The timing is not perfect, but it is close to 10 ms per sample.

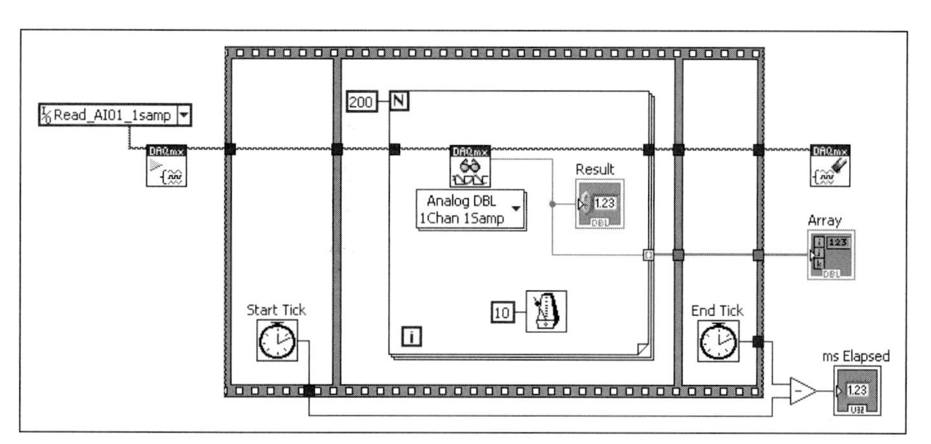

Figure 5.45
Using a sequence structure to test sample timing.

The reason for reading one sample at a time is that we can use the results as they are available. The VI shown in Figures 5.46 and 5.47 monitors a tank level (AI) and writes an AO to 0 or 1 V based on whether the AI value is less than or greater than 5 V. The AO would be used to switch on an outlet pump whenever the tank level is over 5 m.

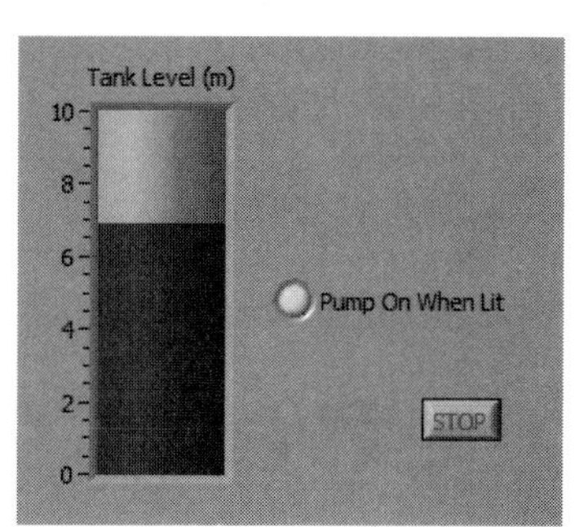

Figure 5.46
VI used to activate a pump when tank level is high, front panel.

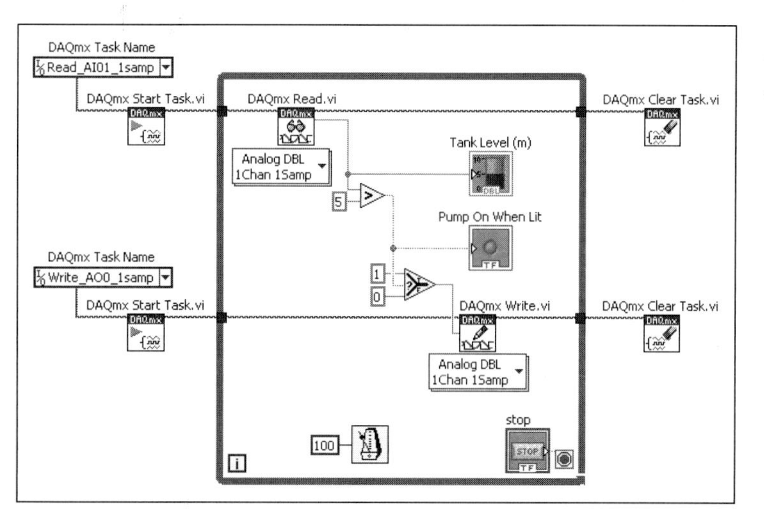

Figure 5.47
VI used to activate a pump when tank level is high, block diagram.

Notice that a While Loop has been used in Figure 5.47 so that tank values are monitored continuously until the **STOP** button is clicked.

This chapter has presented the basics of data acquisition using LabVIEW. Proficiency comes with much practice.

KEY TERMS

AI (analog input)
aliasing
amplification
analog signal
analog-to-digital converter (ADC)
AO (analog output)
band-pass filters
bits
calibration
channel
CMOS (complementary metal-oxide-semiconductor)
DAQ Assistant Express VI
data acquisition
data acquisition system (DAQ)
DI (digital input)
differential input
digital signal
digital-to-analog converter (DAC)
DO (digital output)
dynamic data
electromotive force (emf)
filter
floating
grounded
ground loop
high-pass filters
low-pass filters
Measurement and Automation Explorer
multiplexer (MUX)
offset
oversampling
range
resistance temperature device (RTD)
sampling rate
sensor
signal
signal conditioning
signal-to-noise ratio
single-ended, ground referenced input
single-ended, non-referenced input
signal filtering
source
span
task
thermocouple
transducers
TTL (transistor–transistor logic)
voltage difference
waveform
zero

SUMMARY

Data Acquisition
The process of automatically importing data from an instrument (source) directly into a computer.

Signals

- **Analog**—vary continuously (smoothly)
- **Digital**—take only specific values

Signal Conditioning

- **Filtering**
    - Low-pass filters—remove high-frequency components, often used to remove high-frequency noise
    - High-pass filters—remove low-frequency components
    - Band-pass filters—allow specified range of frequencies to pass
- **Amplification and Offset**
    - **Range**—the expected signal range, minimum to maximum (e.g., 4–10 V)
    - **Span**—the difference between the minimum and maximum expected signal (e.g., 6 V)
    - **Amplification** adjusts the signal span
    - **Offset** slides the entire signal range (a **zero** control is often used to adjust the offset)

Aliasing
Sampling at too low a frequency can create artifacts in the sampled data. *Oversampling* is a technique of sampling fast enough to record the noise as well as the desired signal. When oversampling is used, filtering after data acquisition can be used without danger of aliasing.

Calibration
Relating the measured sensor output to several known values of the quantity to be measured. The known values should cover the entire range of the anticipated measurements.

Data Acquisition Hardware

Channels

- Analog Input (AI)
- Analog Output (AO)
- Digital Input (DI)
- Digital Output (DO)

Wiring

- **Differential Inputs**—measure voltage difference, reject ground loop errors
- **Single-Ended, Ground Referenced Inputs**—measure voltage relative to ground
- **Single-Ended, Non-referenced Inputs**—measure voltage relative to ground

Precision (related to number of bits in ADC)
Sample Rate (the number of samples to be collected in a given period)

Data Acquisition Tasks
Can be created two ways:

- outside of LabVIEW using Measurement and Automation Explorer
- inside of LabVIEW using DAQ Assistant Express VI

A data acquisition task defines:

- what signal(s) to measure
- how the AI connections will be wired
- how often to take readings
- how many readings to collect or, how long to continue reading the signal(s)

Data Acquisition with LabVIEW
There are two basic approaches:

- Collect a batch of data, then process it.
- Collect data point by point and process each value as soon as collected.

Collecting a Batch of Data—task is easy to create from LabVIEW using DAQ Assistant Express VI.
Collecting Data Point by Point—use Measurement and Automation Explorer to create a task that collects one data point on demand. Then, inside LabVIEW:

- identify the task
- start the task
- use the task to read one value—this step is probably in a loop
- stop the task

SELF-ASSESSMENT

1. What are the minimum components needed for automated data acquisition?
   ANS: A signal source, a data acquisition device, and a computer.
2. What is the difference between an analog signal and a digital signal?
   ANS: Digital signals take on only certain values while analog signals can vary continuously (smoothly). Digital signals associated with digital inputs and outputs take on only two values (high and low).

3. Why is signal conditioning sometimes needed before a signal can be connected to a data acquisition system?
ANS: Signals may be noisy, or the signal range may not align with the input range of the data acquisition system.
4. Which type of filter is most commonly used to remove high-frequency noise from a signal?
ANS: A low-pass filter.
5. What is the difference between a signal's range and its span?
ANS: The span indicates the difference between the maximum and minimum values of the signal, but not the minimum and maximum values themselves. The range specifies the minimum and maximum signal values.
6. Describe how amplification and offset are used to align an input signal with the range of a data acquisition device.
ANS: Amplification is used to adjust the span, and offset is used to slide the signal range.
7. What is aliasing?
ANS: Aliasing is the generation of artifacts in a sampled data set because of interplay between sampling rate and signal frequency.
8. How are known values used to calibrate a signal being sent to a data acquisition system?
ANS: Known values allow you to build a relation between the measurement and the acquired data values (called a calibration equation, or calibration curve).
9. Why is it important to know the range, and use values that span the entire range of expected measurements?
ANS: If your measured values are outside the range of your calibration, you are extrapolating the calibration equation and assuming that the extrapolation is valid. Using known values that span the entire range of expected measurements eliminates the need to extrapolate.
10. What is the difference between an analog input and an analog output?
ANS: An analog input receives a signal from a source and sends it to the analog-to-digital converter. An analog output is used to generate an analog signal (typically a voltage) and send it out of the data acquisition system.
11. What is the difference between an analog input and a digital input?
ANS:
    - An analog input is used to receive an analog signal that can vary continuously over a range of signal values (e.g., 2–7 V). The objective of using an analog input is to determine the signal level (e.g., voltage).
    - A digital input is used to receive a digital signal that should have only high- or low-voltage values (e.g., 0 or 5 V). The objective of using a digital input is determining whether the signal level is high or low.
12. What is a ground loop?
ANS: A ground loop is formed when two different devices (e.g., sensor and data acquisition system) are both referenced to ground, but the ground potentials are different. The ground loop is the current path that is formed when current flows from one ground potential to the other.
13. Why are differential inputs generally preferred?
ANS: By measuring the voltage difference between two analog inputs, differential inputs reject ground loop errors.

14. Explain why analog to digital converters with more bits can generate more precise results.
    ANS: More bits allow the input range to be divided into smaller increments (higher precision).
15. Why must you consider the number of samples coming in through all analog inputs when determining the required sample rate?
    ANS: Because most data acquisition systems have multiple analog inputs that are multiplexed through one analog-to-digital converter.
16. What are the two ways to generate a data acquisition task that can be used within LabVIEW?
    ANS: Data acquisition tasks can be created outside of LabVIEW using the Measurement and Automation Explorer, or inside of LabVIEW using the DAQ Assistant Express VI.
17. What information must be supplied to define a data acquisition task?
    ANS:
    - what input(s) to sample (channel(s))
    - how the AI connections will be wired (differential or single-ended)
    - how often to take readings (sample rate)
    - how many readings to collect or (number of samples)

PROBLEMS

1. If the signal being sent from a transducer ranges from 2 to 14 V, what offset and amplification are required to make the signal compatible with
   a. a data acquisition system with an allowable input range of 0–10 V?
   b. a data acquisition system with an allowable input range of −5 to 5 V?
2. How many analog input channels are required to sample two temperature signals and one pressure signal if
   a. All signals are wired using differential inputs.
   b. The temperature signals are wired using differential inputs, and the pressure signal is wired single-ended.
3. The readings listed below were displayed on the screen of an instrument and represent repeated measurements of the same known quantity; the uncertainty in the values represents the limits of the instrument's precision. The instrument is designed to output a voltage equal to the displayed measurement.

6.01342
6.01340
6.01338
6.01344
6.01336
6.01332
6.01351
6.01348
6.01343
6.01336
6.01328
6.01340

If you want to read the values with no less precision than the instrument is capable of supplying, what type of analog-to-digital converter will be required (assume 0–10 V allowed DAQ input range)? Explain how you made your decision.

   a. 12 bit
   b. 14 bit
   c. 16 bit

4. If your data acquisition system uses one analog-to-digital converter for all analog inputs, what sampling rate must the data acquisition system be capable of if you want to sample three temperature channels and two pressure channels at 200 samples per second per channel?
5. The VI shown in Figure 5.48 was created to demonstrate the effect of aliasing.

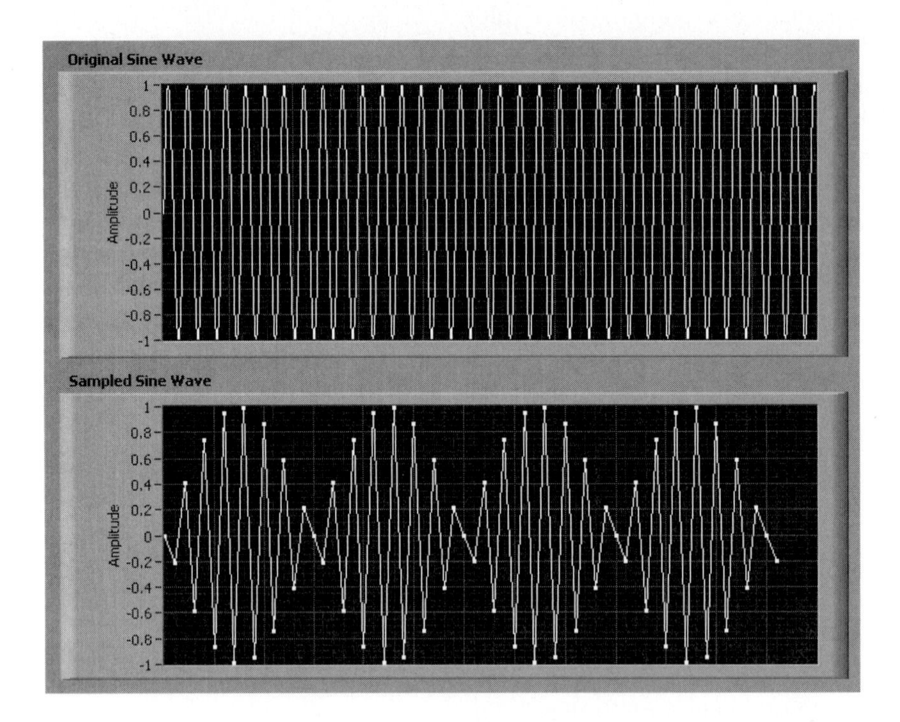

Figure 5.48
Aliasing VI, front panel.

The block diagram for the Aliasing VI is shown in Figure 5.49.

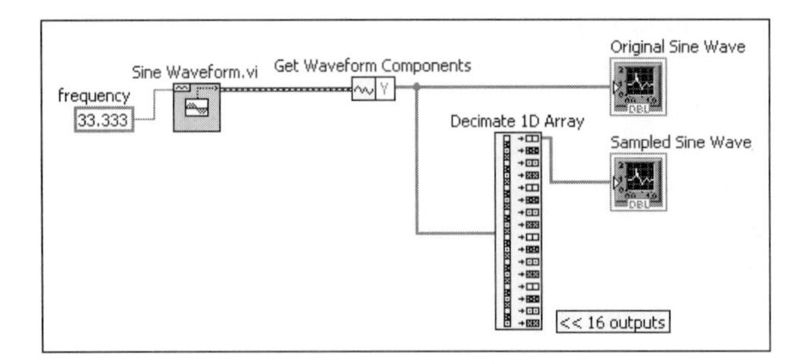

Figure 5.49
Aliasing VI, block diagram.

The VI uses LabVIEW's sine wave generator to create a sine wave at 33.333 Hz. By default, the sine wave generator outputs 1000 samples per second, or 30 dots per cycle of the sine wave. The Get Waveform Components function strips out the time component of the sine waveform and sends only the Y component to the Original Sine Wave graph (Waveform Graph from the Controls Palette / Modern Group/Graph Group).

What causes the aliasing is the Decimate 1D Array function. This function takes the input array (the sine wave with 30 data points per cycle) and divides the array elements between all of the available outputs. We're only using one output to send data to the Sampled Sine Wave graph, but there are 16 outputs displayed in Figure 5.49. LabVIEW is sending one out of every 16 array elements to the Sampled Sine Wave graph. So, instead of 30 dots per original sine cycle, the Sampled Sine Wave graph is only getting less than 2 dots per sine cycle. When you try to sample at about two data points per cycle, some interesting things start happening (aliasing).

Recreate the Aliasing VI shown in Figure 5.49 and run it several times, using a different number of outputs on the Decimate 1D Array function each time. Try 2, 5, 10, 15, 16, 17, and 20 outputs. The relationship between the number of Decimate 1D Array function outputs and samples per sine cycle and effective sampling rate is shown in Table 5.1. The functions required are available at the following locations on the Functions Palette:

Functions Palette / Signal Processing Group / Waveform Generation Group / Sine Waveform VI

Functions Palette / Programming Group / Waveform Group / Get Waveform Components

Functions Palette / Programming Group / Array Group / Decimate 1D Array

Table 5.1 Decimate 1D array outputs and related variables.

Outputs	Data points per cycle	Effective sampling rate* (Hz)	Effective sampling rate/ signal frequency
2	15	500	16.7
5	6	200	6.7
10	3	100	3.3
15	2	67	2.2
16	1.88	63	2.1
17	1.76	59	1.97
20	1.5	50	1.67

*The LabVIEW Sine Waveform VI generates sine waves with 1000 data points per second.

Use your observations to answer the following questions:

a. As you increase the number of outputs, how many outputs are required before you can observe a significant difference between the original and sampled sine waves?

b. What is the effect of aliasing as the number of outputs is increased from 10 to 15?

c. Is aliasing something that should be considered and avoided when preparing to use a data acquisition system?

d. How can you avoid aliasing?

Note: If you have access to a sine wave generator and a data acquisition system, you can sample the sine wave at 1000 Hz and use your acquired data rather than the Sine Waveform function in the Alias VI.

6. Use the DAQ Assistant from LabVIEW to collect 2000 voltage samples through analog input 0 in 5 seconds.
7. Use the Measurement and Automation Explorer to create and save a data acquisition task that will collect 1000 voltage samples through analog input 1 in 2 seconds. Then write a LabVIEW VI that uses the saved task to perform the actual data acquisition.
8. Use the DAQ Assistant from LabVIEW to write 2.3 V to analog output one.
9. Create a LabVIEW VI that uses tasks created in the Measurement and Automation Explorer to send a voltage out through analog output one and read that voltage through analog input zero.

CHAPTER

6

Getting Data Into and Out of LabVIEW without Data Acquisition

Objectives

After reading this chapter, you will know:

- the various types of files that LabVIEW can read from and write to
- how to write LabVIEW data to a file that can be opened in a spreadsheet program
- how to write LabVIEW data to a measurement file
- how to read data from a spreadsheet program in LabVIEW
- how to use spreadsheet data to initialize controls in LabVIEW

6.1 INTRODUCTION

Data acquisition is a very common way of getting data into LabVIEW, but that was the topic of an earlier chapter; the primary focus of this chapter is reading and writing files that can be used by other programs (such as spreadsheets). LabVIEW supports several types of data files, and the type of file format you should use depends on what you want to do with the data.

File Format	Objective
• *Text Files*	Interchange data with other programs
• *Binary Files*	Not compatible with other programs, used with large data sets when speed and size are critical
• *Datalog Files*	LabVIEW data storage files

In this chapter we will focus on getting data into LabVIEW from another program and out of LabVIEW for use in another program. We will use text files for this. LabVIEW provides three functions to write text files:

- Write to Text File (.txt)—used for single values, but not for arrays or matrices
- Write to Spreadsheet File (.txt)—used for 1D or 2D arrays or matrices
- Write to Measurement File (.lvm)—used to send acquired data (waveforms) to text files

The last two are the most useful and will be covered here.

6.2 WRITING LABVIEW DATA TO A SPREADSHEET FILE

The text files created by LabVIEW can be opened in spreadsheet programs such as Excel®. Writing data to a spreadsheet file is the mechanism used to move LabVIEW data into a spreadsheet.

A LabVIEW *spreadsheet file* is a text file that spreadsheets such as Excel can read, but LabVIEW does not create .xls or .xlsx files directly. However, opening a text file in Excel is so straightforward, it makes little difference that LabVIEW creates .txt files instead of .xls files.

The Write to Spreadsheet File function is available in the Function Palette's Programming Group, in the File I/O group. For example, here are the directions to the Write to Spreadsheet File function:

> **Function Palette / Programming Group / File I/O group / Write to Spreadsheet File**

To learn how to send data from a matrix to a text file, we will create a VI that uses the Write to Spreadsheet File function.

We begin on the front panel by creating an array and filling it with some data values (see Figure 6.1). Many significant figures have been used on purpose to illustrate how many digits are saved in the text file.

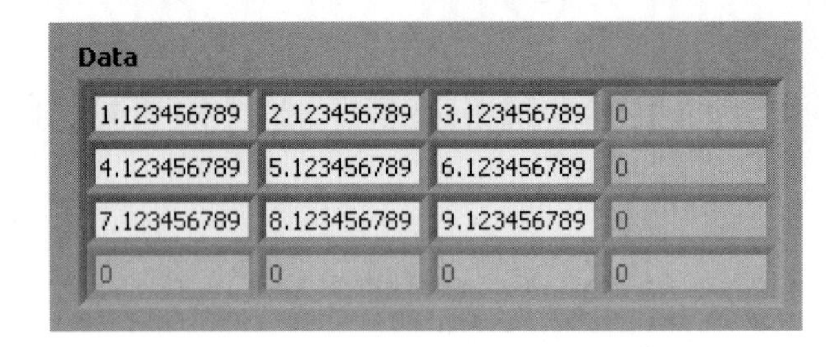

Figure 6.1 Front panel showing data to be saved in each text file format.

Then, we move to the block diagram to use the Write to Spreadsheet File function. The annotated connection pane for the Write to Spreadsheet File function is shown in Figure 6.2.

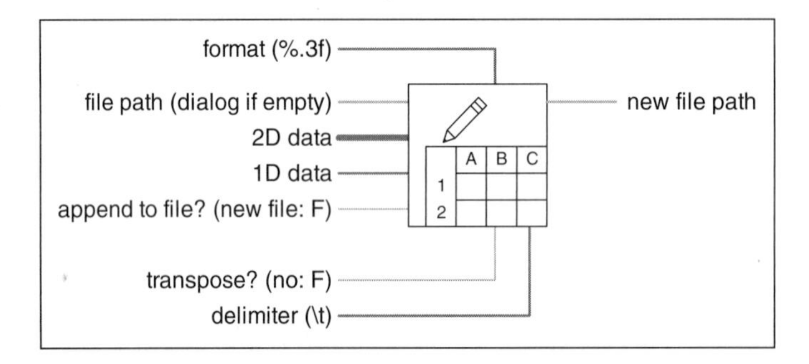

Figure 6.2 Connection pane for Write to Spreadsheet File function.

We will describe the various terminals later, but we can accept the default values for all but one; we must wire the Data matrix output to the 2D data input on the Write to Spreadsheet File function. This is shown in Figure 6.3.

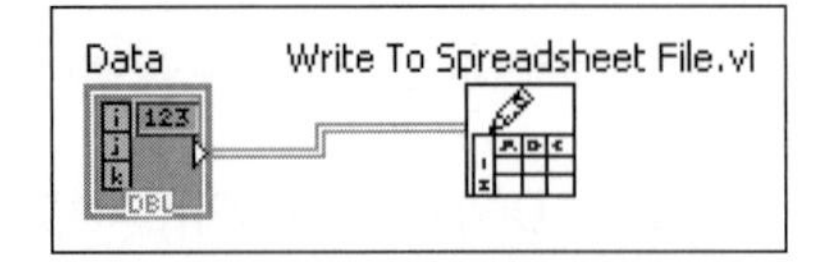

Figure 6.3 Block diagram to send Data array values to a text file.

Because we did not specify a file path, when we run the VI the Choose File to Write dialog (Figure 6.4) opens to allow the user to select a folder and enter a file name. The .txt file extension was used so that the text file will be recognized by other programs. Click **OK** to save the data to the "Write Test.txt" file.

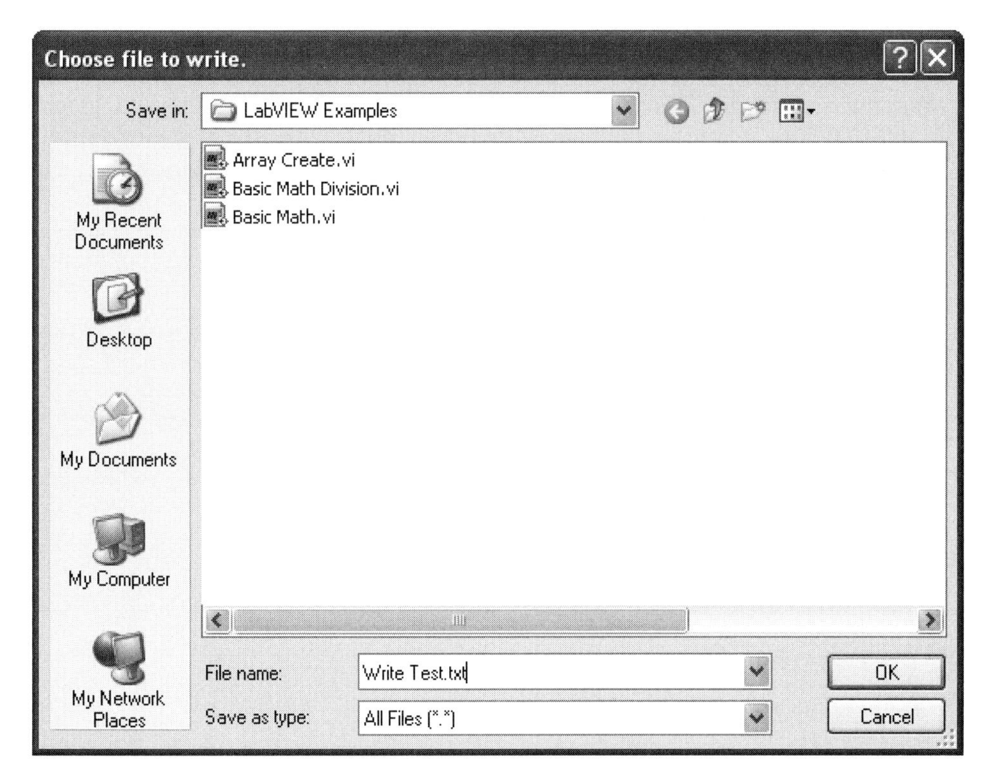

Figure 6.4
The Choose File to Write dialog allows the user to set the file name.

We can open the file in any program that can display a text file (Notepad was used in Figure 6.5) to see how data were saved.

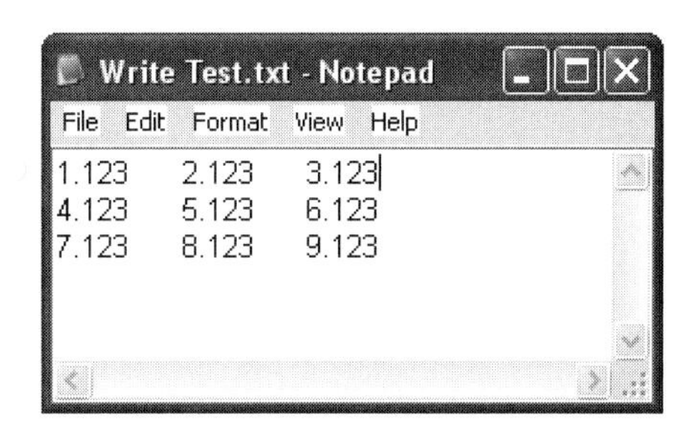

Figure 6.5
The data in the Write Test.txt file.

The values in the Data array were sent to the file, but only three decimal places were saved. This is a result of using the default *format string* (%.3f, as indicated in Figure 6.2.) we can change the format string to save more decimal places, if needed.

The terminals on the Write to Spreadsheet File function block (see Figure 6.2) include

Inputs:

- **format** (%.3f)—the format string used to write the values to the text file.
- **file path** (dialog if empty)—this is a path variable indicating where the file should be stored. If no path is specified, the Choose File to Write dialog is opened to determine the path at run time.

- **2D data**—the terminal used to save a 2D array or matrix of values.
- **1D data**—the terminal used to save single row or column of values.
- **append to file** (new file: F)—a Boolean (True or False) value used to tell LabVIEW how to handle the data if the file already exists. TRUE means append the data to the data already in the file. FALSE (the default) means overwrite the existing data.
- **transpose?** (no: F)—a Boolean (True or False) value indicating whether or not the data should be transposed (rows and columns interchanged) before saving. The default is not to transpose the data.
- **delimiter** (\t)—the *delimiter* is the value placed between numbers to separate them. The default is a tab character ("\t").

Output:

- **new file path**—provides programmatic access to the file path that was used to save the data.

To see how these inputs can be used to modify the way the data are saved, we will change the format string that controls how LabVIEW writes the data to the file.

6.2.1 Format Strings

A format string for writing floating point numbers:

- Begins with the "%" character.
- Sometimes includes "^" (caret, [Shift 6]) to force *engineering notation* (*scientific notation* in multiples of three, e.g., e3, e6.)
- Sometimes a "#" symbol—instructs LabVIEW to drop trailing zeros.
- Sometimes includes a period and a number (e.g., ".3") indicating the number of decimal places to show.
- Ends with a final letter indicating the notation style.
  - f—*floating point notation* (decimal point)
  - e—scientific notation
  - g—LabVIEW uses "f" or "e" depending on size of number

The following table shows how the value 12.3450000 will be saved using various formats.

%.3f (default)	12.345
%.7f	12.3450000
%#.7f	12.345
%e	1.234500e1
%.3e	1.235e1 (rounded)
%^.3e	12.345e0
%g	12.345000
%#g	12.345

Figure 6.6 shows how the format string can be sent to the Write to Spreadsheet File function.

The resulting file is shown in Figure 6.7. Notice that LabVIEW rounds when the "f" format is used.

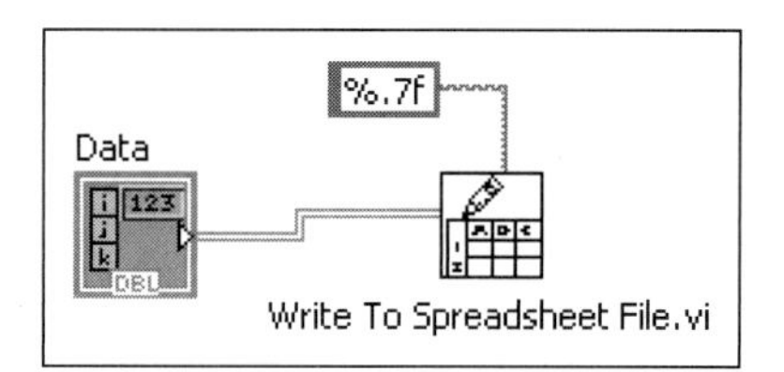

Figure 6.6
Asking for seven decimal places on saved values.

Write Test.txt - Notepad
File Edit Format View Help

1.1234568	2.1234568	3.1234568
4.1234568	5.1234568	6.1234568
7.1234568	8.1234568	9.1234568

Figure 6.7
Data file created using the "%.7f" format.

A good all-purpose format is the "%g" format. The data file created with this format is shown in Figure 6.8. Note that the "%g" format also rounds the values that are written to the file.

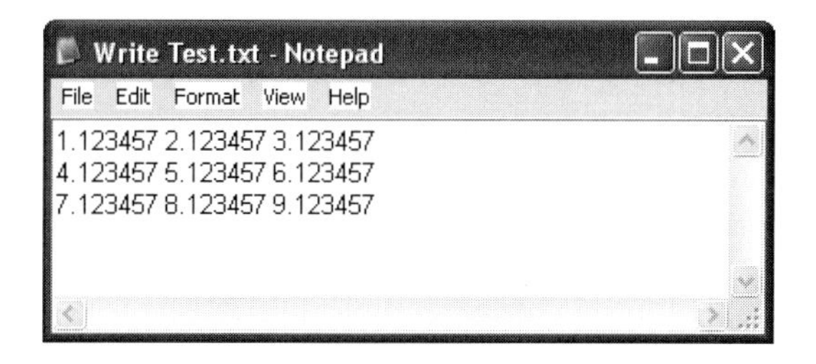
Write Test.txt - Notepad
File Edit Format View Help
1.123457 2.123457 3.123457
4.123457 5.123457 6.123457
7.123457 8.123457 9.123457

Figure 6.8
Data file created using the "%g" format.

6.3 WRITING LABVIEW DATA TO A MEASUREMENT FILE

A LabVIEW *measurement file* (file extension .lvm) is a text file used to save waveforms. A LabVIEW *waveform* is a collection of data values collected over time plus some additional header information. Waveforms are automatically generated when data are acquired using LabVIEW data acquisition functions.

Measurement files are commonly used to save the data collected using a data acquisition system. LabVIEW can read measurement files to reload data for processing and analysis. And, because measurement files are text files, they can be opened by other programs such as word processors and spreadsheets.

A measurement file is created using the Write to Measurement File Express VI. An *Express VI* is a function, or VI, that comes with a dialog box to help the programmer configure the required connections. You can recognize Express VIs by a blue border around the icon in the Function Palette.

The Write to Measurement File Express VI is located in the Programming Group's File I/O Group.

Functions Palette / Programming Group / File I/O Group / Write to Measurement File

When the Write to Measurement File Express VI is placed on the block diagram, the icon for the function is placed on the block diagram and the Configure Write to Measurement File dialog (shown in Figure 6.9) automatically opens.

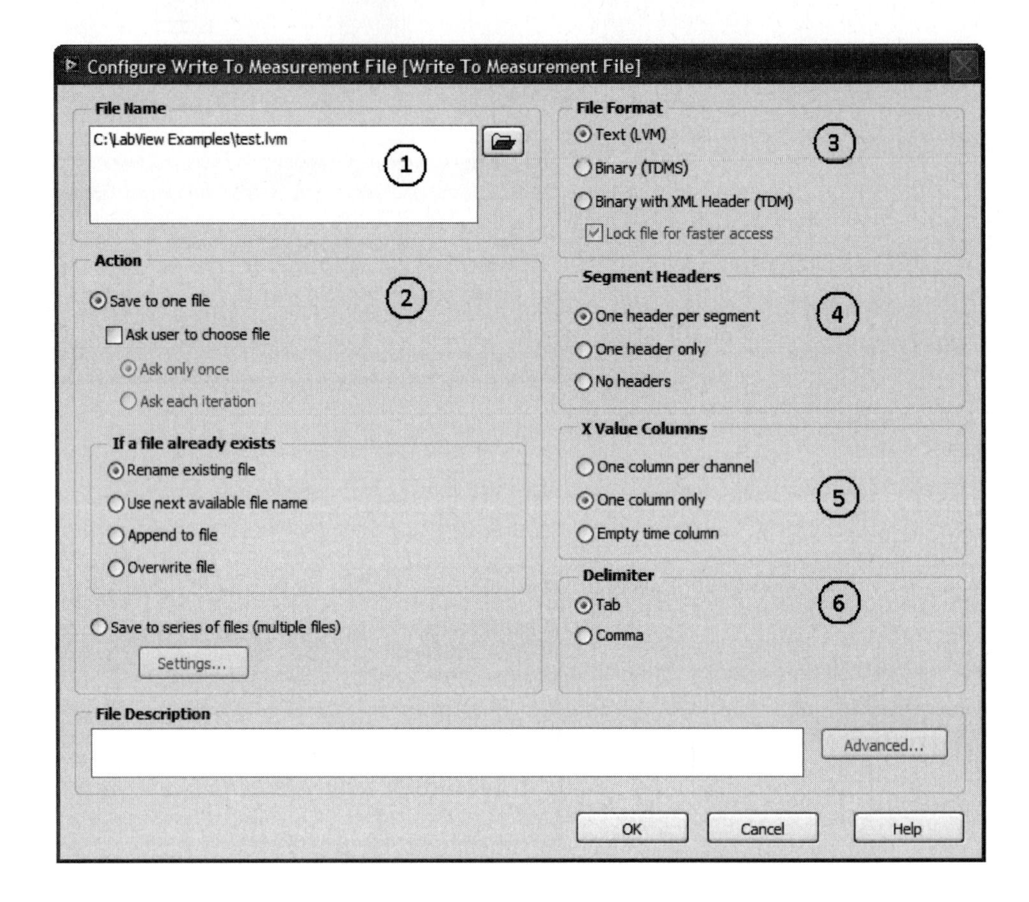

Figure 6.9
The Configure Write to Measurement File dialog.

Six areas have been indicated on the Configure Write to Measurement File dialog (shown in Figure 6.9):

1. **File Name**—you can specify the name of the file to be created.
2. **Action**—instructs LabVIEW what to do if you run the VI multiple times (creating multiple files).
3. **File Format**—the function will output text files (.lvm) and binary files.
4. **Segment Headers**—you can configure the data acquisition functions to take multiple data sets in a single run. Each of these would be considered a segment. The segment header provides information on when the data were collected, who collected the data, and what data were collected.
5. **X Value Columns**—a typical waveform contains data collected over a period of time, and time is the *X* value. You can select how the time values are reported in the measurement file.
6. **Delimiter**—the delimiter is the character that is placed between each numeric value so that the program that reads the file can tell where one number ends and the next starts. Tab delimiters are very commonly used.

When the Configure Write to Measurement File dialog has been completed, click **OK** to return to the block diagram.

The Express VI icon for the Write to Measurement File function can be displayed in two forms, as illustrated in Figure 6.10. Both forms provide the same terminals, but the expanded form is easier to read as long as space on the block diagram is available.

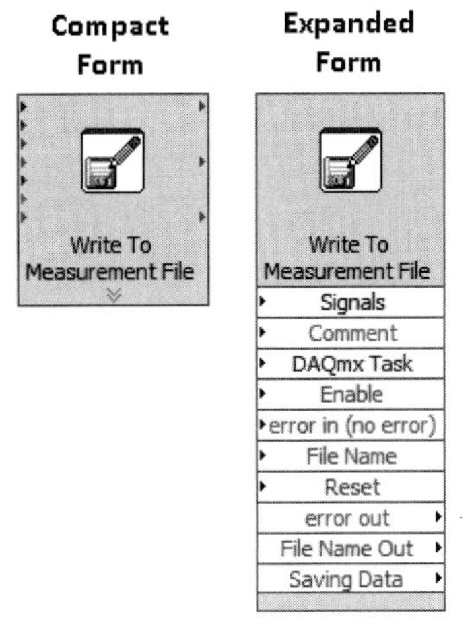

Figure 6.10
Write to Measurement File icon in compact and expanded forms.

The **Signals** input terminal expects to receive a waveform. Waveforms are created by the DAC Assistant Express VI as the result of a data acquisition task. To demonstrate the creation of a waveform, a data acquisition device (NI USB-6009, courtesy of National Instruments) was connected to a simple low-voltage source. The voltage was read using the DAQ Assistant Express VI and the resulting waveform was filtered, plotted, and sent to a LabVIEW measurement file. The complete block diagram is shown in Figure 6.11.

The front panel after collecting the data is shown in Figure 6.12.

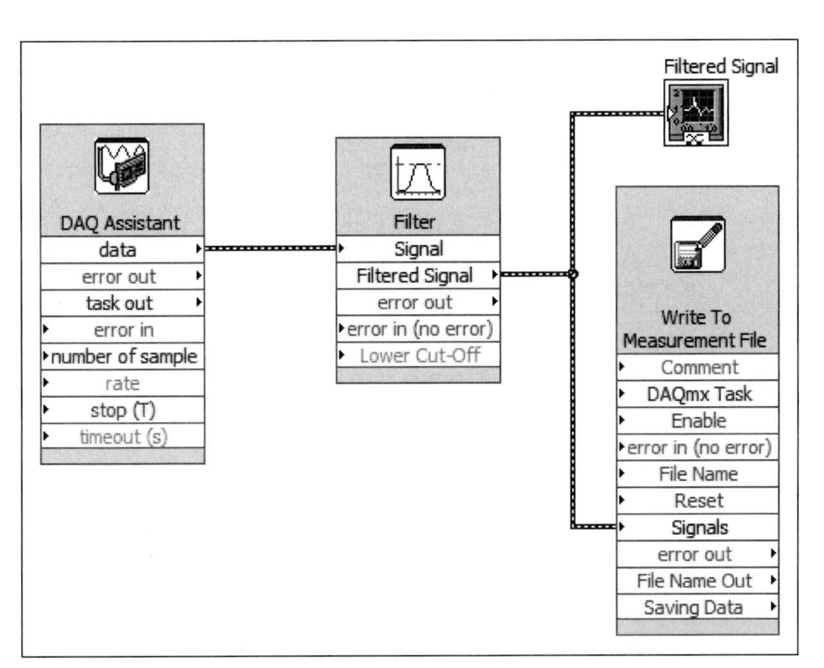

Figure 6.11
Block diagram for data acquisition and writing measurement file.

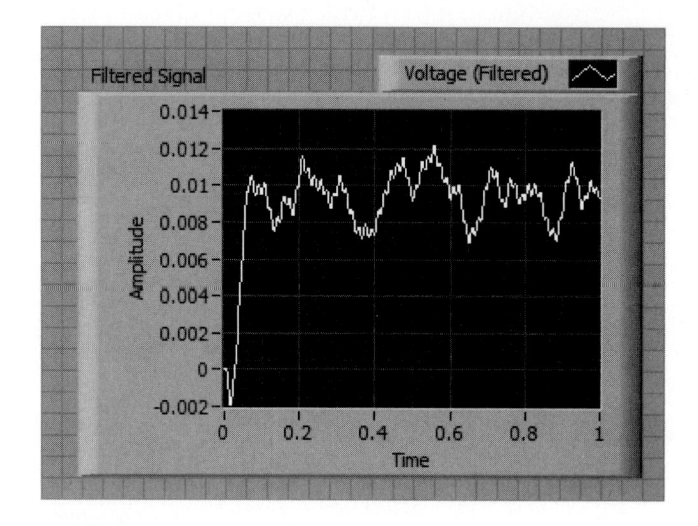

Figure 6.12
Front panel of data acquisition VI showing filtered waveform.

The collected data were automatically saved to the file named in the Configure Write to Measurement File dialog, C:\LabVIEW Examples\test.lvm (see Figure 6.9). Figure 6.13 shows what the measurement file looks like when opened in Excel. (The graph was created in Excel and is not part of the original measurement file.)

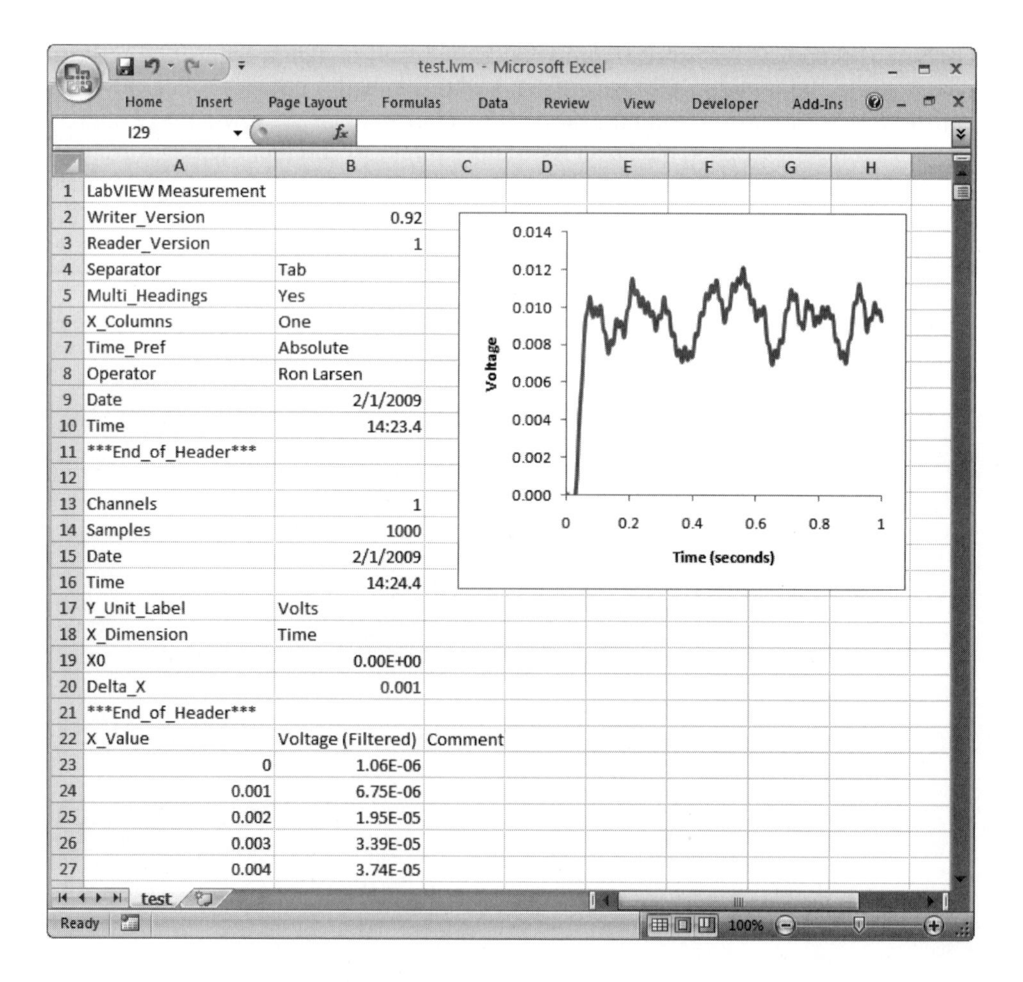

	A	B	C
1	LabVIEW Measurement		
2	Writer_Version	0.92	
3	Reader_Version	1	
4	Separator	Tab	
5	Multi_Headings	Yes	
6	X_Columns	One	
7	Time_Pref	Absolute	
8	Operator	Ron Larsen	
9	Date	2/1/2009	
10	Time	14:23.4	
11	***End_of_Header***		
12			
13	Channels	1	
14	Samples	1000	
15	Date	2/1/2009	
16	Time	14:24.4	
17	Y_Unit_Label	Volts	
18	X_Dimension	Time	
19	X0	0.00E+00	
20	Delta_X	0.001	
21	***End_of_Header***		
22	X_Value	Voltage (Filtered)	Comment
23	0	1.06E-06	
24	0.001	6.75E-06	
25	0.002	1.95E-05	
26	0.003	3.39E-05	
27	0.004	3.74E-05	

Figure 6.13
Measurement file opened in Excel (graph created in Excel).

The measurement file shown in Figure 6.13 contains 21 rows of header information (cells A1:B21) followed by column headers for the data values (cells A22:C22), and 1000 data points in cells A23:B1022. The number of data points collected is set in the DAQ Assistant Express VI's dialog; 1000 data points collected in 1 second is LabVIEW's default for the NI USB-6009 device.

The data in cells A23:B1022 were plotted in Excel using an XY Scatter chart. Once the measurement file has been opened in Excel, all of Excel's capabilities can be used with the data.

Note: When a LabVIEW measurement file is opened in Excel, the .lvm file extension is retained. When new Excel features are added, such as the graph in Figure 6.13, those features are not compatible with the .lvm file format and will be lost if the file format and file extension are not changed when the spreadsheet file is saved from Excel. Excel will warn you (see Figure 6.14) if there is a danger of losing information when you try to save the .lvm file from Excel. To save the graph, we must save the spreadsheet as an Excel file with an .xls or .xlsx file extension.

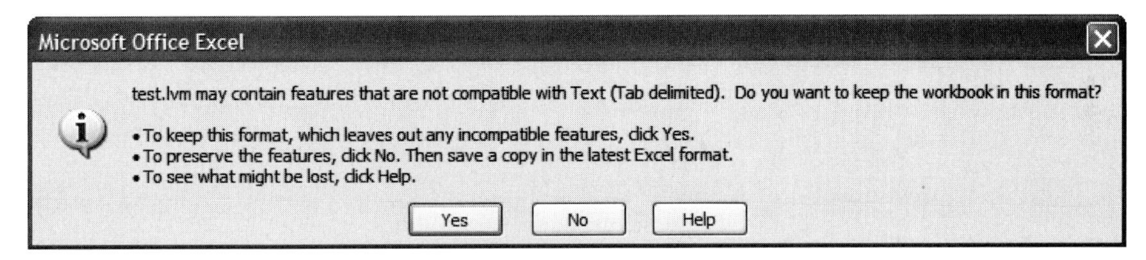

Figure 6.14
Excel warns to change the file type when saving an .lvm file from Excel.

6.4 READING A LABVIEW MEASUREMENT FILE

To read a LabVIEW measurement file, use the Read From Measurement File Express VI (see Figure 6.15) found in the File I/O group:

> **Functions Palette / Programming Group / File I/O Group / Read From Measurement File**

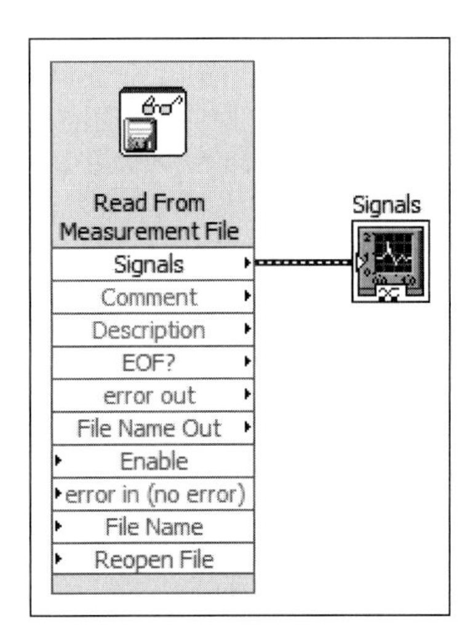

Figure 6.15
Block diagram for a VI to read a measurement file and plot the values.

The Signals waveform graph block was created in the block diagram by right-clicking on the Signals output terminal on the Read From Measurement File icon and selecting **Create / Graph Indicator** from the pop-up menu.

When the Read From Measurement File Express VI is placed on the block diagram, the Configure Read From Measurement File dialog (shown in Figure 6.16) automatically opens. You should ensure that the correct .lvm file is indicated in the **File Name** field, or check **Ask user to choose file** in the **Action** section to select the desired file at run time.

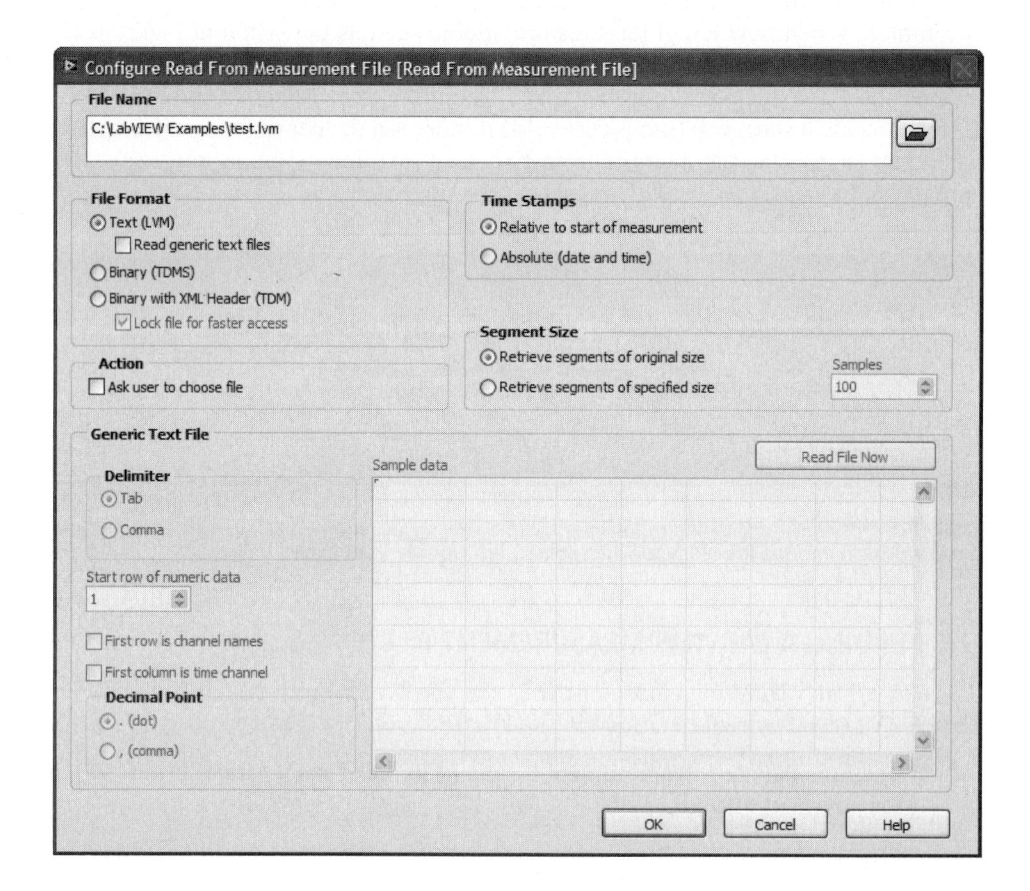

Figure 6.16
The Configure Read From Measurement File dialog.

When the VI is run, the measurement file is read and the results are plotted, as shown in Figure 6.17. The signal is described as Voltage (Filtered) in Figure 6.17 because that was how the data were described in the measurement file (see cell B22 in Figure 6.13).

6.5 READING A SPREADSHEET FILE IN LABVIEW

LabVIEW provides a Read From Spreadsheet File function, which can be used to read values from a text file into LabVIEW.

> **Functions Palette / Programming Group / File I/O Group / Read From Spreadsheet File**

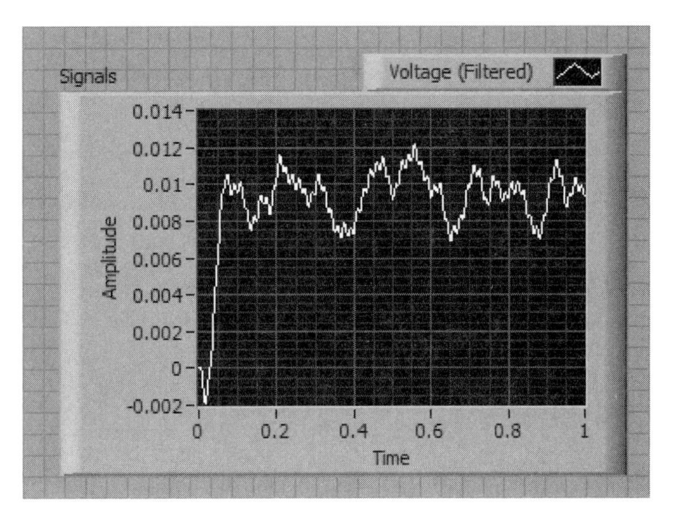

Figure 6.17
The VI front panel after reading the measurement file.

Note: LabVIEW does not read Excel files directly. But an Excel workbook can be saved as a tab-delimited text file (.txt file extension) and LabVIEW can read it. However, this approach will only work for the first worksheet in an Excel workbook. When you try to save an Excel workbook with multiple worksheets as a .txt file, Excel will warn you that only the first worksheet will be saved in the text file.

The general process for getting data from an Excel workbook into LabVIEW is as follows:

1. Get the data values (no text column headers) into an Excel worksheet.
   a. Make sure it is the only worksheet in the workbook.
   b. The top-left value should be in cell A1.
2. Save the Excel file as a .txt file.
3. Place a Read From Spreadsheet File function on a LabVIEW block diagram.
4. Select the data type of the values to be read from the file.
5. Use a string constant to specify the format string that should be used to read the values, if desired. The default format string is "%.3f", which reads floating point numbers with three decimal places.
6. Use a string constant to specify the path name if desired. If the path name is omitted, LabVIEW will ask the user to choose the file at run time.
7. Add an indicator to the block diagram to display the values read from the file.

Step 1: Get the data values into an Excel worksheet
Figure 6.18 shows an Excel file containing temperature data from seven thermocouples collected over a period of 1 hour. It is a typical spreadsheet with

- Title information at the top
- Column headings
- Graph
- Three worksheets in the workbook

All of that information needs to be deleted before saving the workbook as a text file that LabVIEW can read. Figure 6.19 shows the same temperature and time data in a worksheet by itself.

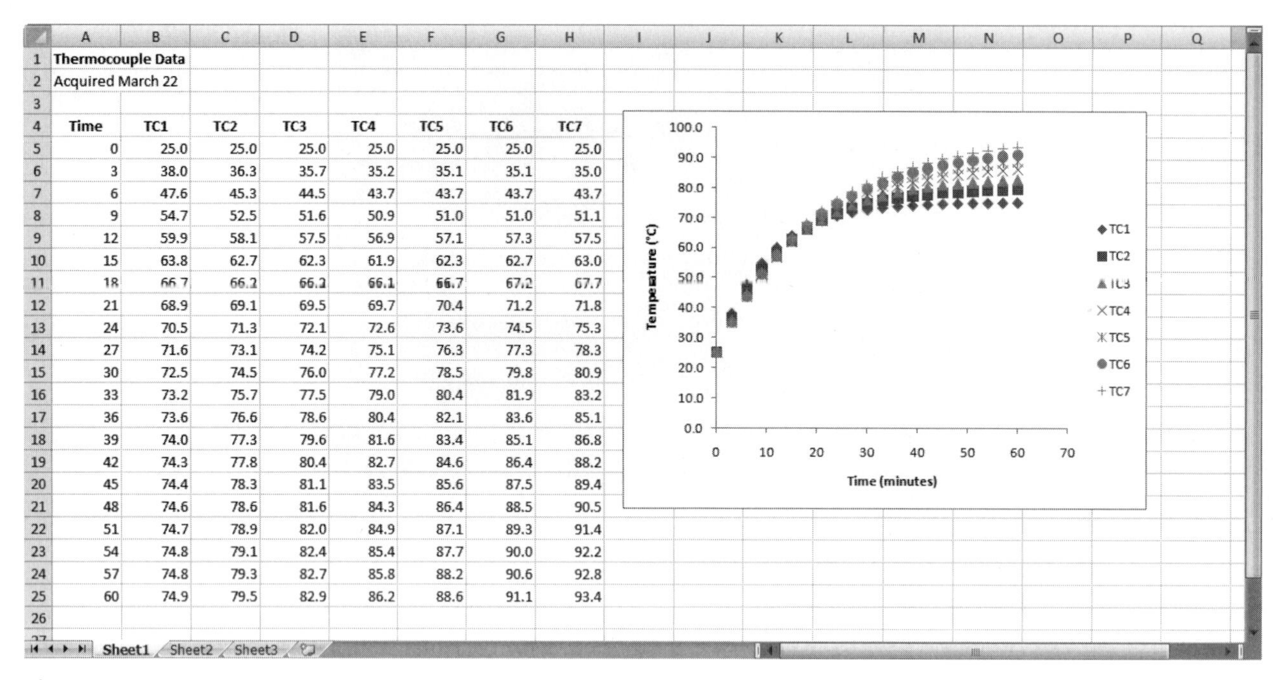

	A	B	C	D	E	F	G	H
1	Thermocouple Data							
2	Acquired March 22							
3								
4	Time	TC1	TC2	TC3	TC4	TC5	TC6	TC7
5	0	25.0	25.0	25.0	25.0	25.0	25.0	25.0
6	3	38.0	36.3	35.7	35.2	35.1	35.1	35.0
7	6	47.6	45.3	44.5	43.7	43.7	43.7	43.7
8	9	54.7	52.5	51.6	50.9	51.0	51.0	51.1
9	12	59.9	58.1	57.5	56.9	57.1	57.3	57.5
10	15	63.8	62.7	62.3	61.9	62.3	62.7	63.0
11	18	66.7	66.2	66.2	66.1	66.7	67.2	67.7
12	21	68.9	69.1	69.5	69.7	70.4	71.2	71.8
13	24	70.5	71.3	72.1	72.6	73.6	74.5	75.3
14	27	71.6	73.1	74.2	75.1	76.3	77.3	78.3
15	30	72.5	74.5	76.0	77.2	78.5	79.8	80.9
16	33	73.2	75.7	77.5	79.0	80.4	81.9	83.2
17	36	73.6	76.6	78.6	80.4	82.1	83.6	85.1
18	39	74.0	77.3	79.6	81.6	83.4	85.1	86.8
19	42	74.3	77.8	80.4	82.7	84.6	86.4	88.2
20	45	74.4	78.3	81.1	83.5	85.6	87.5	89.4
21	48	74.6	78.6	81.6	84.3	86.4	88.5	90.5
22	51	74.7	78.9	82.0	84.9	87.1	89.3	91.4
23	54	74.8	79.1	82.4	85.4	87.7	90.0	92.2
24	57	74.8	79.3	82.7	85.8	88.2	90.6	92.8
25	60	74.9	79.5	82.9	86.2	88.6	91.1	93.4
26								

Figure 6.18
Excel file with extraneous information.

Figure 6.19
Excel worksheet with extraneous information removed.

	A	B	C	D	E	F	G	H	I	J
1	0	25.0	25.0	25.0	25.0	25.0	25.0	25.0		
2	3	38.0	36.3	35.7	35.2	35.1	35.1	35.0		
3	6	47.6	45.3	44.5	43.7	43.7	43.7	43.7		
4	9	54.7	52.5	51.6	50.9	51.0	51.0	51.1		
5	12	59.9	58.1	57.5	56.9	57.1	57.3	57.5		
6	15	63.8	62.7	62.3	61.9	62.3	62.7	63.0		
7	18	66.7	66.2	66.2	66.1	66.7	67.2	67.7		
8	21	68.9	69.1	69.5	69.7	70.4	71.2	71.8		
9	24	70.5	71.3	72.1	72.6	73.6	74.5	75.3		
10	27	71.6	73.1	74.2	75.1	76.3	77.3	78.3		
11	30	72.5	74.5	76.0	77.2	78.5	79.8	80.9		
12	33	73.2	75.7	77.5	79.0	80.4	81.9	83.2		
13	36	73.6	76.6	78.6	80.4	82.1	83.6	85.1		
14	39	74.0	77.3	79.6	81.6	83.4	85.1	86.8		
15	42	74.3	77.8	80.4	82.7	84.6	86.4	88.2		
16	45	74.4	78.3	81.1	83.5	85.6	87.5	89.4		
17	48	74.6	78.6	81.6	84.3	86.4	88.5	90.5		
18	51	74.7	78.9	82.0	84.9	87.1	89.3	91.4		
19	54	74.8	79.1	82.4	85.4	87.7	90.0	92.2		
20	57	74.8	79.3	82.7	85.8	88.2	90.6	92.8		
21	60	74.9	79.5	82.9	86.2	88.6	91.1	93.4		
22										

Sheet1

Step 2: Save the Excel file as a .txt file

To save the Excel worksheet as a tab-delimited text file, start with the following menu options, illustrated in Figure 6.20:

Office button / Save As / Other Formats

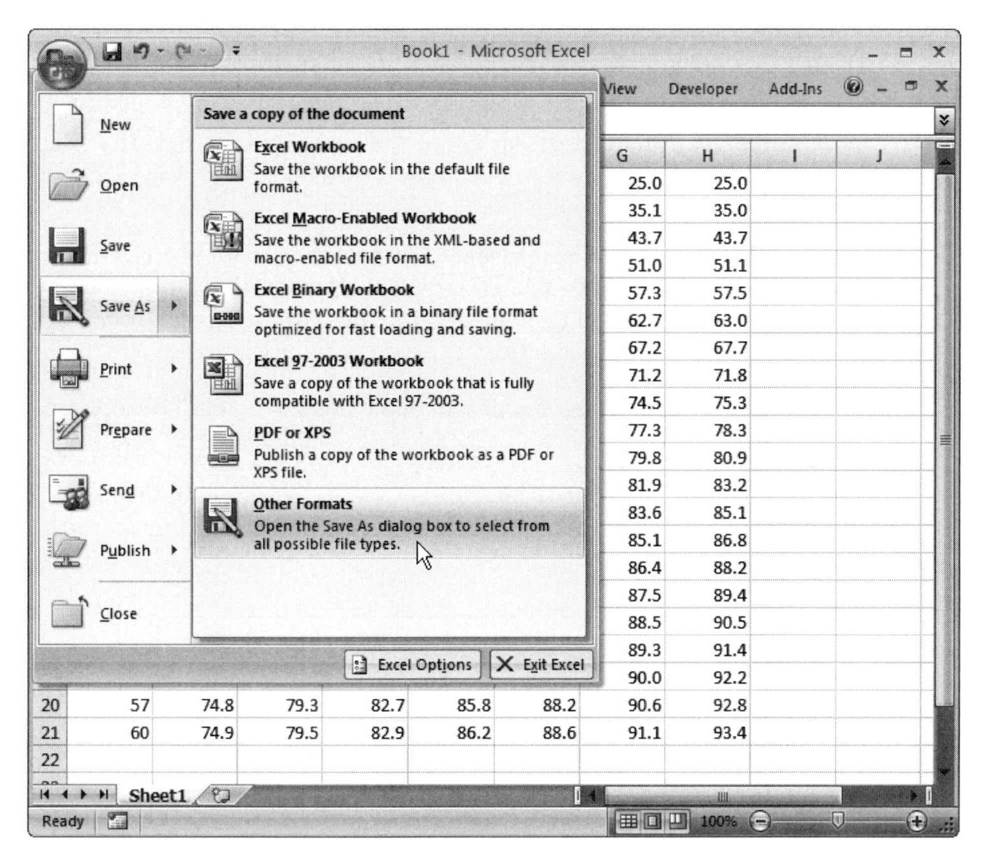

Figure 6.20
Using the Office button to save the Excel worksheet in another format.

This opens the Save As dialog shown in Figure 6.21. Select **Text (Tab delimited) (*.txt)** in the **Save as type:** field.

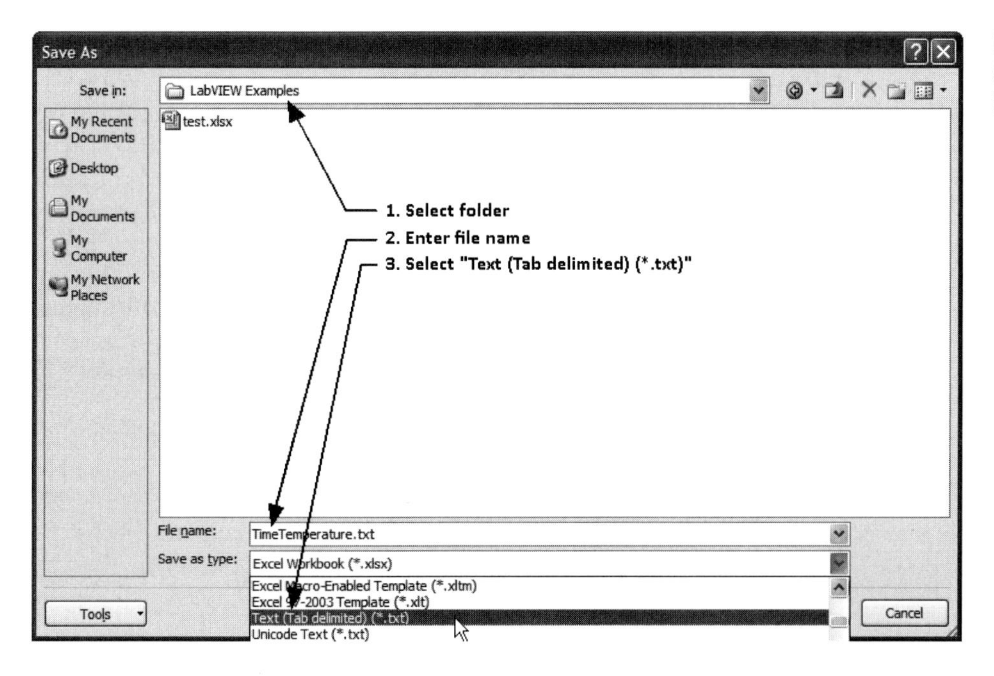

Figure 6.21
Select tab-delimited text file in the Save As dialog.

Excel will likely display a warning box that information may be lost by saving in a text file, but you must use a text file for LabVIEW. The numeric values will not be lost when the file is saved as a text file.

Note: Undisplayed digits are lost when the text file is saved. The Time Temperature.txt file will contain values with only one decimal place (because that's how they were displayed in the Excel worksheet, see Figure 6.19).

Step 3: Place a Read From Spreadsheet File function on a LabVIEW block diagram

The Read From Spreadsheet File function is in the Programming Group's File I/O Group on the Functions Palette.

> **Functions Palette / Programming Group / File I/O Group / Read From Spreadsheet File**

The annotated connection pane for the function is illustrated in Figure 6.22. Most of the terminals have default values that work most of the time.

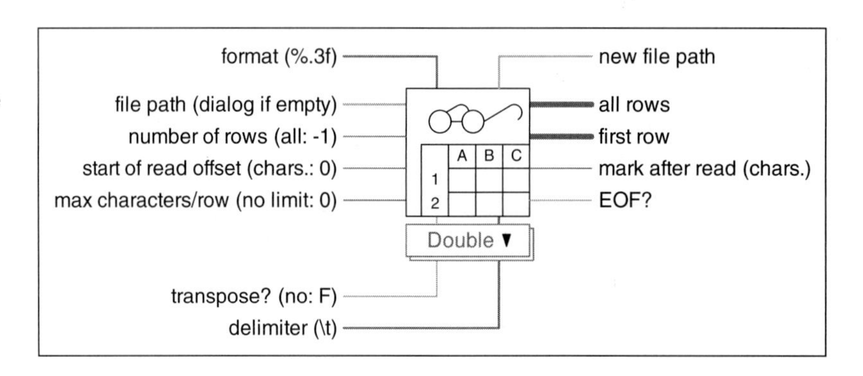

Figure 6.22
Connection pane for the Read From Spreadsheet File function.

Inputs:

- **format** (%.3f)—the format string used to read the values from the text file.
- **file path** (dialog if empty)—this is a path variable indicating where the file should be stored. If no path is specified, the user is asked to choose a file at run time.
- **number of rows (all: −1)**—if only a part of the file is to be read, you must indicate the number of rows to be read. Accept the default to read the entire file.
- **start of read offset (chars.: 0)**—if the top-left corner of the data set is not in cell A1, you must indicate the required offset. Because the offset must be specified in characters, not cells, this is very inconvenient to use.
- **max characters/row (no limit: 0)**—you can set a limit on the number of characters read per row. The default is to read the entire row.
- **transpose?** (no: F)—a Boolean (True or False) value indicating whether or not the data should be transposed (rows and columns interchanged) as it is read. The default is not to transpose the data.
- **delimiter** (\t)—the delimiter is the value that has been placed between numbers to separate them. The default is a tab character ("\t").

Outputs:

- **new file path**—provides programmatic access to the file path that the user may have selected using the Choose File to Write dialog.
- **all rows**—this is the most commonly used output terminal. All rows of data are available through this output.
- **first row**—this terminal provides access to only the first row of values.
- **mark after read (chars.)**—the location of the file marker after the file has been read. Rarely used, this might be useful if you needed to append data to the end of the file.
- **EOF?**—this output is set to TRUE when the end of the file has been read.

In this example, we will set a **format** string constant to increase the number of decimal places slightly, and send **all rows** to an indicator.

Step 4: Select the data type of the values to be read from the file
LabVIEW must know the data type of the values that will be read so that it can read, store, and display the values correctly. The default data type is "Double" (double-precision floating point numbers), as indicated on the *data type selector* shown below the Read From Spreadsheet File function icon in Figure 6.23. In this example, we want to use the Double data type.

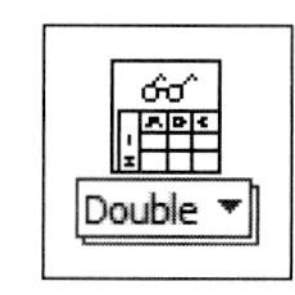

Figure 6.23
The drop-down selector labeled "Double" is used to indicate the data type of the values to be read.

Note: The data type selector can be hidden. Right-click on the node and select **Visible Items** from the pop-up menu if needed.

Step 5: Use a string constant to specify the format string
To attempt to display more decimal places than were saved in the text file, we will instruct LabVIEW to use format string "%.5f" to read the data. This will allow us to observe how LabVIEW handles a request to read more digits than are present in the values in the file.
To set the format string:

1. Right-click on the format input terminal.
2. Select Create/Constant from the pop-up menu.
3. Enter "%.5f" (without the quotes) into the constant.

The result is shown in Figure 6.24.

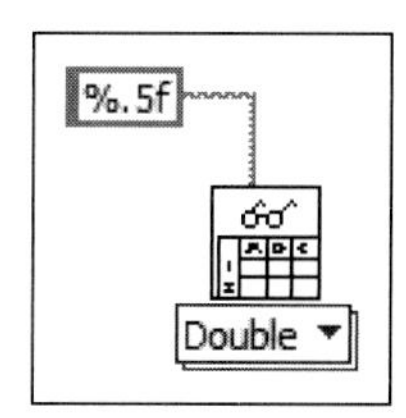

Figure 6.24
Setting the string constant that specifies the read format.

Step 6: Use a string constant to specify the path name
If the path name is not specified, LabVIEW will ask the user to choose the file at run time. In this example, we will not specify the file path name.

Step 7: Add an indicator to the block diagram to display the read values
Right-click on the Read From Spreadsheet File function's **all rows** output and select **Create / Indicator** from the pop-up menu. LabVIEW will add an array indicator to the block diagram and the front panel to display the values read from the file. The complete (for now) block diagram is shown in Figure 6.25.

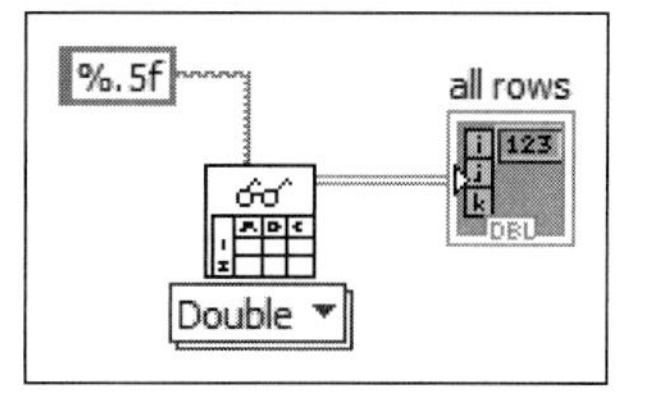

Figure 6.25
Block diagram, ready to read data from text file.

When the VI is run from the front panel, the user is asked to select the file to be read as shown in Figure 6.26.

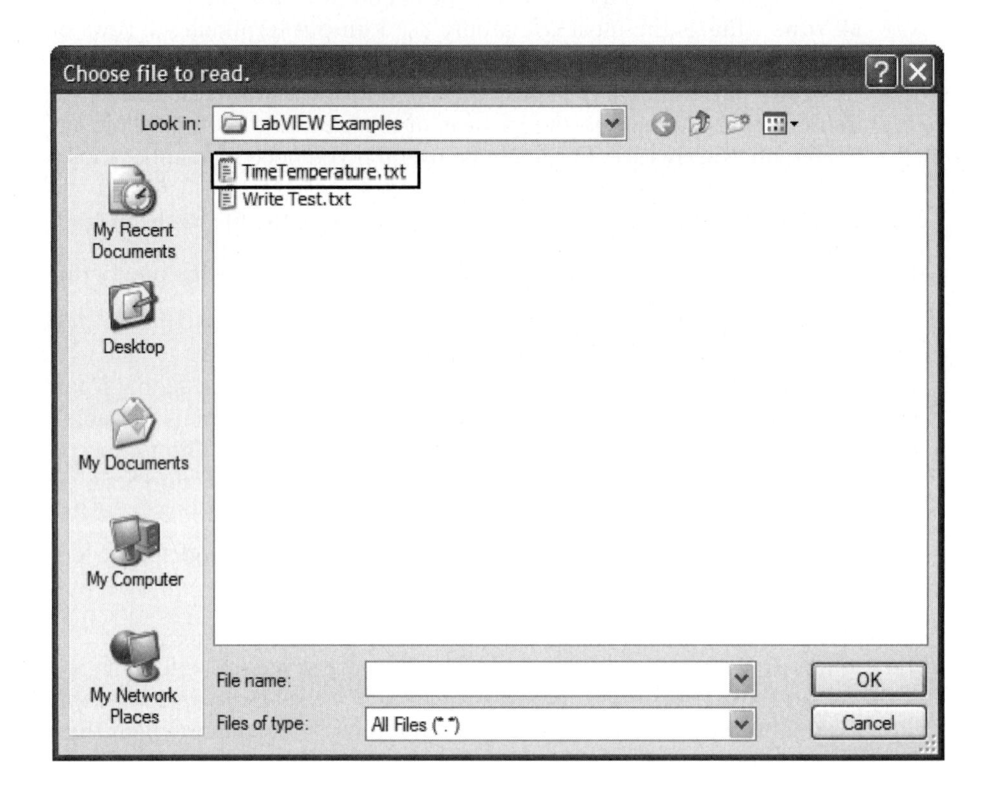

Figure 6.26
The Choose file to read dialog is used to select the text file.

The results are shown in Figure 6.27. The array indicator was resized to show the entire array of values.

Recall that we asked LabVIEW to read five decimal places ("%.5f"), but the values in the text file contained only one decimal place. LabVIEW read the values in the file anyway, with as much precision as the values would allow. The values are displayed on the front panel with three decimal places because of the (default) display format used with the array indicator.

6.5.1 Pulling Single Columns or Rows from 2D Arrays

Once the array values have been read from a file, they can be used for other calculations. It is often handy, and sometimes necessary, to pull a single column or single row from a 2D array. The Index Array function is used for this purpose. The connection pane for the Index Array function is illustrated in Figure 6.28.

Functions Palette / Programming Group / Array Group / Index Array

When you connect a multidimensional array (2D or higher) to the array input (labeled **2D array** in Figure 6.28, but higher order arrays can be connected), the connection pane automatically resizes to display one index input for each array dimension. Figure 6.28 has been labeled for a 2D input array.

all rows

0
0

0.000	25.000	25.000	25.000	25.000	25.000	25.000	25.000	0.000
3.000	38.000	36.300	35.700	35.200	35.100	35.100	35.000	0.000
6.000	47.600	45.300	44.500	43.700	43.700	43.700	43.700	0.000
9.000	54.700	52.500	51.600	50.900	51.000	51.000	51.100	0.000
12.000	59.900	58.100	57.500	56.900	57.100	57.300	57.500	0.000
15.000	63.800	62.700	62.300	61.900	62.300	62.700	63.000	0.000
18.000	66.700	66.200	66.200	66.100	66.700	67.200	67.700	0.000
21.000	68.900	69.100	69.500	69.700	70.400	71.200	71.800	0.000
24.000	70.500	71.300	72.100	72.600	73.600	74.500	75.300	0.000
27.000	71.600	73.100	74.200	75.100	76.300	77.300	78.300	0.000
30.000	72.500	74.500	76.000	77.200	78.500	79.800	80.900	0.000
33.000	73.200	75.700	77.500	79.000	80.400	81.900	83.200	0.000
36.000	73.600	76.600	78.600	80.400	82.100	83.600	85.100	0.000
39.000	74.000	77.300	79.600	81.600	83.400	85.100	86.800	0.000
42.000	74.300	77.800	80.400	82.700	84.600	86.400	88.200	0.000
45.000	74.400	78.300	81.100	83.500	85.600	87.500	89.400	0.000
48.000	74.600	78.600	81.600	84.300	86.400	88.500	90.500	0.000
51.000	74.700	78.900	82.000	84.900	87.100	89.300	91.400	0.000
54.000	74.800	79.100	82.400	85.400	87.700	90.000	92.200	0.000
57.000	74.800	79.300	82.700	85.800	88.200	90.600	92.800	0.000
60.000	74.900	79.500	82.900	86.200	88.600	91.100	93.400	0.000
0.000	0.000	0.000	0.000	0.000	0.000	0.000	0.000	0.000

Figure 6.27
Front panel showing the values read from the text file.

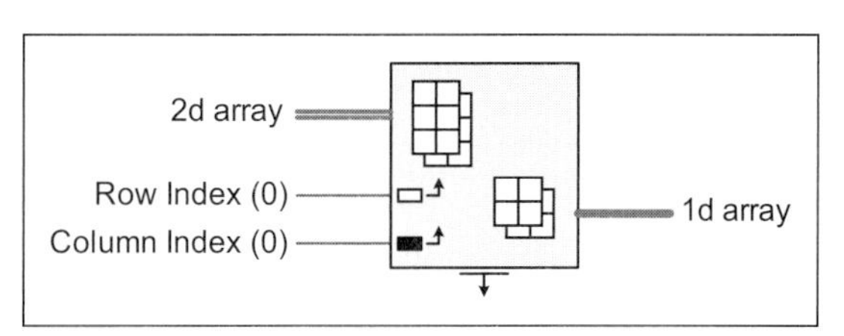

Figure 6.28
Connection pane for Index Array function.

- To select a single row, wire an integer value (the row index value) to the row index input and leave the column index input unwired.
- To select a single column, wire an integer value (the column index value) to the column index input and leave the row index input unwired.

In Figure 6.29 two Index Array functions have been used to pull out the Time and TC1 columns. The Time data are in column 0, and the TC1 data are in column 1. Since we wanted the data from two columns, the row index inputs were not used. The front panel (Figure 6.30) shows the Time and TC1 arrays, and the XY Graph.

Note: By specifying the file path on the block diagram, the Choose file to read dialog is bypassed.

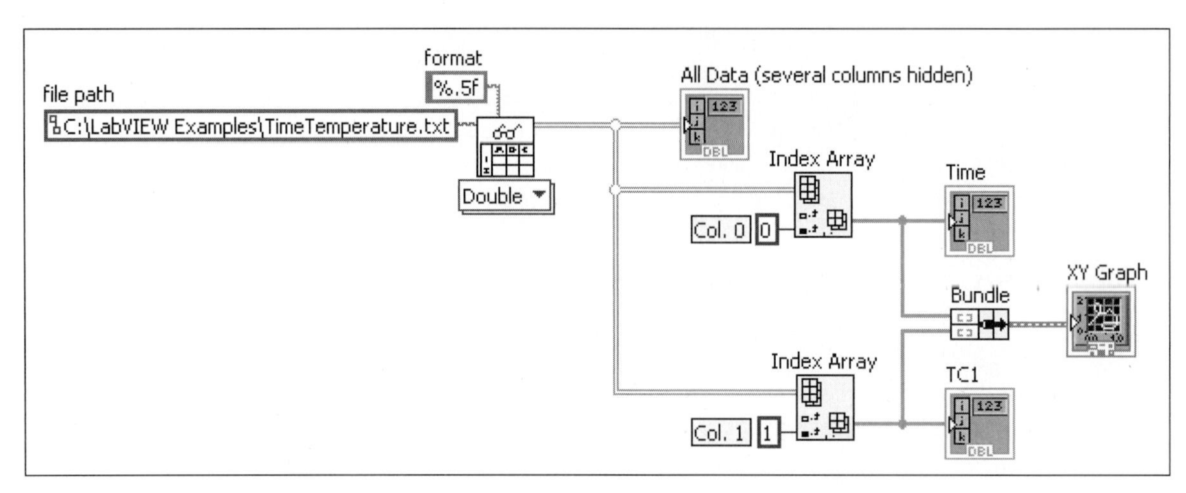

Figure 6.29
Using the data read from the text file.

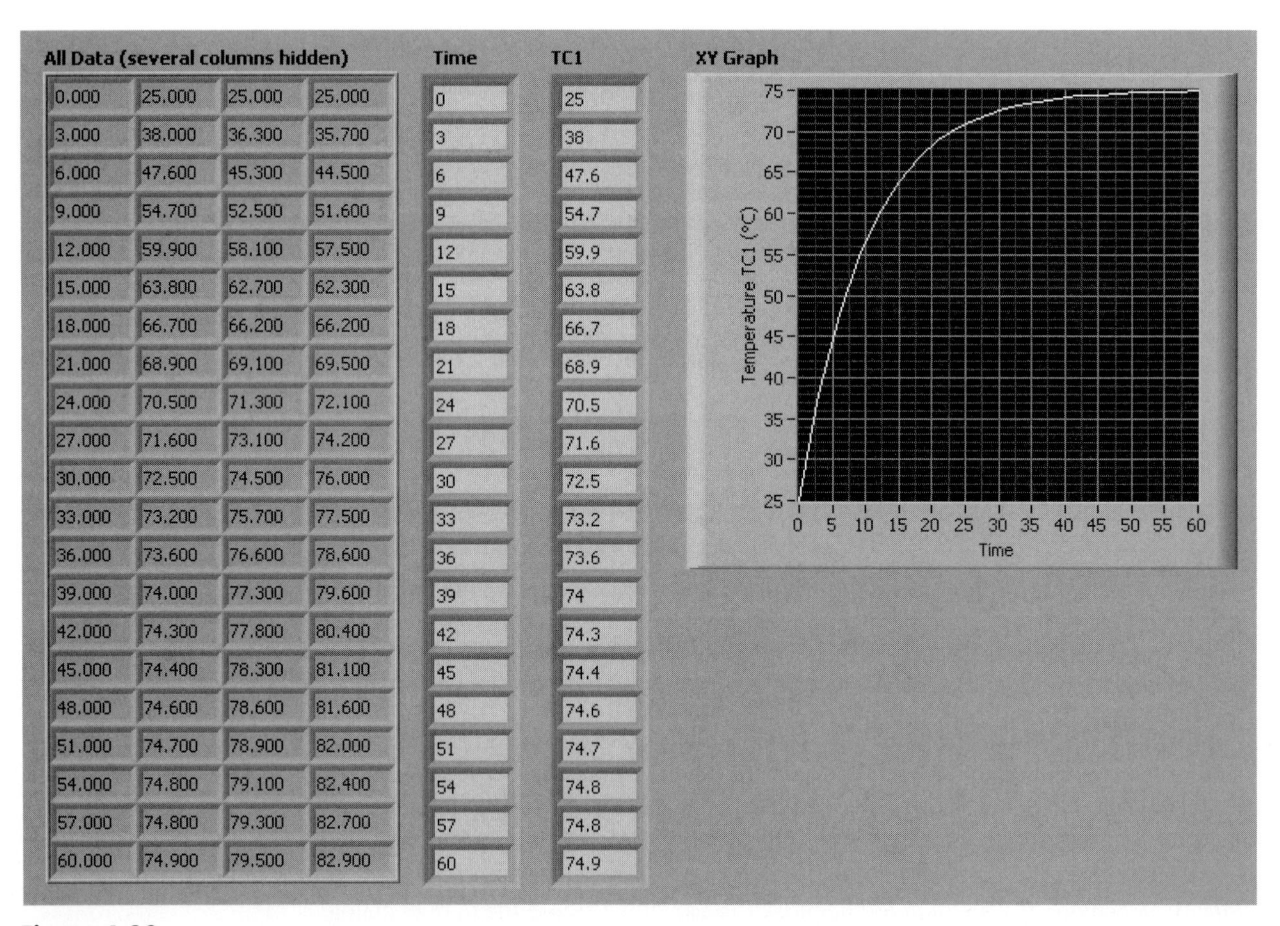

Figure 6.30
Graphing the data read from the text file.

6.6 USING SPREADSHEET DATA TO INITIALIZE A MATRIX CONTROL

Occasionally, you may want to initialize a control using data from a file. For example, you may want to send someone a compiled VI with the required data already loaded into a control (and not attach a text file). This is easy to do:

1. Read the text file and show the results in an array indicator.
2. Change the array indicator to a control.
3. Delete the Read From Spreadsheet File function and any broken wires.
4. Make the current data the default data for the control.

Step 1. Read the text file and show the results in an array indicator
The VI shown in Figure 6.31 reads a text file and sends the data values to an array indicator.

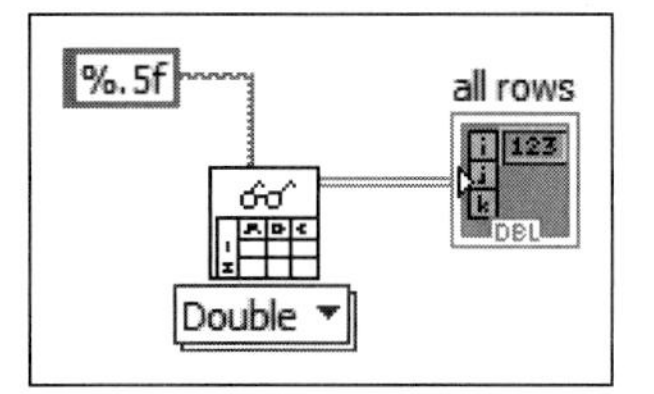

Figure 6.31
Block diagram used to read data from text file.

Step 2. Change the array indicator to a control
To change the array indicator to a control, right-click on the matrix indicator and select **Change to Control** from the pop-up menu as shown in Figure 6.32. The indicator will be turned into a control, but the values that the indicator was displaying will be left in the control.

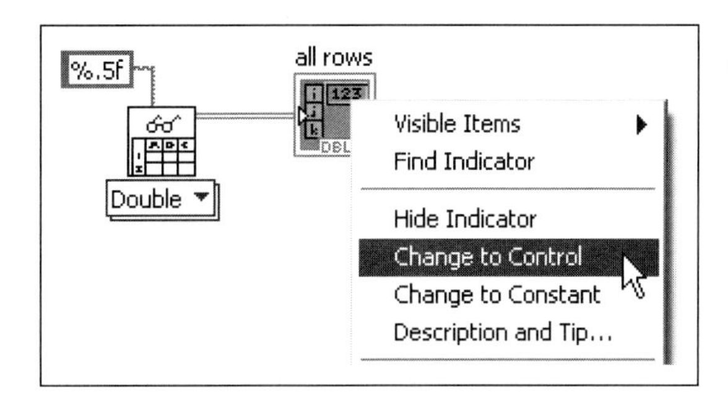

Figure 6.32
Right-click on the array indicator and select Change to Control.

Step 3. Delete the Read From Spreadsheet File function
The wire from the Read From Spreadsheet File function's **all rows** output will be broken (see Figure 6.33) because you cannot sent output to a control, but we don't need the Read From Spreadsheet File function anymore; we've already filled the control with values from the text file. Just delete the Read From Spreadsheet File function and the broken wire.

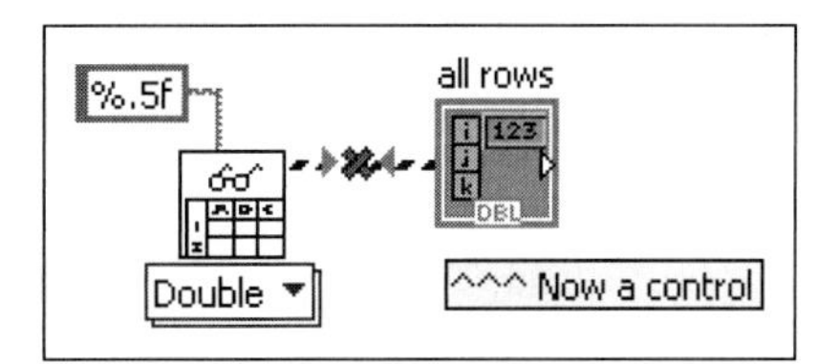

Figure 6.33
Changing the indicator to a control breaks the wire from the Read function.

Step 4. Make the current data the default data for the control

The array control now contains the values read from the file, but if the VI is closed and reopened, default values (zeroes) will be loaded, not the values from the text file. We need to make the current values (read from the text file) the default values before closing the VI so that they will be saved with the VI.

To make the current values the default values, right-click on the array control and select **Data Operations / Make Current Value Default** from the pop-up menu (illustrated in Figure 6.34).

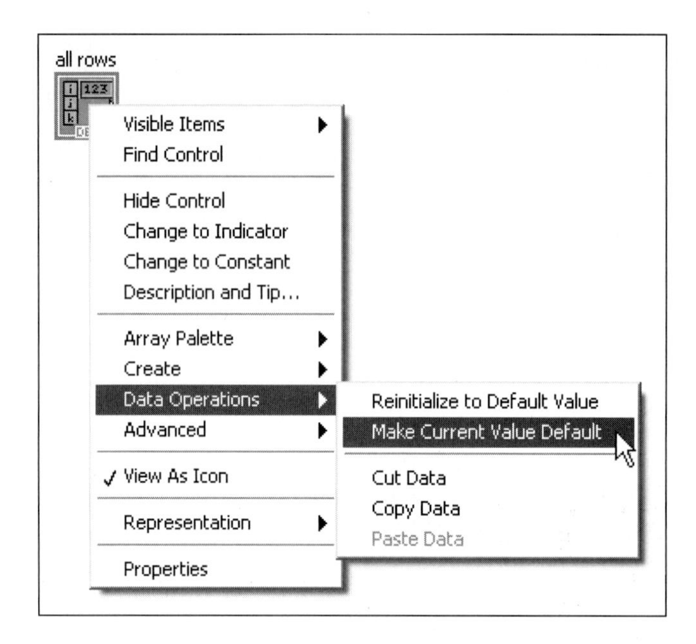

Figure 6.34 Making the current values the default values for the array control.

Once the array control has default values, it can be used like any other control as illustrated in Figure 6.35.

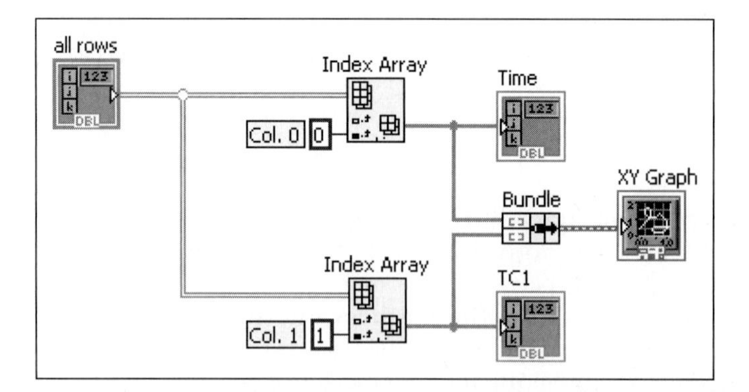

Figure 6.35 Using the array control in a VI.

APPLICATION

Spline Interpolation of Excel Values

One reason to import and export data is to take advantage of the features of various software products. Excel is handy for working with data, but it does not provide a *spline interpolation* function. In this example we import Excel data (via a .txt file) and perform the spline interpolation in LabVIEW, and then export the results back to Excel for plotting.

The original data can be seen in the Excel image shown in Figure 6.36. The shape is a decaying oscillation.

$$y = e^{-ax} \sin(bx)$$

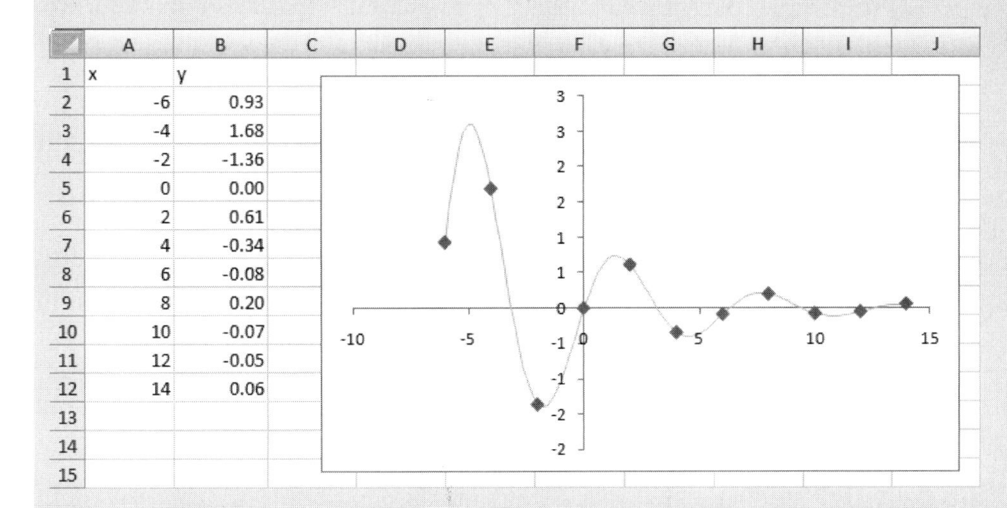

x	y
-6	0.93
-4	1.68
-2	-1.36
0	0.00
2	0.61
4	-0.34
6	-0.08
8	0.20
10	-0.07
12	-0.05
14	0.06

Figure 6.36
The original decaying sine wave data.

The complete curve is shown behind the data points in Figure 6.36, but only the small data set will be sent into LabVIEW for interpolation. We will attempt to use a cubic spline interpolation in LabVIEW to smooth out the curve, and then send the result back to Excel to compare the spline interpolation with the calculated curve.

The data have been saved in a text file, DecayOsc.txt. This file can be read into a LabVIEW VI as shown in Figure 6.37. The X and Y arrays have been separated using Index Array functions.

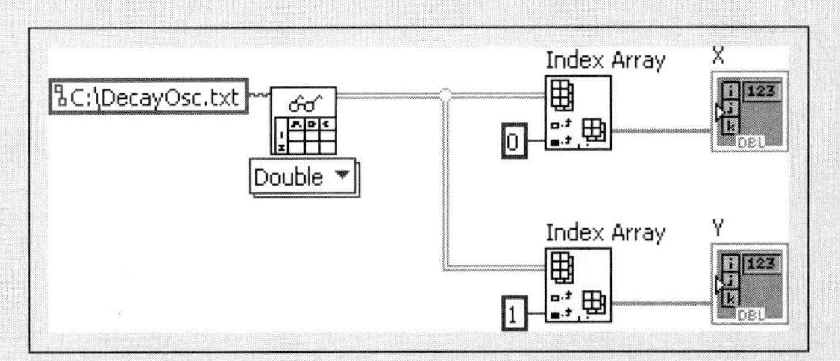

Figure 6.37
VI block diagram for reading the decaying oscillation data file.

Next, the X and Y 1D arrays are combined ("bundled" is the official term) and sent to an XY Graph control as shown in Figure 6.38 (block diagram) and Figure 6.39 (front panel).

The next step is to use LabVIEW's cubic spline interpolation functions to interpolate between each point.

Functions Palette / Mathematics Group / Interpolation & Extrapolation / Interpolate 1D.vi

LabVIEW's Interpolation 1D function is easy to use. The Interpolation 1D function has been placed on the block diagram in expanded form to show the terminals in Figure 6.40. The **method** has been set to "spline" and the number of iteration

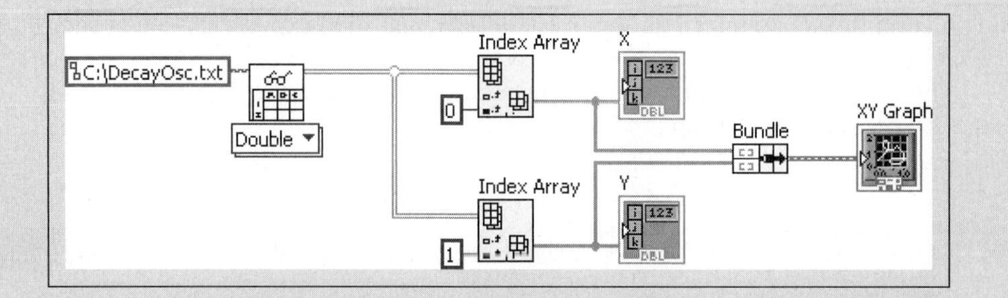

Figure 6.38
The DecayOsc VI (block diagram) with the *X* and *Y* arrays graphed.

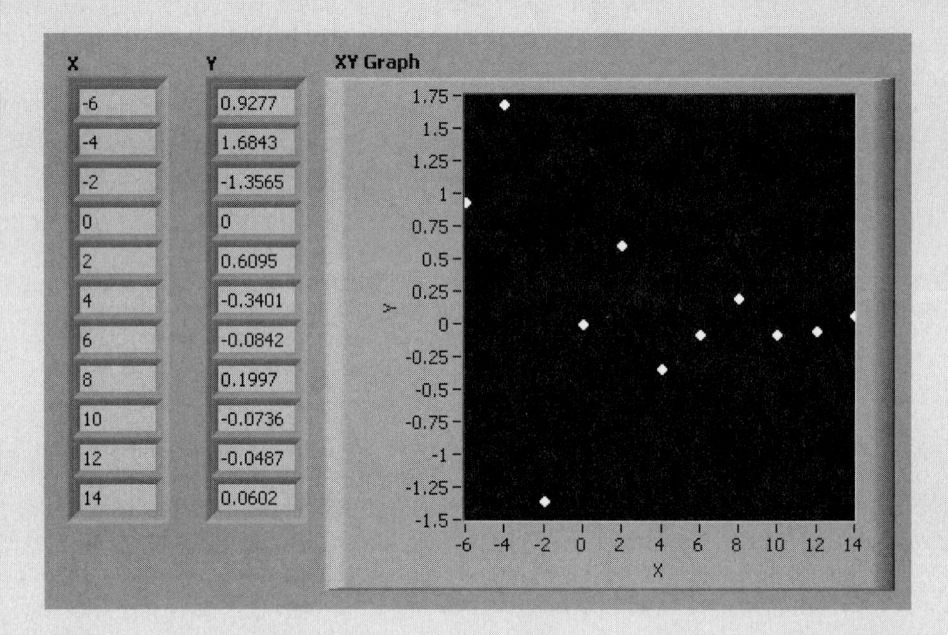

Figure 6.39
The DecayOsc VI (front panel) showing the original *X* and *Y* data.

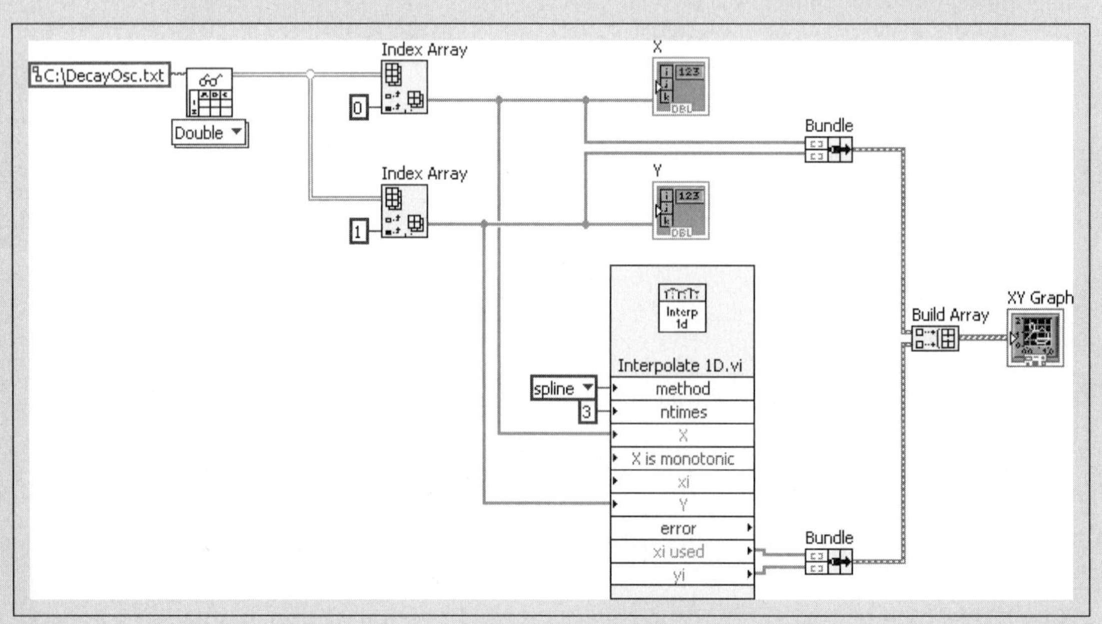

Figure 6.40
Adding spline interpolation to the DecayOsc VI.

passes, **ntimes**, has been set to 3. The original X and Y values are sent into the function as inputs, and the interpolated values (xi *used*, yi) are sent from the function, bundled, and sent to the XY Graph. The interpolated values are plotted on the front panel, as shown in Figure 6.41.

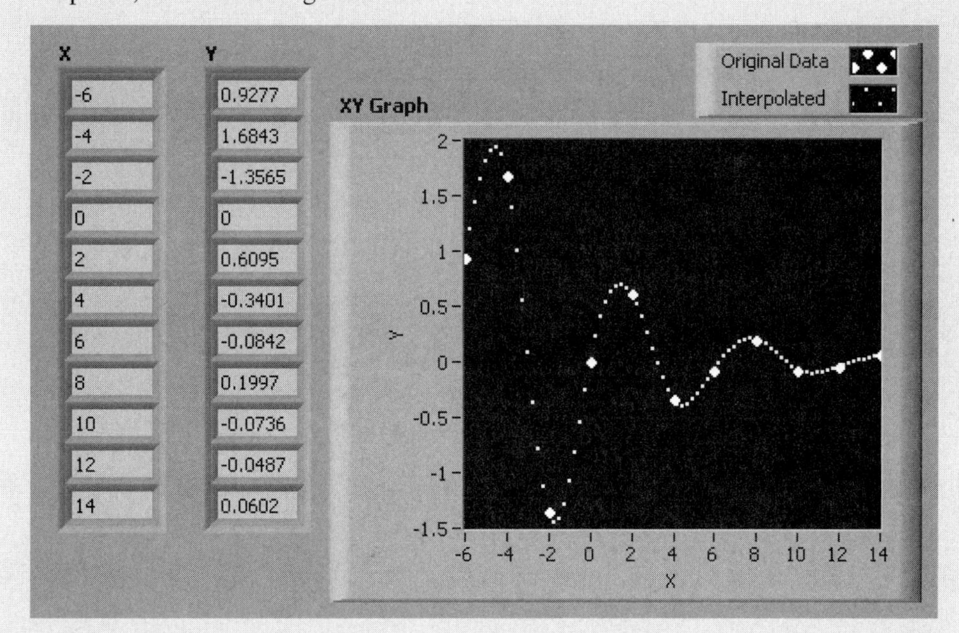

Figure 6.41
The interpolated points superimposed on the original data.

The last step is to use the Write Spreadsheet File function to send the results back to Excel, so that they can be compared to the calculated decaying oscillatory curve. The completed block diagram is shown in Figure 6.42.

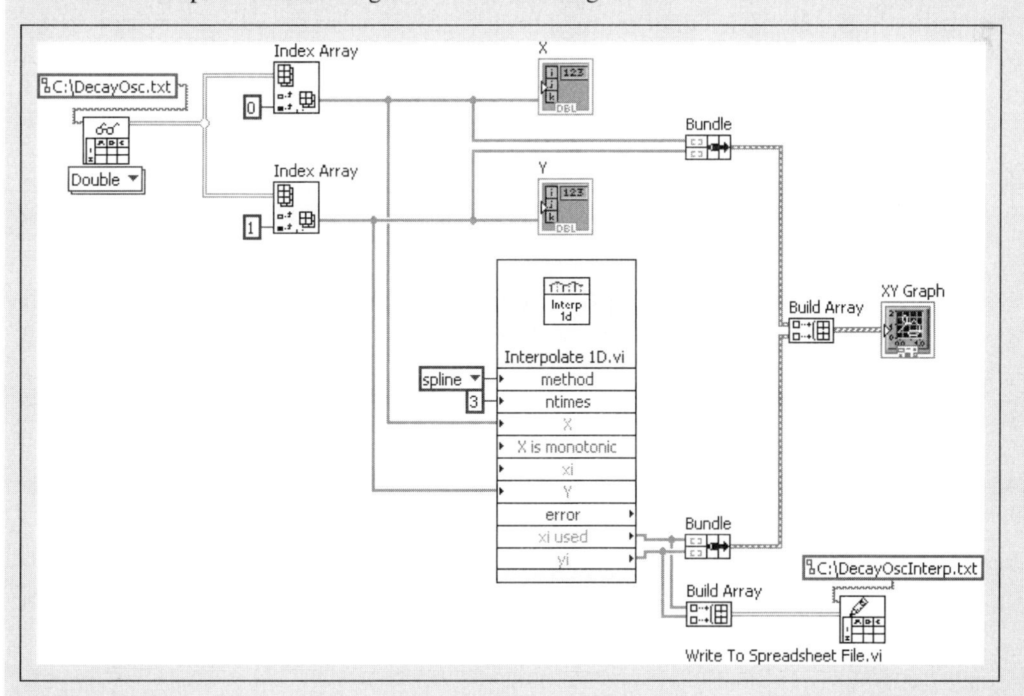

Figure 6.42
The completed DecayOsc VI block diagram with interpolated values sent to .txt file.

All that remains is to open the DecayOscInterp.txt file in Excel, and plot the interpolated results with the calculated curve. The result is shown in Figure 6.43. You can see that the interpolated values fit well where there were at least a couple of original points per peak, but the interpolation missed the first peak badly.

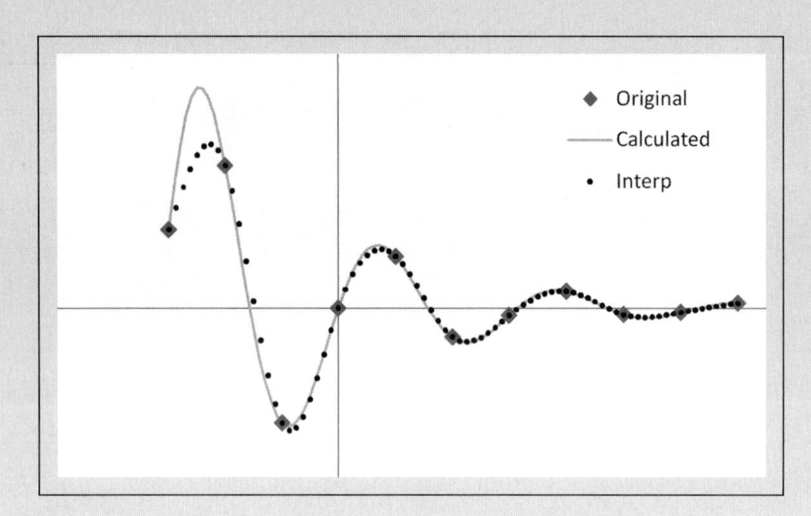

Figure 6.43
Comparing the interpolated results with the calculated decaying oscillation curve.

6.6.1 Reading the First Row or Column from a Text File

Some LabVIEW controls and functions (especially graphing controls) require 1D arrays as inputs. When data are read from a .txt file using the **all rows** output on the Read From Spreadsheet File function, the output is a 2D array. A 2D array is not compatible with controls that require 1D arrays as inputs (but you can use the Index Array function to pull a single column or row from a 2D array, as shown in Figures 6.29 and 6.35).

If you only need the first row or first column from the text file, you can use the **first row** output on the Read From Spreadsheet File function. Using the **first row** output allows you to get a 1D array directly, without needing the Index Array function—but it only works for the first row or column in the text file.

- **Reading One Row**—By default, the **first row** output on the Read From Spreadsheet File function will read one row from the .txt file.
- **Reading One Column**—Send a True into the **transpose** input on the Read From Spreadsheet File function to read one column from the .txt file.

The steps involved in reading a single column of values from a .txt file are as follows:

1. Place the Read From Spreadsheet function on the block diagram.
2. Connect a 1D array indicator to the **first row** terminal.
3. Send True to the **transpose** terminal. (This step is needed to read a column instead of a row.)
4. Run the VI. Select the .txt file containing the values when prompted.

As an example, the spreadsheet shown in Figure 6.44 has a column of values that we want to bring into LabVIEW to chart.

	A	B
1	-74.2032	
2	-45.003	
3	-27.2899	
4	-16.5426	
5	-10.0179	
6	-6.0502	
7	-3.62686	
8	-2.12928	
9	-1.1752	
10	-0.5211	
11	0	
12	0.521095	
13	1.175201	
14	2.129279	
15	3.62686	
16	6.050204	
17	10.01787	
18	16.54263	
19	27.28992	
20	45.00301	
21	74.20321	

Figure 6.44
Spreadsheet data to be read into LabVIEW.

Step 1: Place the Read From Spreadsheet function on the block diagram
The Read From Spreadsheet function is found in the Programming Group:

> **Functions Palette / Programming Group / File I/O Group / Read From Spreadsheet File**

In preparation for reading the values from the spreadsheet (saved as file Sinh Data.txt), the Read From Spreadsheet function is placed on the block diagram. In Figure 6.45 the icon has been expanded to show all terminals. The terminals we will use in later steps have been indicated.

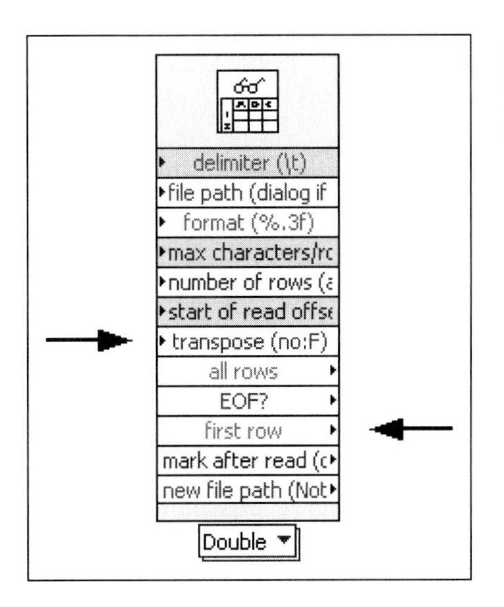

Figure 6.45
Block diagram showing expanded Read From Spreadsheet File icon.

Step 2: Connect a 1D array indicator to the first row terminal
Right-click on the first row terminal and select **Create / Indicator** from the pop-up menu (illustrated in Figure 6.46).

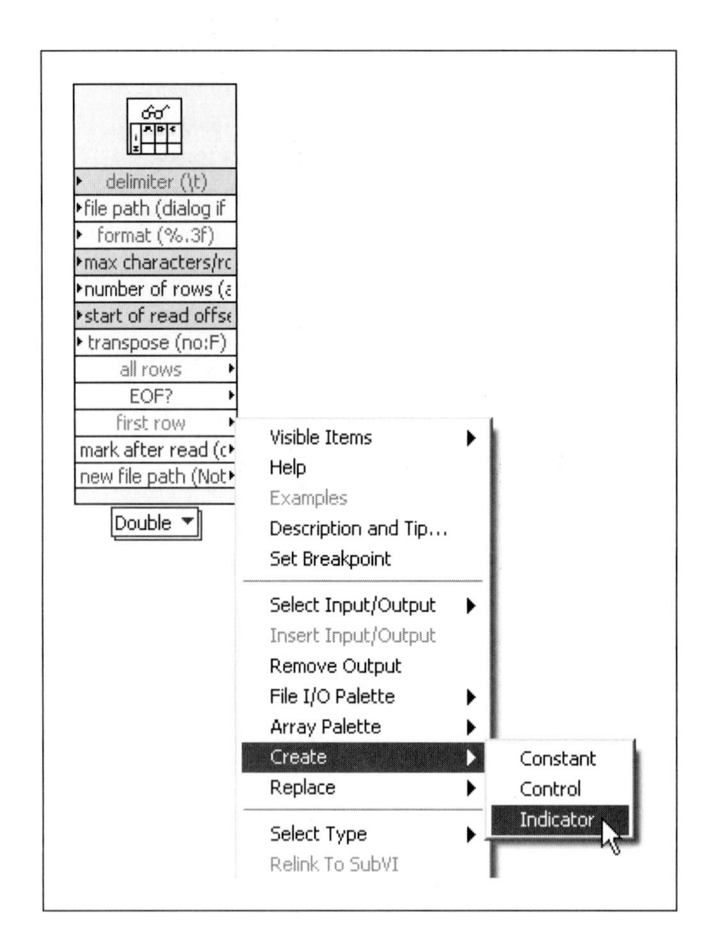

Figure 6.46
Connecting a 1D array indicator to the first row terminal.

Step 3: Send True to the transpose terminal
Since the spreadsheet data are in a column instead of a row, we need to instruct LabVIEW to read a column. This is done by sending a True to the **transpose** terminal.

To connect a Boolean constant to the **transpose** terminal, right-click on the **transpose** terminal and select **Create / Constant** from the pop-up menu. Then slide the switch on the Boolean constant (with the mouse) to output True. The result is shown in Figure 6.47.

Step 4: Run the VI
Run the VI to read the .txt file. LabVIEW will open the Choose File to Read dialog, shown in Figure 6.48. Once the file has been selected, LabVIEW will read the data into a 1D array (as a row). This is illustrated in Figure 6.49.

By default, 1D arrays are imported as rows, but any 1D array in LabVIEW can be presented as either a row or a column. Simply resize the

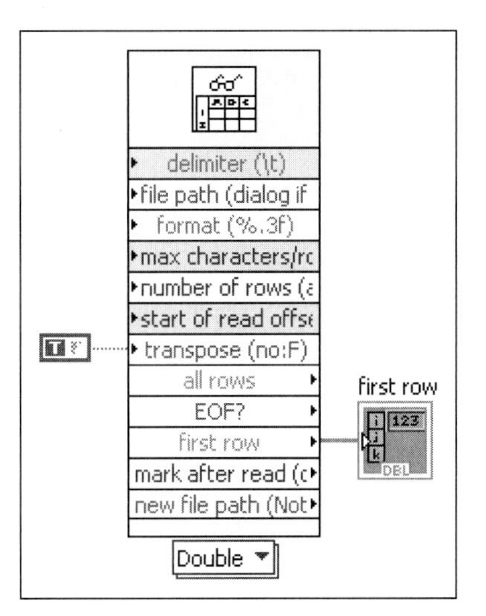

Figure 6.47
A value of True is sent to the transpose terminal.

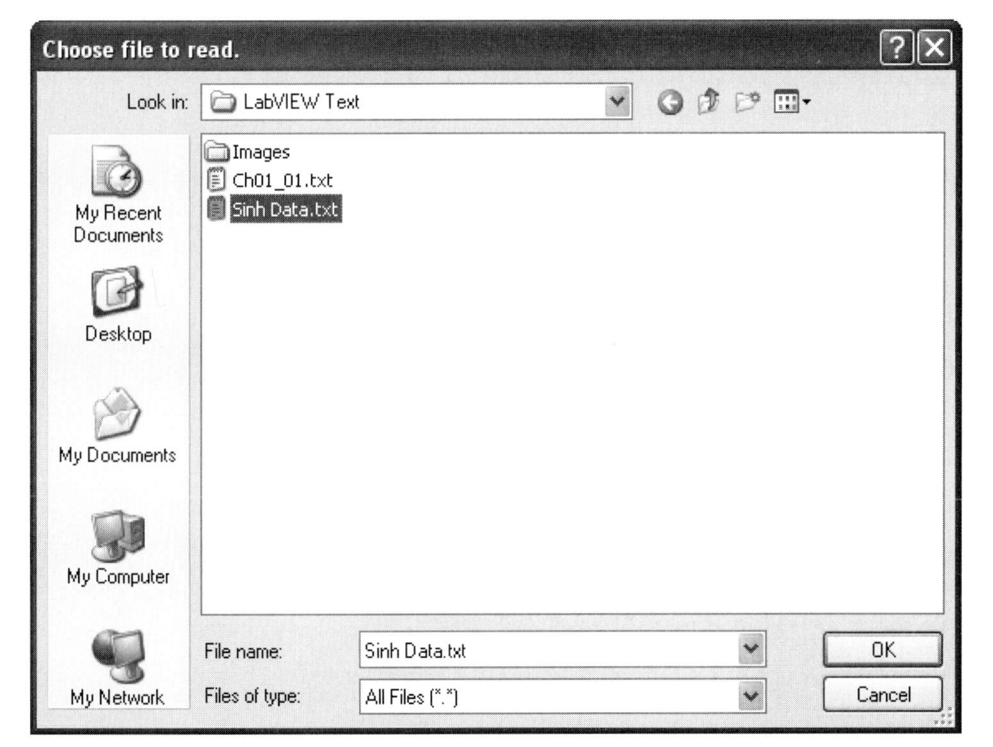

Figure 6.48
The Choose file to read dialog is used to identify the .txt file containing data.

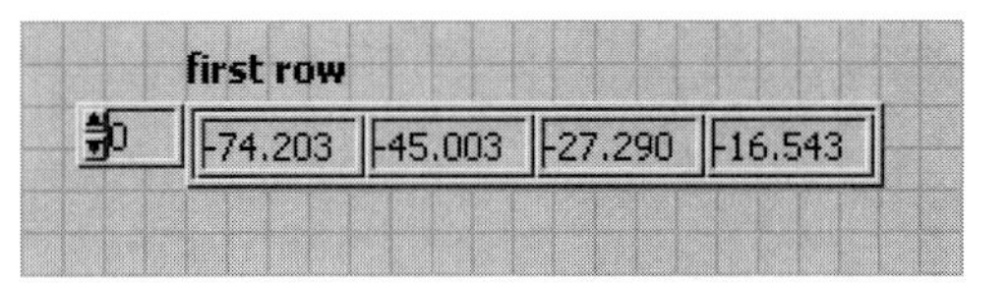

Figure 6.49
The imported data as a 1D (row) array.

array to show the values as a column (optional). In Figure 6.50 the array has been resized to display as a column, and the values have been sent to a Waveform Chart control for plotting.

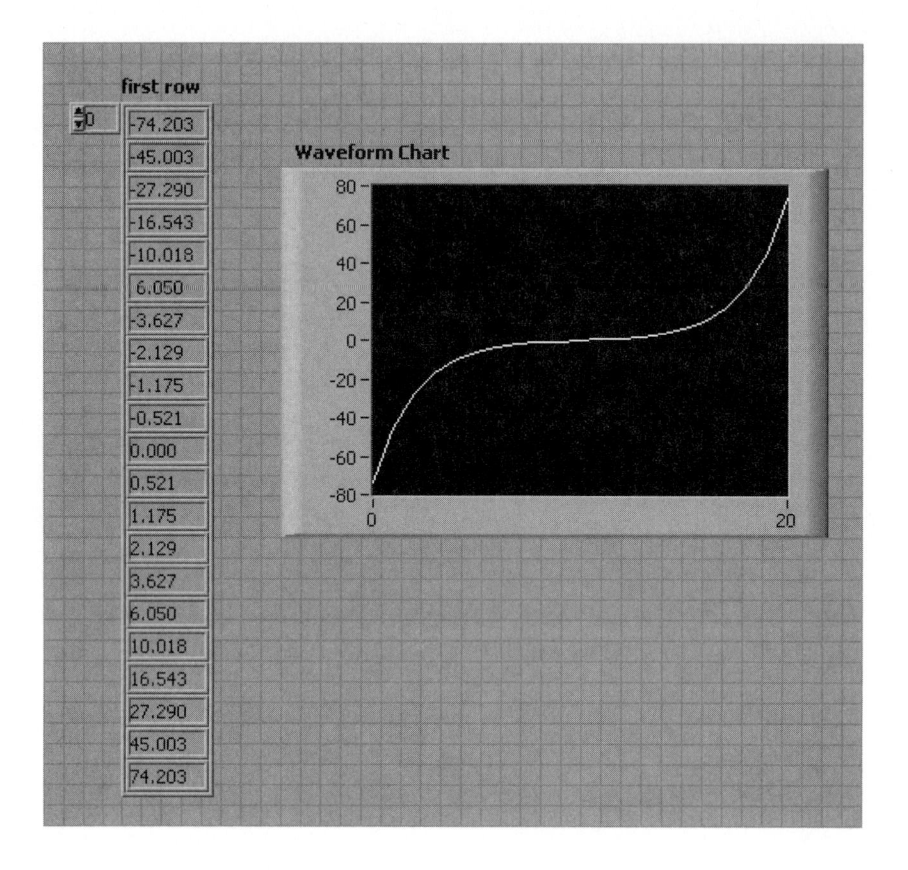

Figure 6.50
The imported data as a 1D (column) array, charted using a Waveform Chart control.

At this point the spreadsheet data are in the VI, but displayed in an indicator. The data are accessible by splicing into the wire running from the first row terminal to the array indicator. This is how the data were sent to the Waveform Chart control in Figure 6.51.

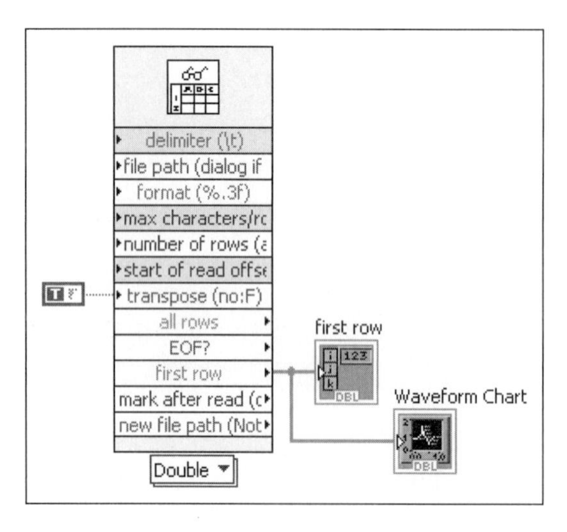

Figure 6.51
Block diagram of the data import and graphing VI.

KEY TERMS

binary files
data type selector
datalog files
delimiter
engineering notation
Express VI
floating point notation
format string
measurement file (.lvm)
scientific notation
spline interpolation
spreadsheet file
text files (.txt)
waveform

SUMMARY

Data File Types

- **Text Files**—stored in standard alphanumeric characters, very interchangeable, and can be large
- **Binary Files**—stored using binary values, fast, but specific to each software product
- **Datalog Files**—LabVIEW data storage files

Writing LabVIEW Data to a Spreadsheet File

Function: **Write to Spreadsheet File** (see Figure 6.52)

Function Palette / Programming Group / File I/O group / Write to Spreadsheet File

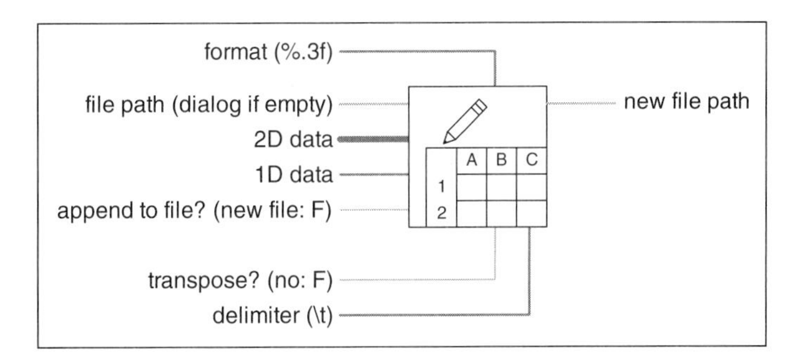

Figure 6.52
Connection pane for Write to Spreadsheet File function.

Format Strings

- Begins with the "%" character
- "^" (caret, [Shift 6])—engineering notation
- "#"—drop trailing zeros
- period and number (e.g., ".3")—number of decimal places to include
- final letter—notation style
  - f—floating point
  - e—scientific notation
  - g—general format

Examples:

%.3f (default)	12.345
%.7f	12.3450000
%#.7f	12.345
%e	1 .234500e1
%.3e	1.235e1 (rounded)
%^.3e	12.345e0
%g	12.345000
%#g	12.345

Writing LabVIEW Data to a Measurement File
Measurement files are text files used to save the data collected using a data acquisition system. Saved data include data values with time information plus header information.

Function: **Write to Measurement File**—an Express VI, opens a dialog box when used

> **Functions Palette / Programming Group / File I/O Group / Write to Measurement File**

Reading a LabVIEW Measurement File
Function: **Read From Measurement File**—an Express VI, opens a dialog box when used

> **Functions Palette / Programming Group / File I/O Group / Read From Measurement File**

Reading a Spreadsheet File in LabVIEW
Function: **Read From Spreadsheet File**

> **Functions Palette / Programming Group / File I/O Group / Read From Spreadsheet File**

Process:

1. Get the data values (no text column headers) into an Excel worksheet.
   (a) Make sure it is the only worksheet in the workbook.
   (b) The top-left value should be in cell A1.
2. Save the Excel file as a .txt file.
3. Place a Read From Spreadsheet File function on a LabVIEW block diagram.
4. Select the data type of the values to be read from the file.
5. Use a string constant to specify the format string that should be used to read the values, if desired. The default format string is "%.3f".
6. Use a string constant to specify the path name if desired. If the path name is omitted, LabVIEW will ask the user to choose the file at run time.
7. Add an indicator to the block diagram to display the read values.

Pulling Single Columns or Rows from 2D Arrays
Function: **Index Array**

> **Functions Palette / Programming Group / Array Group / Index Array**

- To select a single row, wire the row index input and leave the column index input unwired.
- To select a single column, wire the column index input and leave the row index input unwired.

Using Text File Data to Initialize a Control

1. Read the text file and show the results in an array indicator.
2. Change the array indicator to a control.
3. Delete the Read From Spreadsheet File function and any broken wires.
4. Make the current data the default data for the control.

Reading the First Row or Column from a Text File

1. Place the Read From Spreadsheet function on the block diagram.
2. Connect a 1D array indicator to the **first row** terminal.
3. Send True to the **transpose** terminal. (This step is needed to read a column instead of a row.)

SELF-ASSESSMENT

1. By default, how many decimal places are included in values written to text files by LabVIEW?
   ANS: Three (default format string is "%.3f")
2. If you want to save five decimal places when saving data to a text file, what format string should you specify?
   ANS: "%.5f"
3. How do you tell LabVIEW to drop trailing zeros when writing to a file?
   ANS: Include "#" in the format string. For example: "%#.5f"
4. What happens if you use a LabVIEW function that writes values to a text file, but you do not provide a file path?
   ANS: LabVIEW will show the Choose File to Write dialog at run time.
5. What is the difference between a .txt file and a .lvm file?
   ANS: LabVIEW measurement files (.lvm) include a header containing information about how and when the data were collected as well as the data set. Text files (.txt) do not contain the header information.
6. What is a delimiter?
   ANS: A delimited is a character (often a tab, comma, or space) placed between numeric values in a file. Delimiters are used when the file is read to determine where one value ends and the next begins.
7. Can the following file types be opened in Excel?
   - .txt
   - .lvm

   ANS:
   - .txt—YES, automatically
   - .lvm—YES, will probably have to explicitly "open with" Excel
8. Can LabVIEW open an Excel workbook?
   ANS: No. However, the Excel worksheet (first worksheet in the workbook) can be saved as a .txt file from Excel, and LabVIEW can open that .txt file.
9. When LabVIEW reads data using the Read From Spreadsheet function, the data must be assigned a data type. How do you specify the data type associated with the imported data?
   ANS: There is a drop-down selector under the Read From Spreadsheet node that is used to select the data type of the imported values.
10. What steps are required to read one column from a .txt file into a 1D array?
    ANS:
    a. Place the Read From Spreadsheet function on the block diagram.
    b. Right-click on the **first row** terminal and select **Create / Indicator** from the pop-up menu.
    c. Right-click on the **transpose?** terminal and select **Create / Constant** from the pop-up menu. Set the True/False constant to True. (This step is needed to read a column instead of a row.)
    d. Run the VI. Select the .txt file containing the values when prompted.

PROBLEMS

1. Look at the LabVIEW measurement file shown in Figure 6.53 (only a small portion of the file is visible in the image) to answer the following questions:
   a. What are the units on the *Y* values?
   b. When were the data collected?
   c. How many samples were collected?
   d. What is the time interval between samples?
   e. What delimiter separates values in the data list?

Figure 6.53
LabVIEW measurement file opened in Excel.

	A	B	C	D
1	LabVIEW Measurement			
2	Writer_Version	0.92		
3	Reader_Version	1		
4	Separator	Tab		
5	Multi_Headings	Yes		
6	X_Columns	One		
7	Time_Pref	Absolute		
8	Operator	Ron Larsen		
9	Date	2/1/2009		
10	Time	14:23.4		
11	***End_of_Header***			
12				
13	Channels	1		
14	Samples	1000		
15	Date	2/1/2009		
16	Time	14:24.4		
17	Y_Unit_Label	Volts		
18	X_Dimension	Time		
19	X0	0.00E+00		
20	Delta_X	0.001		
21	***End_of_Header***			
22	X_Value	Voltage (Filtered)	Comment	
23	0	1.06E-06		
24	0.001	6.75E-06		
25	0.002	1.95E-05		
26	0.003	3.39E-05		
27	0.004	3.74E-05		
28	0.005	1.63E-05		
29	0.006	-3.92E-05		

2. Use LabVIEW to convert the following temperature values to °C, and then export the values to Excel and create a time (*x* axis) and temperature (*y* axis) plot.

Start Time: 0
Time Interval: 5 seconds

T (°F)
75
109
133
152
166
176
183
189
193
196
198

3. A typical Excel data file might look something like Figure 6.54 , with headings, units, and graphs. The contents must be stripped down to nothing but data values before saving as a .txt file and importing the values into LabVIEW.
The Excel file (CalibData.xls) shown in Figure 6.54 is available at the text's website: www.chbe.montana.edu/LabVIEW

   a. Download the Excel file (or create something similar).

   b. Create a .txt file suitable for transferring the data into LabVIEW.

   c. Import the data into a matrix or array control in LabVIEW.

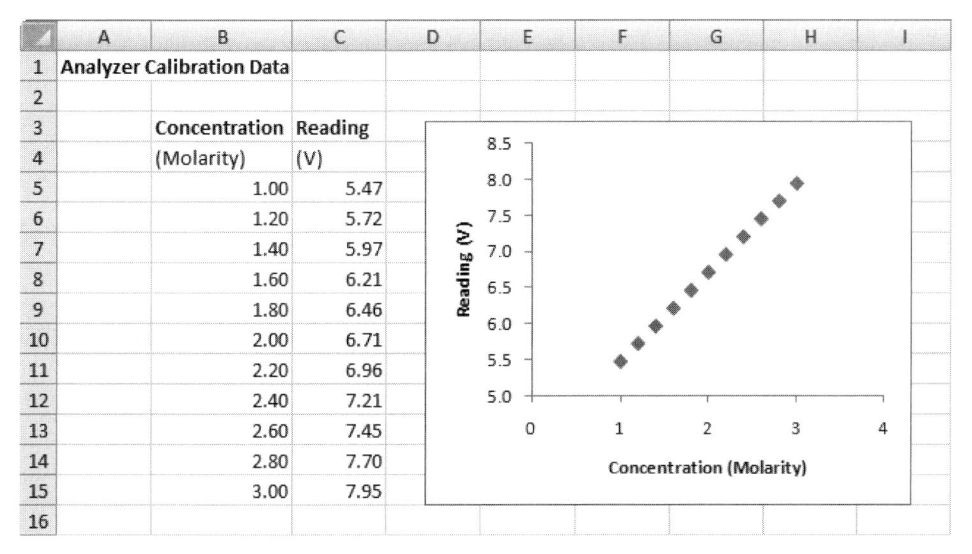

	A	B	C
1	Analyzer Calibration Data		
2			
3		Concentration	Reading
4		(Molarity)	(V)
5		1.00	5.47
6		1.20	5.72
7		1.40	5.97
8		1.60	6.21
9		1.80	6.46
10		2.00	6.71
11		2.20	6.96
12		2.40	7.21
13		2.60	7.45
14		2.80	7.70
15		3.00	7.95
16			

Figure 6.54
Calibration data in Excel.

4. The Excel file shown in Figure 6.55 contains a system of simultaneous linear equations in matrix form. Prepare two .txt files from the Matrices.xls file, one for the coefficient matrix and one for the right-hand-side vector. Import the matrices into LabVIEW and solve the equations simultaneously.

	A	B	C	D	E	F	G	H	I	J
1	Coefficient Matrix							Right-Hand-Side Vector		
2										
3		1	3	5	7	4		15		
4		2	7	4	1	5		-22		
5		3	1	7	9	2		-10		
6		5	1	6	8	4		17		
7		8	2	5	3	1		-31		
8										

Figure 6.55
Excel file containing coefficient and right-hand-side matrices.

5. The VI shown in Figure 6.56 (front panel) and Figure 6.57 (block diagram) calculates sine data in the range of $0-2\pi$. Recreate the VI, and then save the calculated sine values to a .txt file using a format that displays five decimal places and does not truncate trailing zeros.

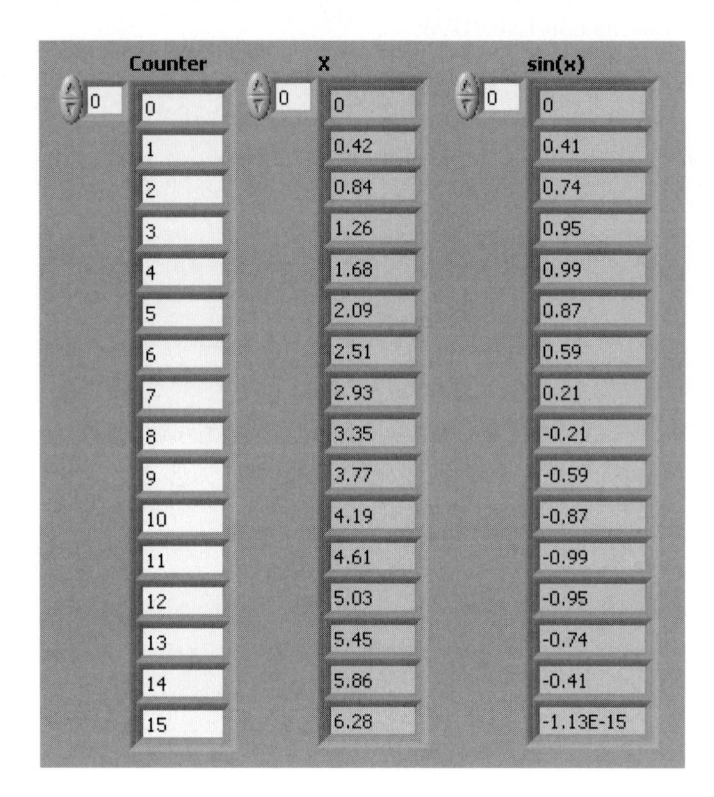

Figure 6.56
Front panel of a VI that calculates sine values.

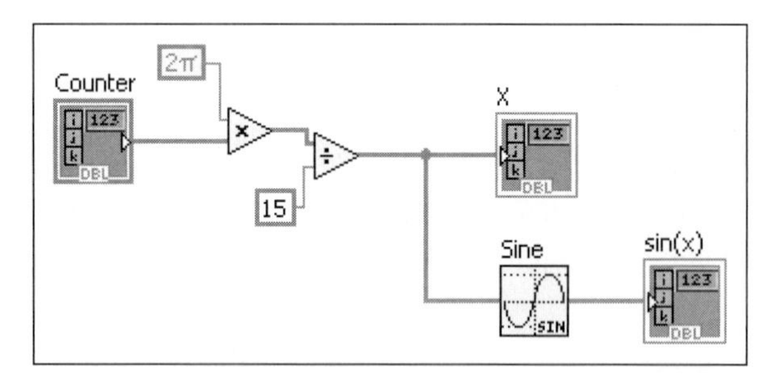

Figure 6.57
Block diagram of Sine VI.

6. The VI shown in Figure 6.58 calculates Bessel function (jn(x)) data and writes the data to a .txt file, specifically, Bessel jn.txt. The file is available on the text's website.

   Download the Bessel jn.txt file, and then create a VI that reads the file and sends the output to a Waveform Graph control. The Waveform Graph control can be found at

   Controls Palette / Modern Group / Graph Group / Waveform Graph

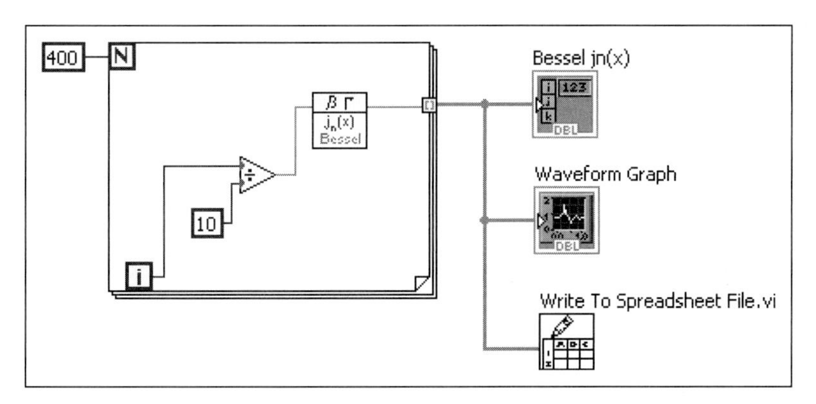

Figure 6.58
VI to generate and save Bessel function values.

From the Waveform Graph, what is the value of jn(x) when $x = 100$? Check your answer by evaluating jn(100) using the Bessel jn(x) function located at

> **Functions Palette / Mathematics Group / Elementary & Special Functions Group / Spherical Bessel Function jn(x).vi**

CHAPTER 7

Graphing with LabVIEW

Objectives

After reading this chapter, you will know:

- the difference between a chart and a graph in LabVIEW
- how to use Waveform Charts to display LabVIEW waveform data
- how to use Waveform Graphs to display array data
- how to modify the features of a LabVIEW graph
- how to create a data array for graphing
- how to use LabVIEW XY Graphs for 2D plotting
- how to use LabVIEW 3D graphs
- how to get an image of a LabVIEW graph into a report

7.1 INTRODUCTION

LabVIEW provides a number of charting and graphing options for producing data plots; only some of the more common will be presented in this chapter.

Chart Types	Graph Types
• Waveform Chart	• Waveform Graph
• Intensity Chart	• XY Graph
	• Intensity Graph
	• Digital Waveform Graph
	• Mixed Signal Graph
	• 3D Surface Graph
	• 3D Parametric Graph
	• 3D Curve Graph

The chart and graph indicators are available on the Controls Palette, either in the Modern Group or the Express Group.

In LabVIEW,

- A *chart* is able to receive individual data points and continuously update the presentation of the data. A *Waveform Chart*, for example, is typically used during data acquisition to monitor the data as they are being collected. (Waveform Charts can also receive entire arrays.)
- A *graph* receives a complete data set (as an array) before preparing the graphical display.

LabVIEW provides

- Graph and chart indicators for 1D plotting (assuming uniformly spaced x values)
- Graph indicator for 2D plotting (XY Graph)
- Graph indicators for 3D plotting

7.2 USING WAVEFORM CHARTS

The Waveform Chart is LabVIEW's basic charting indicator for data acquisition. In its most fundamental form, the Waveform Chart receives data one point at a time and displays the data values on a graph. This is termed **point-by-point** plotting.

Controls Palette / Modern Group / Graph Group / Waveform Chart

Controls Palette / Express Group / Graph Indicators Group / Waveform Chart

7.2.1 Waveform Charts—Point-by-Point Plotting

In the block diagram shown in Figure 7.1, every time the For Loop cycles a new iteration value is divided by 100 and sent to the Sine function. The Sine function sends one sin(x) value to the Waveform Chart each time the For Loop cycles.

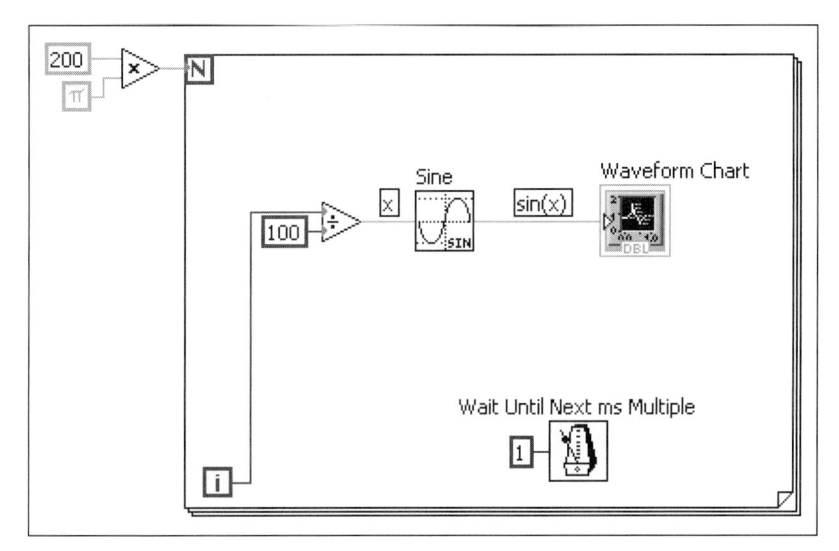

Figure 7.1 The Waveform Chart indicator used for point-by-point plotting.

When the VI is run, the Waveform Chart starts displaying values, as illustrated in Figure 7.2. Here, the image was captured less than half way through the For Loop, so less than a full sine cycle is shown.

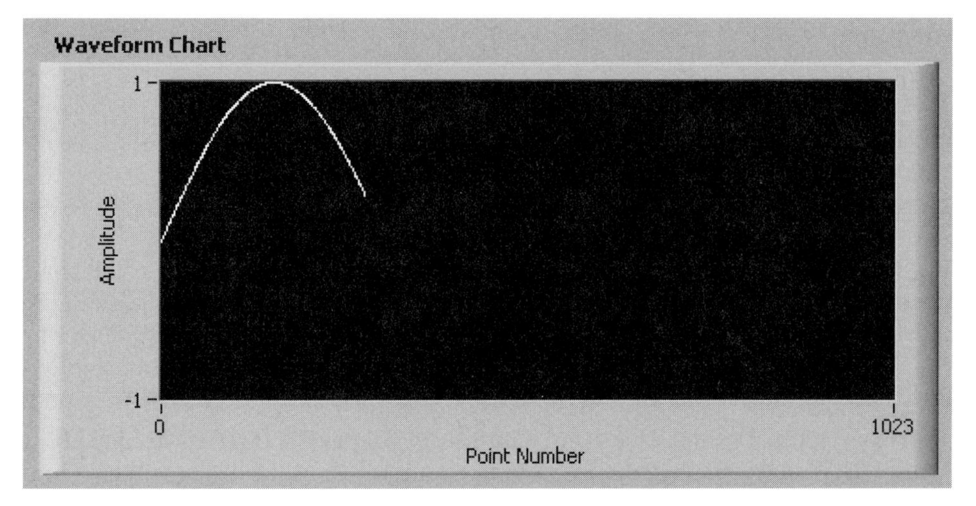

Figure 7.2 The Waveform Chart—For Loop interrupted.

If the For Loop is allowed to complete (the VI terminates), a full sine cycle is graphed as shown in Figure 7.3.

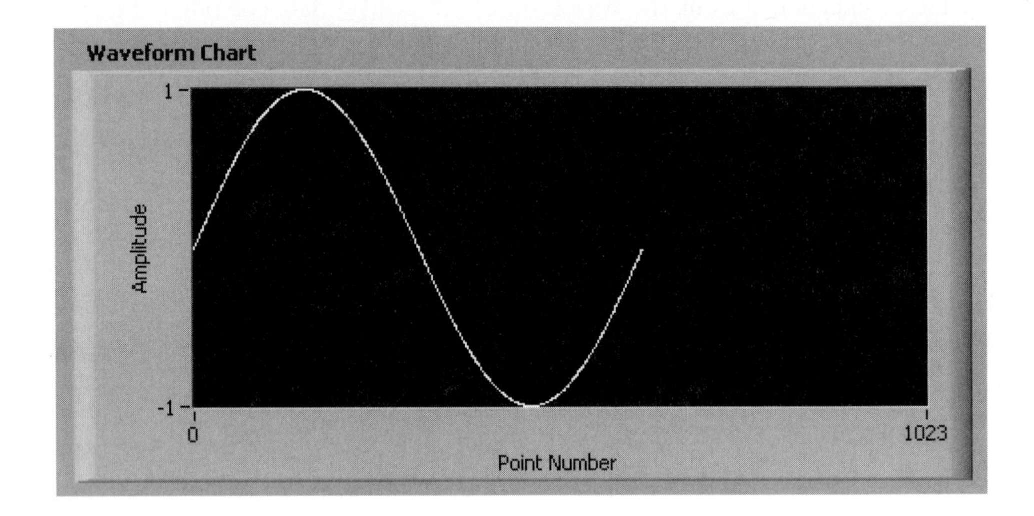

Figure 7.3
Waveform Chart—one complete For Loop.

If the VI is run again, the Waveform Chart continues to display the sine wave, as shown in Figure 7.4. This is because all but the last value plotted are stored in the *chart history*, an array of previously plotted values; only the last point is sent to the chart each time the For Loop cycles.

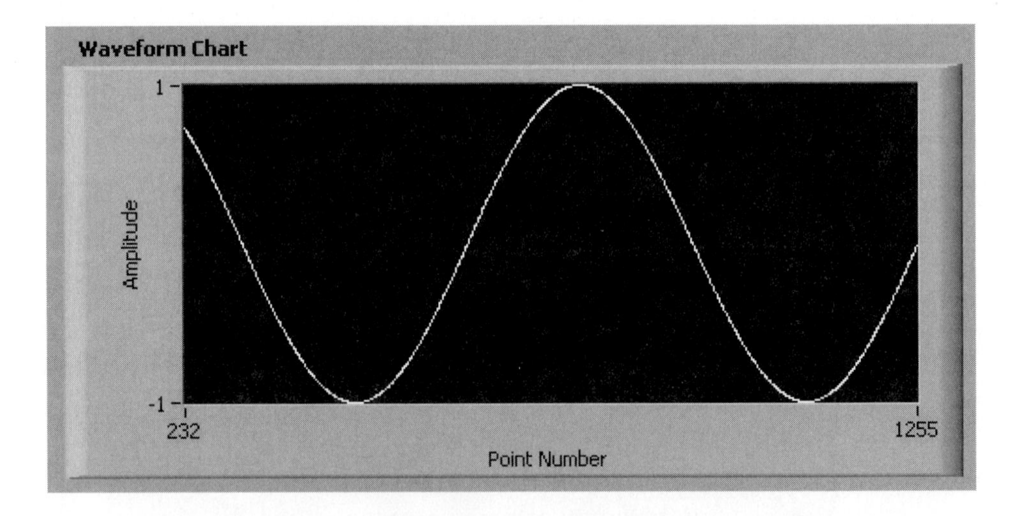

Figure 7.4
Waveform Chart—running the VI again continues the curve.

By default the chart history stores 1024 points. This is why the X axis limits in Figure 7.2 were set from 0 to 1023, to show as many values as stored in the chart history. When the VI was run the second time, the number of points sent to the Waveform Chart exceeded 1024 and the curve started scrolling to the left. That is why the X axis limits in Figure 7.4 are shown as 232–1255. Only the most recent 1024 values are displayed; the original 232 values (0–231) have scrolled off the display.

Note: You can change the size of the chart history. Right-click on the chart and select **Chart History Length . . .** from the pop-up menu.

By default, Waveform Charts scroll to the left when the chart history is full, but this action is based on the *update mode*, and there are three options:

- **Strip Chart** (default)—the data scrolls smoothly to the left. The current value is always at the right end of the curve.
- **Scope Chart**—the chart display fills, then clears, and fills again as often as needed.
- **Sweep Chart**—the chart does not clear, but a sweep line moves across indicating the new data plotted to the left of the sweep and the old values remaining on the right side of the sweep line.

The three update options are shown in Figure 7.5. The block panel is shown in Figure 7.6.

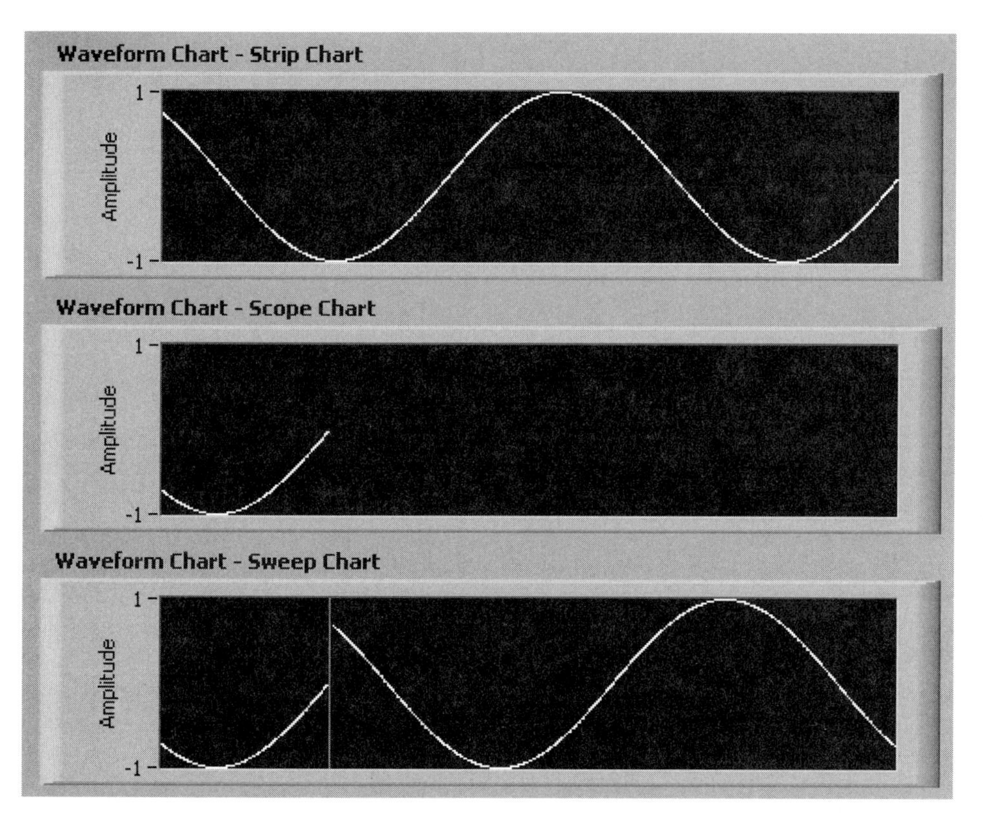

Figure 7.5
Waveform Chart update options.

The Update Modes can be selected in two ways:

- From the chart pop-up menu: **Advanced / Update Mode**
- From the Chart Properties dialog (right-click on chart, select **Properties** to open the dialog): **Appearance Panel / Update Mode** drop-down list

You can right-click on the chart and change the Update Mode when a waveform chart is running, too.

7.2.2 Waveform Charts—Array Plotting

The Waveform Chart can also be used for *array plotting*. With array plotting, an entire array of values is sent to the Waveform Chart for plotting (not point by point).

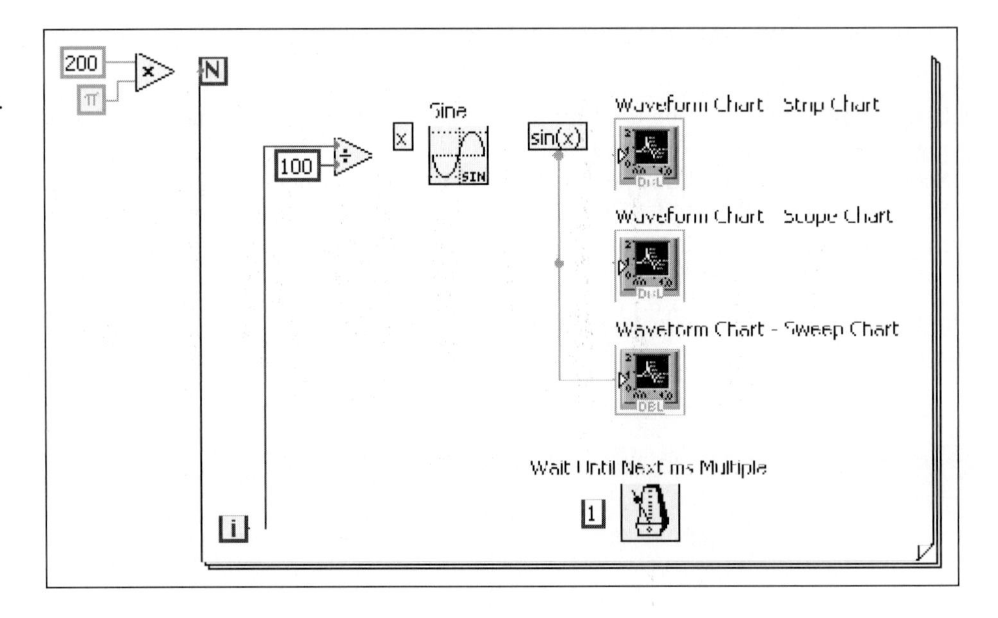

Figure 7.6
Block diagram used to demonstrate update modes.

Note: Waveform Charts require 1D arrays rather than matrices. This is because a matrix in LabVIEW is always 2D, even if you only use one column or one row. If you attempt to send a matrix into a LabVIEW graphing function, you will generate a data type mismatch error.

In the VI shown in Figure 7.7 the Waveform Chart has been moved outside the For Loop boundary. The sin(x) values are built into an array at the For Loop boundary, and then the entire array is sent to the Waveform Chart when the For Loop terminates.

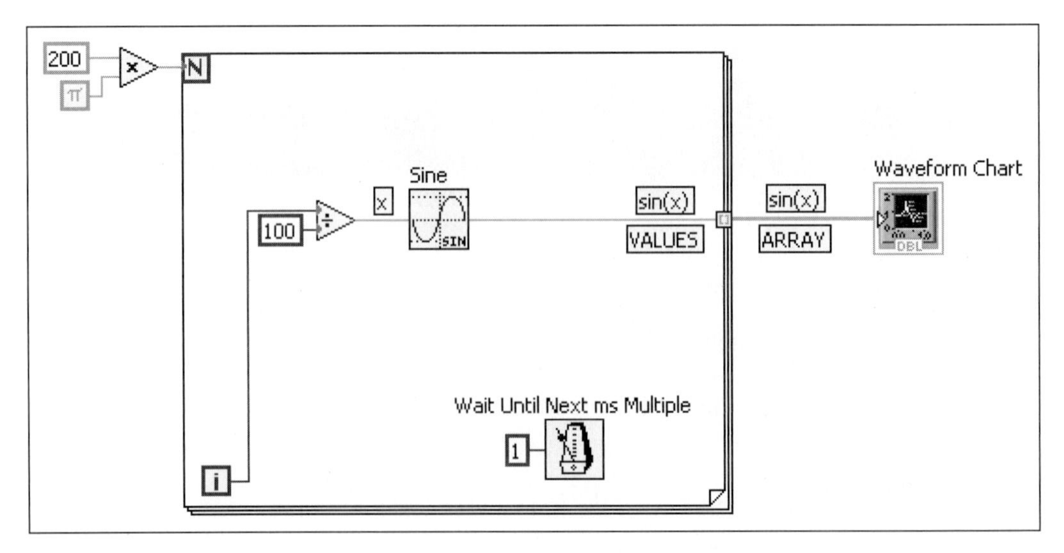

Figure 7.7
Waveform Chart used to display an array.

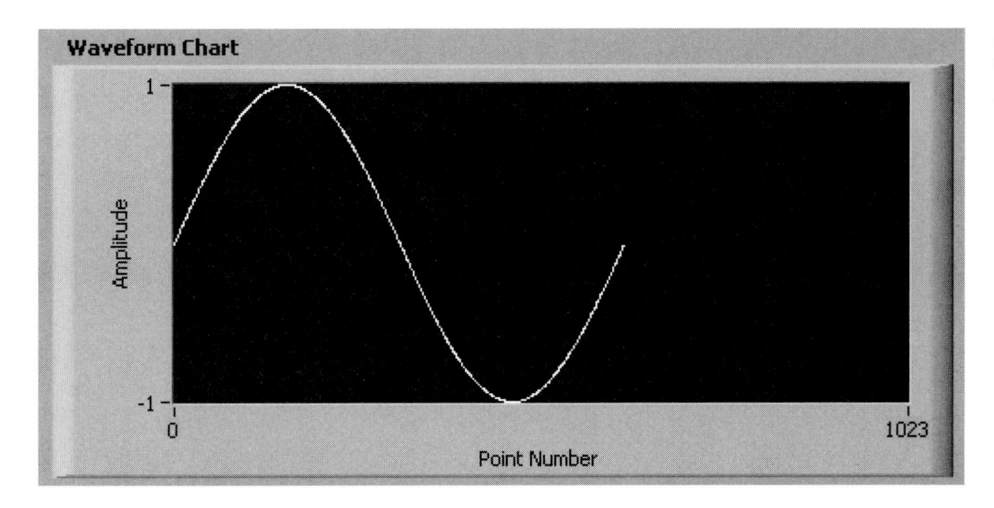

Figure 7.8
Waveform Chart display—array plotting.

The graph in figure 7.8 looks very similar to what was observed with point-by-point plotting except that when the VI is run you won't see the plot being drawn; it just pops on the screen when the array is sent to the Chart. The chart history is still the same, and the curve will scroll (in batches, not smoothly) when the chart history is full.

So, the Waveform Chart works for either point-by-point plotting, or array plotting.

7.2.3 Using the Waveform Chart with Data Acquisition

The two common modes of data acquisition are reading analog input values point by point, and reading in an entire array of values. These coincide with the two ways that the Waveform Chart can display data—this is probably not a coincidence.

In Figure 7.9, a data acquisition system is being used to acquire data point by point, and a Waveform Chart is used to display the values as they are collected.

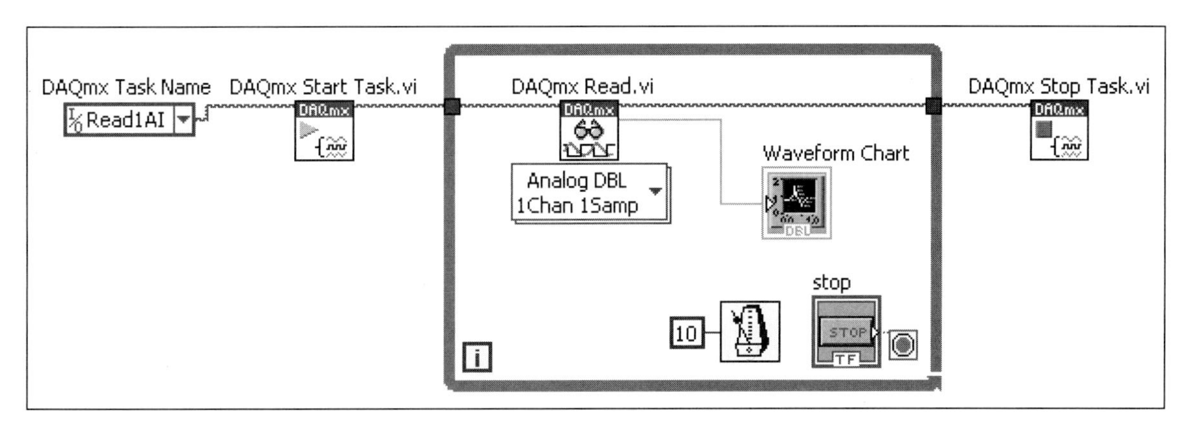

Figure 7.9
Block diagram of point-by-point data acquisition and plotting.

(Each time the While Loop cycles, another value is read from the data acquisition system.) The resulting chart is shown in Figure 7.10.

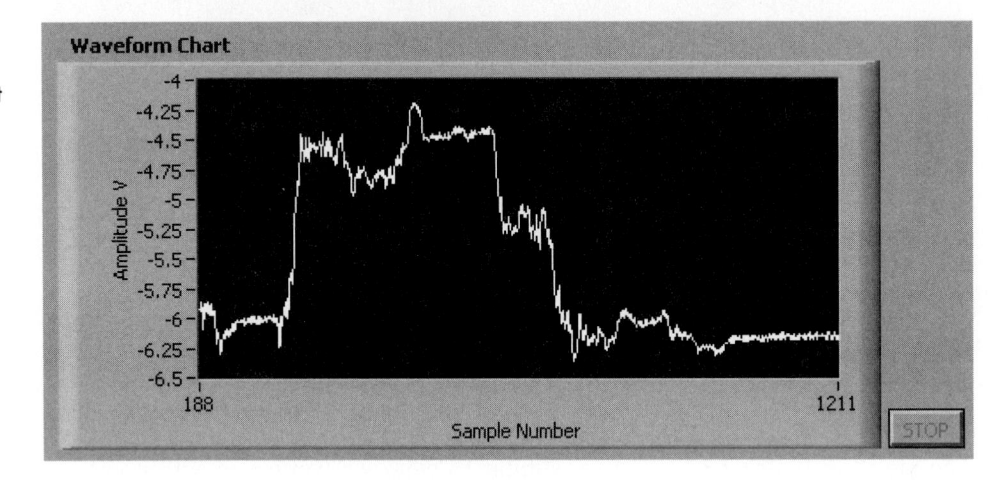

Figure 7.10
Waveform Chart used to display acquired data point by point.

Alternatively, data can be acquired in a batch and then plotted using a Waveform Chart, as shown in Figure 7.11 (block diagram) and Figure 7.12 (front panel).

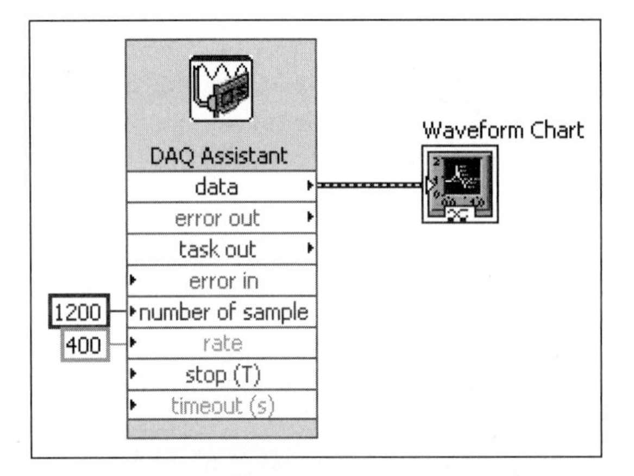

Figure 7.11
Block panel for data acquisition, batch mode, with array plotting.

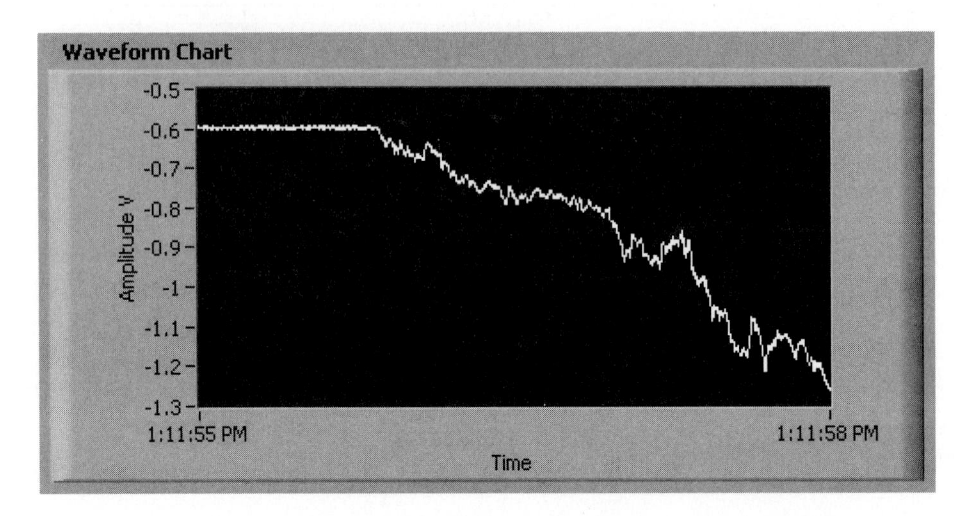

Figure 7.12
Acquired data, plotting with Waveform Chart using array plotting.

Notice, in Figure 7.12, that the X axis now shows sample time, not sample number. This is the result of sending a *signal* (includes time information) to the Waveform Chart instead of just the numerical values. The DAQ Assistant automatically outputs a signal.

The Waveform Chart can receive data in two ways: point by point, and array. Those two options correspond with the two common ways of acquiring data using data acquisition devices.

7.2.4 Displaying Multiple Curves on a Waveform Chart

It is common to acquire multiple channels of data, and the Waveform Chart can display multiple inputs in two ways:

- **Overlay Plots**—all curves shown on the same plot
- **Stacked Plots**—each curve shown in its own plot

The VI shown in Figure 7.13 shows the same data plotted both ways. The block diagram used to generate the plots is shown in Figure 7.14.

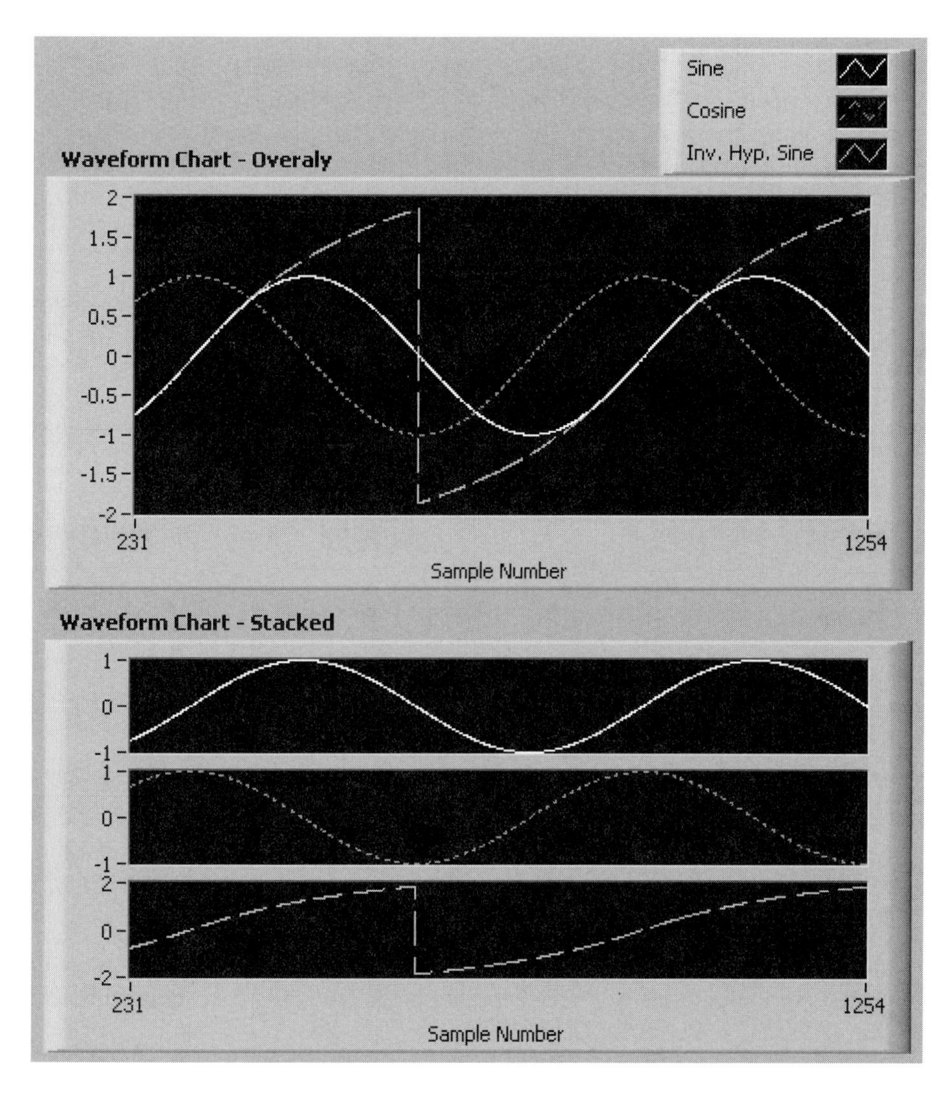

Figure 7.13
Waveform Chart options: Overlay Plots (top), Stacked Plots (bottom).

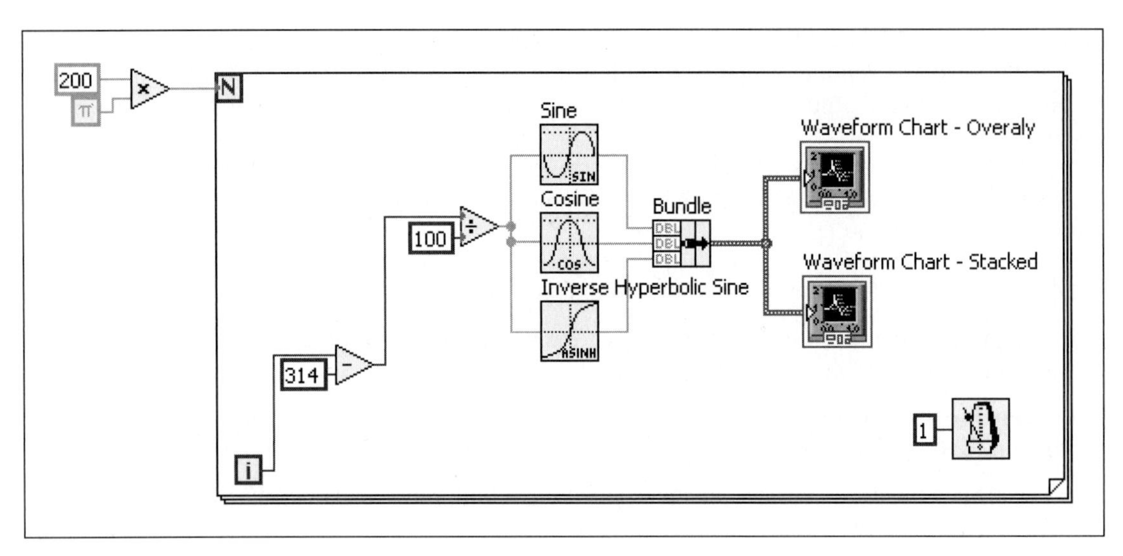

Figure 7.14
Block diagram for generating three curves for plotting.

You can pull the Waveform Charts outside the For Loop to attempt to plot using arrays instead of point by point. The block diagram for this is shown in Figure 7.15. As the front panel (Figure 7.16) shows, array plotting works for the Overlay Plot (build an array of arrays for plotting, and deselect transpose array option), but stacked plots are not an option when plotting arrays.

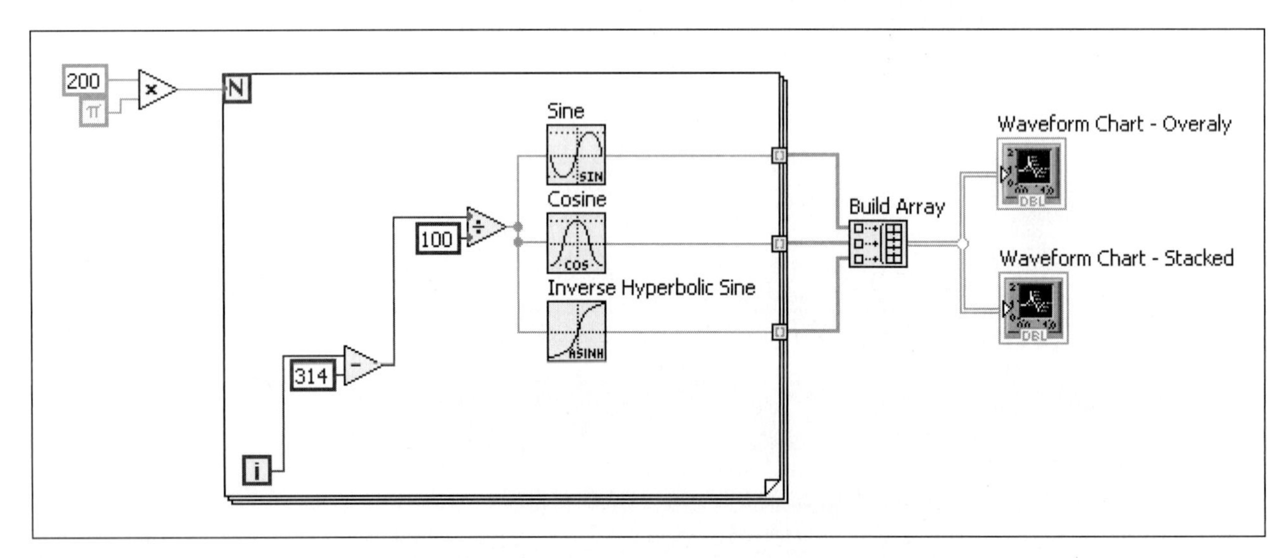

Figure 7.15
Block diagram for multi-curve plotting using arrays.

Conclusion: Put the Waveform Chart inside the loop for point-by-point plotting if you want stacked plots. If you need separate plots and you are working with arrays, use three separate Waveform Charts.

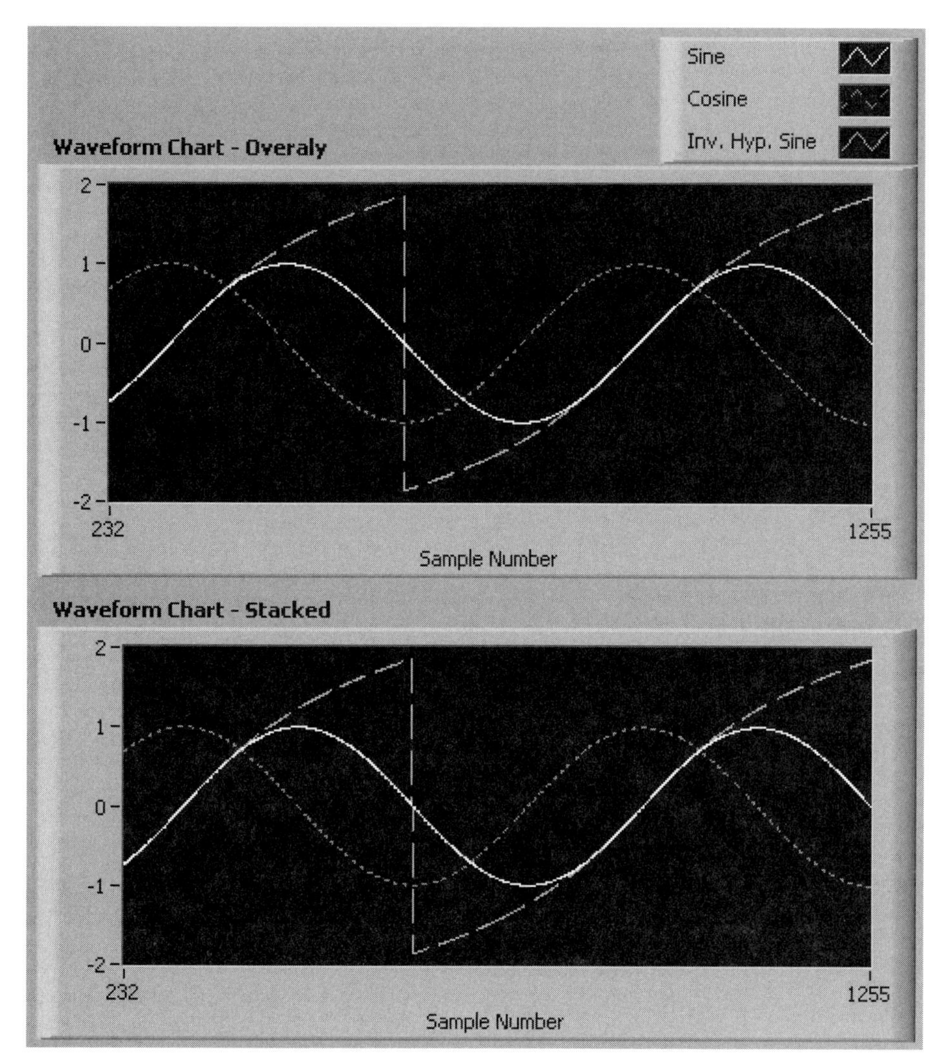

Figure 7.16
Stacked plots only work with point-by-point plotting (arrays used here).

APPLICATION

Build a LabVIEW Function Generator

To demonstrate the use of the Waveform Chart function, we will create a function generator that can create any of the following waveforms at run time.

- sine wave (signal type = 0)
- triangle wave (signal type = 1)
- square wave (signal type = 2)
- sawtooth wave (signal type = 3)

The LabVIEW function that creates these waveforms is called the Basic Function Generator, and it is available deep inside the Programming Group's Waveform Group:

> **Functions Palette / Programming Group / Waveform Group / Analog Waveform Group / Waveform Generation Group / Basic Function Generator**

The connection pane for the Basic Function Generator function is illustrated in Figure 7.17.

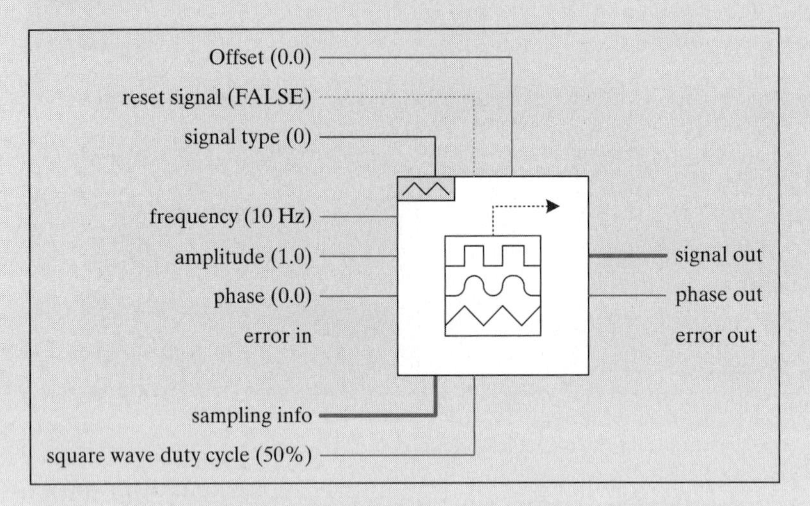

Figure 7.17
Connection pane for Basic Function Generator.

The front panel of the Function Generator VI is shown in Figure 7.18, and the block diagram is shown in Figure 7.19.

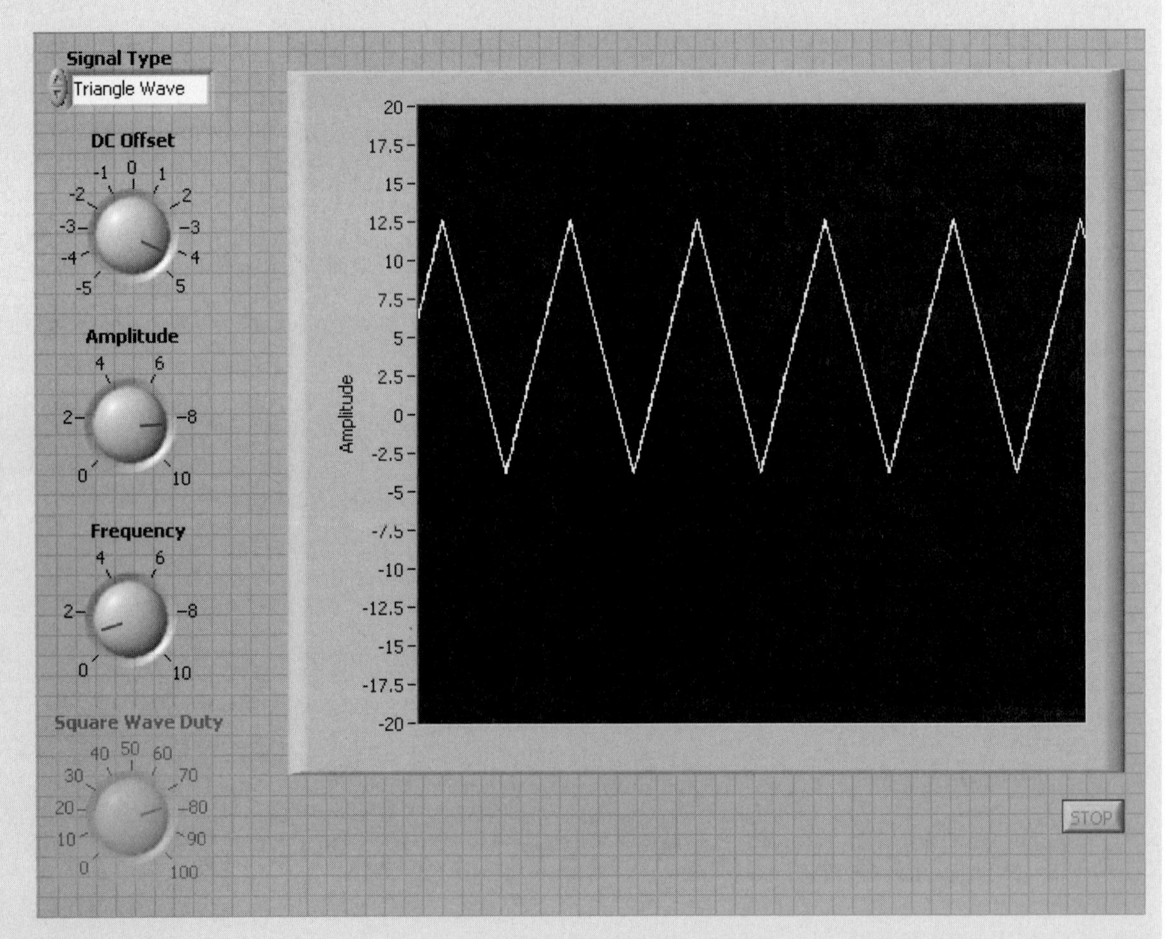

Figure 7.18
Function Generator VI front panel.

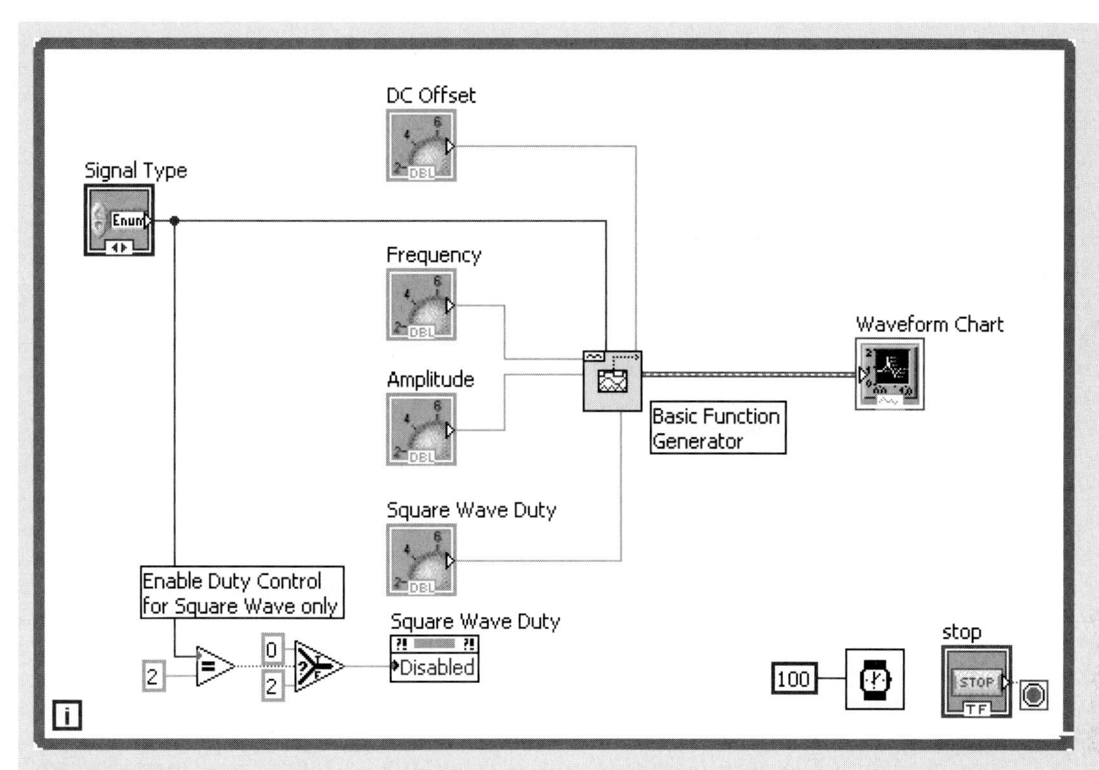

Figure 7.19
Block diagram for the Function Generator VI.

Notice, in Figure 7.19, that the Waveform Chart indicator has been used and the entire block diagram has been enclosed in a While Loop. This program runs continuously once it has started, and the Basic Function Generator continues to generate signal values appropriate for the selected waveform (e.g., sine, square, or triangle wave). The Waveform Chart indicator is designed to receive a waveform point by point and (when built into a While Loop) continuously update the display of the waveform.

Note: A Waveform Graph indicator cannot be used in this situation because the signal coming out of the Basic Function Generator is a series of point values, not the 1D array required by the Waveform Graph indicator.

The Square Wave Duty control (bottom-left corner of Figure 7.18) is used to adjust the percentage of time that the square wave is in the high position. The *square wave duty* only has meaning when the square wave has been selected. The portion of the block diagram reproduced in Figure 7.20 is responsible for enabling the Square

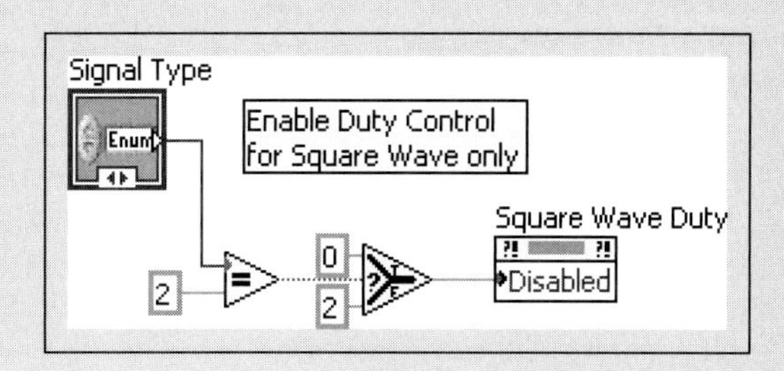

Figure 7.20
The portion of the Function Generator block diagram that enables the Square Wave Duty control.

Wave Duty control when “square wave” has been selected and disabling the control otherwise. Here’s how it works:

- First the Signal Type is compared against a constant (“2” is the signal type for “square wave”) and a TRUE is sent to the Select function when Signal Type = 2.
- The Select function sends either a “0” or a “2” to the Square Wave Duty control’s Disabled property node. Setting the Disabled property value to “0” enables the control, and setting the property value to “2” disables and grays the control.

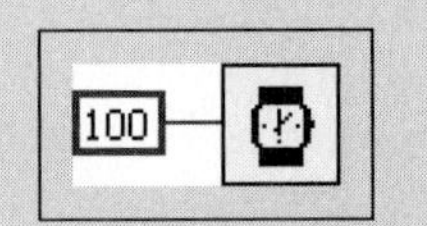

Figure 7.21
The Timer function is used to slow the While Loop.

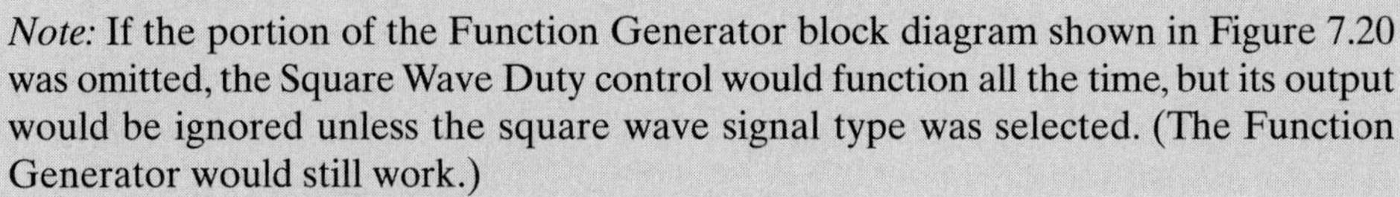

Note: If the portion of the Function Generator block diagram shown in Figure 7.20 was omitted, the Square Wave Duty control would function all the time, but its output would be ignored unless the square wave signal type was selected. (The Function Generator would still work.)

The Timer function (shown in Figure 7.21) puts a 100 ms wait inside the While Loop shown in Figure 7.19. This slows down the execution of the While Loop to improve the display of the waveform.

The Waveform Chart function is commonly used in LabVIEW to display a waveform that is changing with time. That waveform could be coming from a function generator (as in this example) or it could be a signal coming in from a data acquisition system.

7.3 USING WAVEFORM GRAPHS

LabVIEW’s *Waveform Graph* indicator is distinct from the Waveform Chart indicator in that Waveform Graphs must receive a complete array of values for plotting. There is no chart history, Waveform Graphs just display the array values that they receive as input. Waveform Graphs are never used for point-by-point plotting.

Controls Palette / Modern Group / Graph Group / Waveform Graph

Controls Palette / Express Group / Graph Indicators Group / Waveform Graph

Note: Waveform Graphs require 1D arrays rather than matrices. This is because a matrix in LabVIEW is always 2D, even if you only use one column or one row. If you attempt to send a matrix into a LabVIEW graphing function you will generate a data type mismatch error.

EXAMPLE 7.1

Calculate *y* array values given polynomial coefficients and *x* array values. Graph the *y* array.

Polynomial: $y = A + Bx + Cx^2 + Dx^3$
Coefficients: $A = 4, B = 3.7, C = -1.9, D = 0.17$
x array values: uniformly spaced integers between 0 and 10

First, the x array values are entered into a 1D array on the front panel, and a Waveform Graph indicator is placed on front panel, as shown in Figure 7.22.

The block diagram is shown in Figure 7.23.
The result is shown in Figure 7.24.

In Figure 7.23 you can see that most of the VI is dedicated to solving for the Y values. There are a number of ways that the Polynomial VI could be programmed. The next two figures show a couple of possible modifications to simplify the block diagram.

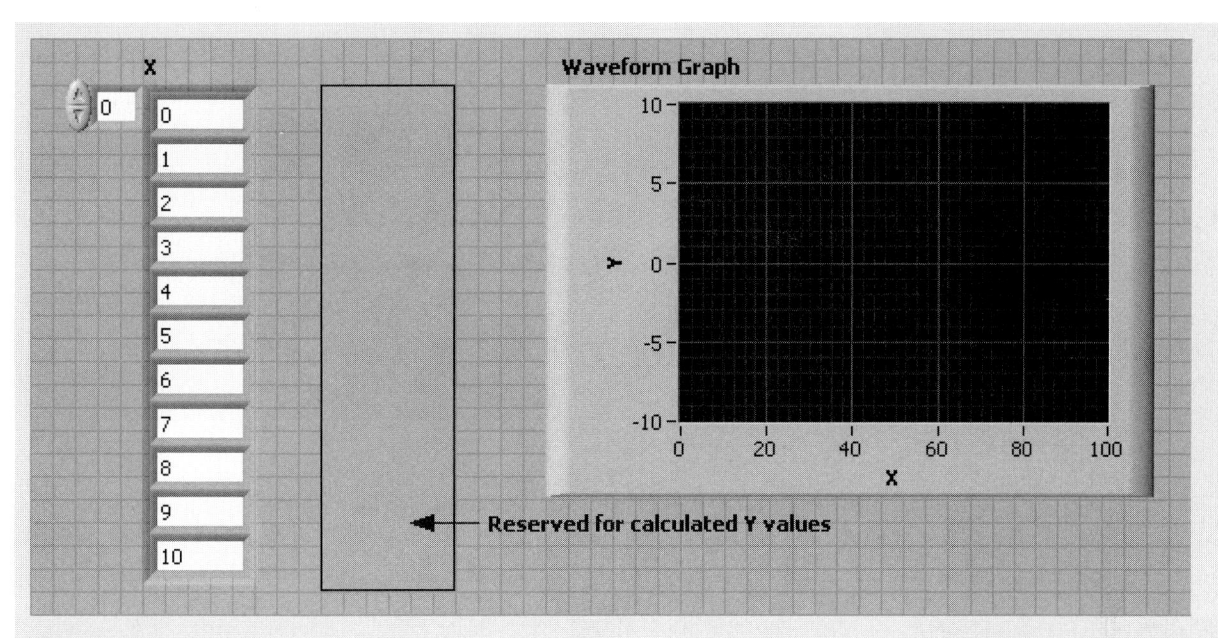

Figure 7.22
Front panel for Polynomial VI.

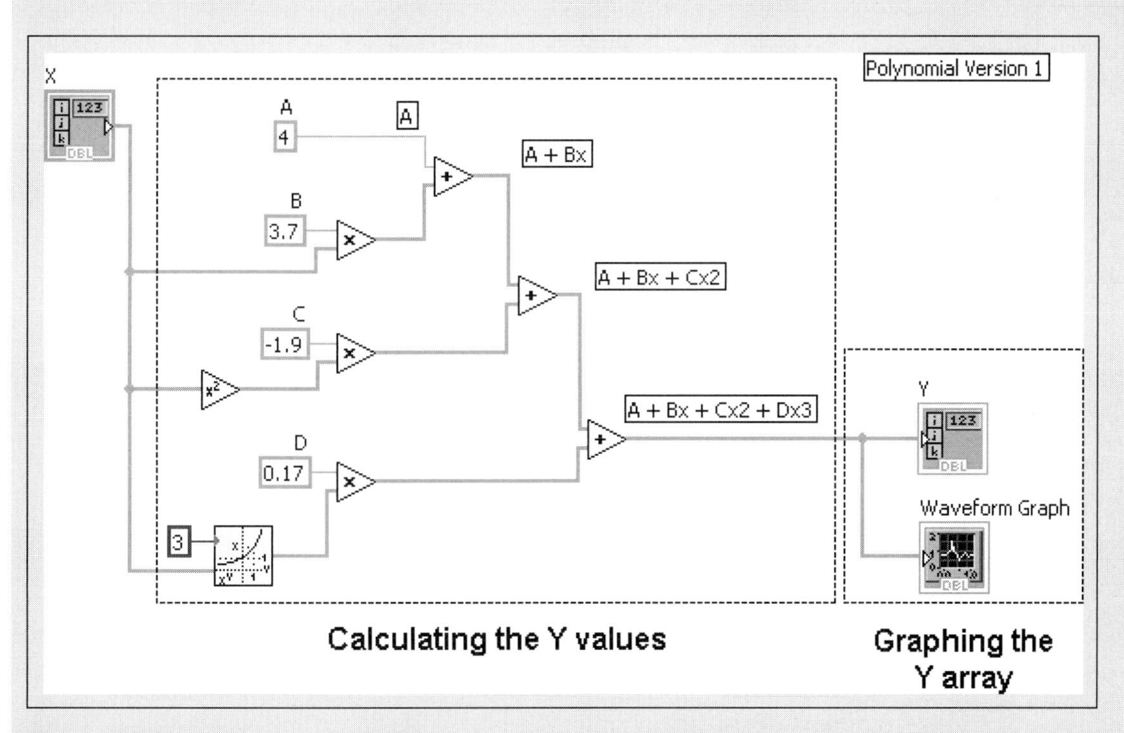

Figure 7.23
Block diagram of Polynomial VI (version 1, annotated).

In the second version of the Polynomial VI (see Figure 7.25), we have replaced all of the individual summations with a Compound Arithmetic function. The Compound Arithmetic function can be found in either of these locations:

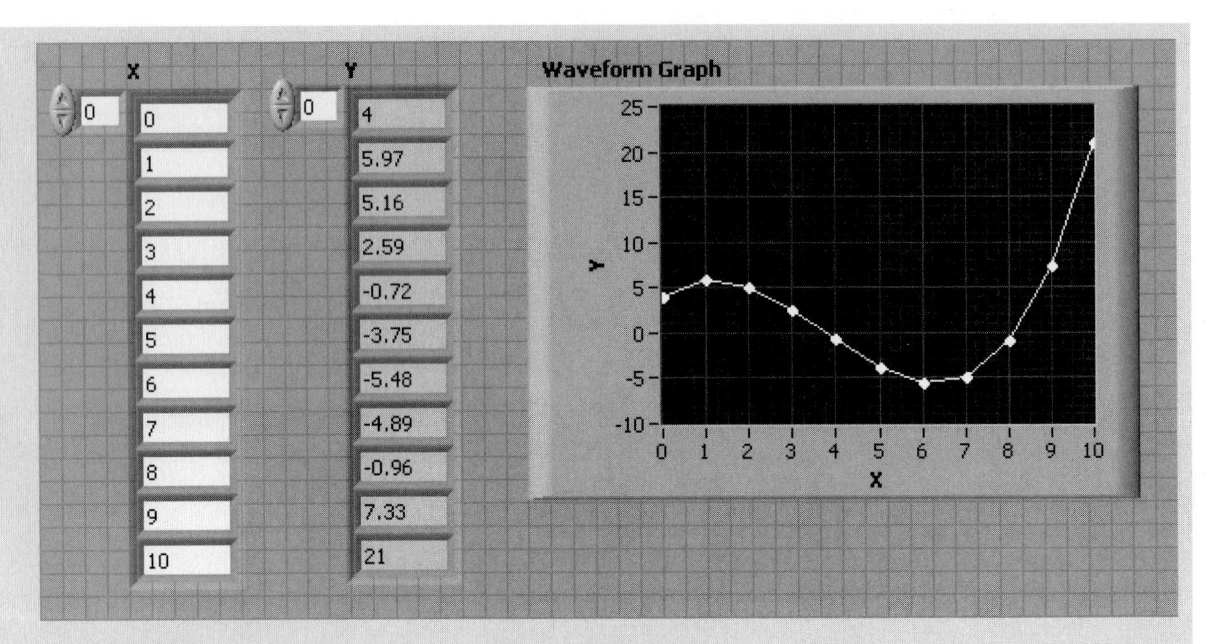

Figure 7.24
The calculated polynomial values and graph of the *Y* array values.

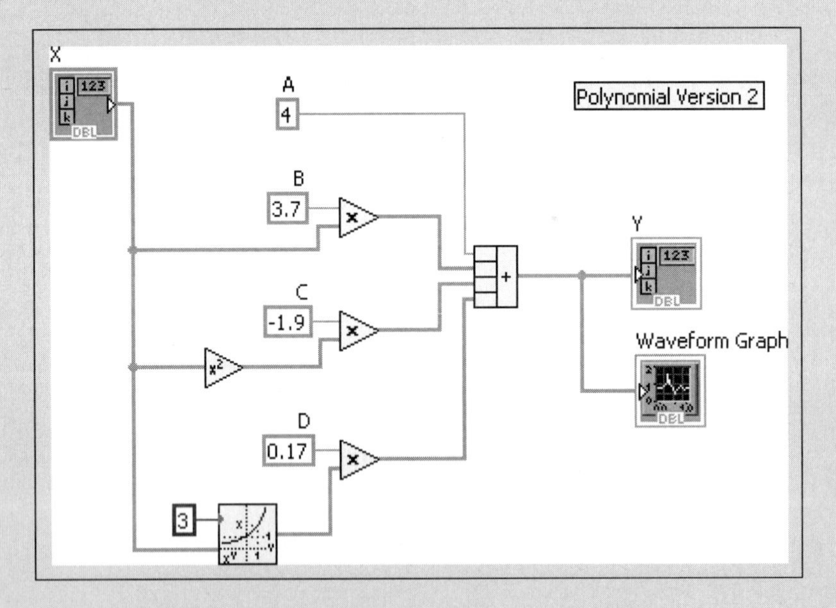

Figure 7.25
Second version of Polynomial VI using Compound Add function.

Functions Palette / Programming Group / Numeric Group / Compound Arithmetic function

Functions Palette / Mathematics Group / Numeric Group / Compound Arithmetic function

In the third version of the Polynomial VI (see Figure 7.26), we have replaced all of the math functions with a Formula Express VI. The formula was entered as 4 + 3.7* X − 1.9* X **2 + 0.17* X **3. The Formula Express VI can be found in

Functions Palette / Mathematics Group / Scripts and Formulas Group / Formula Express VI

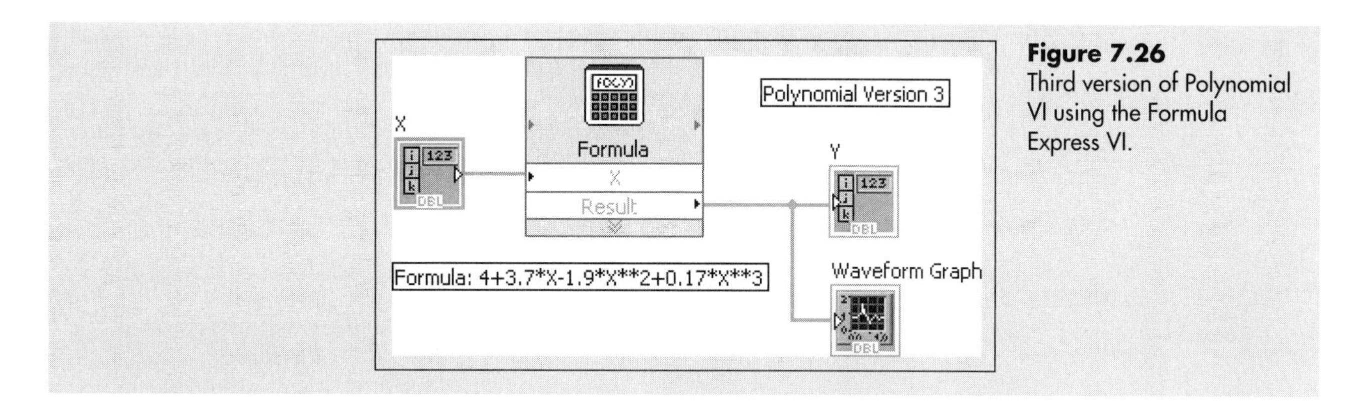

Figure 7.26
Third version of Polynomial VI using the Formula Express VI.

7.3.1 Comparing the Waveform Charts and Waveform Graphs

In the next VI, we will send the same array data to both a Waveform Chart indicator and a Waveform Graph indicator to observe how each handles the same data. The array contains only four values [2, 4, 4, 2]. The font panel (before running the VI) is shown in Figure 7.27, and the block diagram is shown in Figure 7.28.

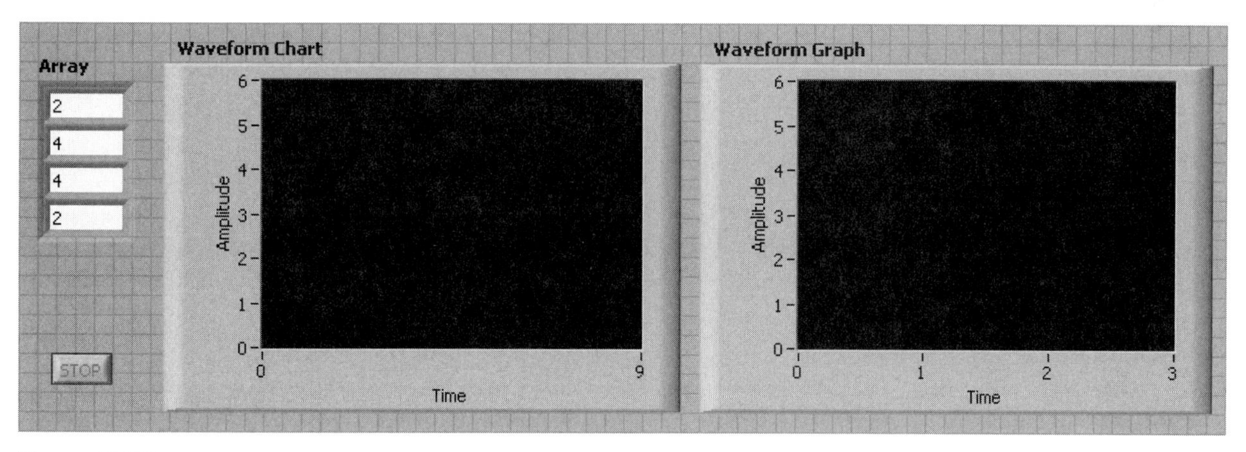

Figure 7.27
Front panel before running the VI.

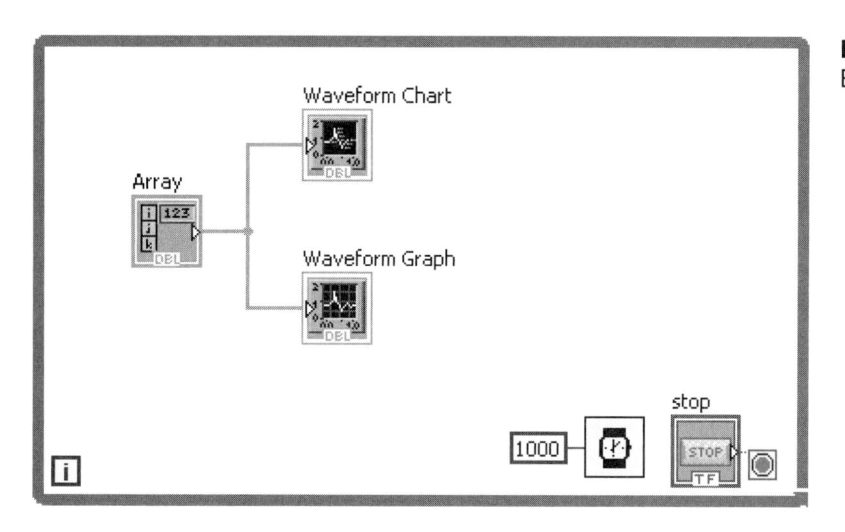

Figure 7.28
Block diagram of the VI.

As you can see in Figure 7.28, the 1D array is wired directly into both the Waveform Chart indicator and the Waveform Graph indicator; they both get exactly the same input.

Now, we run the VI. After one cycle through the While Loop, the front panel looks as shown in Figure 7.29; both the chart and the graph are presenting the same information.

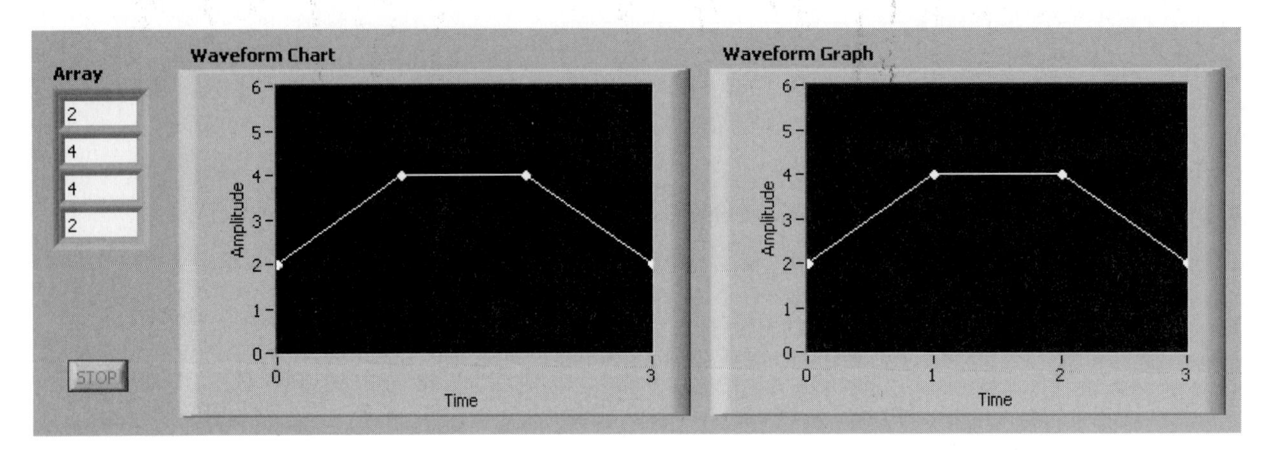

Figure 7.29
The front panel after one cycle through the While Loop.

The VI continues to run, and the next time through the While Loop, the same array data are sent to both the chart and the graph indicators. The result is shown in Figure 7.30. Both the chart and the graph have received the array values as input twice, but the chart shows the history (the array data twice) while the graph only displays the most recently received array values.

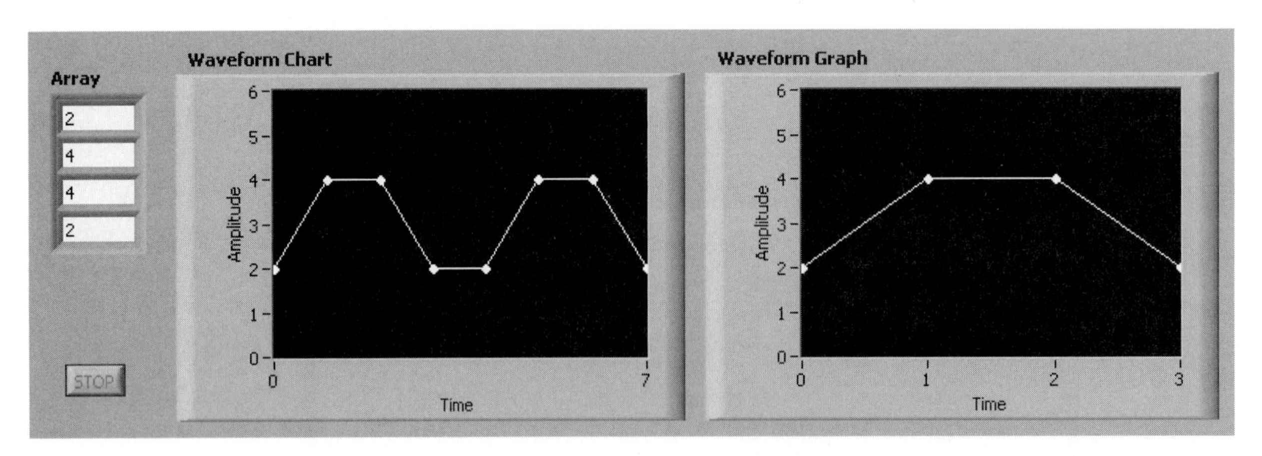

Figure 7.30
The front panel after two cycles through the While Loop.

The Waveform Chart indicator is unique in that it keeps and displays a history of the data received for plotting rather than just showing the most recent values. If we continue to run the VI for a few more cycles of the While Loop (see Figure 7.31),

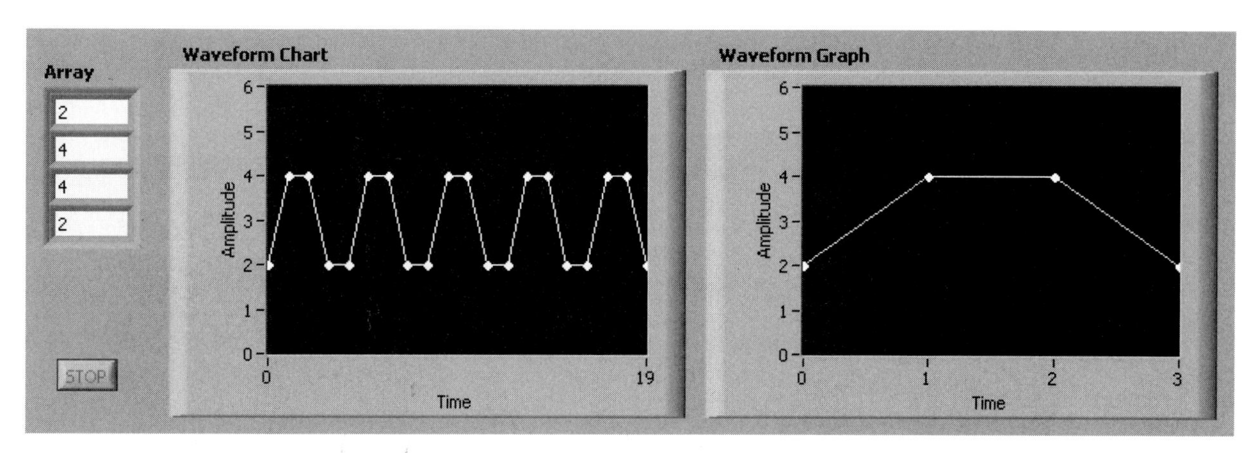

Figure 7.31
The front panel after five cycles through the While Loop.

the chart continues to display all of the values received, while the graph will always display only the most recently received array values.

The number of values stored in the Waveform Chart indicator's history list is 1024 by default. To change the value, right-click on the indicator and select **Chart History Length . . .** from the pop-up menu.

7.3.2 Plotting Multiple Curves Using Waveform Graphs

If you want to plot multiple arrays on a Waveform Graph indicator, you must send a 2D array to the input of the Waveform Graph indicator. Use the Build Array function to combine the X and Y arrays into a 2D array that is sent to the Waveform Graph indicator. The Build Array function is found in the Programming group's Array group:

Functions Palette / Programming Group / Array Group / Build Array

In Figure 7.32 the Polynomial VI has been modified to build the 2D array for graphing. The Transpose 2D Array function was used to convert the two-row array

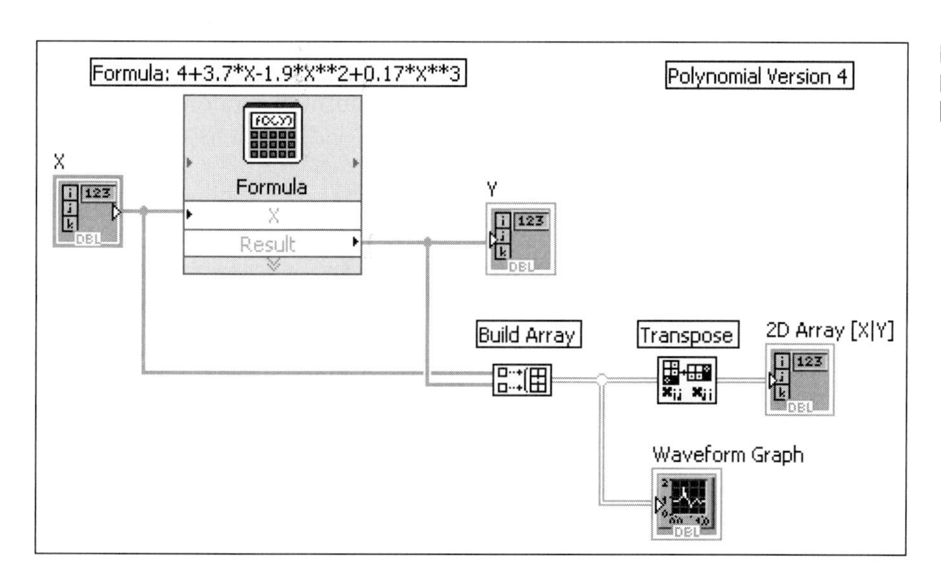

Figure 7.32
Block diagram that plots both X and Y arrays.

created by the Build Array function into a two-column array. The result is shown in Figure 7.33.

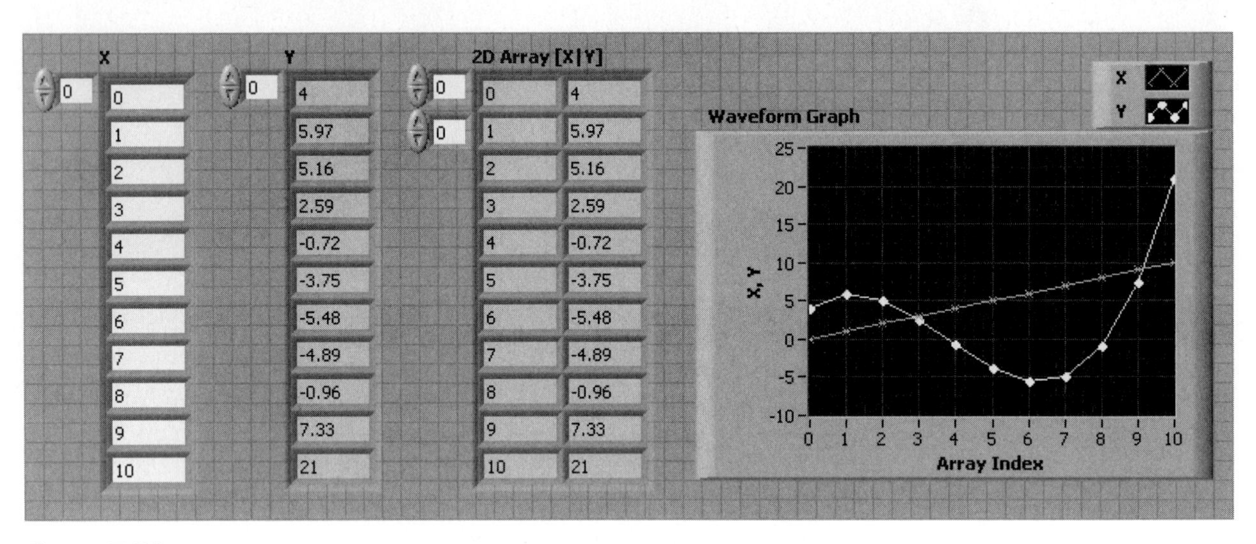

Figure 7.33
Plotting both *X* and *Y* arrays using a Waveform Graph indicator.

7.3.3 Data Acquisition and Waveforms

When a data acquisition system is used, the system is configured to record one or more measurements at a specified time interval. The data set that comes from a data acquisition system often contains the following items:

- Start time
- Time interval between readings
- Array of recorded values

In LabVIEW, these three pieces of information are bundled together in a cluster, and called a *waveform*. There are several ways to obtain a waveform in LabVIEW:

- A data set collected with a data acquisition system will be available as a waveform in LabVIEW.
- LabVIEW provides functions that act as waveform generators (simulated data).
- You can create a waveform in LabVIEW by bundling (start time, time interval, data array) values into a cluster.

EXAMPLE 7.2

Plotting array data and waveform data using LabVIEW's Waveform Graph indicator

LabVIEW's Waveform Graph indicator will accept a waveform as input (so will a Waveform Chart). To illustrate the difference between plotting arrays and plotting waveforms, consider a data set that was collected starting at 11:00 AM with an interval between measurements of 3 minutes (180 seconds). The measured values were: 10, 12, 16, 22, and 30. In Figure 7.34 both the waveform and the 1D array have been plotted using Waveform Graph indicators.

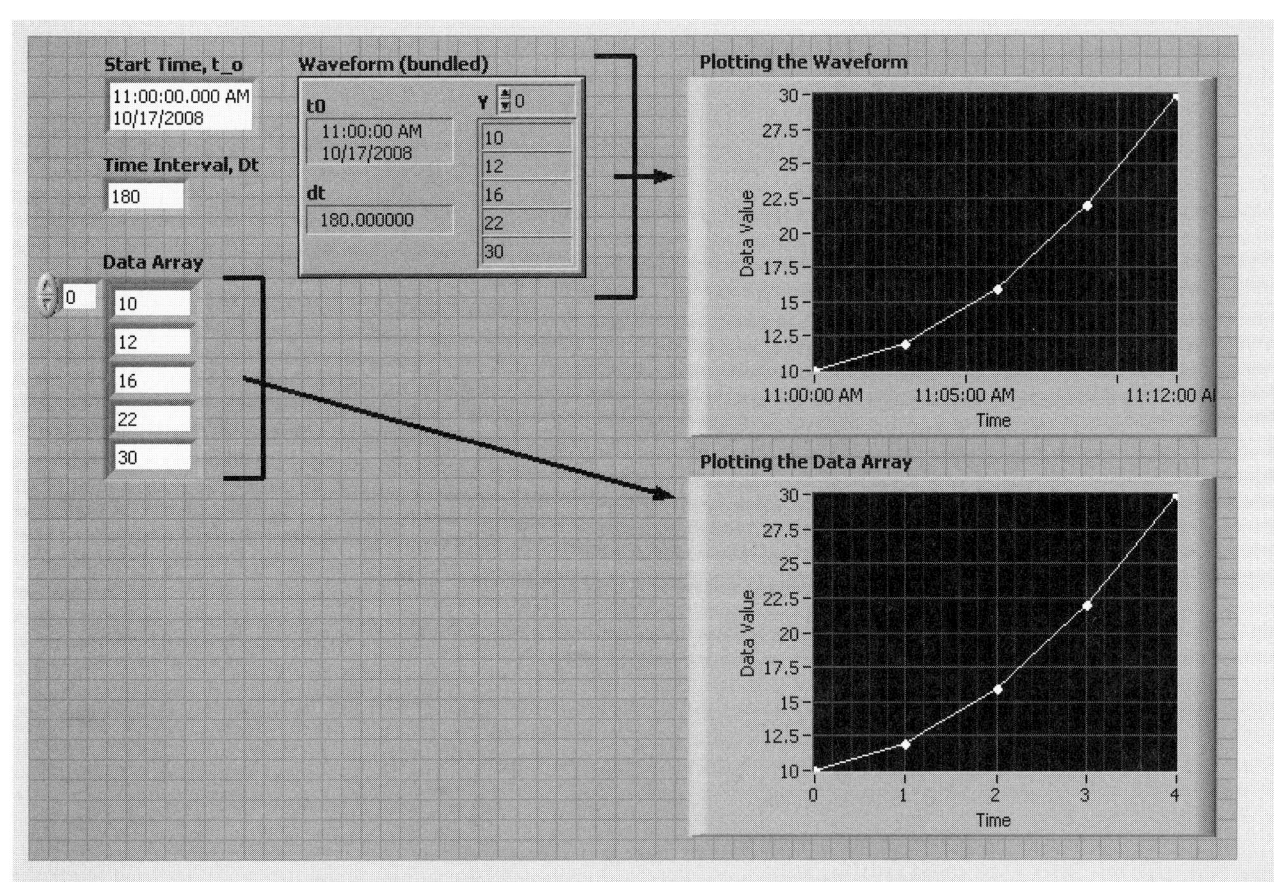

Figure 7.34
Two Waveform Graph indicators, one plotting a LabVIEW waveform and one plotting a 1D array.

The graphs are pretty much the same, except for the values on the Time axes.

- **Waveform**—When the waveform was sent to the Waveform Graph (upper graph in Figure 7.34), start time and time interval information was available to the graph indicator, so actual time values were plotted on the x axis.
- **1D Array**—When the data array was plotted (lower graph in Figure 7.34) the Waveform Graph received no information about time, so the markers on the Time axis just represent array index values.

Note: When plotting waveforms in LabVIEW, plotting actual times on the Time axis is not the default; the default is to plot times relative to the beginning of data acquisition, in seconds. You must request that actual times be plotted by right-clicking on the graph and clearing the check mark before the **Ignore Time Stamp** item in the pop-up menu.

A waveform provides all of the data values, plus information on the timing of the data acquisition. When a waveform is available, it gives you more options for presenting the data graphically.

The block diagram that was used to bundle the waveform information and create the graphs is shown in Figure 7.35.

The Build Waveform function is available in the Waveform group:

Functions Palette / Programming Palette / Waveform Group / Build Waveform

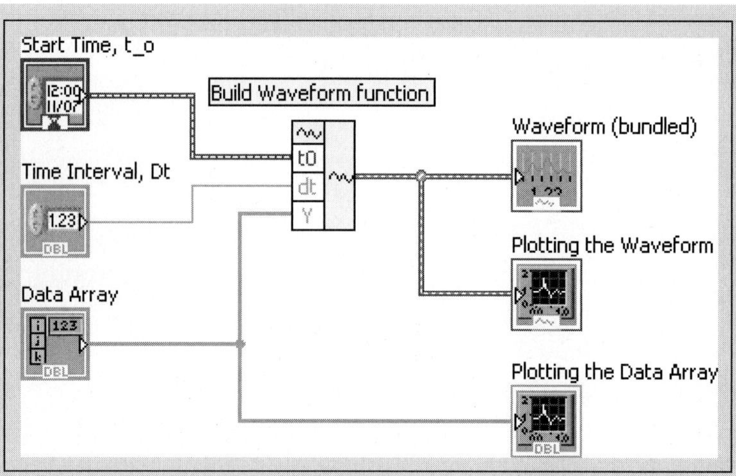

Figure 7.35
Block diagram of VI that bundles and plots a waveform.

7.4 MODIFYING GRAPH FEATURES

Some of the features of a graph, such as the axis labels, can be modified just by double-clicking on the displayed text string. Features can be activated and deactivated by using the graph's pop-up menu. Right-click on the graph to see the pop-up menu options. This is illustrated in Figure 7.36.

Many of the options under **Visible Items** will be presented later, in Section 7.3.2, but a few comments on the **X Scale** and **Y Scale** menu options are in order.

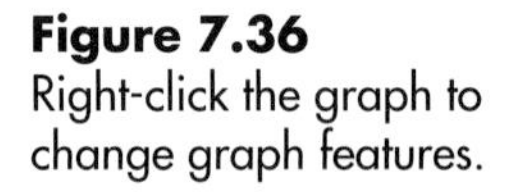

Figure 7.36
Right-click the graph to change graph features.

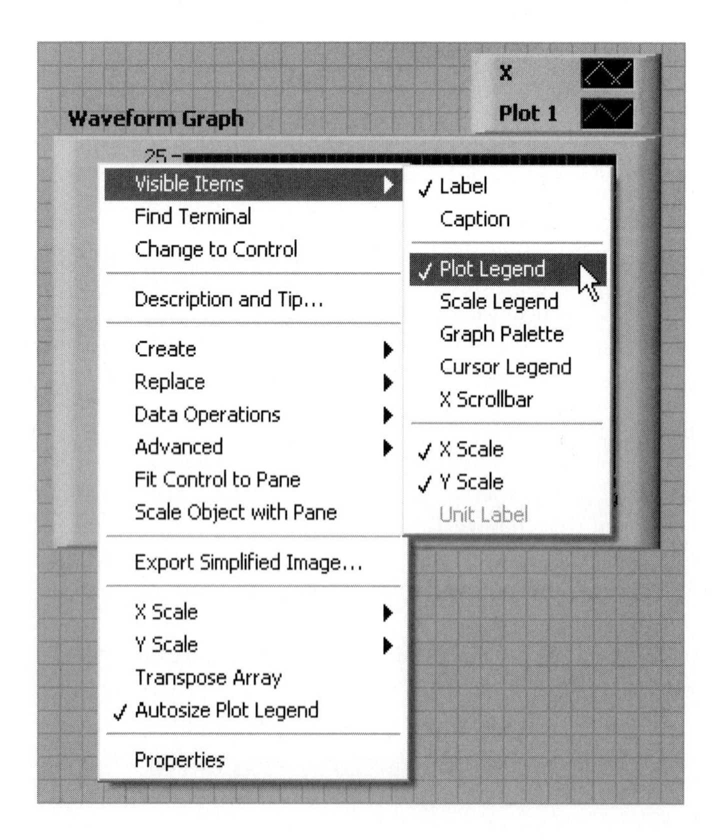

If you select the X Scale menu option (the options for the Y Scale are the same), the options shown in Figure 7.37 appear on the submenu.

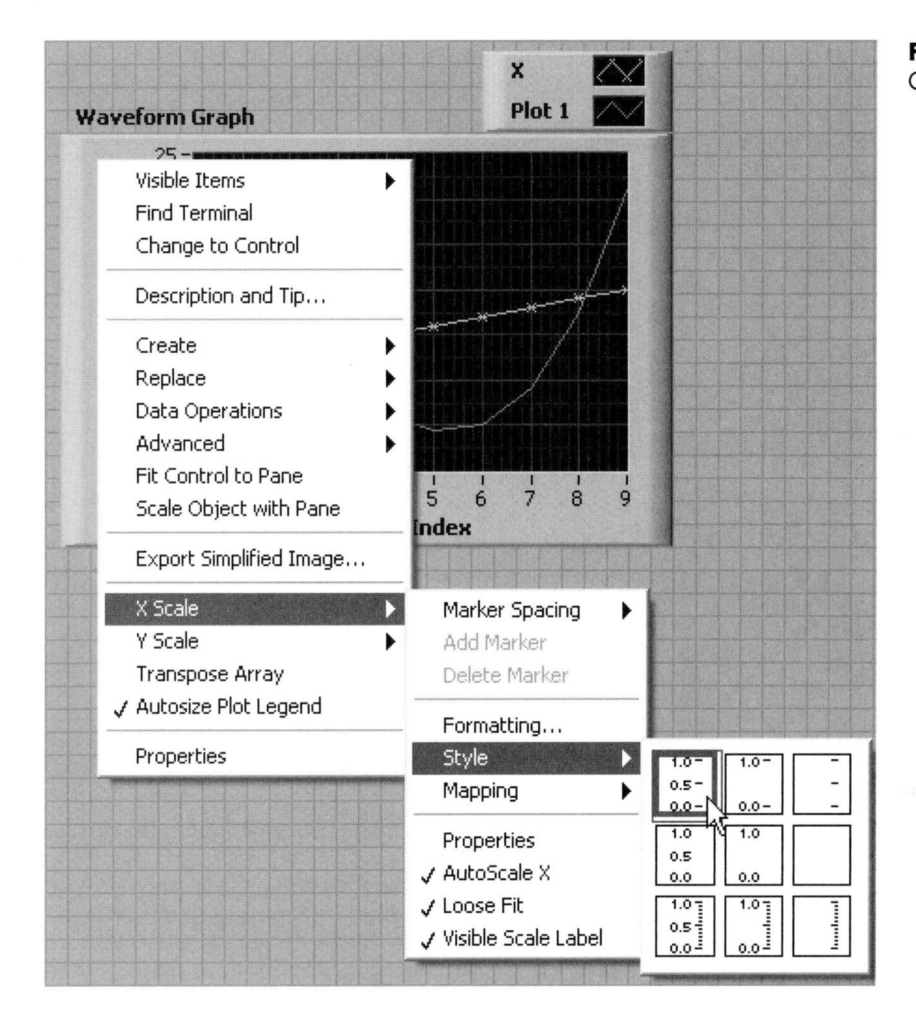

Figure 7.37
Options for the *X* axis.

- **Marker Spacing**—options are Uniform and Arbitrary. Uniform is the most commonly used. Arbitrary markers allow you to indicate specific levels.
- **Formatting . . .** —opens the Graph Properties dialog to the Display Format panel shown in Figure 7.38. This is described in more detail at the end of this list.
- **Style**—allows you to select how much information (scale values, tick marks) you want to see on the axis.
- **Mapping**—allows you to select whether the axis is Linear or Logarithmic.
- **Properties**—opens the Graph Properties dialog to the Scales panel shown in Figure 7.39. The Scales panel allows you to activate or deactivate autoscaling, and enter minimum and maximum scale values if autoscaling is deselected.
- **AutoScale X**—allows you to activate or deactivate autoscaling from the pop-up menu. When *autoscale* is active, LabVIEW automatically adjusts the axis range to fit the displayed values. If you deactivate autoscaling, the axis limits that were in place when autoscaling was deactivated will continue to be used.

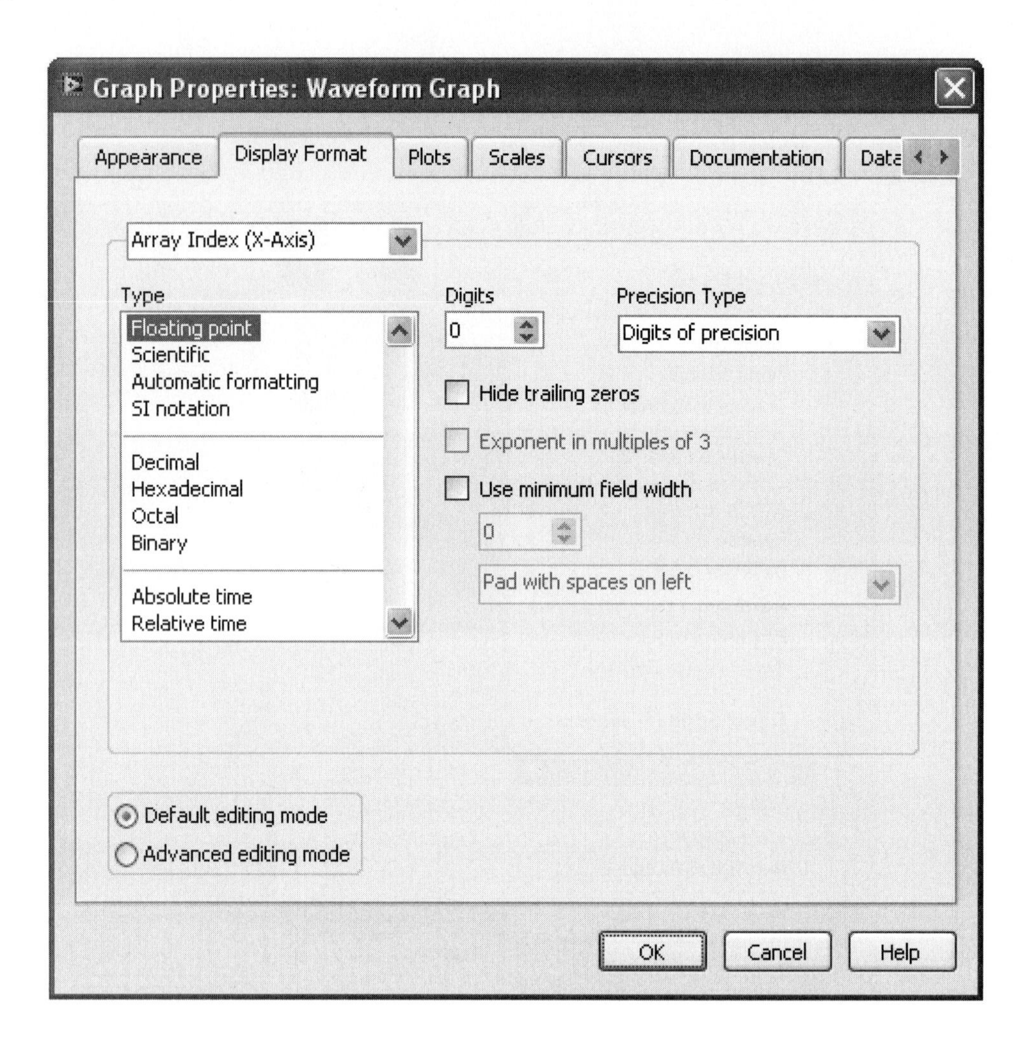

Figure 7.38
Graph Properties dialog, Display Format panel.

- **Loose Fit**—When checked, the minimum and maximum scale values will always be a whole multiple of the axis value interval; this makes your axis look cleaner.
- **Visible Scale Label**—allows you to activate or deactivate the display of the axis label (not the numbers). To hide both the axis label and the scale values, set the **Style** to **blank**.

A *Properties dialog* is available for any control that displays numeric values, but some of the panels change depending on the type of control or indicator. To open the Properties dialog, right-click on a control or indicator (including a graph) and select Properties from the pop-up menu.

For a graph, the Display Format panel (Figure 7.38) allows you to select either axis using the drop-down list near the top of the dialog, and then

- Set the number of Digits desired.
- Select whether the number of digits represents *Digits of precision* or *Significant figures*.
- Choose whether or not to display *trailing zeroes* (Hide trailing zeroes checkbox).

The Properties dialog Scales panel (Figure 7.39) allows you to adjust a number of graph elements. First, select an axis using the drop-down list near the top of the panel, and then you can

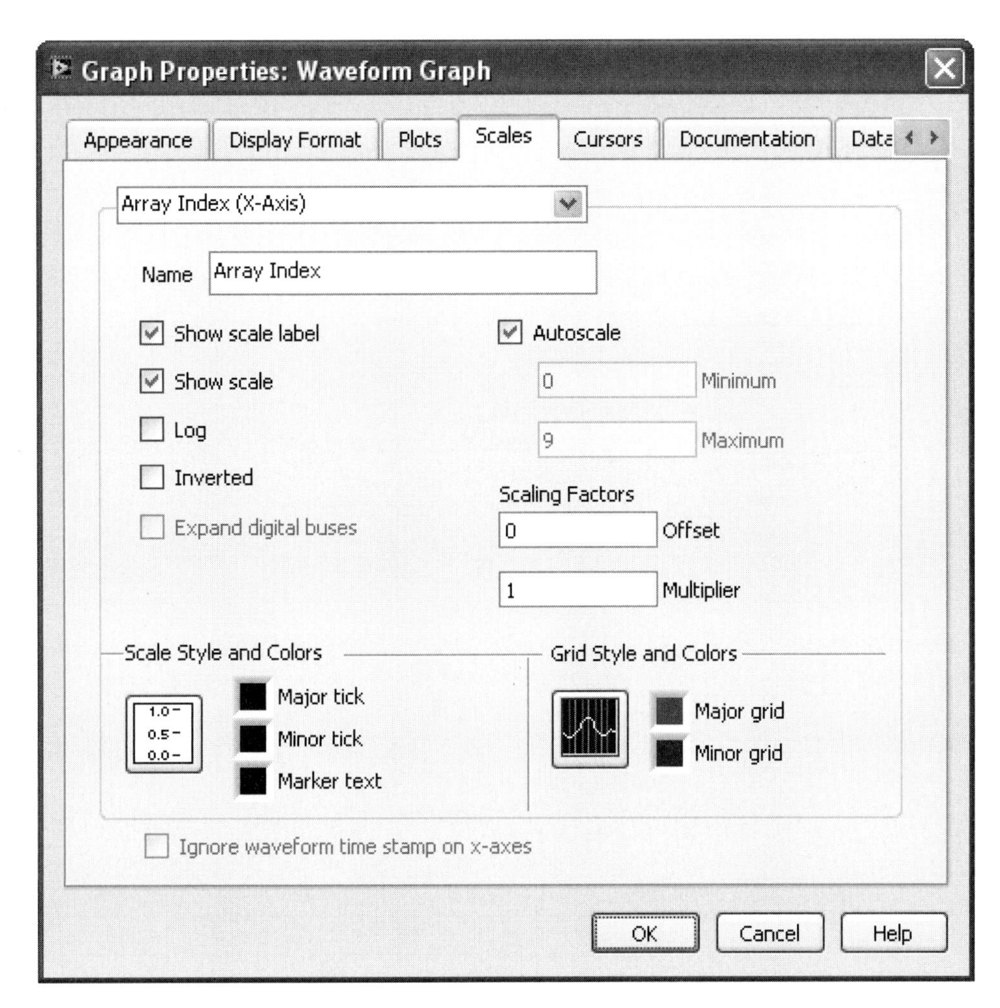

Figure 7.39
Graph Properties dialog, Scales panel.

- Assign or edit the display name (Name field).
- Indicate whether or not the label will be displayed (Show scale label checkbox).
- Indicate whether or not the numeric scale will be displayed (Show scale checkbox).
- Select a Log or Inverted scale, activate or deactivate autoscaling.
- Set display colors for the tick marks, marker values, and gridlines.

The other commonly used panel when working with graphs is the Plots panel, shown in Figure 7.40.

Graph features that can be changed from the Plots panel are as follows:

1. Select the plot (if multiples plots are shown in the graph).
2. Enter or modify the display name for the plot.
3. Choose the line style (solid, dashed, etc.).
4. Choose the line thickness.
5. Choose the marker style.
6. Choose how the plot will be displayed.
    - Markers only
    - Markers with lines, no smoothing, horizontal lines at right marker value
    - Markers with smoothed lines (selected in Figure 7.40)

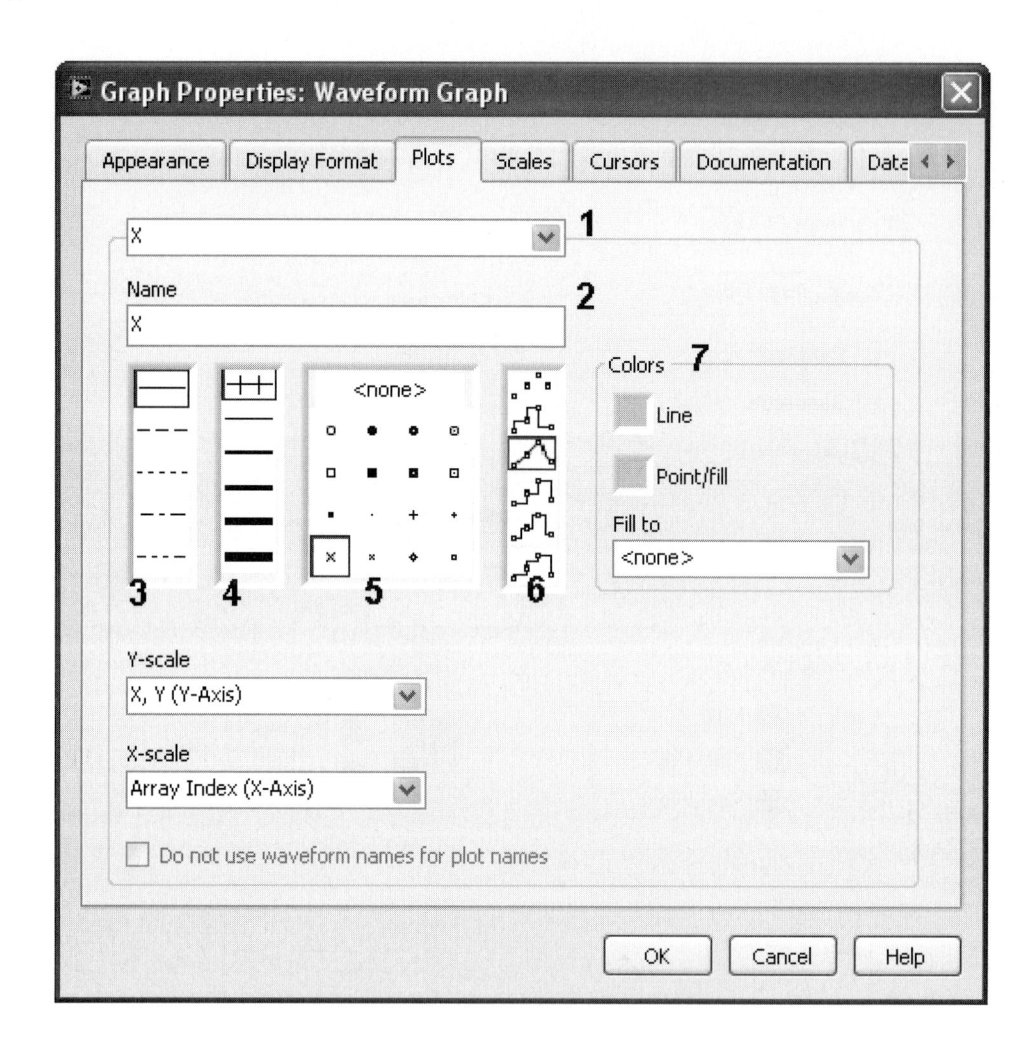

Figure 7.40
Graph Properties dialog, Plots panel.

- Markers with lines, no smoothing, horizontal lines at left marker value
- Markers with lines, no smoothing, markers between horizontal lines
- Markers with lines, no smoothing, markers between vertical lines

7. Choose the colors of the lines and markers.

7.5 GENERATING 1D ARRAYS FOR GRAPHING

Because most LabVIEW graphing controls require 1D arrays as inputs, you need to be careful to generate 1D arrays prior to graphing.

When data are read from a .txt file using the **all rows** output on the Read From Spreadsheet File function, the data are placed in a 2D array, which is not compatible with graphic controls that require 1D arrays as inputs. But you can use the Index Array function to extract individual 1D arrays (either rows or columns) from a 2D array. The Index Array function is located at

Functions Palette / Programming Group / Array Group / Index Array

Note: If your .txt file only has one row or column, use the Read From Spreadsheet File function's **first row** output and a 1D array will be created automatically.

To get data from a .txt file into a 1D array, use the following steps:

1. Place the Read From Spreadsheet function on the block diagram.
2. Wire the **all rows** output to an Index Array function's **2D array** input (use one Index Array function for each row or column needed).
3. Wire the Index Array's **row index** or **column index** to indicate which row to pull out (the top row or left column is identified as 0).
4. Wire the **1D array** output of the Index Array to an array indicator to see the values.

We will use these steps in the following example.

EXAMPLE 7.3

Plotting the involute of a circle

The *involute* of a circle is the curve that results by following the end of a string as it is being unwound from a circle. Data for the involute of a circle are available in the text files Involute.txt. The first step is to use the Read From Spreadsheet function to read the files.

> **Function Palette / Programming Group / File I/O Group / Read From Spreadsheet.vi**

In Figure 7.41, the Read From Spreadsheet function is shown expanded so that the terminals can be seen. The **file path** and **all rows** terminals are in use.

- **file path**—indicates the location of the .txt file
- **all rows**—reads the entire .txt file

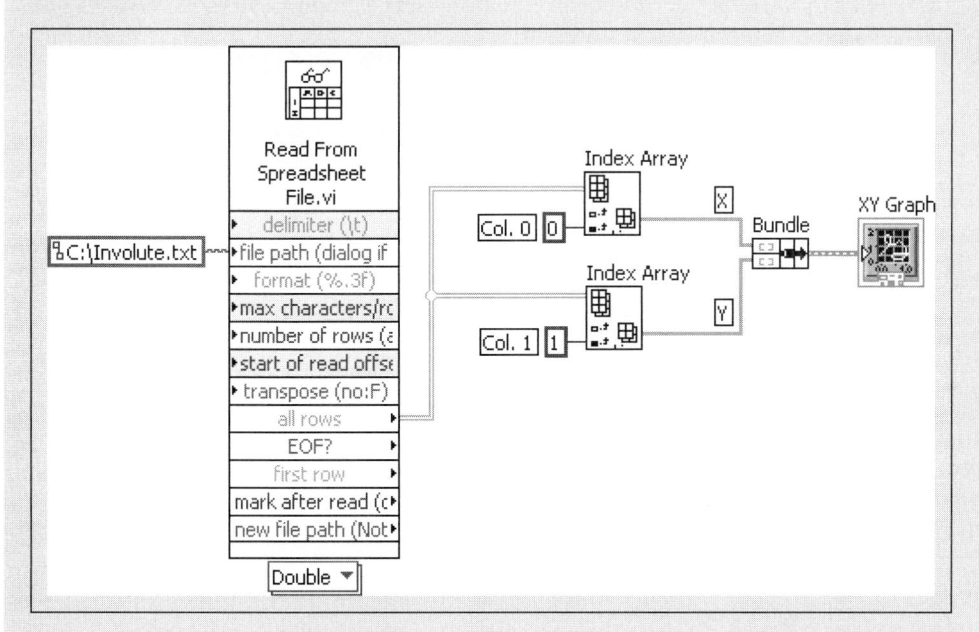

Figure 7.41
Using Read From Spreadsheet function to read data from .txt file.

Once the data values are read from the Involute.txt file, two Index Array functions are used to grab the first and second columns as 1D arrays. Then, the 1D arrays (of *X* and *Y* values) are bundled and sent to an XY Graph control. The result is shown in Figure 7.42.

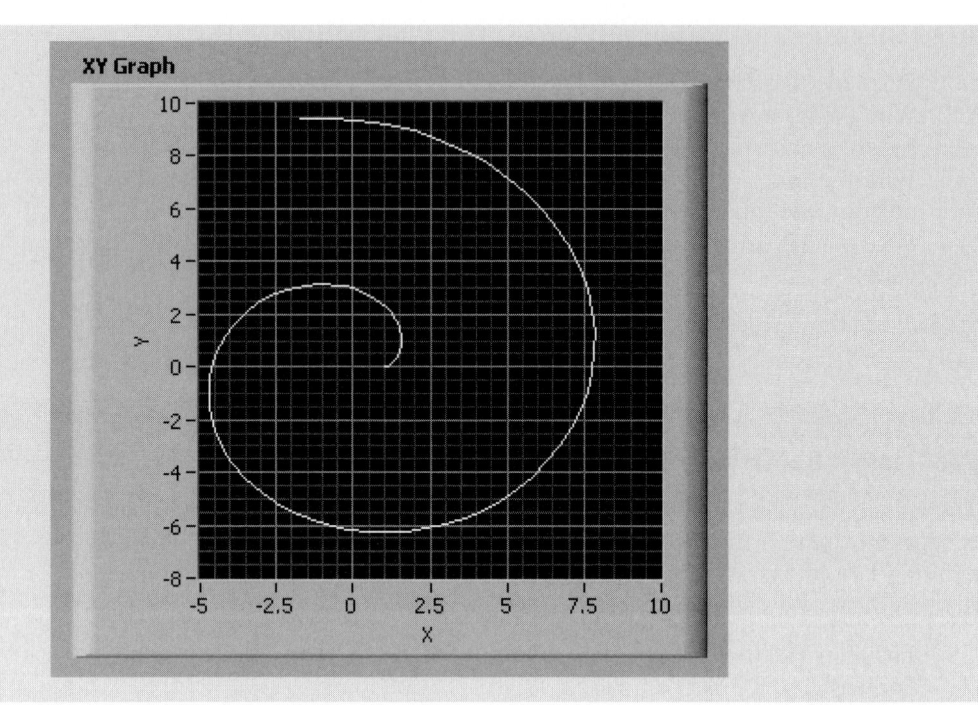

Figure 7.42
The involute of a circle.

7.6 PUTTING LABVIEW GRAPHS TO WORK

LabVIEW provides some tools that can help get values from graphs. We will use those tools to solve for the *roots* of an equation using a graph. (The roots are the values of the variable that satisfy the equation.)

As an example, we'll solve for the values of x that satisfy the equation

$$x^2 - 9.4x^2 + 22.95x - 10.602 = 0$$

Since the greatest power on x is 3, we expect three roots.

This equation is entered into the Formula Express VI as shown in Figure 7.43.

The VI shown in Figure 7.44 uses a For Loop to evaluate the formula for x values between 0 and 10. The roots are the values of x that generate y values equal to 0.

By plotting the X and Y arrays using an XY Graph (as shown in Figure 7.45), we can see that the roots are close to 0.5, 3, and 5.5. We'll use some features of LabVIEW's graphics to zoom in and get more precise values for the roots.

If you right-click on the XY Graph and select Visible Items from the pop-up menu, the following options are displayed:

1. **Label**—(defaults to graph type, XY Graph in this example)
2. **Caption**—(defaults to graph type, changed to "Plotting a Polynomial" in this example)
3. **Plot Legend**—identifies the curves displayed on the plot (defaults to Plot 0, Plot 1, etc.).
4. **Scale Legend**—shows the text strings used as axis labels (X and Y here); buttons allow axes to be unlocked.
5. **Graph Palette**—provides access to cursor, zoom, and pan features. Axes must be unlocked before cursor, zoom, and pan features are useful.

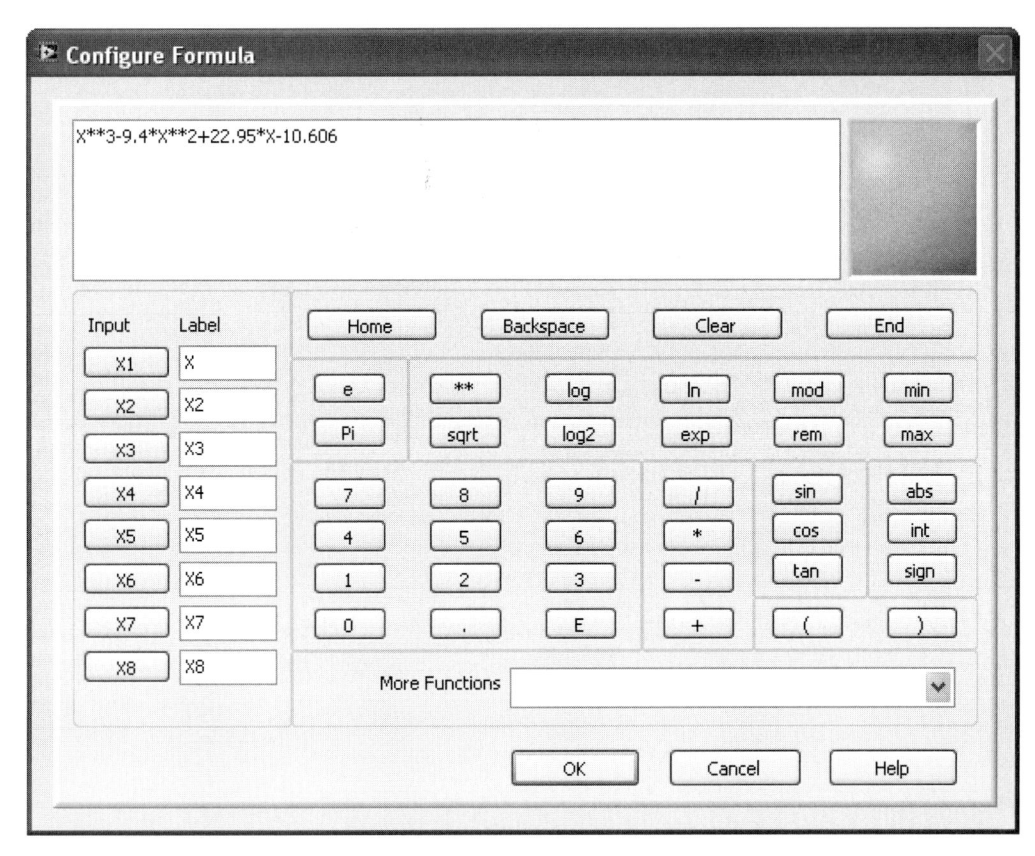

Figure 7.43
Entering the polynomial into the Configure Formula dialog.

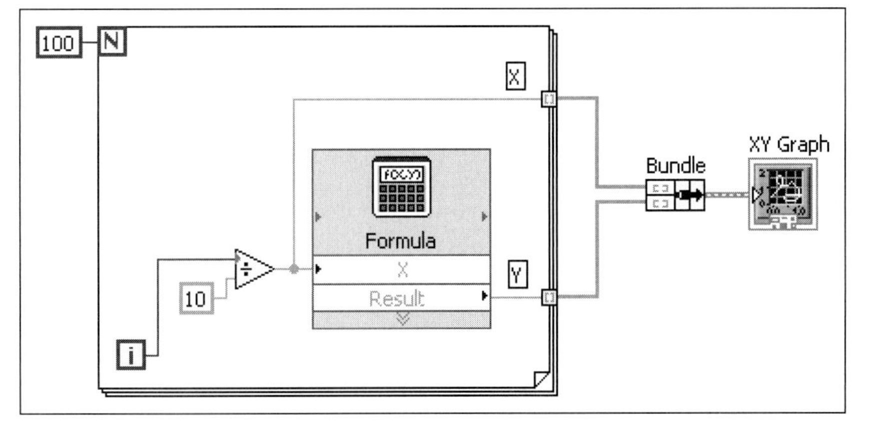

Figure 7.44
Block diagram of VI used to evaluate and graph polynomial.

6. **Cursor Legend**—once a cursor has been defined, the *cursor legend* shows the current location on the axes. The diamond controller allows you to use the mouse to move the active cursor in four directions.
7. **X Scrollbar**—if the graph has been zoomed or panned, the *X Scrollbar* can be used to slide the graph back and forth.

Figure 7.46 shows each of these items on the polynomial graph.

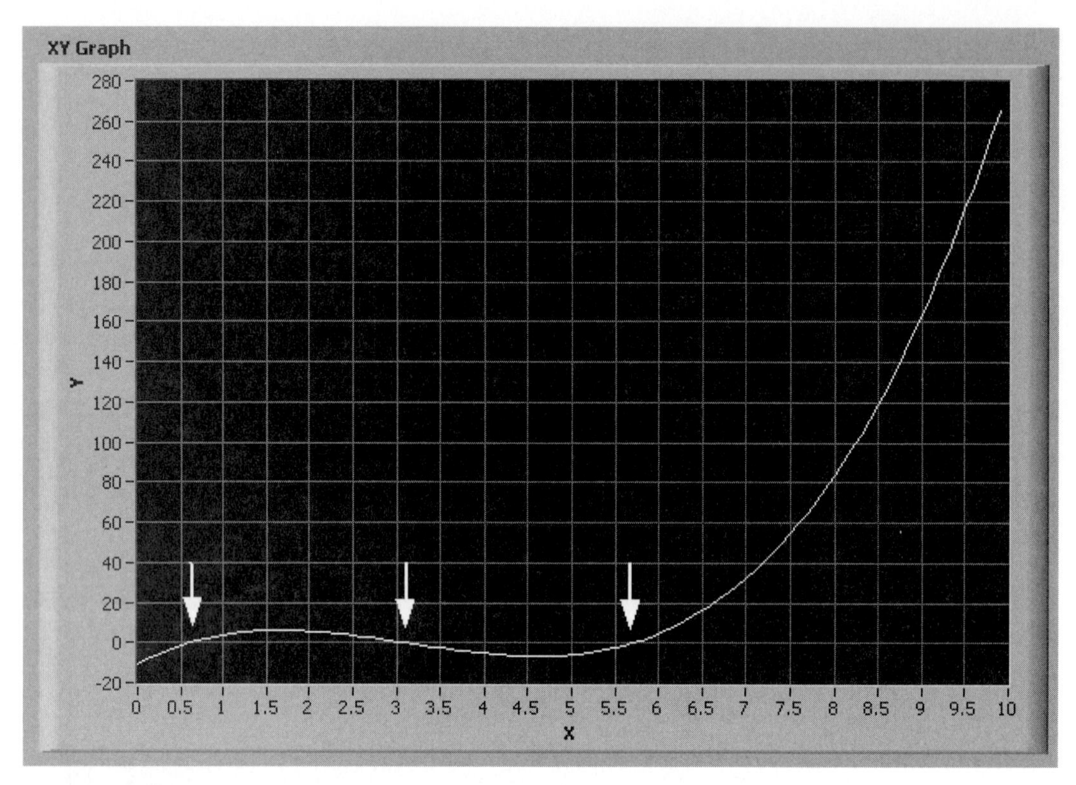

Figure 7.45
XY graph of the polynomial showing approximate locations of roots.

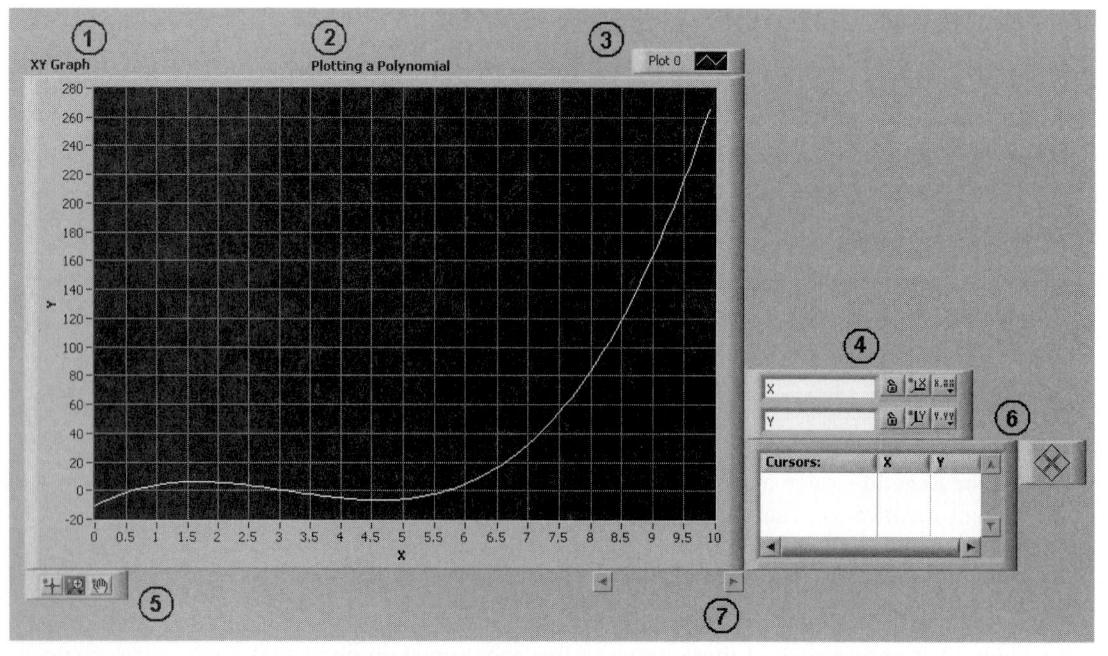

Figure 7.46
Polynomial graph with all graphing features made visible.

Zoom In

To try to more precisely determine the root values, we will first zoom in on the portion of the graph in the lower-left corner. The following steps are needed:

1. Unlock the X and Y axes using the Scale Legend buttons (if needed), in Figure 7.47 the axes are shown already unlocked.

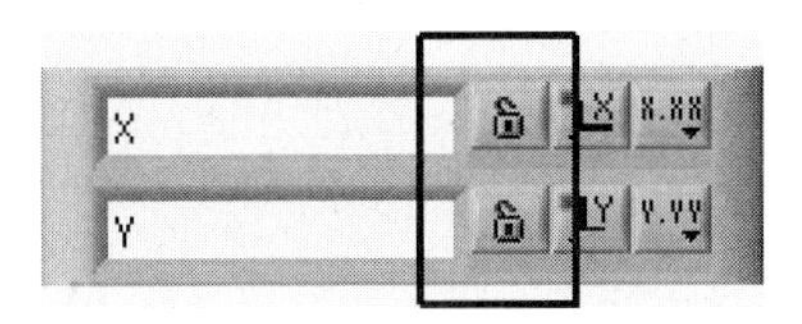

Figure 7.47
The axes are unlocked on the Scale Legend.

2. Click the Zoom (magnifying glass) button on the Graph Palette and select a zoom option (see Figure 7.48).
   - Top-left (selected)—zoom to an area (to be) indicated with the mouse.
   - Top-middle—use full Y axis, zoom to selected region of X axis.
   - Top-right—use full X axis, zoom to selected region of Y axis.
   - Bottom-left—restore full graph (unzoom).
   - Bottom-middle—zoom in on a selected point.
   - Bottom-right—unzoom at a selected point.

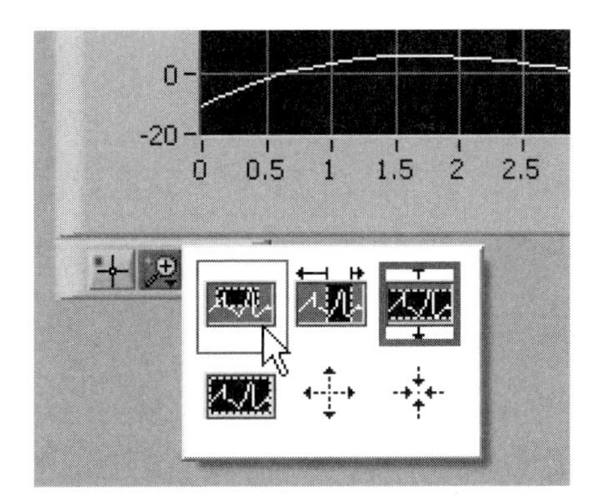

Figure 7.48
The Zoom button opens a menu of zoom options.

3. Select the bottom-left corner of the graph, from about (0, −20) to (6.5, 20). The displayed portion of the graph will zoom in on the selected region as shown in Figure 7.49. Use the panning tool (hand) on the Graph Palette to slide the graph around if needed.

Activate a Cursor

Next, we will activate a *cursor* to help us read values off the graph very precisely. To activate a cursor, right-click in the Cursor legend and select **Create Cursor / Single-Plot** from the pop-up menu as shown in Figure 7.50.

- A **Free** cursor can be moved anywhere on the graph.
- A **Single-Plot** cursor can be moved around, but the cursor intersection will always be on the displayed plot (that's what we want for finding roots).
- The **Multi-Plot** cursor is not available because there is only one plot displayed on the graph.

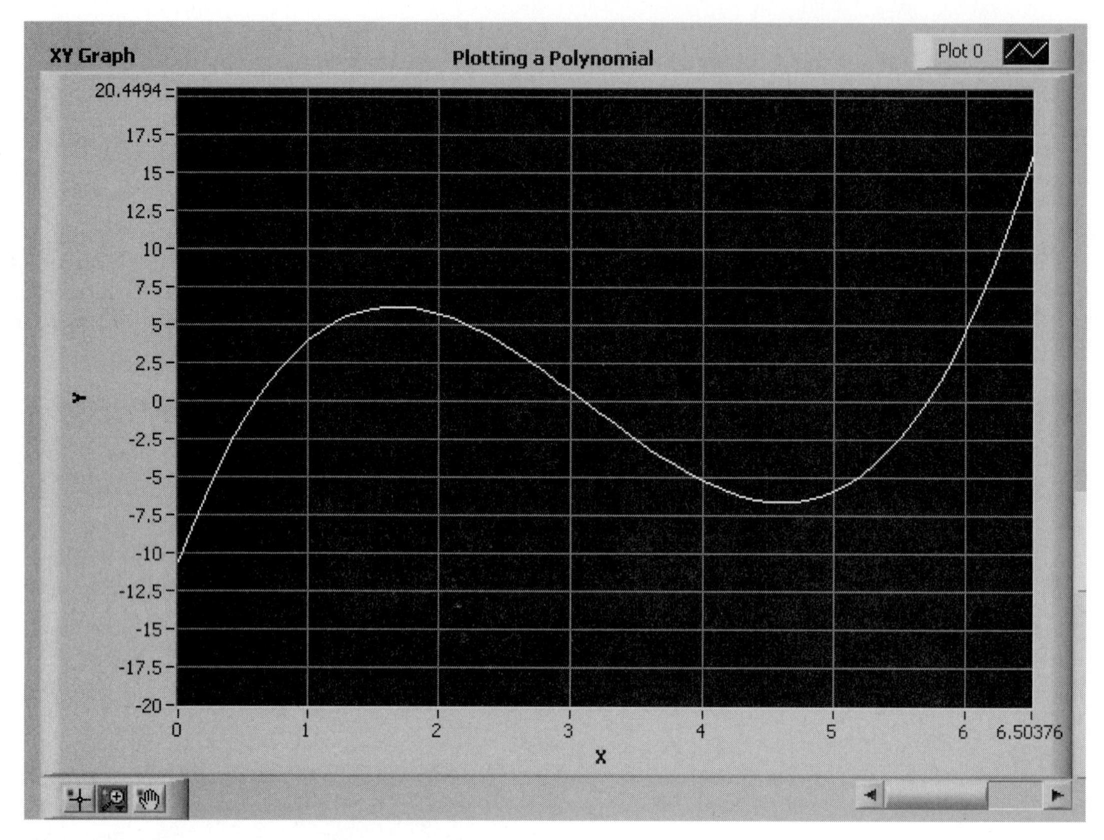

Figure 7.49
The zoomed graph.

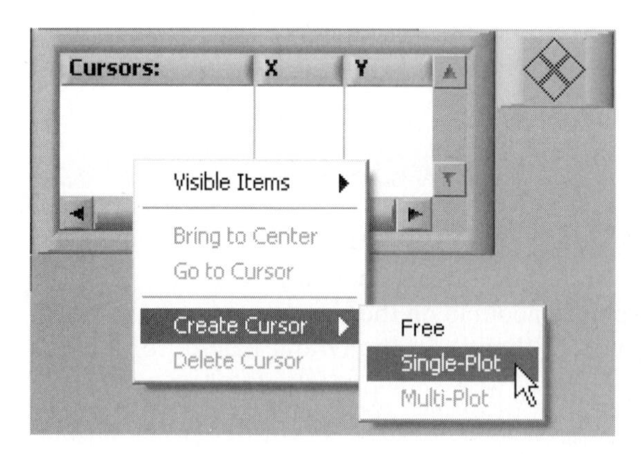

Figure 7.50
Creating a Single-Plot cursor.

The cursor is made up of horizontal and vertical lines as shown in Figure 7.51. You can move around with the mouse. Click on the cursor button on the Graph Palette if needed to activate the cursor tool. In Figure 7.51, the cursor has been placed very close to $Y = 0$ at $X = 0.6$. This tells us that $x = 0.6$ is one of our roots.

Simply move the cursor to the other two locations where the polynomial crosses $Y = 0$ to find the other two roots. One is at $x = 3.1$ as shown in Figure 7.52. The other is at $x = 5.7$ (not shown).

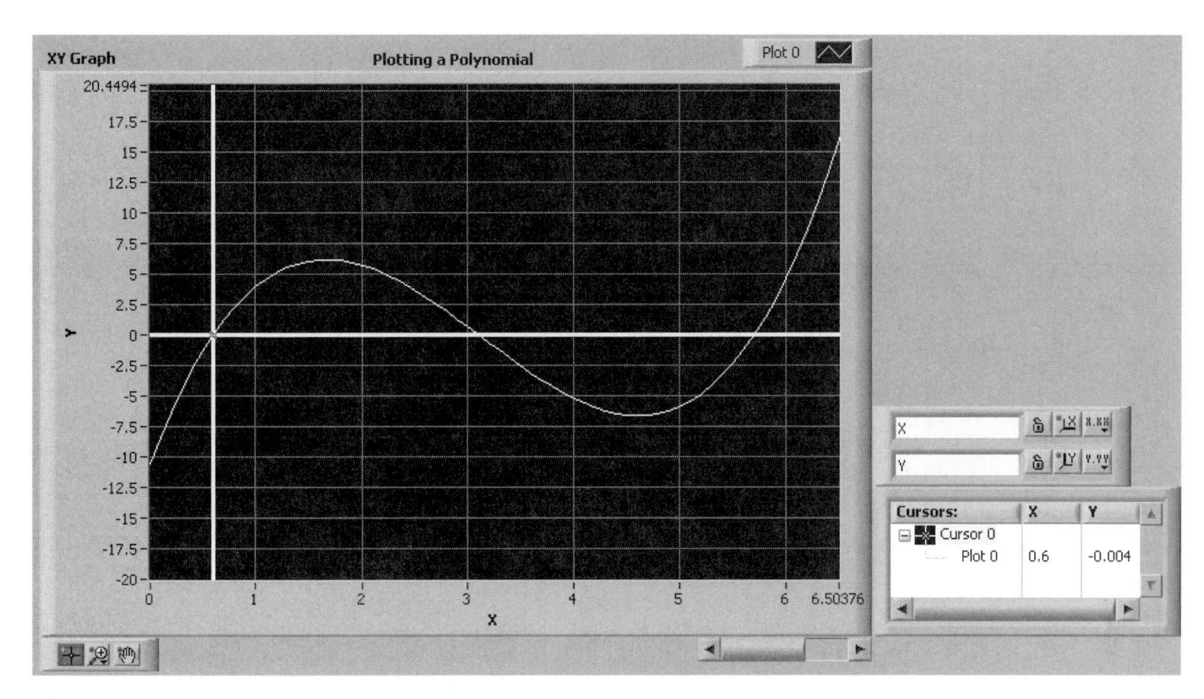

Figure 7.51
Move the cursor to $Y = 0$ to find a root.

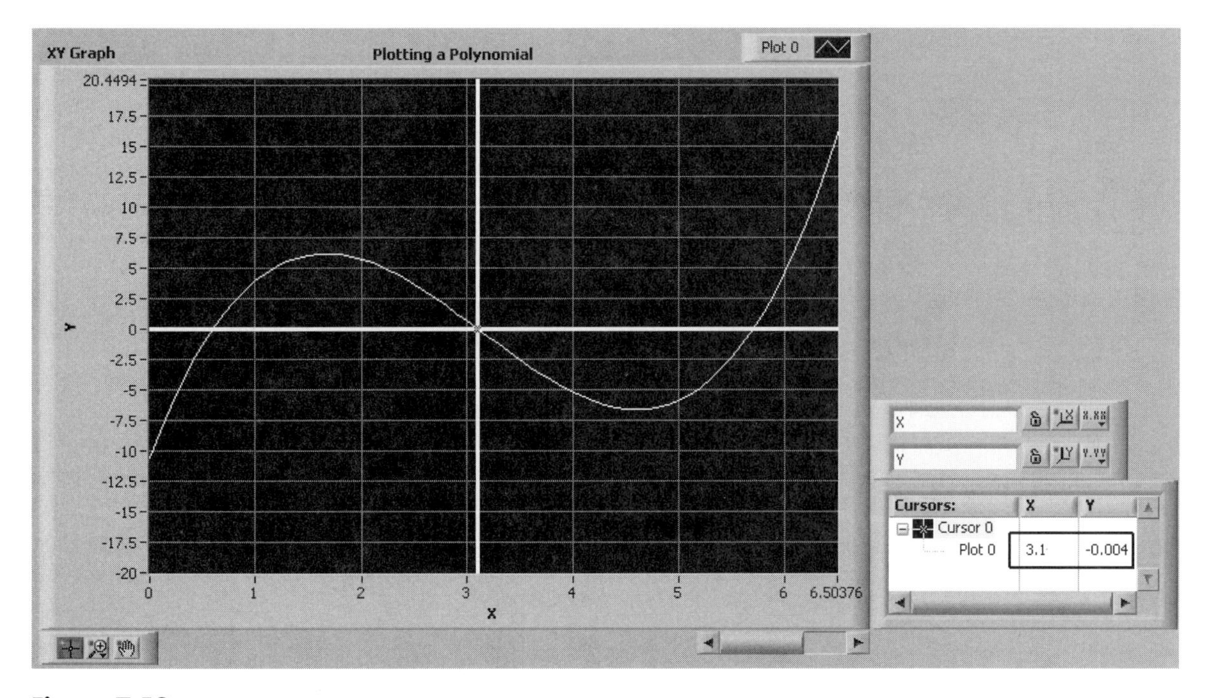

Figure 7.52
The second root is located at $X = 3.1$.

The ability to use a cursor to find values on a graph is very convenient.

7.7 USING XY GRAPHS—2D PLOTTING

To this point, all of the graphing examples in this chapter have been plotted with uniform spacing on the x axis; no x values have been sent to either the graph or chart indicators. Even when a waveform was plotted, a constant time interval was assumed so that the points were evenly spaced across the x axis.

Collecting data using an automated data acquisition system commonly produces values uniformly spaced through time. But there are many data analysis situations (and some data acquisition situations) where the data to be plotted do not have uniformly spaced points in the x direction. When this is the situation, you must provide both an X array and a Y array, and use LabVIEW's *XY Graph* indicator to accurately plot non-uniformly spaced values.

Controls Palette / Modern Group / Graph Group / XY Graph

EXAMPLE 7.4

Lab testing a new medical diagnostic tool

A new medical test is being developed that uses an enzyme to convert a blood chemical into a form that is easy to measure using a test strip. The problem is that the enzyme slowly begins to lose potency as soon as the test kit is opened to air. A researcher is testing to see if a new formulation will allow the concentration to reach 0.37 μg/L (this is the critical value for a successful test) before the enzyme activity has fallen to 20% of its original level. The data sheet is shown in Figure 7.53.

Ultimately, we will want to plot both the product concentration and the enzyme activity as functions of time, but we will begin with just the product concentration. The front panel of a VI that plots the product concentration on the y axis and the sample time on the x axis is shown in Figure 7.54. The block diagram used to generate Figure 7.54 is shown in Figure 7.55.

Figure 7.53
Data sheet for medical diagnostic study.

	A	B	C	D	E
1	Test Product Concentration and Enzyme Activity Study				
2	22-May-2009				
3					
4	Time	Product Concentration	Enzyme Activity		
5		μg/L	% of initial		
6	11:30 AM	0	100		
7	11:45 AM	0.07	98		
8	12:40 PM	0.18	88		
9	1:35 PM	0.26	72		
10	2:25 PM	0.32	55		
11	3:20 PM	0.37	36		
12	4:10 PM	0.40	24		
13	5:15 PM	0.42	14		
14					

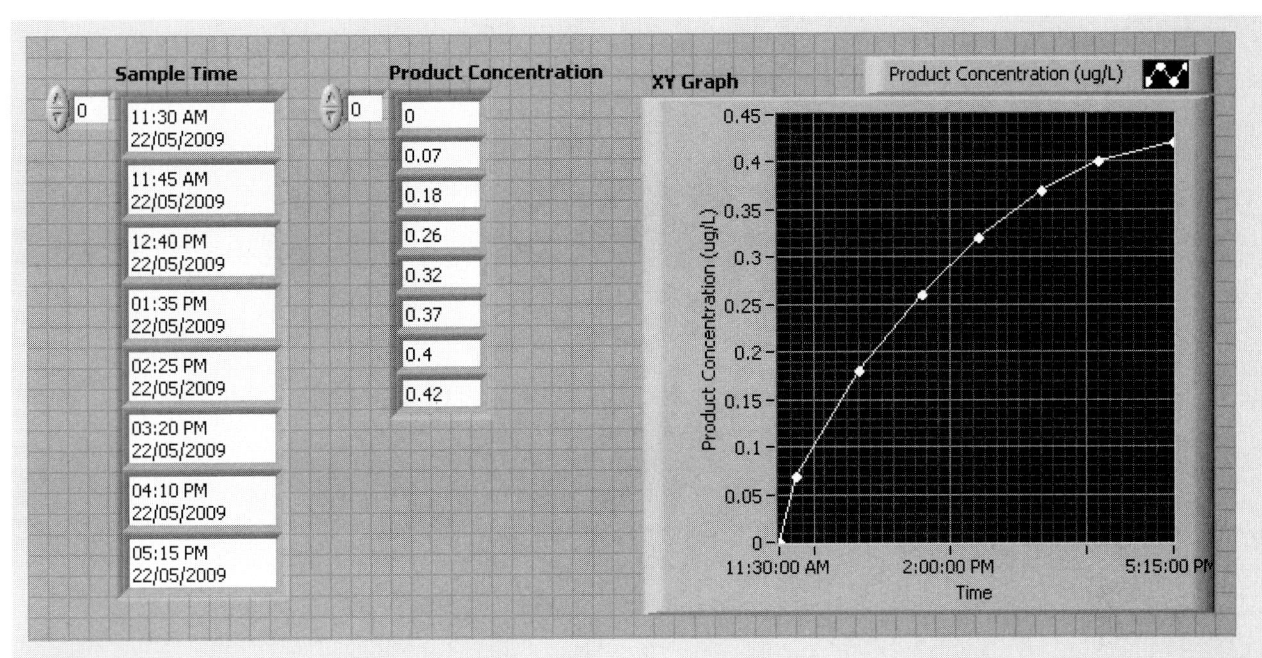

Figure 7.54
Front panel showing XY plot of Product Concentration vs. Time.

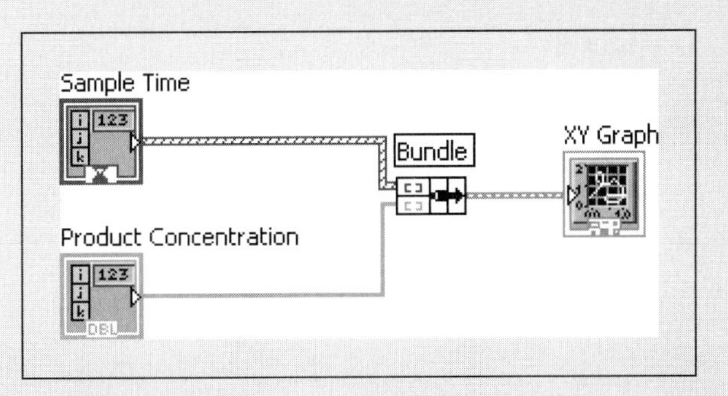

Figure 7.55
Block diagram used to create the XY Graph of Product Concentration vs. Time.

The array or sample times and the array or product concentration values must be bundled before wiring to the XY Graph input. The Bundle function is available at

Functions Palette / Programming Group / Cluster, Class & Variant Group / Bundle

The enzyme activity can be plotted as well, as shown in Figure 7.56. The units on enzyme activity were changed from percent of initial activity level to fraction of initial level. This was done to get both the product concentration and the enzyme activity on the same scale for plotting. From the graph, you can see that the product concentration reaches 0.37 μg/L when the enzyme activity level is still 0.36 or 36% of the initial activity level. Since this is well over the target of 20%, the new formulation is a success!

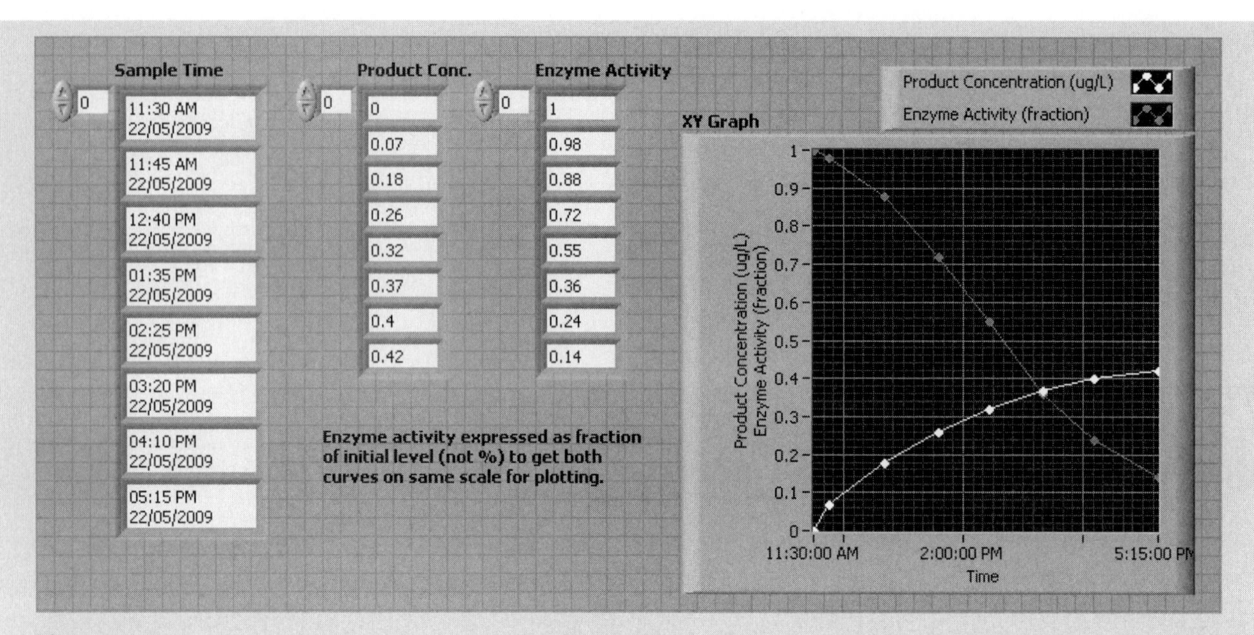

Figure 7.56
Plotting two curves on the XY Graph.

The block diagram required to plot two curves on an XY plot is shown in Figure 7.57.

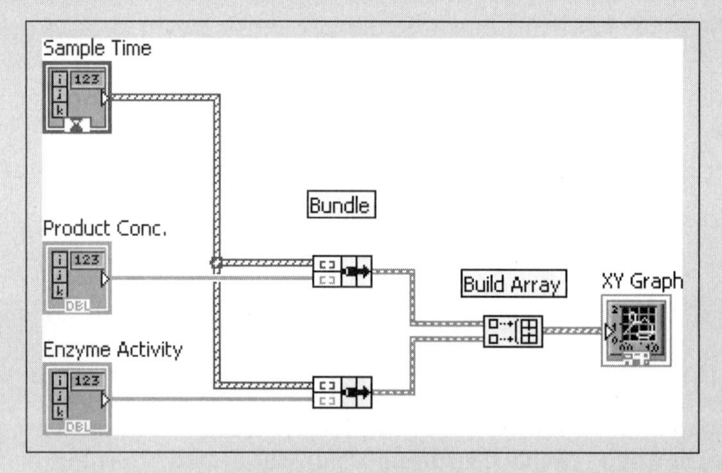

Figure 7.57
Block diagram to plot two curves on an XY plot.

Preparing the data for plotting takes a few steps:

- The product concentration and sample time arrays are bundled.
- The enzyme activity and sample time arrays are bundled.
- The two bundles are appended using the Build Array function.
- The 2D array of bundles of X and Y values is sent to the XY Graph indicator.

The Build Array function is available at

Functions Palette / Programming Group / Array Group / Build Array

In this example the product concentration and enzyme activity measurements were made at the same time, but this is not required. Because the X and Y values for each curve are bundled before being built into the final array, each curve can use a different array of X values.

PRACTICE

Bundling and (Array) Building for Multi-Curve XY Graphs

Learning when to bundle and when to build an array will make working with LabVIEW XY Graph controls easier. Practice by creating a VI that plots the following data:

X	*Y*1	*Y*2
1	1	49
2	4	36
5	25	9
6	36	4
7	49	1

The *X* values are non-uniformly spaced to make it obvious that the XY Graph control uses the *X* values to locate the markers.

Note: Your VI should have three 1D arrays:

- *X*
- *Y*1
- *Y*2

Part 1: Create two XY Graphs

First, practice bundling (*X*, *Y*) pairs by bundling (*X*, *Y*1) and (*X*, *Y*2). Send each pair to a separate XY Graph Control. The bundle function and XY Graph control are located at

Functions Palette / Programming Group / Clusters, Class & Variant Group / Bundle

Controls Palette / Modern Group / Graph Group / XY Graph

The front panel should look something like Figure 7.58.

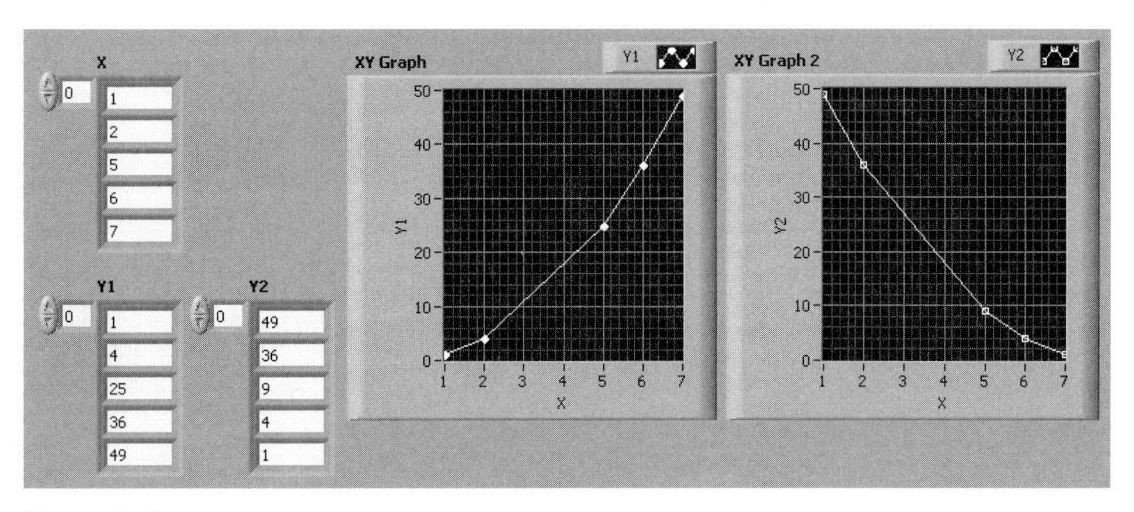

Figure 7.58
Part 1, Create two XY graphs.

Part 2: Create one XY Graph with two curves

Modify your VI as follows:

- Delete one XY Graph control
- Build the two bundles into an array, as $\begin{bmatrix} (X,Y1) \\ (X,Y1) \end{bmatrix}$
- Send the array of bundles into the XY Graph control

Your result should look like Figure 7.59.

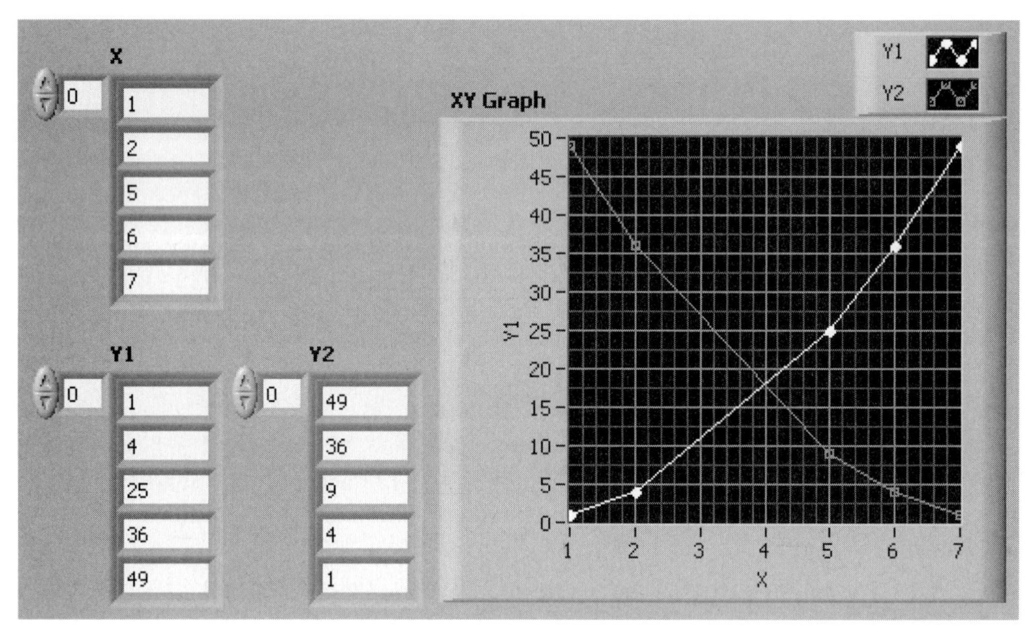

Figure 7.59
Part 2, Create one XY graph with two curves.

The block diagram for part 1 is shown in Figure 7.60, and for part 2 in Figure 7.61. Notice, in part 2, that bundling comes first; then the array (of bundles) is built.

Figure 7.60
Block diagram for part 1.

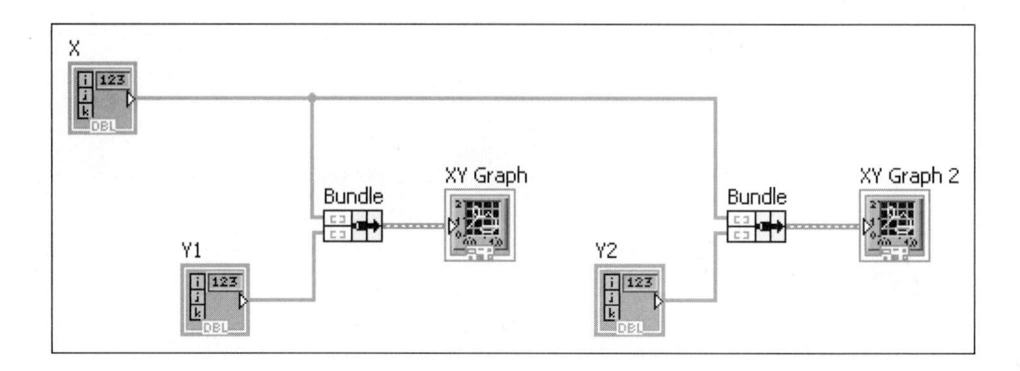

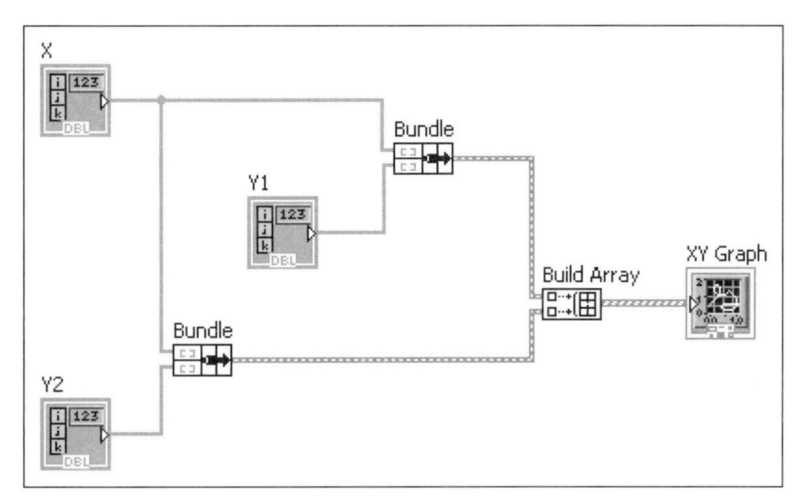

Figure 7.61
Block diagram for part 2.

APPLICATION

XY Graphics Demonstration—Spinning Sine Waves

This demonstration just shows what can be done with a For Loop, a While Loop, a couple of trig functions, three knobs and an XY plot. If you run the program at a public gathering, people will stop by to turn the knobs. The front panel is shown in Figure 7.62, and the block diagram in Figure 7.63.

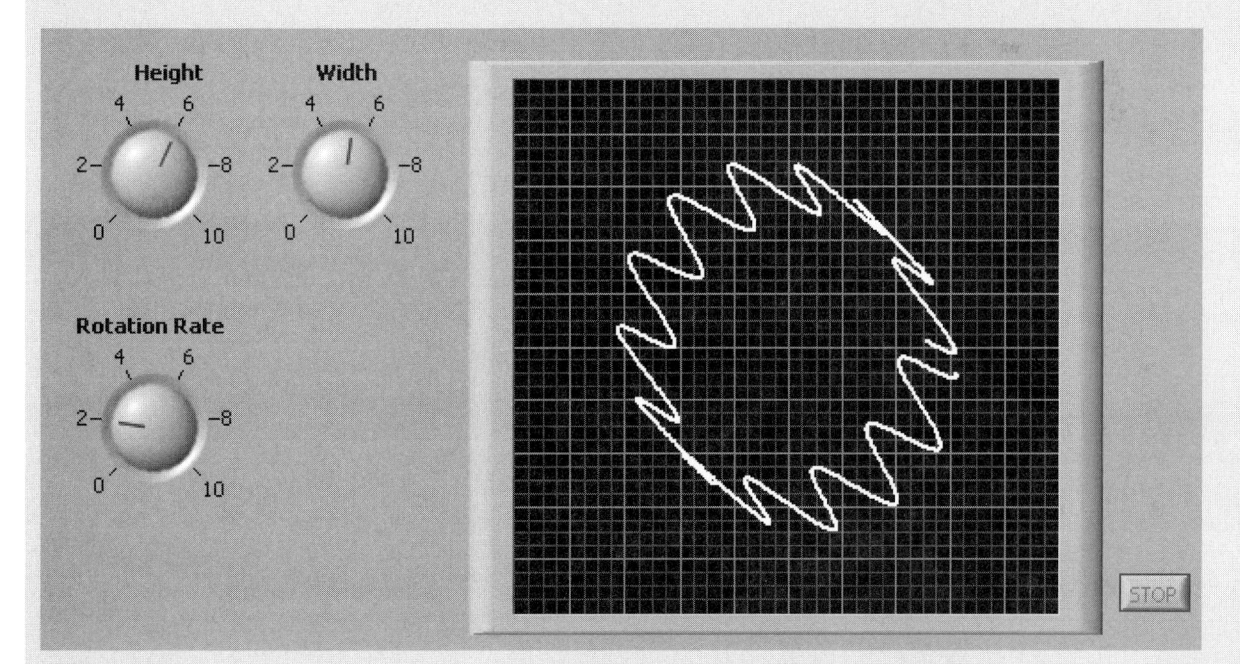

Figure 7.62
Spinning Sine Waves VI.

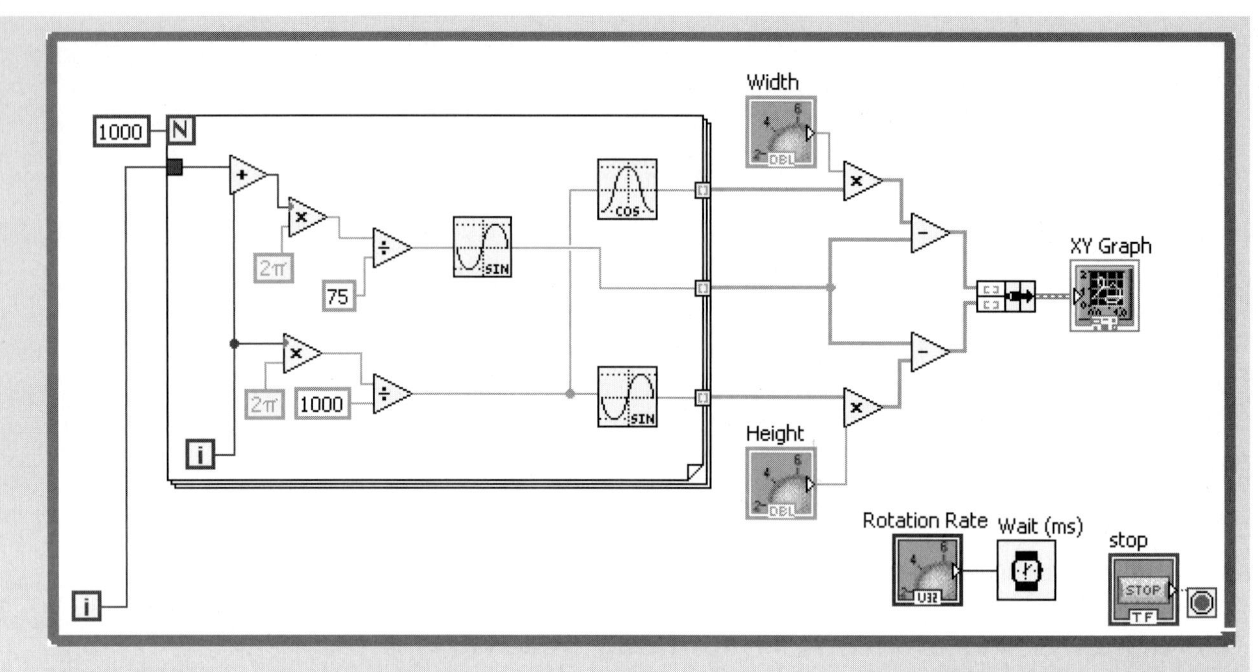

Figure 7.63
Spinning Sine Waves, block diagram.

It's a simple application with an interesting graphic display.

7.8 3D GRAPHING

LabVIEW provides three indicators for presenting 3D graphs:

- 3D Surface Graph
- 3D Parametric Graph
- 3D Curve Graph

The icon for the *3D Parametric Graph* indicator is a *torus* (doughnut shape). We will generate that shape to demonstrate the use of the 3D graphing indicators in LabVIEW.

EXAMPLE 7.5

Creating a 3D Parametric Graph of a torus

To help define terms, consider the torus shown in Figure 7.64.

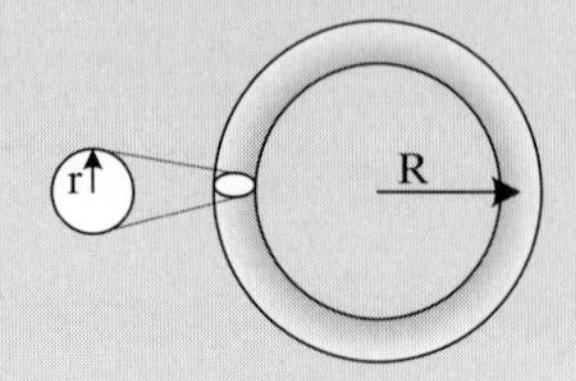

Figure 7.64
Defining the radius values for the torus.

We will call the diameter of the smaller circle "r" and the diameter from the center of the torus to the center of the tube "R". With these definitions, the surface of a torus can be described with the following equations:

$$x = (R + r\cos(u))\cos(v)$$
$$y = (R + r\cos(u))\sin(v)$$
$$z = r\sin(v)$$

Where u and v are working variables that each range between 0 and 2π radians. The number of increments used for u and v is arbitrary, but a smaller step size creates a smoother surface when the torus is plotted. In this example we will use 30 steps for both u and v. The step values for u and v were calculated in Excel, and are shown in Figure 7.65. Values of u are in column B, and values for v are in row 7.

The values of x, y, and z are then calculated using the surface equations for the torus. The calculations for x values are illustrated (partially) in Figure 7.65. Ultimately the size of the 2D x array will be 31×31 elements.

C8 fx =(C3+C4*COS(C$7))*COS($B8)

	A	B	C	D	E	F	G
1	Toroid Array Data - X						
2							
3		R:	10				
4		r:	2				
5							
6			v -->				
7			0.000	0.209	0.419	0.628	0.838
8	u	0.000	12.000	11.956	11.827	11.618	11.338
9	|	0.209	11.738	11.695	11.569	11.364	11.090
10	|	0.419	10.963	10.923	10.805	10.614	10.358
11	V	0.628	9.708	9.673	9.568	9.399	9.173
12		0.838	8.030	8.000	7.914	7.774	7.587
13		1.047	6.000	5.978	5.914	5.809	5.669
14		1.257	3.708	3.695	3.655	3.590	3.504
15		1.466	1.254	1.250	1.236	1.214	1.185
16		1.676	-1.254	-1.250	-1.236	-1.214	-1.185

Figure 7.65
Calculating ***X*** values for a torus in Excel.

To prepare for importing the values into LabVIEW, the x, y, and z arrays are each moved to the top-left corner of a worksheet (no headings) and then stored in separate tab-delimited text (.txt) files.

Once the three text files have been prepared, the Read From Spreadsheet function can be used three times to read the x, y, and z arrays into LabVIEW as three 2D arrays.

The Read From Spreadsheet function is available at

Functions Palette / Programming Group / File I/O Group / Read From Spreadsheet

The Read From Spreadsheet functions are used to read each of the three text files. Then the 2D arrays are displayed, and sent to the 3D Parametric Graph function for plotting. The complete block diagram is shown in Figure 7.66, and the resulting plot on the front panel in Figure 7.67.

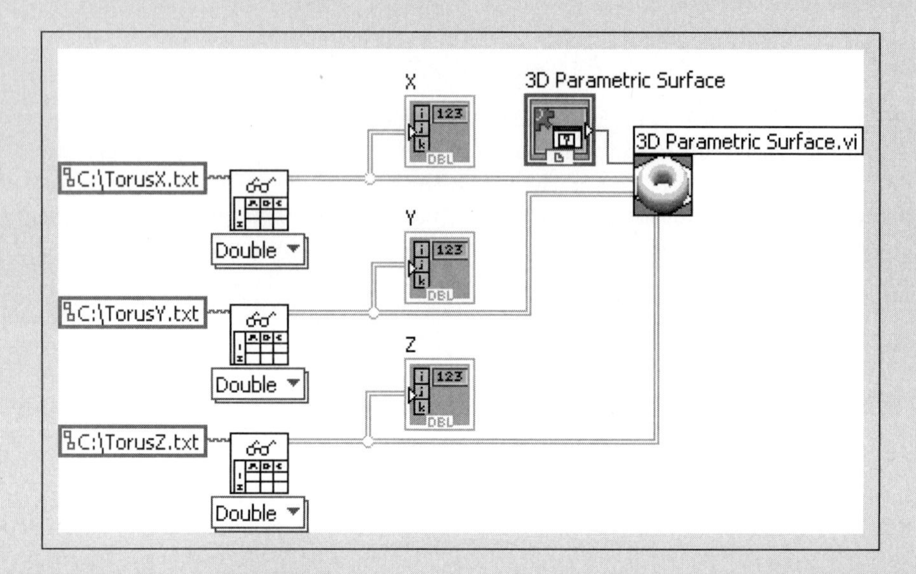

Figure 7.66
Final block diagram for Torus VI.

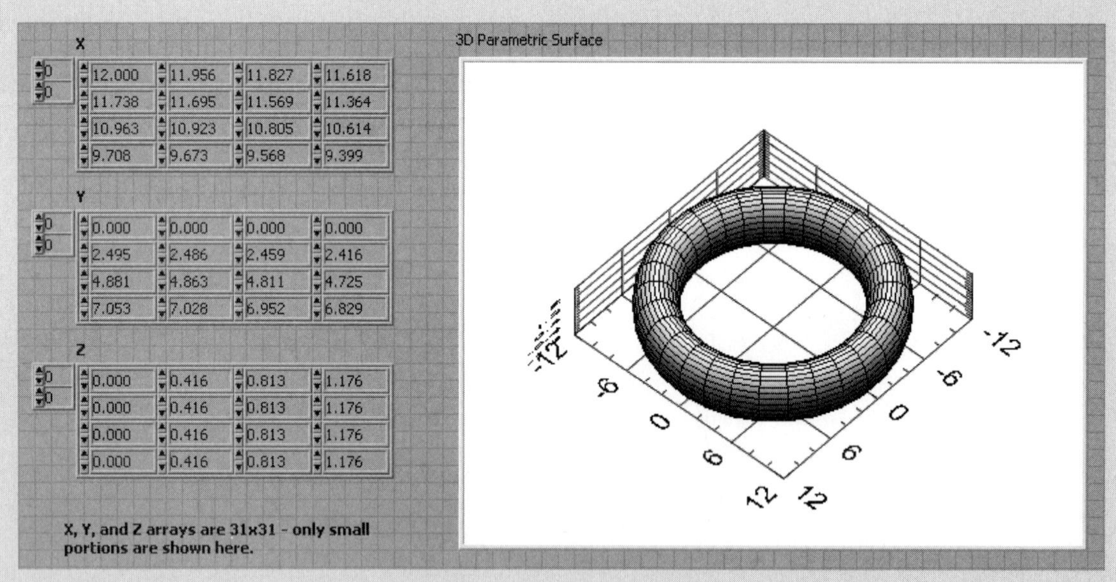

Figure 7.67
Front panel of the Torus VI.

To adjust the display properties of the 3D graph, right-click on the graph and select **CWGraph3D / Properties . . .** from the pop-up menu. The CWGraph3D Control dialog will open, as illustrated in Figure 7.68.

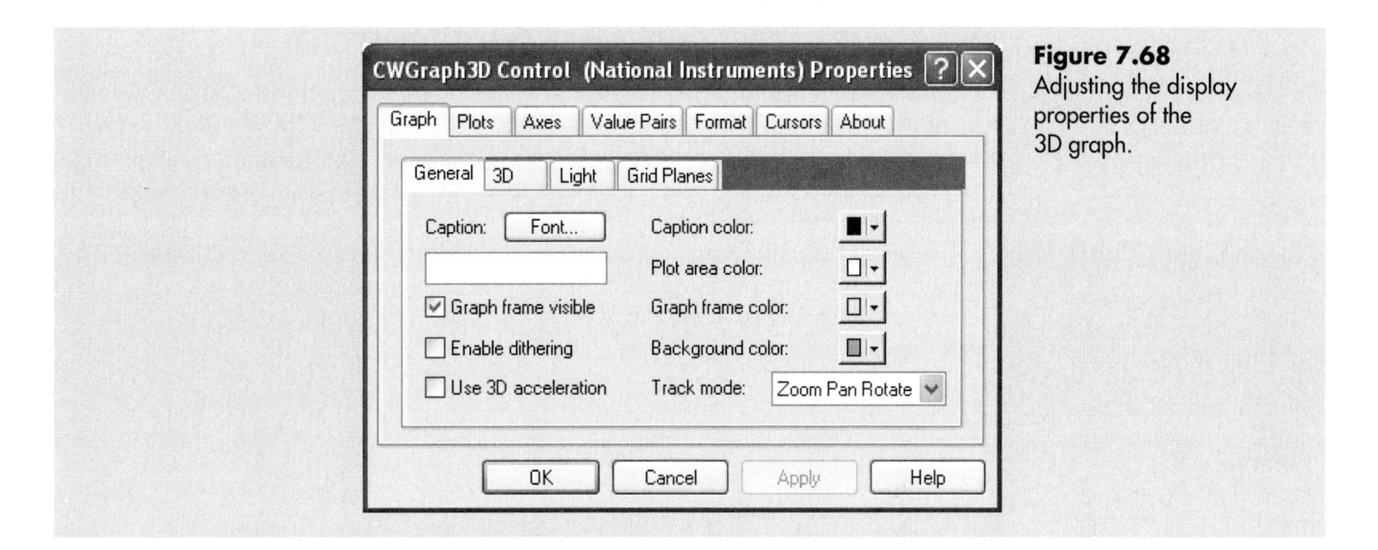

Figure 7.68
Adjusting the display properties of the 3D graph.

7.8.1 A Look Ahead

It is not necessary to create the *x*, *y*, and *z* values in Excel and import them into LabVIEW. The same calculations that were performed in Excel can be done in LabVIEW; it requires two nested For Loops. The block diagram for the LabVIEW Torus VI is shown in Figure 7.69 as a preview of what LabVIEW can do.

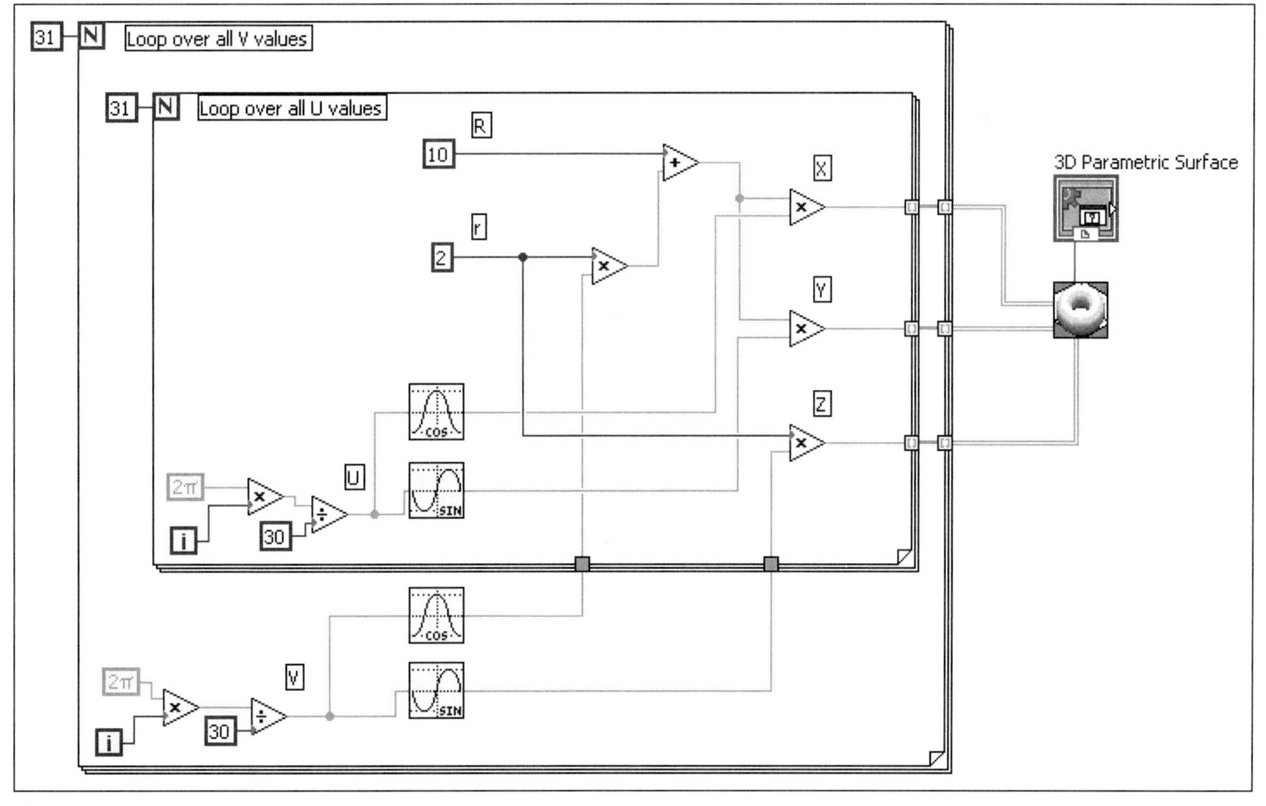

Figure 7.69
Block diagram for the LabVIEW Torus VI.

7.9 GETTING GRAPHS ONTO PAPER AND INTO REPORTS

One way to get a printout of a LabVIEW graph is simply to print the VI front panel that shows the graph. While that works much of the time, LabVIEW will also *export* a simplified version of the graph that can be either pasted or inserted into a document. To export an image of a graph, right-click on the graph and select **Export Simplified Image . . .** from the pop-up menu.

None of the export methods is perfect. The .bmp (bitmap) export in Figure 7.70, for example, is a little pixilated and truncates the x-axis value label on the right. Still, the export method provides a way to get a LabVIEW graph into a report when needed.

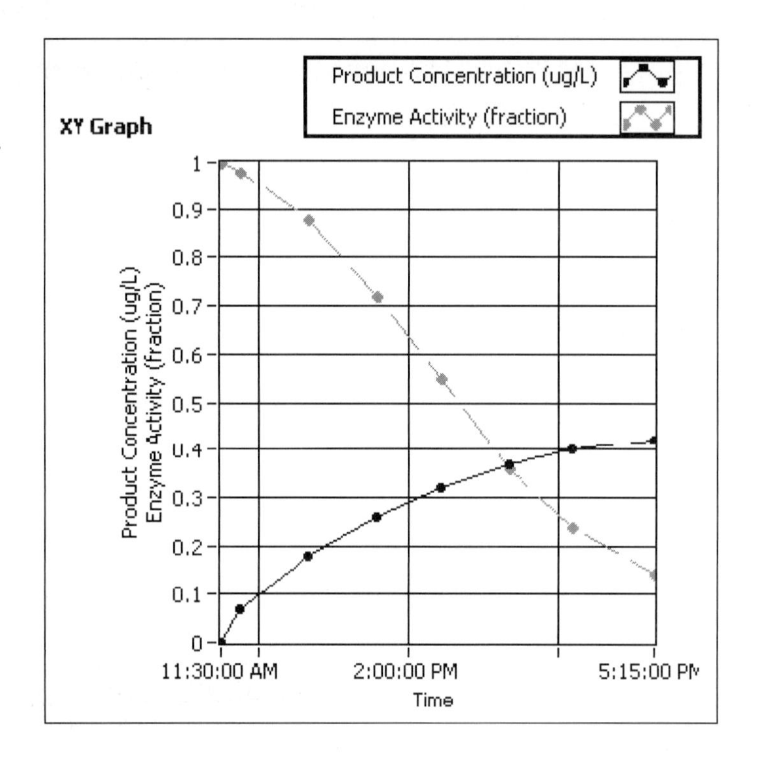

Figure 7.70
Graph from Figure 7.56 exported as .tif file, then inserted into this document.

KEY TERMS

3D parametric graph
array plotting
autoscale
chart
chart history
cursor
cursor legend
digits of precision
export
graph
Graph Palette
involute
overlay plots (waveform chart)
Plot Legend
Properties dialog
root (of polynomial)
sawtooth wave
Scale Legend
Signal
significant figures
sine wave
square wave
square wave duty
stacked plots (waveform chart)
torus
trailing zeroes
triangle wave
update mode (strip chart, scope chart, sweep chart)
waveform
Waveform Chart
Waveform Graph
X Scrollbar
XY Graph

SUMMARY

Chart—receives individual data points and continuously update the presentation of the data.
Graph—receives a complete data set (as an array) before preparing the graphical display.

Chart Types

- Waveform Chart
- Intensity Chart

Graph Types

- Waveform Graph
- XY Graph
- Intensity Graph
- Digital Waveform Graph
- Mixed Signal Graph
- 3D Graph (surface, parametric, curve)

1D Graphing and Charting

- **Waveform Chart**—receives single values, 1D array, or waveform
  - Keeps a chart history (right-click on chart, choose **Chart History Length** . . . to change size)
  - Update Modes (Strip Chart, Scope Chart, Sweep Chart)
  - Stack or Overlay Plots
- **Waveform Graph**—receives a 1D array, or a waveform

Controls Palette / Modern Group / Graph Group / Waveform Chart
Controls Palette / Modern Group / Graph Group / Waveform Graph

2D Plotting

XY Graph

- Receives a bundle of 1D arrays (*X* values array and *Y* values array are bundled)
- For two curves, build an array of two (X|Y) bundles

Plotting Multiple Curves

- Waveform Chart receives a bundle of values to plot multiple curves point by point (stack or overlay).
- Waveform Graph receives a 2D array to plot multiple curves (use Build Array function to create 2D array from two 1D arrays).
- XY Graph receives an array of (*X*|*Y*) bundles to plot multiple curves.

Modifying Graph Features

Pop-Up Menu

- **Marker Spacing**—options are Uniform and Arbitrary
- **Style**—select scale values, tick marks
- **Mapping**—linear or Logarithmic scale
- **Properties**—opens the Graph Properties dialog

- **AutoScale X**—activate or deactivate autoscaling
- **Loose Fit**—When checked, minimum and maximum scale values will be a whole multiple of the axis value interval
- **Visible Scale Label**—activate or deactivate the display of the axis label
- **Visible Items**
  - **Label** (defaults to graph type)
  - **Caption** (defaults to graph type)
  - **Plot Legend** (defaults to Plot 0, Plot 1, etc.)
  - **Scale Legend**—shows the text strings used as axis labels; buttons allow axes to be unlocked
  - **Graph Palette**—access to cursor, zoom, and pan features; axes must be unlocked
  - **Cursor Legend**—once a cursor has been defined, shows the current cursor location
  - **X Scrollbar**—used to slide the graph back and forth

Properties Dialog **(right-click on graph, choose Properties to open the dialog)**

Display Format Panel

- Set the number of Digits desired.
- Select whether the number of digits represents Digits of precision or Significant figures.
- Choose whether or not to display trailing zeroes.

Scales Panel

- Assign or edit the display name.
- Indicate whether or not the label will be displayed.
- Indicate whether or not the numeric scale will be displayed.
- Select a Log or Inverted scale; activate or deactivate autoscaling.
- Set display colors for the tick marks, marker values, and gridlines.

Plots Panel

- Select the plot (if multiples plots are shown in the graph).
- Enter or modify the display name for the plot.
- Choose the line style (solid, dashed, etc.).
- Choose the line thickness.
- Choose the marker style.
- Choose how the plot will be displayed.
  - Markers only
  - Markers with lines, no smoothing, horizontal lines at right marker value
  - Markers with smoothed lines
  - Markers with lines, no smoothing, horizontal lines at left marker value
  - Markers with lines, no smoothing, markers between horizontal lines
  - Markers with lines, no smoothing, markers between vertical lines
- Choose the colors of the lines and markers

Outputting Graphs

- Print the front panel
- Export Simplified Image . . . —right-click on the graph and select **Export Simplified Image . . .**

SELF-ASSESSMENT

1. What is the difference between a chart and a graph in LabVIEW?
   ANS: Chart controls can continuously receive data and update as needed. Graph controls receive an entire data set before creating the plot.
2. Are graphic displays (charts or graphs) added to the front panel or the block diagram?
   ANS: Graphic controls (e.g., Waveform Chart, XY Graph) only appear on the Controls Palette and must be placed on the front panel. When a graphic control is placed on the front panel, a node for the graphic control appears on the block diagram as well.
3. Does LabVIEW expect data for graphic controls in the form of arrays, or matrices?
   ANS: With one exception, LabVIEW expects data sent into graphic controls to be in arrays, not matrices. The exception is the Waveform Graph which will accept a 2D matrix (each row contains data for one curve).
4. How are "bundles" used with XY graphs?
   ANS: A bundle is a grouping of related items. The LabVIEW XY Graph control wants the *X* and *Y* values to be bundled before plotting. This ensures that there are the same number of *X* values as *Y* values being plotted. Because bundles can contain different data types, bundling (as opposed to building an array) allows, for example, integer *X* values to be bundled with double-precision *Y* values.
5. When do you "bundle" and when do you "build an array" when working with XY Graph controls?
   ANS: (*X*, *Y*) pairs are bundled. Bundles are built into arrays when you want to plot multiple curves on the same XY Graph.
6. What is a "waveform" in LabVIEW?
   ANS: A waveform is a data set with header information that includes the following items:
   - Start time
   - Time interval between values
7. What is the difference between the inputs required by a Waveform Graph and an XY Graph?
   ANS: A waveform graph receives a 1D array of *Y* values and plots the points evenly distributed across the horizontal (*X*) dimension of the graph. An XY graph receives a bundle of two 1D arrays, one array contains *X* values, one contains *Y* values.

   While a waveform graph will accept a 2D array of values, it will plot multiple curves; it will not use any input values as *X* values. An XY graph will also plot multiple curves, but the data must be sent in as an array of (*X*, *Y*) bundles.
8. How can you create an image of a LabVIEW plot (to put into a report, for example)?
   ANS: Right-click on the graph and select **Export Simplified Image ...** from the pop-up menu.

PROBLEMS

1. Enter the following values into a 1D array and graph them using a Waveform Graph control. Adjust the plot characteristics to show the plot with solid squares, connected by a dashed line.

Y
2
4
3
5
4
6

2. Plot the following values and use the plot to estimate the curve's maximum value.

Y
1.0
2.2
3.4
4.1
4.0
2.0
1.1

3. Enter the following values into two 1D arrays. Bundle the (X, Y) values and send the bundle to an XY Graph control. Show the data points with markers connected with a dashed line.

X	*Y*
1.2	3.5
2.1	4.0
3.5	4.3
4.2	4.2
4.6	4.1
5.6	3.7
7.2	2.7

4. Enter the following values into two 1D arrays. Bundle the (X, Y) values and send the bundle to an XY Graph control. Estimate the X value where the curve crosses $Y = 0$.

X	*Y*
1.2	9.1
2.3	12.4
3.5	10.5
4.2	6.7
5.8	−9.3
7.6	−39.5
8.2	−52.5

5. Enter the following values into three 1D arrays and create an XY graph with two curves: (X, $Y1$) and (X, $Y2$). Use the Plot Legend and set the curve properties so that the curves are easily distinguished when printed on a black and white printer.

X	$Y1$	$Y2$
1.2	9.1	7.1
2.3	12.4	9.6
3.5	10.5	8.5
4.2	6.7	6.1
5.8	−9.3	−4.4
7.6	−39.5	−24.1
8.2	−52.5	−32.5

6. Enter the following values into four 1D arrays and create an XY graph with two curves: ($X1$, $Y1$) and ($X2$, $Y2$). Set the curve properties so that the curves are easily distinguished when presented in a classroom using a color projection system. Estimate the point of intersection of the two curves.

$X1$	$Y1$	$X2$	$Y2$
1	12	1	1.5
2	9.0	3	4.6
3	7.0	5	7.0
4	5.4	7	8.4
7	1.9	9	9.0
8	1.3		
9	1.0		

7. The data required to generate a plot of Bessel function J0(x) are available in files BesselX.txt and BesselY.txt at the text's website.

www.chbe.montana.edu/LabVIEW

Download the files, then create an XY graph of the Bessel function. What is the value of J0(x) when $x = 22$?

Hint: You can check your result using LabVIEW's function: Bessel Function Jv.vi.

8. A first-order response is a common model for a variety of processes in engineering (e.g., mixing in a tank, warming a thermocouple, charging a capacitor). A first-order response has a time constant, τ, that is indicative of the speed of the response.

- After a time equal to 1τ, 63.2% of the total change in Y will have taken place.
- After a time equal to 2τ, 86.4% of the total change in Y will have taken place.
- After a time equal to 3τ, 95.0% of the total change in Y will have taken place.

Data for a first-order response are available in files FirstOrderTime.txt and FirstOrderY.txt at the text's website. Download the files, and then create an XY graph of the response. What is the value of the time constant, τ, for the process?

CHAPTER

8

Data Analysis Using LabVIEW VIs

Objectives

After reading this chapter, you will know:

- how to use LabVIEW functions to calculate basic statistical quantities
- how to interpolate data tables using a LabVIEW function
- how to fit an interpolated curve through each data point in a data set
- how to use linear regression to finding the best-fit regression model for a data set

8.1 INTRODUCTION

This chapter focuses on the routine calculations that are basic to working with data sets, including

- Calculating means and standard deviations
- Interpolating between data points in a data set
- Fitting a curve to a data set
- Determining regression coefficients for the best-fit curve

The goal of this chapter is not to write elaborate VIs, but to demonstrate how to use the data analysis tools that are built into LabVIEW.

We will use a number of data sets in the examples in this chapter; these data sets are available on the text's website www/chbe/montana.edu/LabVIEW.

8.2 BASIC STATISTICS

LabVIEW provides functions to calculate basic descriptive *statistic* values on a single data set such as

- Maximum
- Minimum
- Mean
- Median
- Standard deviation
- Variance

For two data sets LabVIEW provides functions to compute correlation coefficients.

There are random number generators that can generate sets of random values with various distributions (uniform, normal, etc.)

Additional functions provide for

- Creating histograms
- Hypothesis testing
- Analysis of variance (ANOVA)

We will begin with a VI that calculates and displays the basic descriptive statistics for a data set. To start with a data set very familiar to students, we will begin by looking at scores from an exam.

EXAMPLE 8.1

Descriptive Statistics for a Set of Exam Scores

Given the exam scores in Table 8.1, calculate and display the basic descriptive statistics, including

- Maximum
- Minimum
- Mean
- Median
- Standard deviation
- Variance

Then create a histogram showing how the scores were distributed.

Table 8.1 Data set 1: Exam scores

77	93	87	92
91	83	87	86
85	75	92	90
97	74	85	82
83	53	74	71
85	93	78	78
92	87	92	84
55	82	73	
83	86	81	

The scores were imported from a tab-delimited text (.txt) file using the Read From Spreadsheet function in the Programming Group.

Functions Palette / Programming Group / File I/O Group / Read From Spreadsheet

Since the scores ultimately need to be in a 1D array (for compatibility with the functions in the Probability & Statistics Group), the **first row** output terminal on the Read From Spreadsheet function must be used, and the **Transpose** input must receive a TRUE in order to read a column of values. The block diagram used to read the exam scores is shown in Figure 8.1.

Note: Alternatively, the data can be assigned to a 2D array using the Read From Spreadsheet function's **all rows** output, and then the first column pulled out using the Index Array function. Since there is only one column of values in the .txt file, using the **first row** output is more efficient.

The first row array indicator was changed to a control (current values saved as default), just to save space on the block diagrams.

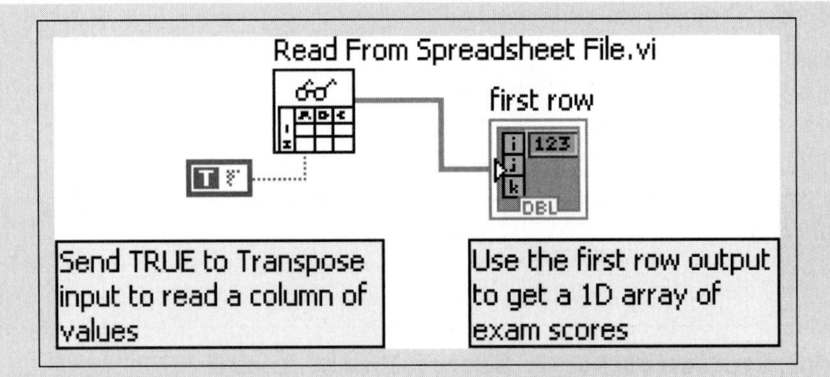

Figure 8.1
Block diagram used to read exam scores from text file.

Note: In these examples the scores are presented as a column array, but the mathematical calculations in the rest of the example will work equally well with either a row array or a column array.

The statistics were calculated using functions from either the Array Group or Probability & Statistics Group:

Functions Palette / Programming Group / Array Group / Array Max & Min

Functions Palette / Mathematics Group / Probability & Statistics Group

- Mean.VI—arithmetic average value
- Median.VI—central value
- Std Deviation and Variance.VI

Definitions

- **Mean**—arithmetic average value
- **Median**—the *median* is the central value when values are arranged in a sorted list
- **Standard Deviation**—the *standard deviation* describes the extent of variability in the values
- **Variance**—the *variance* is the square of the standard deviation

The block diagram for calculating these statistics is shown in Figure 8.2, and the results are shown on the front panel in Figure 8.3.

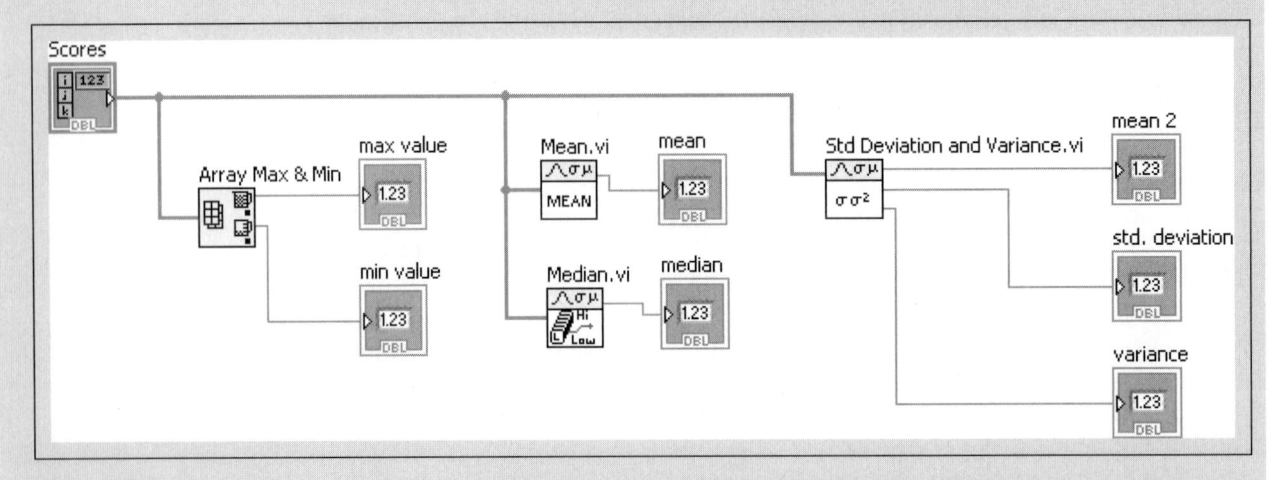

Figure 8.2
Block diagram of Descriptive Statistics VI.

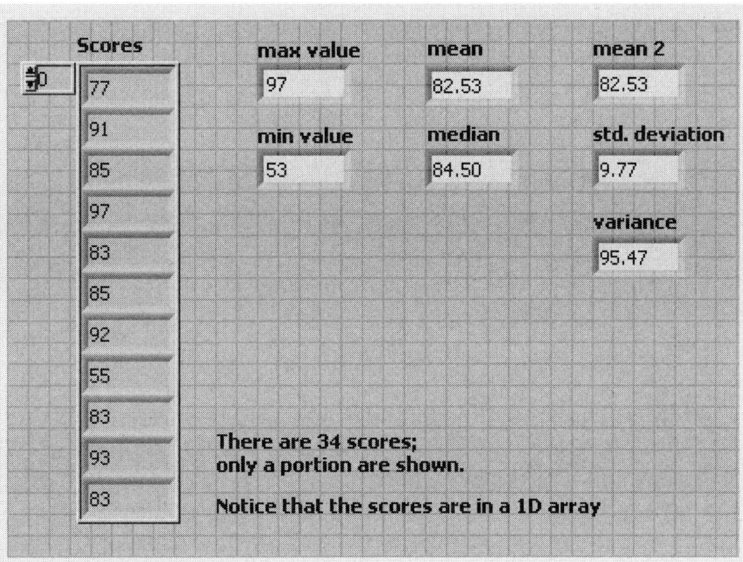

Figure 8.3
Front panel of Descriptive Statistics VI.

Notice that the mean (arithmetic mean) was calculated twice, once using the Mean function and once as one of three outputs from the Std Deviation and Variance function. Both calculations gave the same result (mean = 82.5) and either can be used.

PRACTICE

Find the Mean and Standard Deviation of a Data Set

The mean and standard deviation are commonly used to provide a sense of the average value in a data set and the extent of the spread of the values in the data set. For example, if you take the same measurement several times, you can use the standard deviation to get an idea of the uncertainty associated with the measurement.

The values in the following list represent repeated diameter measurements taken using a caliper with a digital readout. The caliper shows four decimal places, but how accurate are the readings? Calculate the standard deviation to find out.

Diameter (mm)
451.0063
453.5625
451.1954
455.2409
453.4645
453.6030

The mean and standard deviation were determined using LabVIEW's Std Deviation and Variance.VI, and the results are shown in Figure 8.4. The mean is 453 mm, but the standard deviation is 1.6 mm. This indicates that there is uncertainty in the "ones" position, so it is absurd to report these values with seven significant digits.

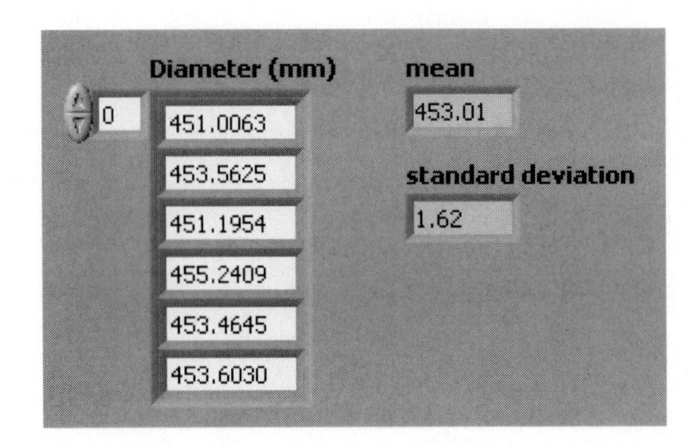

Figure 8.4
Calculating the mean and standard deviation of a data set.

Note: This does not imply that the calipers are unable to deliver a more precise value in a different situation. This result could be imprecise because the measurement is being taken on something that is not a true circle, or perhaps it is in a difficult location and the calipers are not being positioned accurately.

Once the mean and standard deviation have been calculated, they are typically reported as mean ± std. dev., or 453 ± 1.6 mm in this example.

The block diagram used to find the mean and standard deviation of the data set is shown in Figure 8.5.

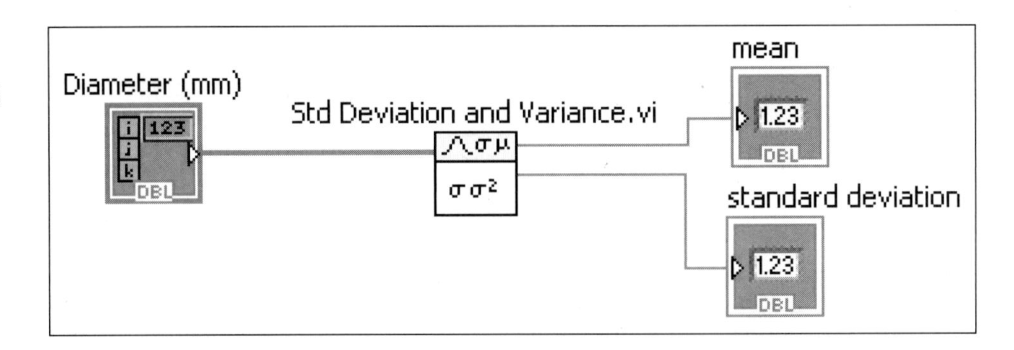

Figure 8.5
Block diagram of VI used to find mean and standard deviation.

As a final step, we will modify the VI to display the exam scores as a *histogram*. LabVIEW provides three functions for creating histograms:

- **Histogram.vi**—you can set the number of bins, but not the bin ranges.
- **General Histogram.vi**—you can specify the bin ranges.
- **Create Histogram.vi**—designed to create a histogram from a signal, although it will accept an array as input.

Functions Palette / Mathematics Group / Probability & Statistics Group

Definitions

- **Histogram**—a graph that shows how values in a data set are distributed. Most students are familiar with grade distribution graphs, which are a type of histogram.
- **Bins**—the values in a data set are sorted into categories, called *bins*. Bins for a grade distribution are typically A, B, C, and so on.

In this example we will use the General Histogram function so that we can specify the bin range values to see how many scores are in the following ranges:

- 50–60
- 60–70
- 70–80
- 80–90
- 90–100

We also need to decide how to handle scores that fall on a bin boundary. For example, should a score of 80 fall into the 80–90 bin, or the 70–80 bin? That is, should someone who got an 80 be in the "B" bin or the "C" bin? The usual way of thinking about grades is to include a score on the lower bin limit in the bin (they get the higher grade). So, the bin limits are now:

- 50–60, include score on the **lower** bin limit in the bin
- 60–70, include score on the **lower** bin limit in the bin
- 70–80, include score on the **lower** bin limit in the bin
- 80–90, include score on the **lower** bin limit in the bin
- 90–100, include scores on **both** the lower bin and upper bin limit in the bin

The bins are specified for the General Histogram function by defining a cluster consisting of (lower bin limit, upper bin limit, boundary inclusion code) where the boundary inclusion codes are as follows:

- 0–lower-score on lower bin boundary is included in bin
- 1–upper-score on upper bin boundary is included in bin
- 2–both-score on either bin boundary is included in bin
- 3–neither-score on either bin boundary is not included in bin

The array of clusters that needs to be sent to the General Histogram function's **Bins** input looks like this:

$$\begin{bmatrix} (50, 60, 0) \\ (60, 70, 0) \\ (70, 80, 0) \\ (80, 90, 0) \\ (90, 100, 2) \end{bmatrix}$$

Fortunately, we don't have to create that array of clusters ourselves. We can right-click on the **Bins** input and select **Create / Control** from the pop-up menu. A control will be placed on the front panel that can be expanded to five array elements to allow us to set the required values for the five bins. This is shown near the bottom of the final front panel of the Descriptive Statistics VI, in Figure 8.6.
The block diagram for the final Descriptive Statistics VI is shown in Figure 8.7.

8.3 INTERPOLATION

When you are working with tabulated data, as often as not it seems like, the value you need is in between the values in the table. For example, steam tables are tabulated values of the thermodynamic properties of water. The example shown in

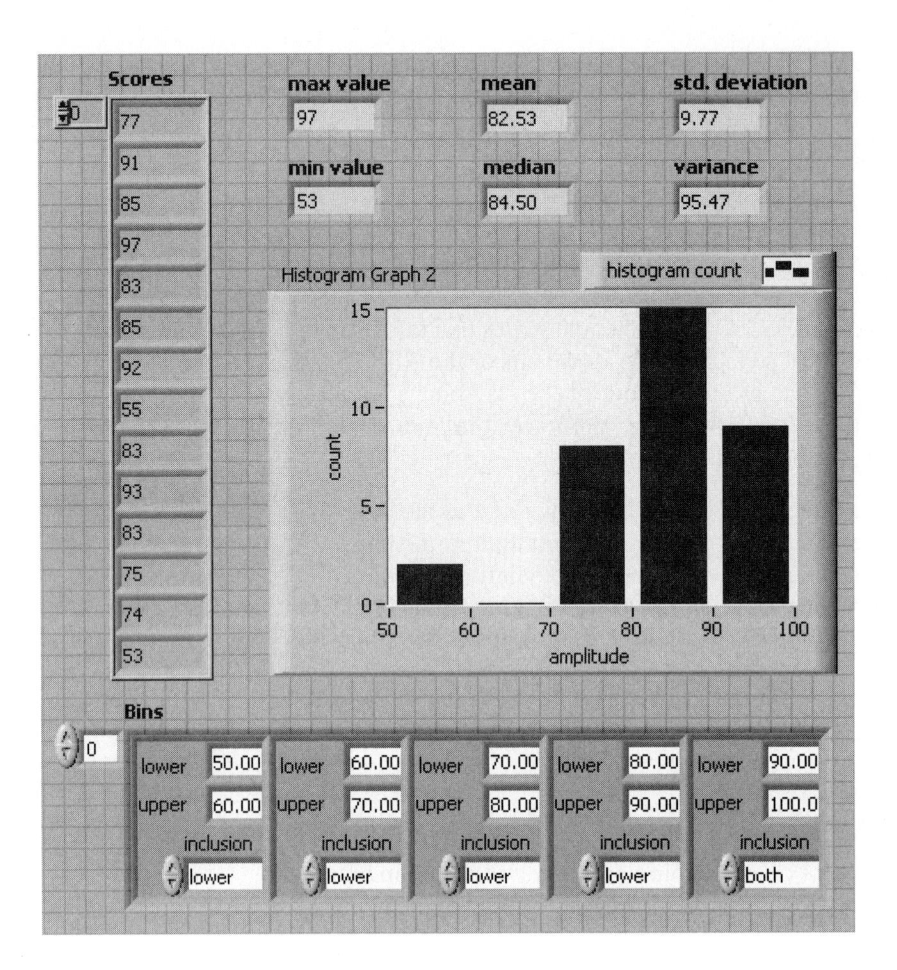

Figure 8.6
Front panel of Descriptive Statistics VI with histogram.

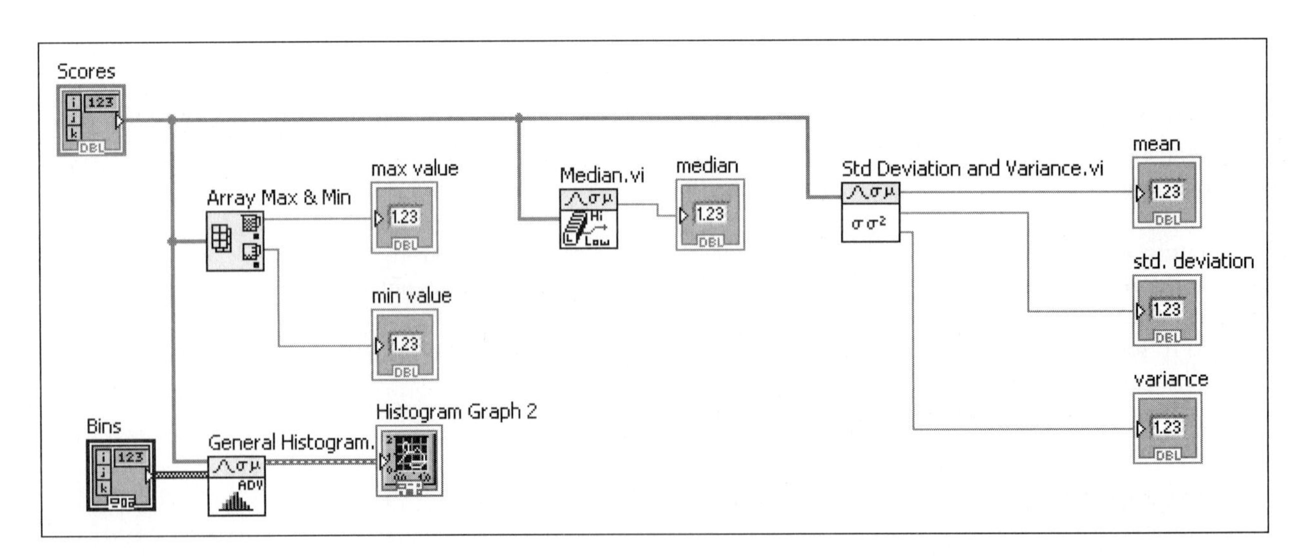

Figure 8.7
Block diagram for the final Descriptive Statistics VI.

Table 8.2 has values of enthalpy at six different temperature values, but not at 230°C. How can we estimate the enthalpy of saturated steam at 230°C using the data in Table 8.2? We need to *interpolate* to find the answer; the process is termed *interpolation*.

Table 8.2 Properties of saturated steam

Temperature (°C)	Internal energy (kJ/kg)	Enthalpy (kJ/kg)	Entropy (J/g*K)
100	2506.0	2675.6	7.3541
150	2559.1	2745.9	6.8371
200	2594.2	2792.0	6.4302
250	2601.8	2800.9	6.0721
300	2563.6	2749.6	5.7059
350	2418.1	2563.6	5.2110

Source: E. W. Lemmon, M. O. McLinden, and D. G. Friend, "Thermophysical Properties of Fluid Systems" in NIST Chemistry WebBook, NIST Standard Reference Database Number 69, Eds. P. J. Linstrom and W. G. Mallard, National Institute of Standards and Technology, Gaithersburg MD, 20899, http://webbook.nist.gov (retrieved February 4, 2009).

Definitions

- **Internal Energy**—When energy is added to an object (heat transfer), it is stored as internal energy. The usual evidence that internal energy is increasing is an increase of temperature. Commonly used with closed (non-flow) systems.
- **Enthalpy**—Accounts for not only the internal energy change when energy is added, but also changes in pressure and volume. Commonly used with open (flow) systems.
- **Entropy**—a measure of randomness, sometimes described as the amount of energy in a system that cannot be used for mechanical work.

EXAMPLE 8.2

Interpolate to Find the Enthalpy of Saturated Steam at 230°C

Interpolation involves using neighboring data values to estimate an unknown quantity. One method is to look at the data and estimate. From Table 8.2 we know that the enthalpy is 2792.0 kJ/kg at 200°C, and 2800.9 kJ/kg at 250°C. A quick estimate would be 2798 kJ/kg at 230°C. In the following paragraphs, we will look at calculation methods for interpolation, and we can test the estimate.

If we plot the temperature and enthalpy data, we get the plot in Figure 8.8. The dashed line is at 230°C from the graph it is apparent that the enthalpy value is near 2800 kJ/kg, but the graph is hard to read much more accurately than that.

Zooming in on the graph and connecting the dots in a couple of ways will provide some insight into how some interpolations methods work.

In Figure 8.9 the graph has been zoomed in to the region around 230°C, and the data values have been connected with a straight line. From Figure 8.9 it appears that the enthalpy should be about 2797 kJ/kg.

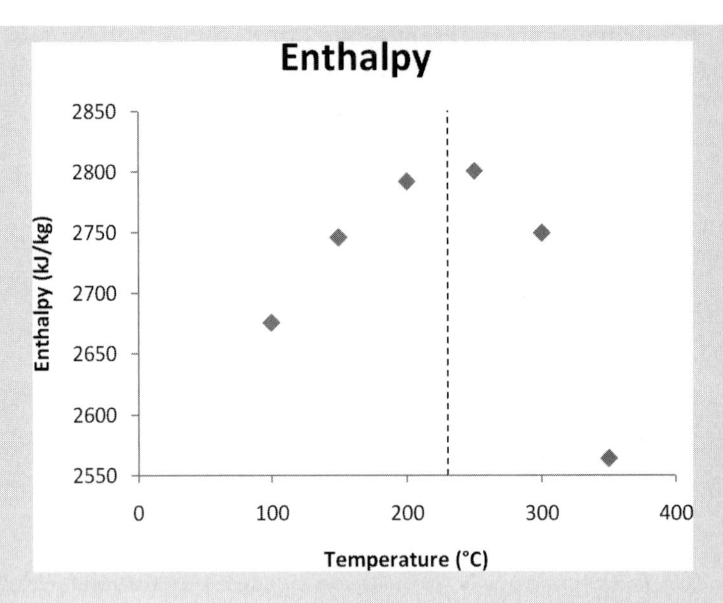

Figure 8.8
Enthalpy of saturated steam.

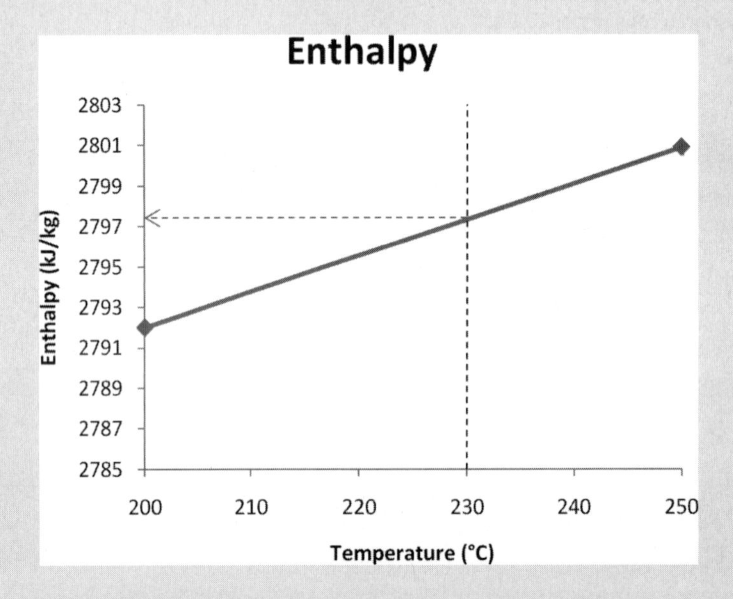

Figure 8.9
Zooming in on the region between 200 and 250°C—straight line.

But, the spreadsheet that was used to create the enthalpy graphs will also connect the data points with "smoothed" lines, as shown in Figure 8.10.

The smoothed lines through the data take into account the data points at 150 and 300°C as well. The result is shown in Figure 8.10, and this curve suggests that the enthalpy value should be about 2804 kJ/kg.

These last two graphs illustrate *linear interpolation* (assuming the unknown value is on a straight line between data points, Figure 8.9) and *cubic spline interpolation* (assuming the unknown value is on a smooth curve between data points, Figure 8.10). Both types of interpolation are available using LabVIEW's Interpolate 1D function.

Note: LabVIEW provides a number of interpolation functions. We are focusing on the Interpolate 1D function because it can be used for several different types of interpolation.

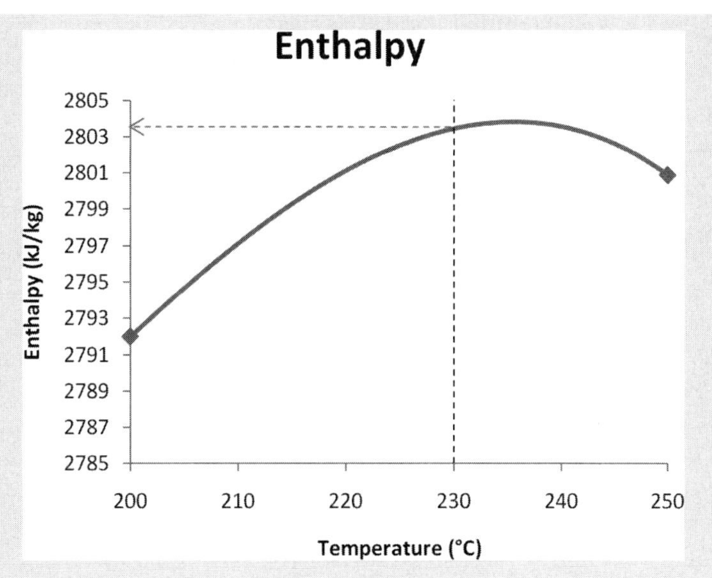

Figure 8.10
Zooming in on the region between 200 and 250°C—smoothed line.

The Interpolate 1D function is located at (not available in Base LabVIEW Package):

Functions Palette / Mathematics Group / Interpolation & Extrapolation / Interpolate 1D

The connection pane for the Interpolate 1D function is illustrated in Figure 8.11.

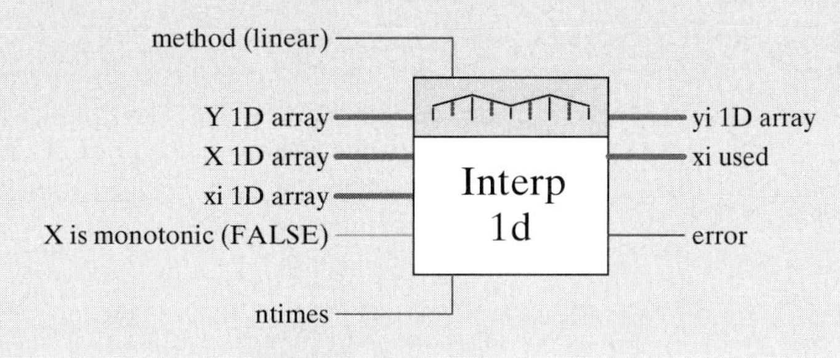

Figure 8.11
Connection pane info for Interpolate 1D function.

The **method** terminal is used to select the interpolation method:

- 0—nearest (no interpolation, just the closest value in the data set)
- 1—linear (default)
- 2—cubic spline
- 3—cubic Hermite
- 4—Lagrange

The ***X*** and ***Y*** input terminals are used to supply the data set values used in the interpolation. The ***xi*** input is used to indicate the value(s) at which the interpolation should be performed (230°C in this example).

Note: The ***xi*** input requires a 1D array. In this example we are interpolating at a single temperature, but we must supply that temperature value to the ***xi*** input as a 1D array containing one value.

If you know that the X values in your data set have been sorted and increase *monotonically* (always increasing, not jumping around), then you can input a TRUE to the ***X* is monotonic** input. This speeds up the interpolation routine since it does not need to sort the *X* array. The **ntimes** input is not used if ***xi*** values are specified.

The ***yi*** output provides our desired solution. The ***xi* used** output provides an array of the *xi* values used to compute *yi* values. This will be useful when the **ntimes** input was used (see next section), otherwise the *xi* values are already known.

To use the Interpolate 1D function, we will need to provide the following:

- the method indicator (we will use several)
- known (data set) Temperature values, as a 1D array of *X* values
- known enthalpy values that correspond to the temperature values, as a 1D array of *Y* values
- the temperature at which we want to determine the enthalpy, as a 1D array of *xi* values

The front panel of the Interpolation VI is shown in Figure 8.12, and the corresponding block diagram is shown in Figure 8.13. In Figure 8.12 the "linear" interpolation method was used, and the interpolated enthalpy value was found to be 2797.34 kJ/kg, very close to the value obtained from Figure 8.9.

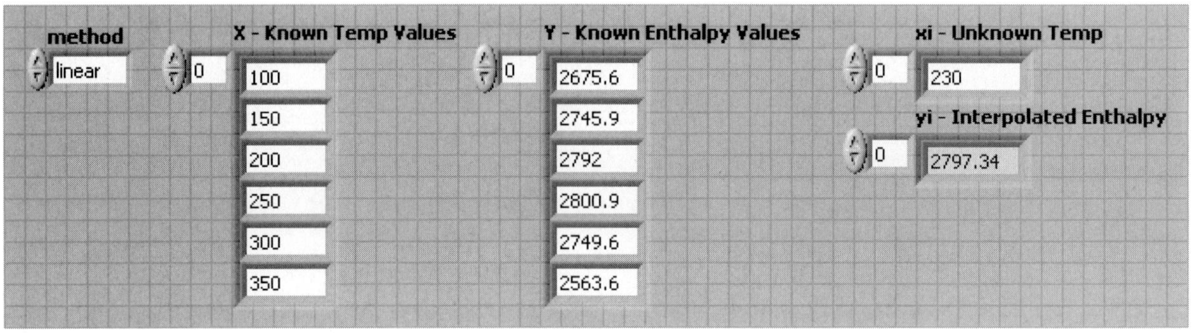

method
linear
X - Known Temp Values
0
100
150
200
250
300
350
Y - Known Enthalpy Values
0
2675.6
2745.9
2792
2800.9
2749.6
2563.6
xi - Unknown Temp
0
230
yi - Interpolated Enthalpy
0
2797.34

Figure 8.12
Interpolation VI, using Linear method.

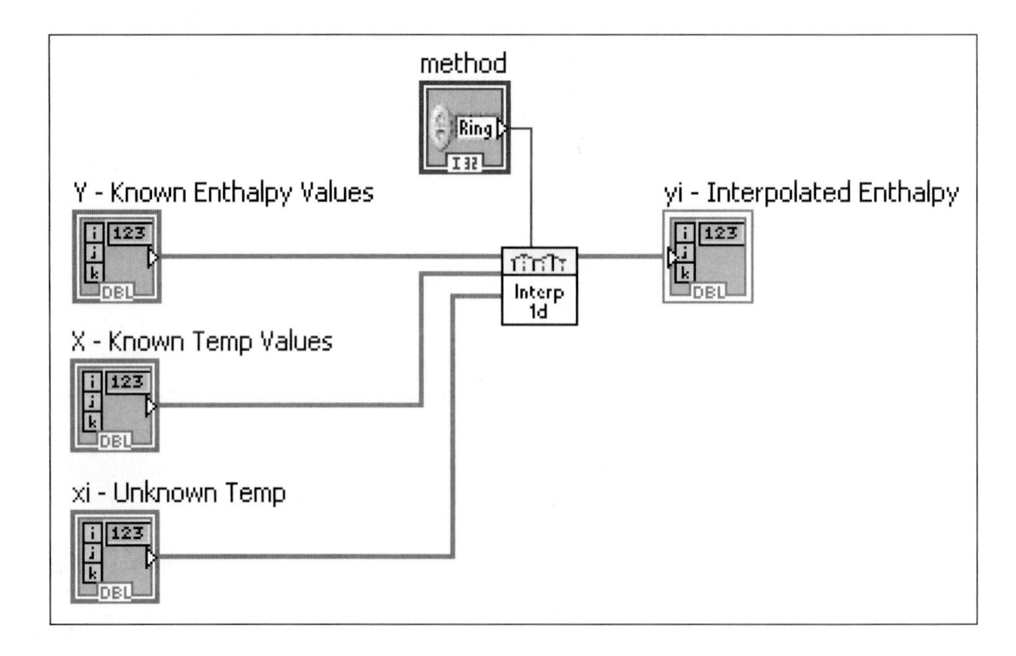

Figure 8.13
Block diagram for Interpolation VI.

To use the other interpolation methods, simply change the **method** control value and run the VI again. The results of the various interpolation methods are

- Linear: 2797.3 kJ/kg
- Spline: 2801.7 kJ/kg—this is the cubic spline result, called "spline" in the method list
- Hermite: 2799.2 kJ/kg—this is the cubic Hermite result, called "cubic" in the method list
- Lagrange: 2802.57 kJ/kg

Which one is "correct"? They are all correct to about three significant digits, and none is likely to be correct to five.

PRACTICE

Interpolating Sine Data

LabVIEW's Interpolate 1D function is powerful and easy to use. Try using both linear and cubic spline interpolation to find the sine of 45° given:

a. Sine values at 40° and 50° [sin(40°) = 0.6428, sin(50°) = 0.7660]
b. Sine values at 30°, 40°, 50° and 60° [sin(30°) = 0.5000, sin(60°) = 0.8660]

Compare the interpolated values with the known result, sin(45°) = 0.7071

Part a—one point on either side. The interpolated result is 0.7044 with either method. There is no difference between the linear interpolation (see Figure 8.14) and the cubic spline interpolation (see Figure 8.15) because there are not enough data points to fit a cubic spline curve.

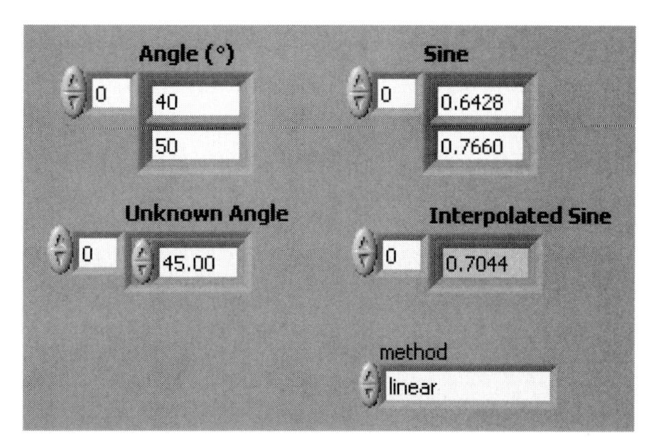

Figure 8.14
Linear interpolation with one point on either side of unknown angle.

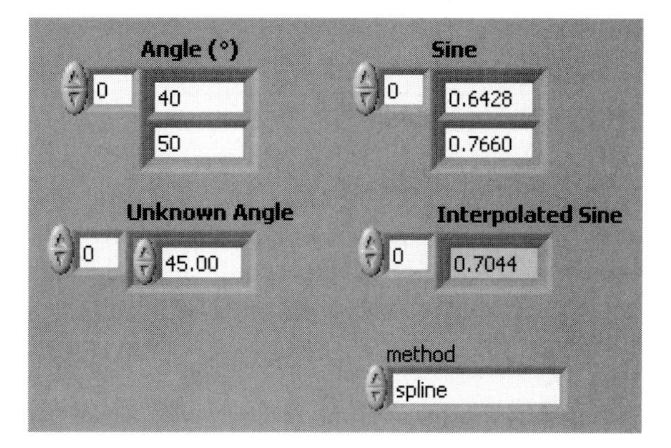

Figure 8.15
Spline interpolation with one point on either side of unknown angle.

Part b—two points on either side. The interpolated result is still 0.7044 with the linear method (see Figure 8.16). This is because linear interpolation only uses the adjacent points; the additional known values were ignored.

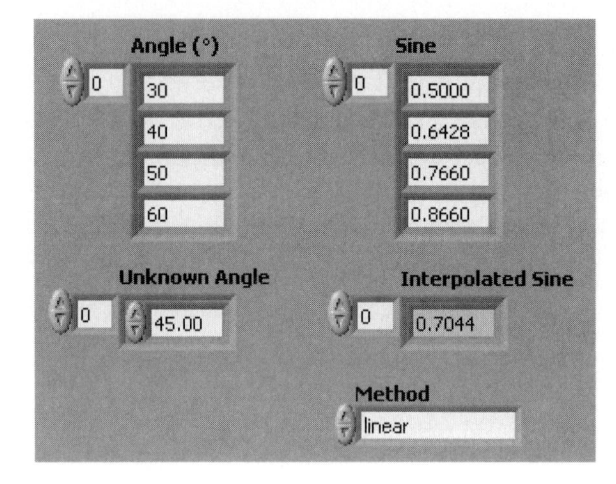

Figure 8.16
Linear interpolation with two points on either side of unknown angle.

The cubic spline interpolation does use the additional known values and generates a more accurate interpolated result of 0.7076 (see Figure 8.17). This result is accurate to 5 parts in about 7000, or 0. 07% error.

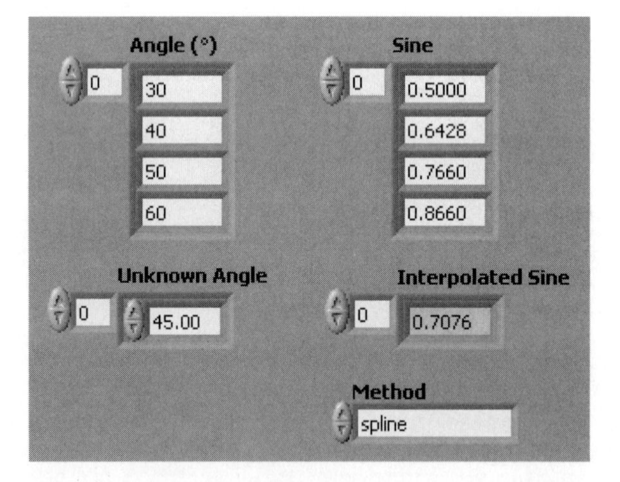

Figure 8.17
Spline interpolation with two points on either side of unknown angle.

8.4 CURVE FITTING

The general term *curve fitting* means getting some sort of a curve through a set of data points. There are two general approaches to curve fitting, with different goals:

- **Method 1**—Get a curve on a graph that goes through each data point; no model (mathematic equation) is needed.

- **Method 2**—Find the coefficients of a mathematical model (regression equation) that best fits the model to the data. It is not required that the plotted regression equation go through each data point.

Method 2 (Regression) is the subject of the next section. This section on curve fitting focuses on Method 1. Method 1 is intrinsically related to interpolation, because the same methods are used for interpolation and for Method 1 curve fitting.

To use LabVIEW's Interpolation 1D function for curve fitting, you have two options:

- Supply an array of ***xi*** values that spans the range of the *X* values in the data set.
- Do not supply the *xi* array at all, but set the **ntimes** value to tell LabVIEW how many times to interpolate between the values.

The meaning of **ntimes** may not be immediately obvious; it indicates how many times LabVIEW will go through the data set interpolating between each current value of *X*. This is illustrated in Figure 8.18. The black circles represent the original data values, and the open circles represent interpolated values.

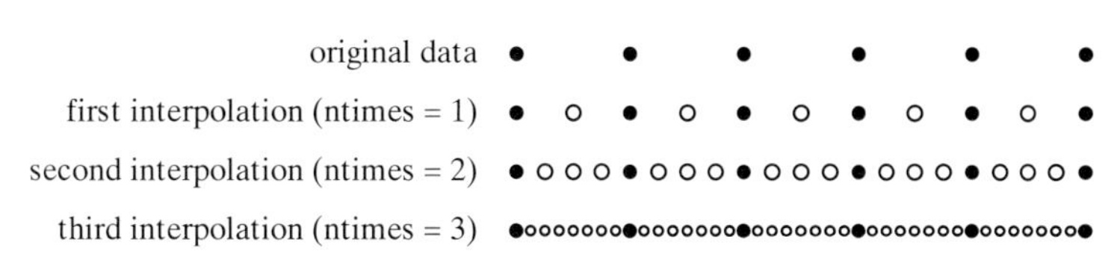

Figure 8.18
The ntimes value indicates how many interpolation passes will be made through the data.

- **ntimes = 1**—LabVIEW makes one interpolation pass through the data and adds one interpolated point between each original data point.
- **ntimes = 2**—LabVIEW makes two interpolation passes through the data. In the first pass one interpolated point is added between each original data point (same as when ntimes = 1). In the second pass, another interpolated point is added between each point after the first pass (original and interpolated). The result is three interpolated points between each original data point.
- **ntimes = 3**—LabVIEW makes three interpolation passes through the data and ultimately leaves seven interpolated points between each original data point.

Figure 8.19 shows the Curve Fitting VI's block diagram. It is similar to the Interpolation VI, but the *xi* array is gone, and a constant with a value of 3 has been wired to the **ntimes** input terminal. The *xi* and *yi* outputs have been wired to array indicators to show the calculated *xi* (Temperature) values and the interpolated *yi* (Enthalpy) values.

The two Bundle functions are used to collect *x* and *y* values (i.e., temperature and enthalpy values) for plotting. The Bundle on the left combines the temperature and enthalpy values from the data set, while the Bundle on the right combines the calculated temperature values and the interpolated enthalpy values. The Build Array function is used to combine the two bundles so that two enthalpy vs. temperature

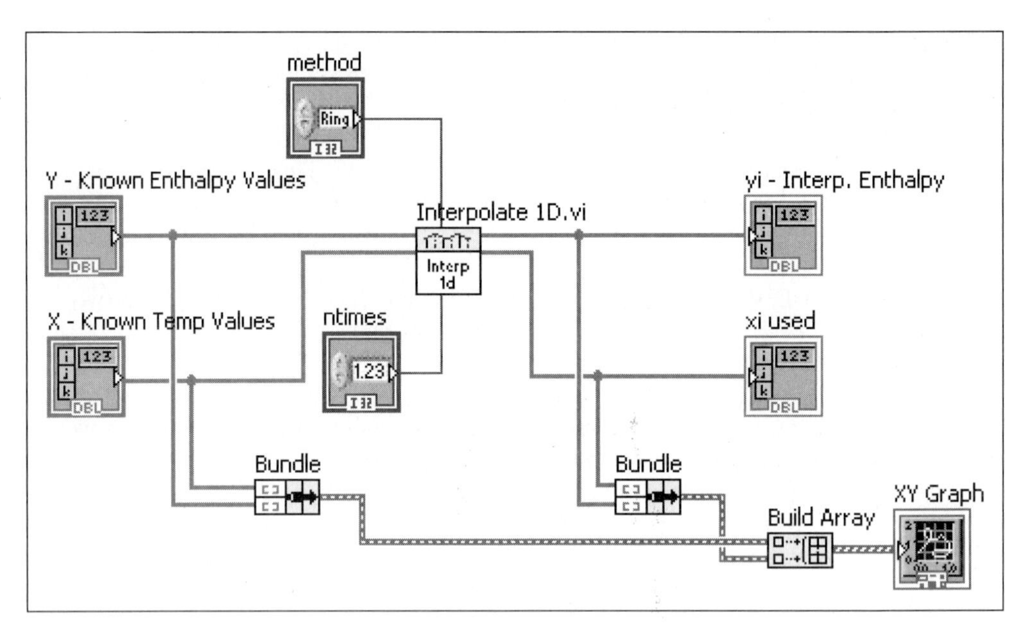

Figure 8.19
Block diagram of the Curve Fitting VI.

curves can be plotted on the XY Graph. Figure 8.20 shows the result of calculating three intermediate points (ntimes = 2) between each original data point using a spline fit.

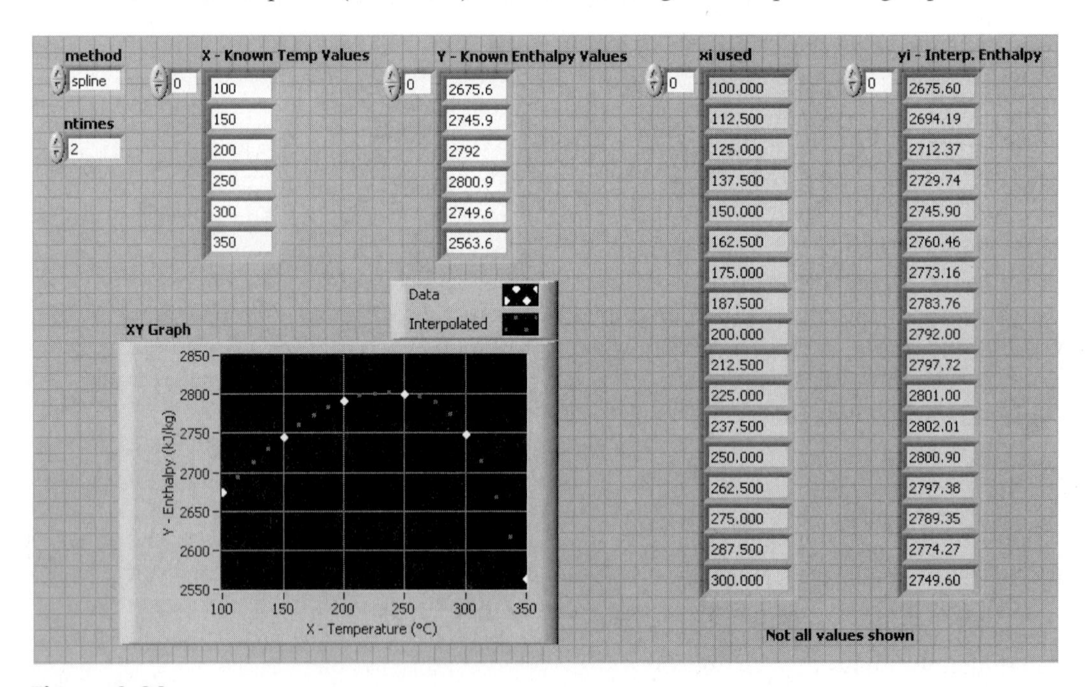

Figure 8.20
Front panel of Curve Fitting VI.

Some of the other method options (cubic spline, cubic Hermite, and Lagrange) can also be used for curve fitting, but when there is a curvature (as in Figure 8.21) the fitted curves can be quite dissimilar. Figure 8.21 shows the result of fitting with a spline curve, while Figure 8.22 shows the Lagrange fit (same data).

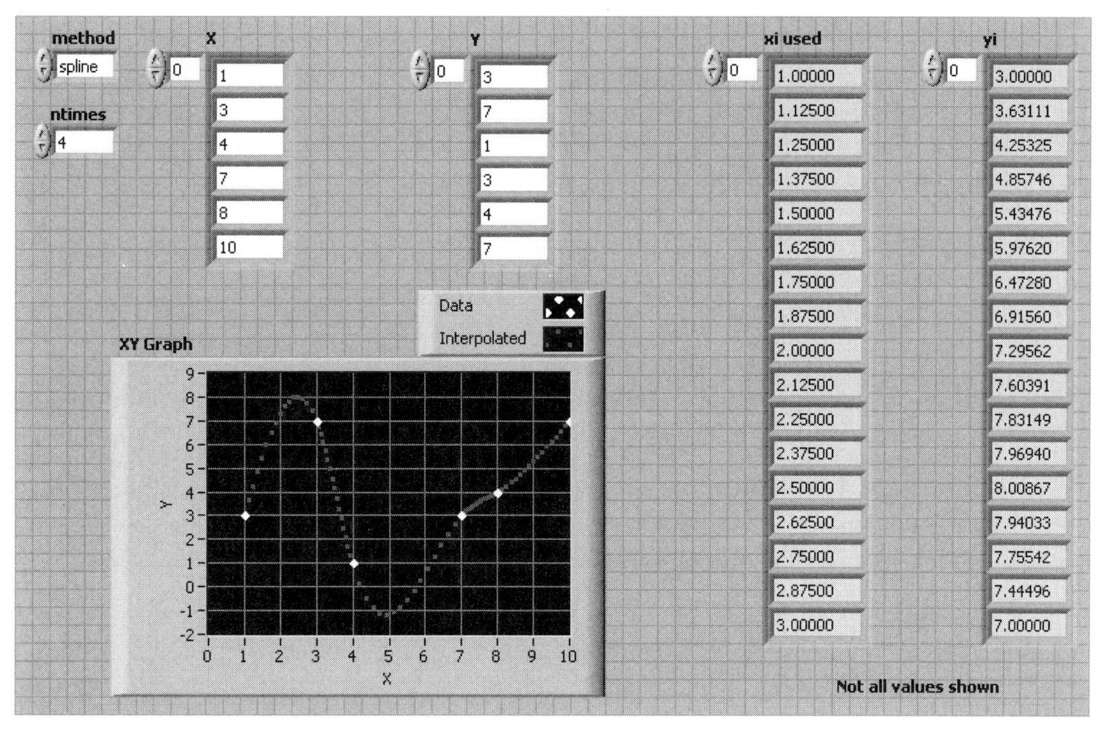

Figure 8.21
Spline fit to data set with lots of curvature.

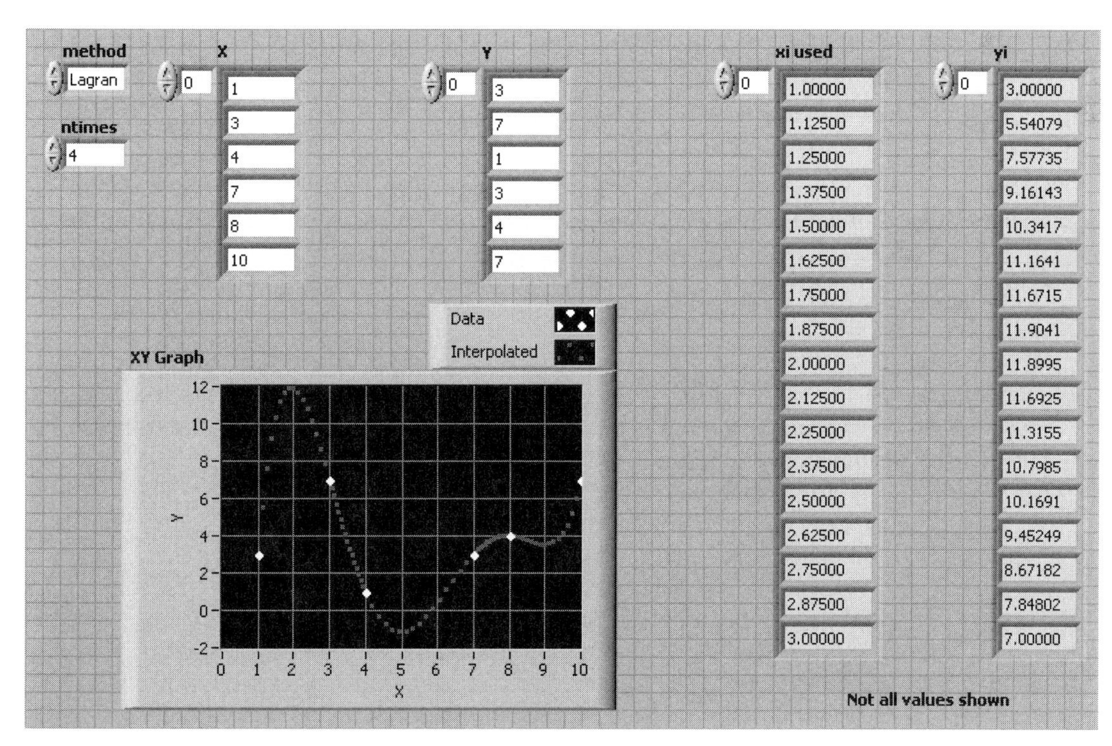

Figure 8.22
Lagrange fit to data set with lots of curvature.

8.5 REGRESSION

Regression involves finding the coefficients that cause a mathematical model (*regression model*) to best fit the values in a data set. "Best" means the coefficients that minimize the *sum of the squared error (SSE)* between the y values predicted by the regression model and the y values in the data set.

Quick Explanation

In Figure 8.23, the regression goal is to find the coefficients of the regression model—the "b" values in $y_{pred} = b_0 + b_1x + b_2x^2$—that cause the regression curve to "best fit" the data values. The distance between any data point and the regression curve at that point is called the *error* at that point (see e_i in Figure 8.23). You might try minimizing the overall error for all data points, but that won't work because some errors are positive and some are negative. To keep positive errors from canceling with negative errors, we use the squared error at each point. Squaring errors makes all of the values positive. By minimizing the sum of the squared error (SSE) for all points, we can find the "b" values that best fit the data points, where "best" means minimum overall SSE.

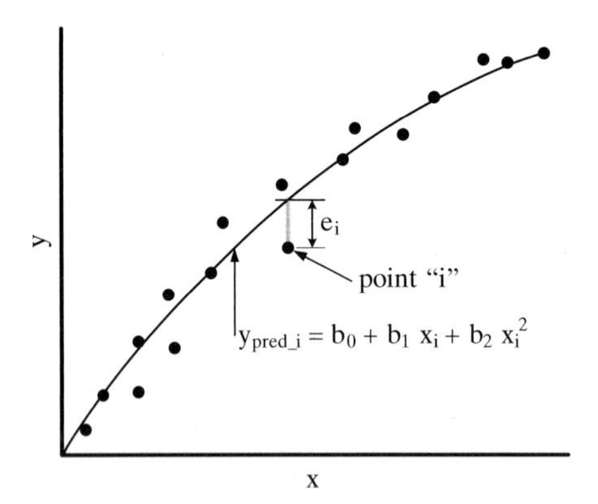

Figure 8.23 How least squares regression works.

Note: Actually LabVIEW allows you to select between three different definitions of "best" for many of the regression functions:

- Least Square
- Least Absolute Residual
- Bisquare

The Least Square method is the common method, and we will use that method in the VIs developed here.

The most common type of regression is *linear regression*. With linear regression the coefficients can be determined directly (perhaps with some significant matrix math). *Non-linear regression* is also possible, but requires an iterative solution of trying various coefficient values until the sum of the squared error is minimized. LabVIEW provides functions for both linear and non-linear regression. All of the regression functions are found in the Fitting Group (not available in Base LabVIEW Package):

Functions Palette / Mathematics Group / Fitting Group

Linear Regression Functions

- Linear Fit $f = ax + b$
- Exponential Fit $f = ae^{bx}$
- Power Fit $f = ax^b$
- Gaussian Peak Fit $f = a \exp\left[-\frac{(x-\mu)^2}{2\sigma^2}\right]$
- Logarithm Fit $f = a \log_c(bx)$ — c (logarithm base) is a function parameter
- General Polynomial Fit $f = \sum_{j=0}^{m} a_j x_i^j$ — $1 \le m \le 25$
- General LS (Least Squares) Linear Fit—fits any linear model that you provide

Non-Linear Regression Functions

- Non-Linear Curve Fit
- Constrained Non-Linear Curve Fit

8.5.1 Linear Fit

A linear fit uses the following regression model:

$$y_p = b_0 + b_1 x$$

where

b_0 is the intercept
b_1 is the slope

The connection pane for the Linear Fit function is illustrated in Figure 8.24.

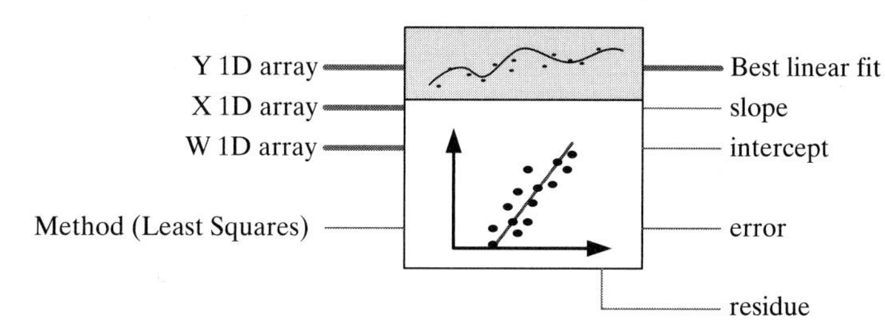

Figure 8.24 Linear Fit function connector pane (showing Least Squares terminals only).

We need to supply 1D arrays of X and Y values. The W (weight) array is optional and rarely used. The slope and intercept will be displayed, and the Best Linear Fit values (y_p values) will be used to create a graph showing the original values and the best fit line.

A common feature of linear regressions is reporting the R^2 value. The Linear Fit function does not provide that value, but it can be calculated by the Goodness of Fit function, which receives the Y (data) and y_p (regressed) values as inputs, and outputs the R^2 value. The Goodness of Fit function is available in the Advanced Curve Fitting Group:

Functions Palette / Mathematics Group / Fitting Group / Advanced Curve Fitting

Definition

- R^2—the *coefficient of determination*. If the best fit regression curve goes through every data point perfectly, $R^2 = 1$. The more distance there is between the data points and the regression curve, the smaller the value of R^2.

The Slope and Intercept VI is shown in Figure 8.25 (front panel) and Figure 8.26 (block diagram). The R^2 value of 0.97 indicates a good fit between the regression curve and the data points.

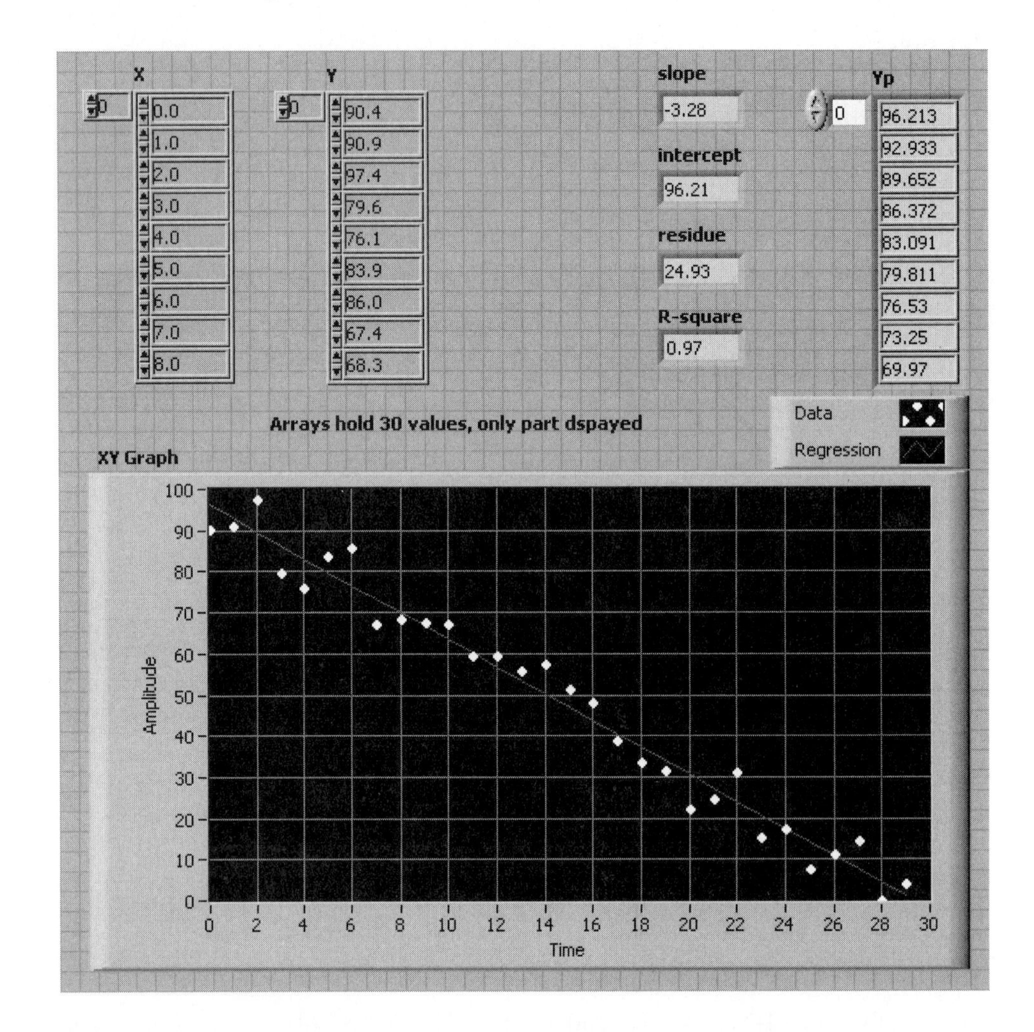

Figure 8.25 Slope and Intercept VI, front panel.

PRACTICE

Linear Regression for a Linear Fit

Regression for a slope and intercept is a very common data analysis task. Practice using LabVIEW for this by regressing the following data using LabVIEW's Linear Fit function. Also use the Goodness of Fit function to calculate R^2.

X	Y
1	2.0
2	7.5
3	8.2
4	12.8
5	15.9
6	19.1

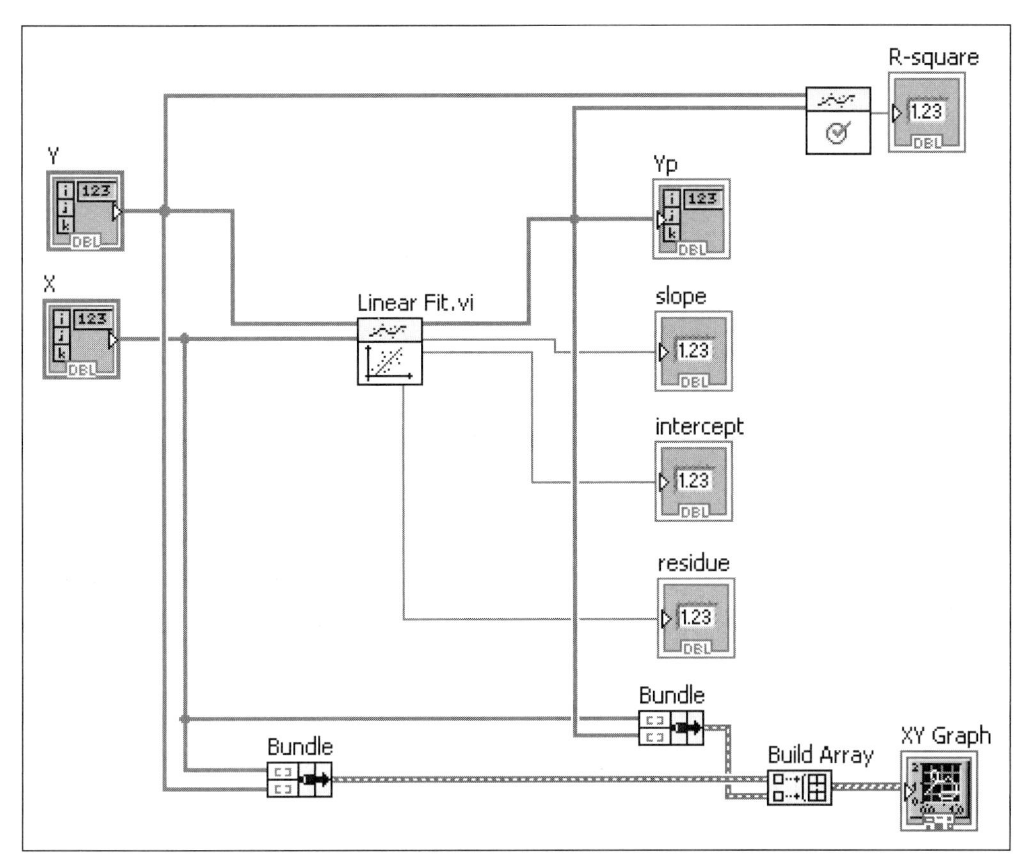

Figure 8.26
Slope and Intercept VI, block diagram.

The intercept, slope, and R^2 values were found to be (see Figure 8.27):

- Intercept −0.61
- Slope 3.29
- R^2 0.979

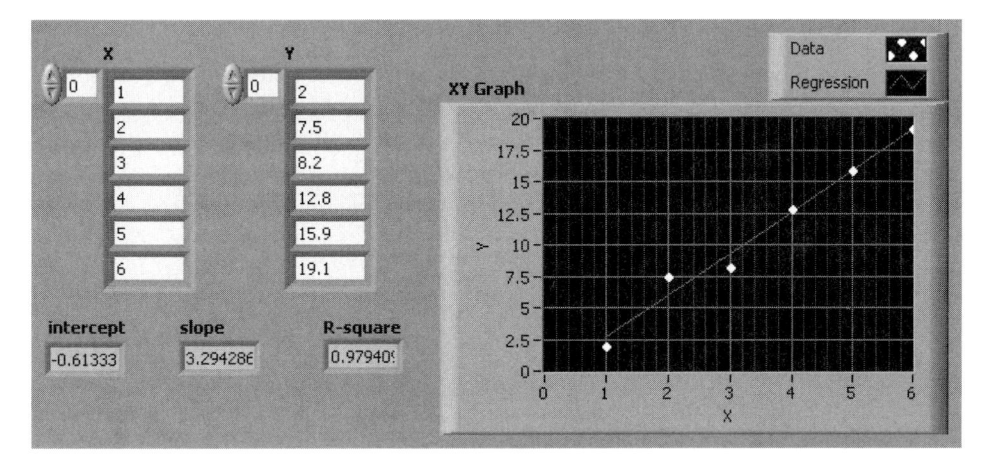

Figure 8.27
Finding slope and intercept.

The block diagram used to perform the regression is shown in Figure 8.28.

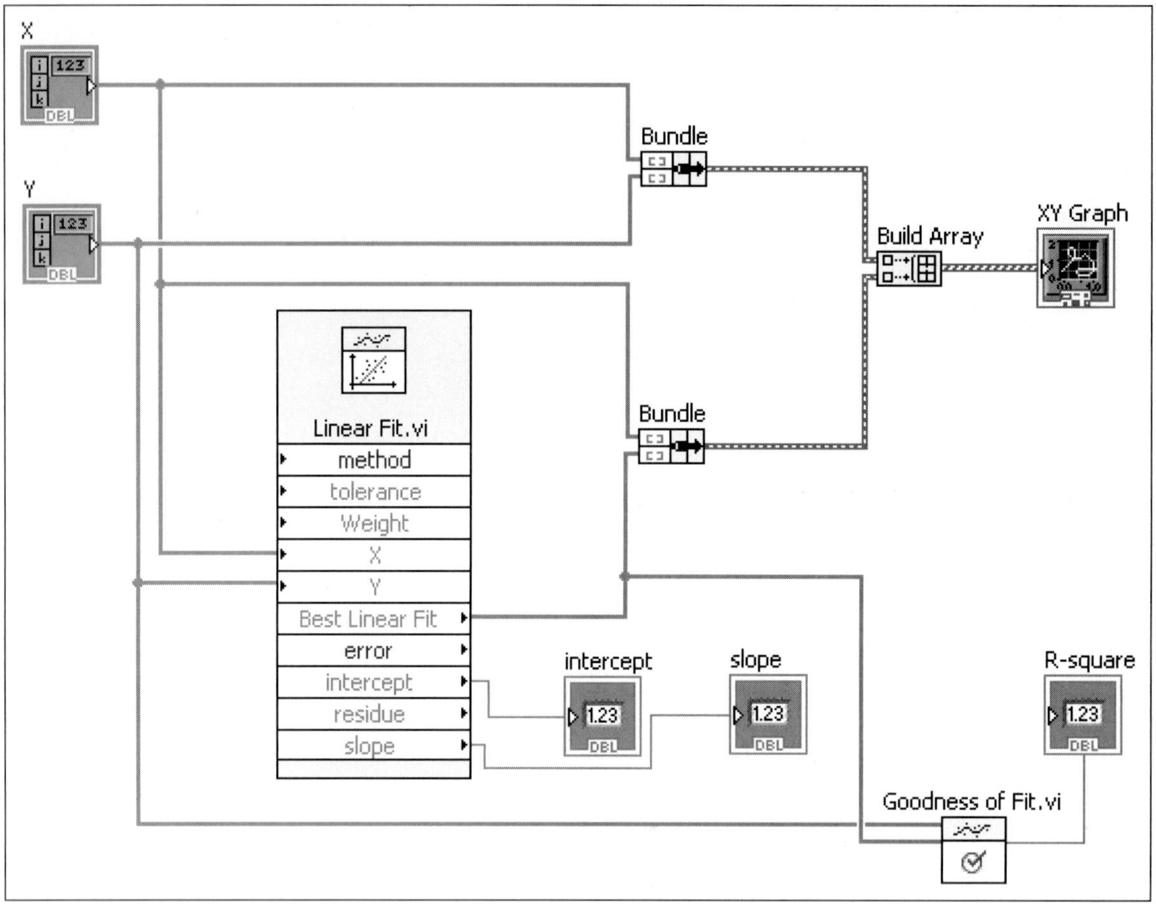

Figure 8.28
Block diagram used for Linear Fit regression.

8.5.2 Gaussian Fit

The various linear regression functions are all similar in layout and function. The Gaussian Fit function is an unusual regression function in that it is not available in common math software products. A *Gaussian curve* is the "bell curve" or normal distribution curve.

The regression model for the Gaussian Fit is

$$y_p = a \exp\left[-\frac{(x-\mu)^2}{2\sigma^2}\right]$$

Where the regression coefficients are

a, the amplitude
μ, the center of the peak
σ, the standard deviation (controls the spread of the bell)

The connection pane for the Gaussian Fit function is illustrated in Figure 8.29.

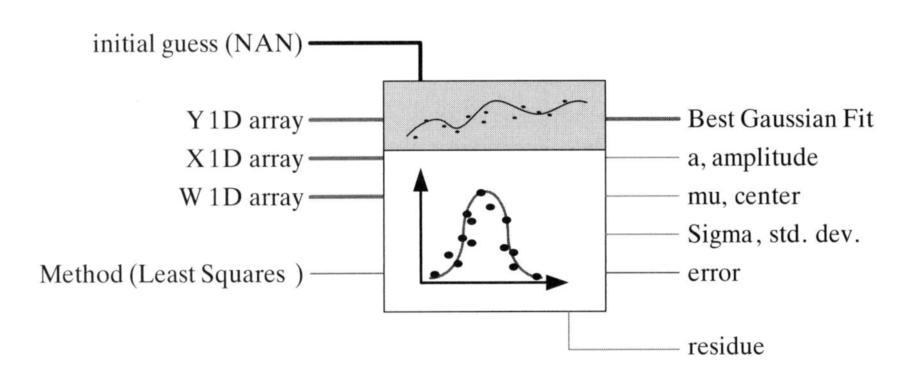

Figure 8.29
Connection pane for the Gaussian Fit function (simplified for Least Squares).

We need to supply 1D arrays of X and Y values. The W (weight) array is optional and rarely used.

The regression coefficients (a, μ, σ) will be displayed, and the Best Gaussian Fit values (Y_p values) will be used to create a graph showing the original values and the best fit line. The Goodness of Fit function will be used to determine the R^2 value.

The results are shown in Figure 8.30 (front panel) and Figure 8.31 (block diagram).

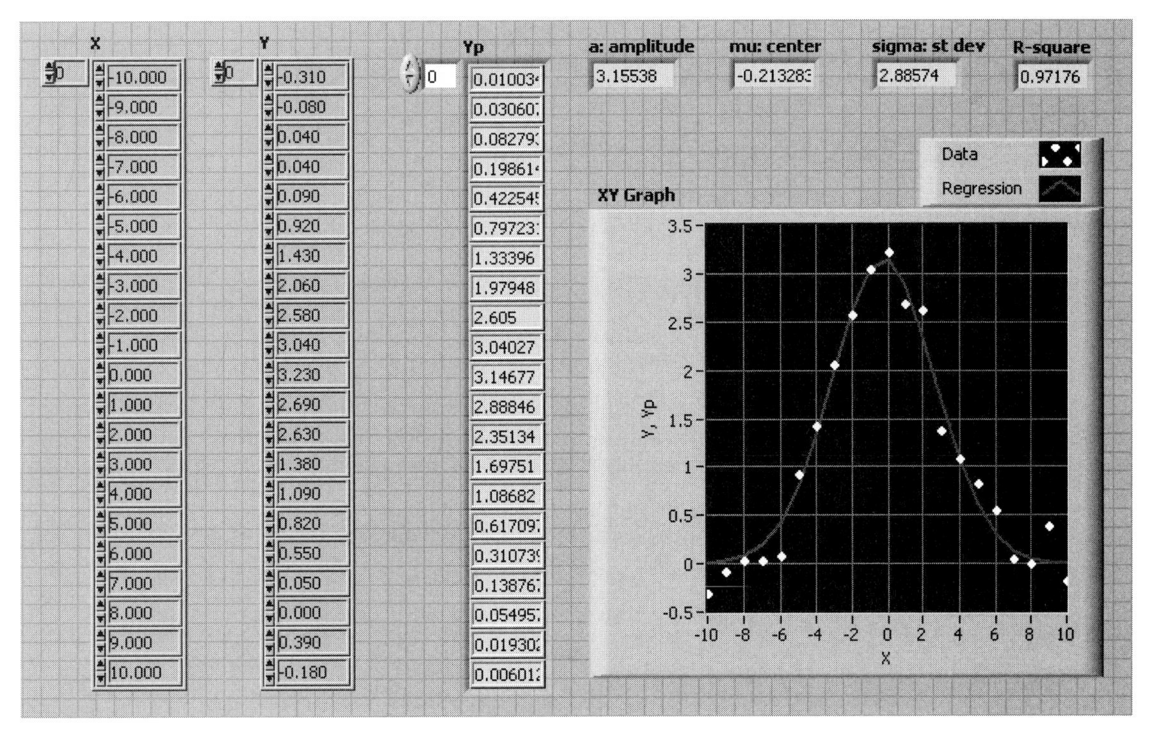

Figure 8.30
Gaussian Regression VI, front panel.

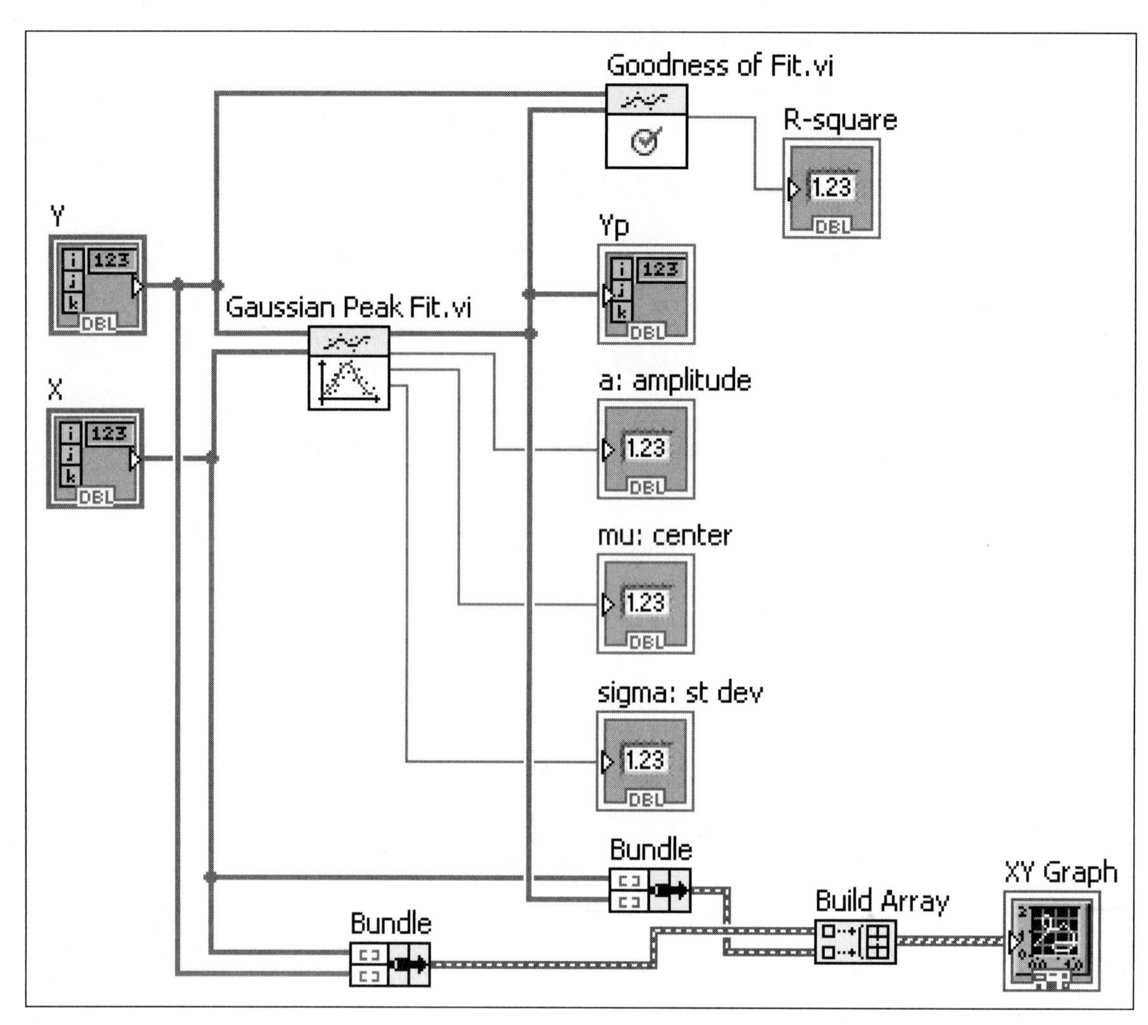

Figure 8.31
Gaussian Regression VI, block diagram.

8.5.3 Polynomial Regression

Polynomial regression is a commonly used approach because the flexibility in setting the order of the polynomial model allows the regression curve to bend. LabVIEW provides the General Polynomial Fit function, which allows polynomial regression up to 25th order.

The regression model for the General Polynomial Fit is

$$y_{pi} = \sum_{j=0}^{m} a_j \, x_i^j$$

Where the regression coefficients are the a_j values.

The connection pane for the General Polynomial Fit function is illustrated in Figure 8.32.

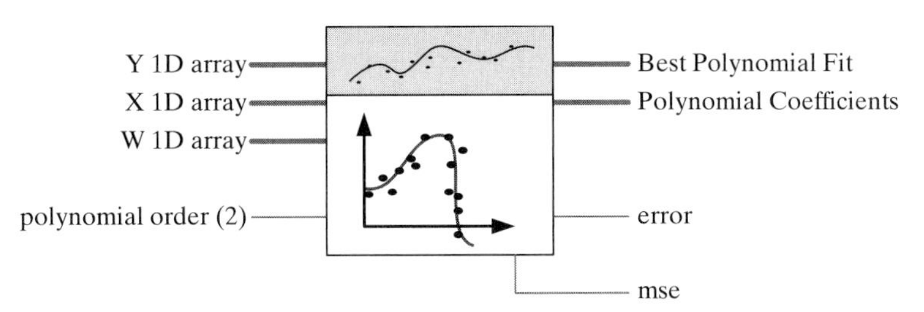

Figure 8.32
Connection pane for the General Polynomial Fit function (simplified for Least Squares).

We need to supply 1D arrays of X and Y values. The W (weight) array is optional and rarely used.

The regression coefficients will be returned as an array, and the Best Polynomial Fit values (Y_p values) will be used to create a graph showing the original values and the best fit line. The Goodness of Fit function will be used to determine the R^2 value.

The results are shown in Figure 8.33 (front panel) and Figure 8.34 (block diagram).

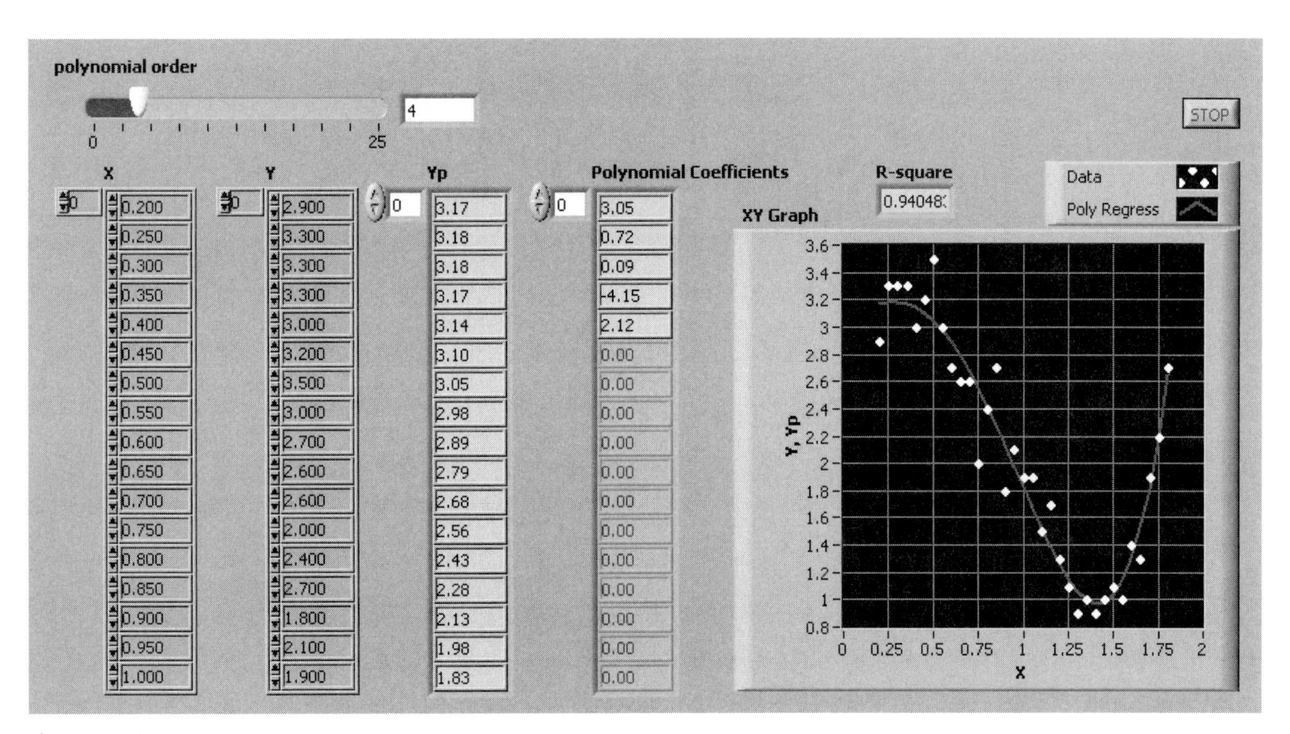

Figure 8.33
Polynomial Regression VI, front panel.

Notice that a While Loop was added to the block diagram so that the Polynomial Regression VI will keep running so that the polynomial order can be varied to observe the impact on the regression result.

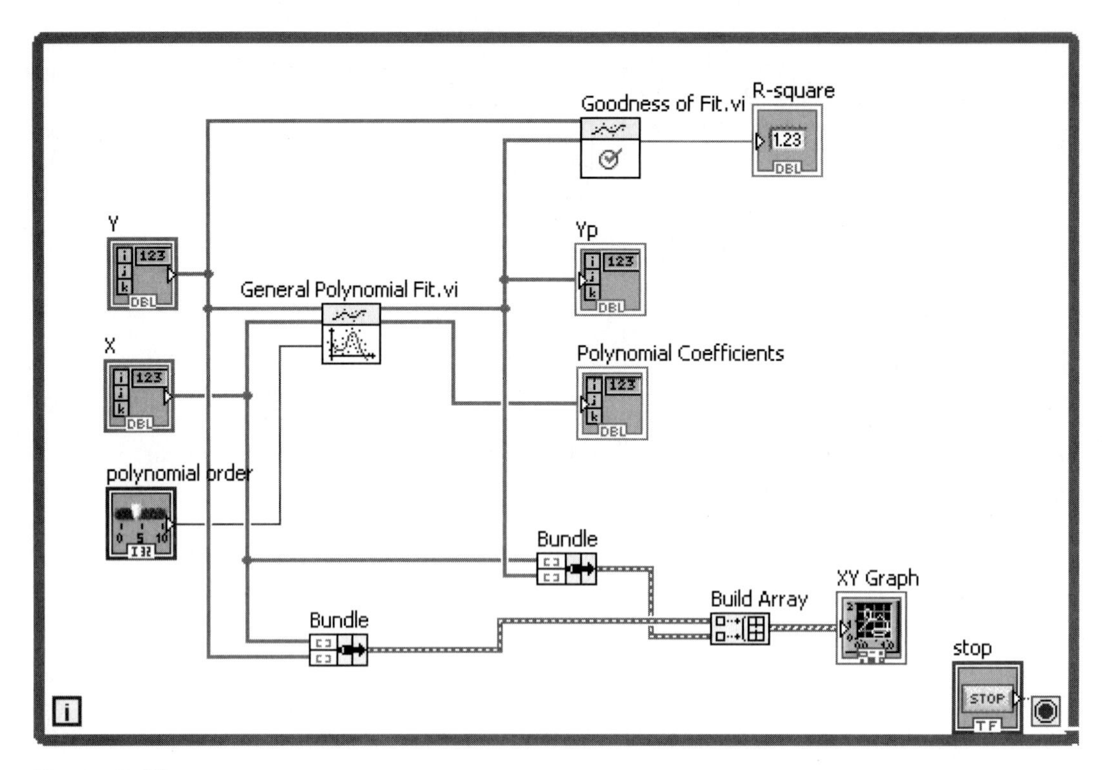

Figure 8.34
Polynomial Regression VI, block diagram.

APPLICATION

Determining Heat Transfer Coefficient from Experimental Data

Heat transfer coefficients are used to predict the amount of energy transferred from a hot surface to a moving fluid. The heat transfer coefficients are obtained from experimental data.

An experiment was conducted using a small (1.0 cm diameter) spherical heater in a tube filled with flowing air (see Figure 8.35.) Two thermocouples are used to record the heater surface temperature, T_S, and the wall temperature, T_W.

The rate of heat transfer from the heater to the fluid is described by the equation

$$\dot{Q} = hA(T_s - T_w)$$

Where

h is the heat transfer coefficient
A is the surface area of the heater

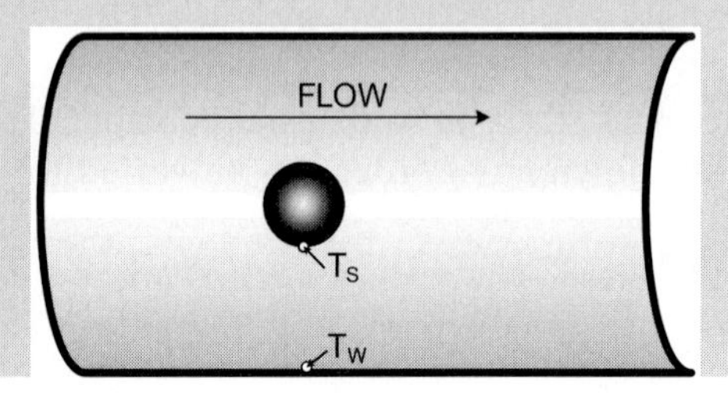

Figure 8.35
Experimental setup.

$\dot{Q}$ can be measured, A can be calculated, and T_S and T_W can be measured. The only unknown is h, and we can determine the heat transfer coefficient as the slope of the plot shown in Figure 8.36.

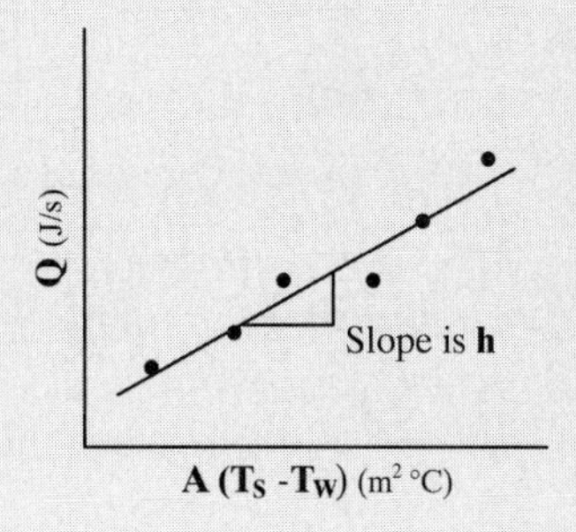

Figure 8.36
Finding the heat transfer coefficient.

The following files contain the experimental data:

- Qdot.txt heat transfer rate, J/s
- DeltaT.txt $(T_S - T_W)$, °C

We can read the data into LabVIEW, and perform a regression to determine the heat transfer coefficient. The results are shown in Figure 8.37. The block diagram is shown in Figure 8.38. The heat transfer coefficient value was found to be 48.4 J/(s m^2 °C).

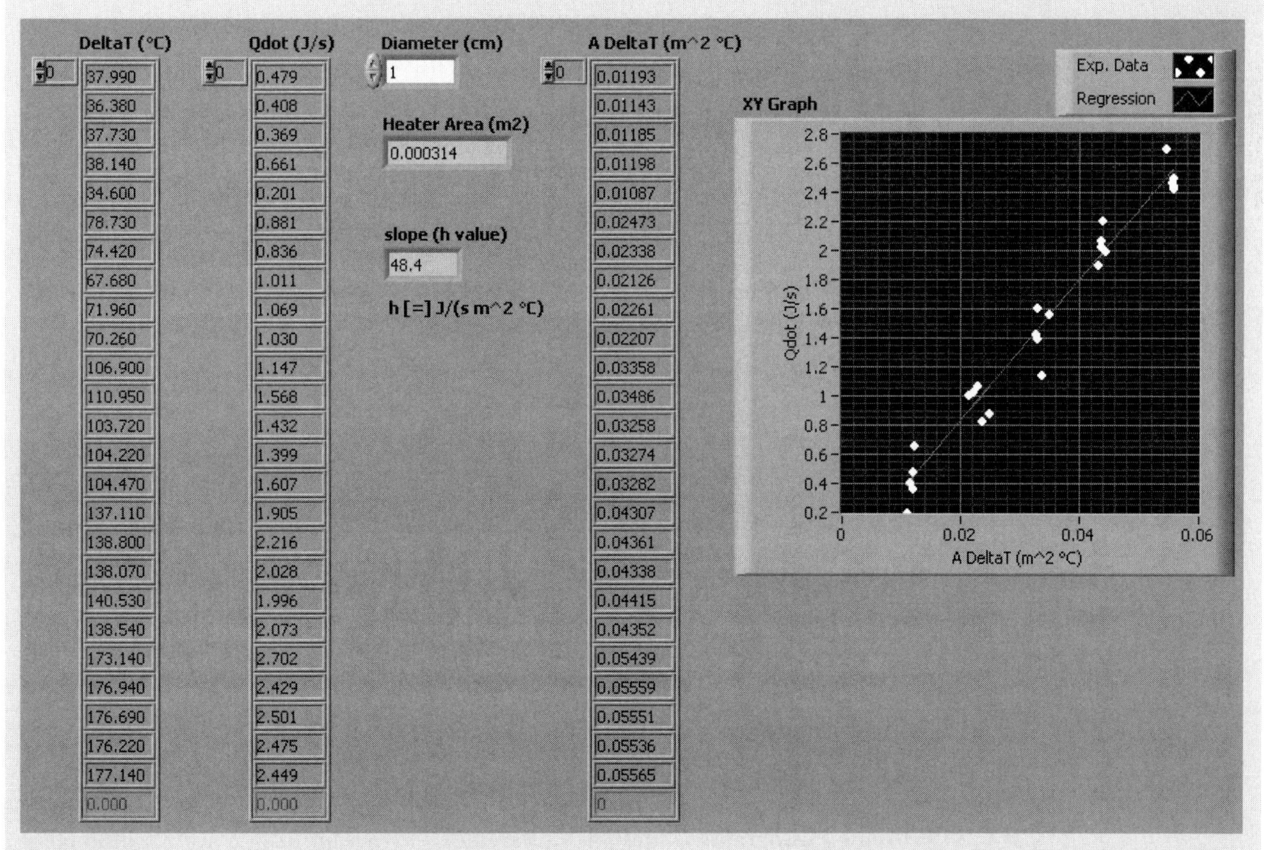

Figure 8.37
Experimental Heat Transfer Coefficient VI, front panel.

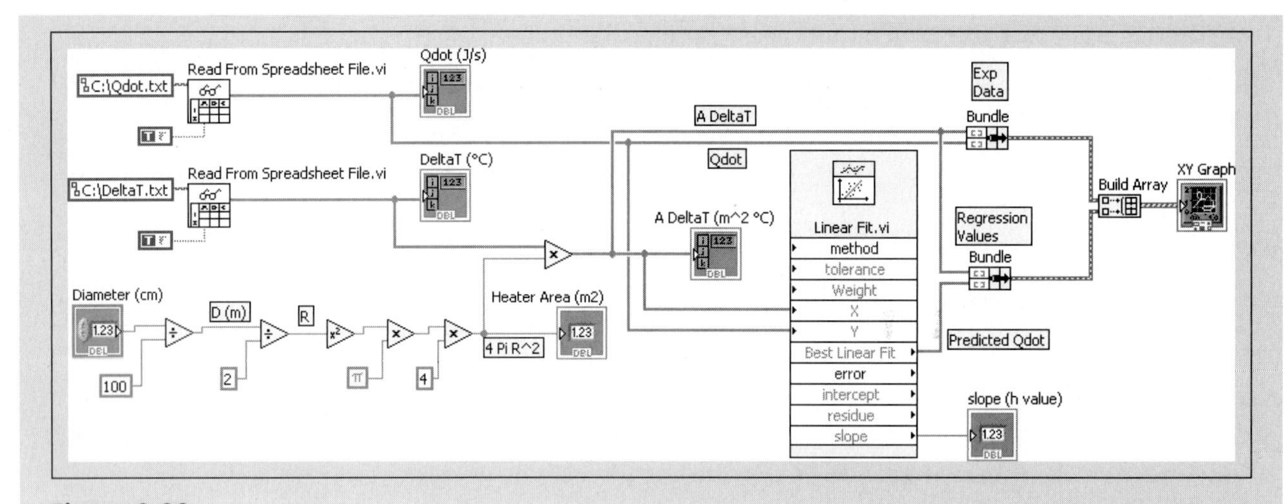

Figure 8.38
Experimental Heat Transfer Coefficient VI, block diagram.

KEY TERMS

- bins
- coefficient of determination (R^2)
- cubic spline interpolation
- curve fitting
- error
- gaussian curve (bell curve, normal distribution)
- heat transfer coefficients
- histogram
- interpolate
- interpolation
- linear interpolation
- linear regression
- mean
- median
- monotonically
- non-linear regression
- polynomial regression
- regression
- regression model
- standard deviation
- statistics
- sum of the squared error (SSE)
- variance

SUMMARY

Basic Statistics

- Maximum—Array Max & Min
- Minimum—Array Max & Min
- Mean—Mean.VI
- Median—Median.VI
- Standard deviation—Std Deviation and Variance.VI
- Variance—Std Deviation and Variance.VI

Functions Palette / Programming Group / Array Group / Array Max & Min

Functions Palette / Mathematics Group / Probability & Statistics Group

Histograms

Functions Palette / Mathematics Group / Probability & Statistics Group

- **Histogram.vi**—you can set the number of bins, but not the bin ranges.
- **General Histogram.vi**—you can specify the bin ranges.
- **Create Histogram.vi**—designed to create a histogram from a signal, although it will accept an array as input.

Interpolation

Functions Palette / Mathematics Group / Interpolation & Extrapolation / Interpolate 1D

You supply an *X* array, *Y* array, and known *x* value. The Interpolate 1D function returns the interpolated *y* value corresponding to the known *x* value.

Methods

- 0—nearest (no interpolation)
- 1—linear (default)
- 2—cubic spline
- 3—cubic Hermite
- 4—Lagrange

Curve Fitting

Functions Palette / Mathematics Group / Interpolation & Extrapolation / Interpolate 1D

You supply an *X* array, *Y* array, and an array of known *x* values. The Interpolate 1D function returns the interpolated *y* array corresponding to the known *x* values.

ntimes input

1—adds one interpolated point between each original data point
2—adds three interpolated point between each original data point
3—adds seven interpolated point between each original data point

Regression
Finding the coefficients that cause a regression model to best fit the values in a data set.

Functions Palette / Mathematics Group / Fitting Group

Linear Regression Functions

- Linear Fit $f = ax + b$
- Exponential Fit $f = ae^{bx}$
- Power Fit $f = ax^b$
- Gaussian Peak Fit $f = a\exp\left[-\frac{(x-\mu)^2}{2\sigma^2}\right]$
- Logarithm Fit $f = a\log_c(bx)$ c (logarithm base) is a function parameter
- General Polynomial Fit $f = \sum_{j=0}^{m} a_j x_i^j$ $1 \le m \le 25$
- General LS (Least Squares) Linear Fit—fits any linear model that you provide

Non-Linear Regression Functions

- Non-Linear Curve Fit
- Constrained Non-Linear Curve Fit

***R*²—Coefficient of Determination**

Functions Palette / Mathematics Group / Fitting Group / Advanced Curve Fitting / Goodness of Fit

SELF-ASSESSMENT

1. Where can you find the data sets used in this chapter (so that you can follow along without having to type in all the array values)?
   ANS: At the text's website: www.chbe.montana.edu/LabVIEW
2. What are the descriptive statistics that are commonly used with a single data set?
   ANS:
   - Mean
   - Standard deviation
   - Variance
   - Median
3. What is a histogram?
   ANS: A graph that shows how the values in a data set are distributed.
4. What is interpolation?
   ANS: A method of using nearby data values to estimate an intermediate value.
5. What methods of interpolation are available using LabVIEW's Interpolate 1D function?
   ANS:
   - Nearest (no interpolation)
   - Linear
   - Cubic Spline
   - Cubic Hermite
   - Lagrange
6. What is the difference between linear and cubic spline interpolation?
   ANS: With linear interpolation a straight line is drawn between adjacent points and the interpolated value is assumed to be on the line. With cubic spline interpolation, a smooth curve is drawn through all of the data points and the interpolated value is assumed to line on the curve.
7. What are the two basic approaches to curve fitting?
   ANS:
   - Find a curve that must go through every data point; no mathematical equation for the curve is required.
   - Find the mathematical equation for the best fit (regression) curve through the data point; the curve does not need to go through each data point.
8. What linear regression functions are available in LabVIEW's Fitting Group on the Functions Palette?
   ANS:
   - Linear Fit
   - Exponential Fit
   - Power Fit
   - Gaussian Fit
   - Logarithm Fit
   - General Polynomial Fit
   - General Least Squares Fit

PROBLEMS

1. Determine the mean and standard deviation of the following values:

1.29
1.32
1.28
1.30
1.33
1.32
1.30
1.31

2. Determine the average (i.e., arithmetic mean) and median values of the following set of test scores. When there are a few very low scores (such as zeros, when people fail to take the exam), which measurement is a better indicator of class performance?

88
95
97
87
100
80
0
78
96
97
62
75
0
90
85
0
0
68

3. Create a histogram showing how many scores in the data set in the previous problem fit into the following bins:

- 90–100
- 80–90
- 70–80
- 60–70
- 0–60

State how you are handling scores that fall on the bin boundaries.

4. A company has advertised an updated instrument that now comes with a digital readout and an enhanced price. The old analog display could be read to two decimal places, but the digital display shows four. To see if the new instrument is truly more precise, an old sample was re-run on the new instrument. The old

and new data sets are shown below. Calculate the mean and standard deviation to find out:

a. If the two instruments give essentially the same mean value.
b. if the new instrument produces results with a smaller standard deviation than the old instrument.

Old	New
11.55	11.5308
11.43	11.5310
11.54	11.5268
11.49	11.5348
11.56	11.5203
11.51	11.5276
11.67	11.5361
11.52	11.5447
11.66	11.5360
11.44	11.5234

5. An assumption of "normally distributed errors" is common in statistical analysis, but the assumption is frequently not tested. To test that assumption, a researcher connected two meters to a data acquisition system and tested the same samples 1000 times for each meter. The data files are available at the text's website: www.chbe.montana.edu/LabVIEW in files Meter1.txt and Meter2.txt.

Read each file into LabVIEW, and create a histogram for each data set. Do the values appear to be normally distributed (do the histograms look like a "bell curve")?

6. Create a VI that will allow you to interpolate the steam table data in Table 8.2 for the following values:

a. Internal Energy at 140°C
b. Enthalpy at 280°C
c. Entropy at 330°C

Compare the results using linear interpolation and spline interpolation.

7. The following table shows how the population of Ireland has varied over time. Create a VI using LabVIEW's Interpolate 1D function that will allow you to

a. Interpolate the data to determine the population in 2000.
b. Extrapolate the data to predict the population in 2015.

Try various methods for part b. How widely do the extrapolated values vary?

Year	Population
1901	3,221,823
1911	3,139,688
1926	2,971,992
1936	2,968,420
1946	2,955,107

1951	2,960,593
1961	2,818,341
1971	2,978,248
1981	3,443,405
1991	3,525,719
1996	3,626,087
2002	3,917,203
2006	4,239,848

Source: Central Statistics Office Ireland, http://www.cso.ie/statistics/Population1901-2006.htm

8. Create a VI using LabVIEW's Interpolate 1D function that will allow you to create an XY Graph of the population of Ireland between 1901 and 2006 with several interpolated points between each value in the data set shown in the previous problem.

9. Analytical instruments typically need to be calibrated for one or both of the following reasons:

a. To ensure that the instrument is still reporting correct values.
b. To correlate the measurement units to the units output by the instrument.

As an example of the latter, consider a meter that measures pH but outputs a voltage signal suitable for automated data acquisition. A calibration curve is used to correlate pH and voltage.

To calibrate the pH meter, solutions of known pH are prepared and used with the meter. The output voltages (in triplicate) corresponding to the known pH values are recorded. The calibration data are listed below, and available in files pHData.txt and VoltData.txt.

pH	Voltage		
1.5	1.21	1.14	1.16
2.3	1.76	1.70	1.76
3.3	2.32	2.39	2.40
4.1	2.89	2.91	2.90
5.6	3.91	3.87	3.89
6.9	4.68	4.72	4.68
7.4	4.98	5.05	5.05
8.2	5.57	5.47	5.57
10.4	7.01	6.98	6.95
11.8	7.96	7.88	7.91

Read the data into LabVIEW, plot it to see if a linear fit is reasonable, and then regress the data to determine the calibration equation. The calibration equation should allow you to calculate a pH given a voltage value.

10. When data have a significant number of bends, many people think of polynomial regression as the only option for trying to get a fit. This problem is designed to demonstrate that polynomial regression has limitations.

Two text files on the text's website contain data for a decaying oscillation: DecayOscXLarge.txt and DecayOscYLarge.txt. Use LabVIEW's General Polynomial Fit function and vary the order of the polynomial (LabVIEW allows orders between 1 and 25.) Use LabVIEW's Goodness of Fit function to determine the R^2 value as well. Which order's provide a "good" fit to the decaying oscillation function, and what happens when the order is too small or too large?

CHAPTER

9

Programming in LabVIEW

Objectives

After reading this chapter, you will know:

- more about controlling display options with LabVIEW controls
- how to use LabVIEW block diagram options, like automatic wiring, to create VIs more efficiently
- how to create SubVIs to allow you to reuse program elements
- how to use LabVIEW projects to store related VIs and other files
- how to use LabVIEW programming structures for more powerful VIs
  - Loop structures
  - Case structures
  - Sequence structures

9.1 INTRODUCTION

Since LabVIEW is an object-oriented graphical programming language, each time you place a control (an object) on the front panel, and every time you connect two terminals with a wire, you are programming. We've included programming examples in every chapter. This chapter on "Programming in LabVIEW" is about taking LabVIEW programming to the next level, about creating more advanced VIs, and about some features that LabVIEW provides that can help you stay organized and work more efficiently.

9.2 LABVIEW PROGRAMMING BASICS, EXPANDED

Here, we'll revisit the basics of LabVIEW programming, but push the boundary of each section a little.

- **Front Panel**—getting greater control over controls and indicators
- **Block Diagrams**—working more efficiently with nodes and wiring
- **SubVIs**—Packaging commonly used code pieces for reuse
- **Projects**—Collecting VIs and other files that work together

9.2.1 Front Panel: Controls and Indicators

The LabVIEW user interface is the *Front Panel*, which holds the controls and indicators needed to set required values, and display calculated results. But LabVIEW gives you much control over how controls and indicators function. For example, you can

- Adjust the display format of displayed values
- Restrict allowed data entry values
- Change the data type of the output value
- Use a logarithm scale (on controls with scales)
- Set default initial values

Adjust the Display Format of Displayed Values
To change the display format on a control or indicator, right-click on the object and select **Display Format . . .** from the pop-up menu. The Numeric Properties dialog will open with the Display Format panel visible as shown in Figure 9.1.

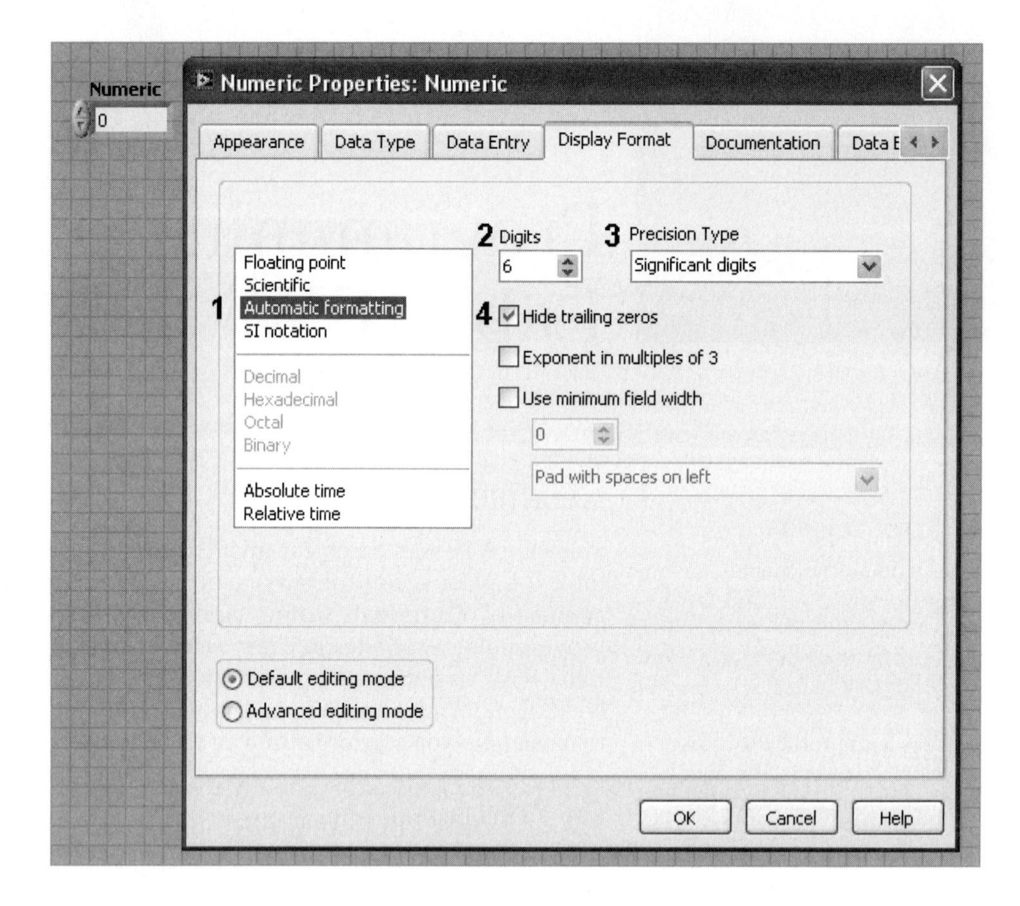

Figure 9.1
Changing the format used to display numbers.

The display options that can be changed are numbered in Figure 9.1:

1. While automatic formatting is usually a good option, you can force the display to use only floating point or scientific notation, or use SI notation which uses text prefixes (e.g., milli, nano) to indicate magnitude.
2. You can change the number of displayed Digits (six is the default).
3. You can change the Precision Type between "Significant digits" and "Digits of precision". To understand the difference, consider the value 102.331. This value is shown with six significant digits, but three digits of precision.
4. You can show or hide trailing zeroes.

Restrict Allowed Data Entry Values
By default, a numeric control will accept any numeric value that will fit within the limits of the control's assigned data type. You can restrict the values that a control will accept. For example, if a math operation is going to take the logarithm of the value entered into the control, you can restrict the data entry to positive values.

Right-click on the control and select **Data Entry . . .** from the pop-up menu. The Numeric Properties dialog will open with the Data Entry panel visible as shown in Figure 9.2. Change the Minimum field to a very small positive number.

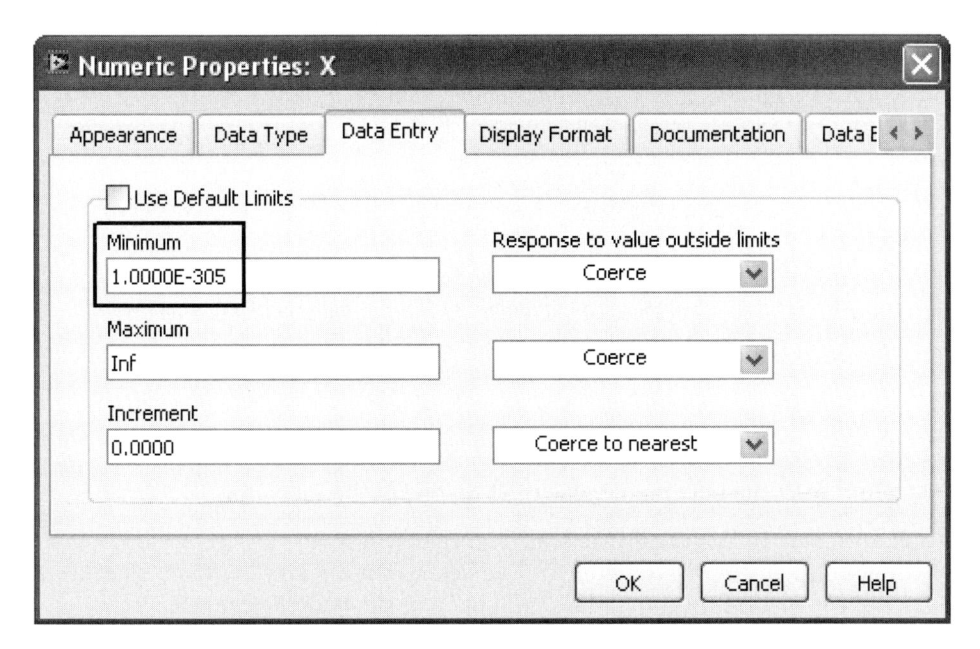

Figure 9.2
Preventing a control from accepting non-positive values.

PRACTICE

Controlling User Inputs

The VI shown in Figure 9.3 displays the volume of N moles of gas at temperature T and pressure P. The temperature Dial control has already been modified to allow high temperatures to be selected. But the VI uses the ideal gas law to solve for volume, so pressures ought to be kept fairly low, perhaps below 8 atm. Create the VI shown in Figure 9.3, but change the upper limit on the P Dial to restrict the allowable range of pressure settings.

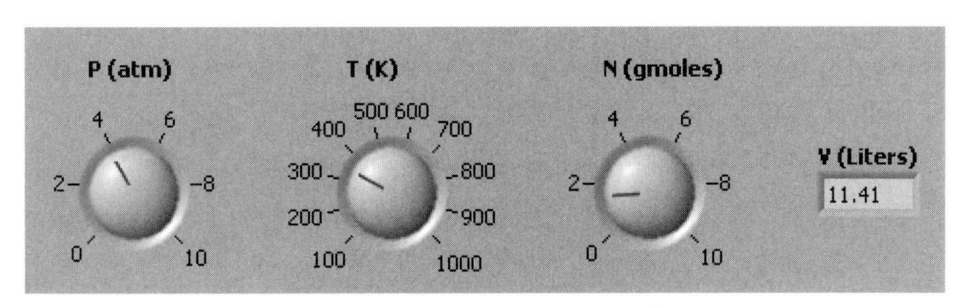

Figure 9.3
Ideal Gas Solver VI.

The block diagram of the VI is shown in Figure 9.4.

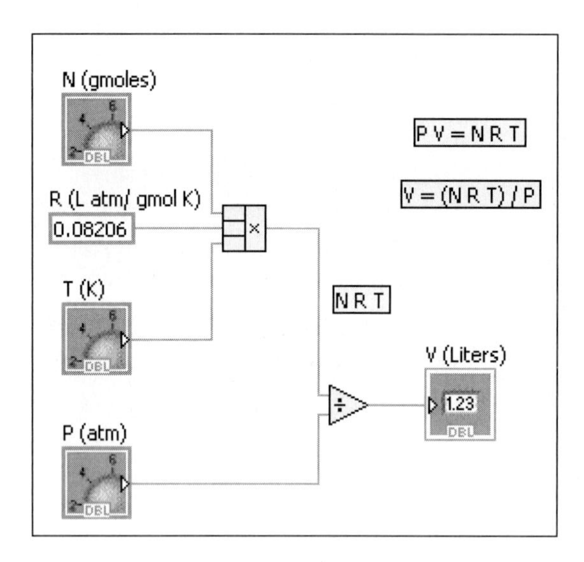

Figure 9.4
Ideal Gas Solver VI, block diagram.

To restrict the maximum value of *P*, do one of the following:

- Click on the "10" on the P Dial scale and change it to an "8".
- Right-click on the P Dial control and select **Data Entry . . .** from the pop-up menu. The Knob Properties dialog will open as shown in Figure 9.5.
  - Clear the Use Default Limits box.
  - Enter "8" in the Maximum field.

Note: The latter approach does not change the upper limit on the Dial control scale; it simply prevents the Dial control from moving past 8.

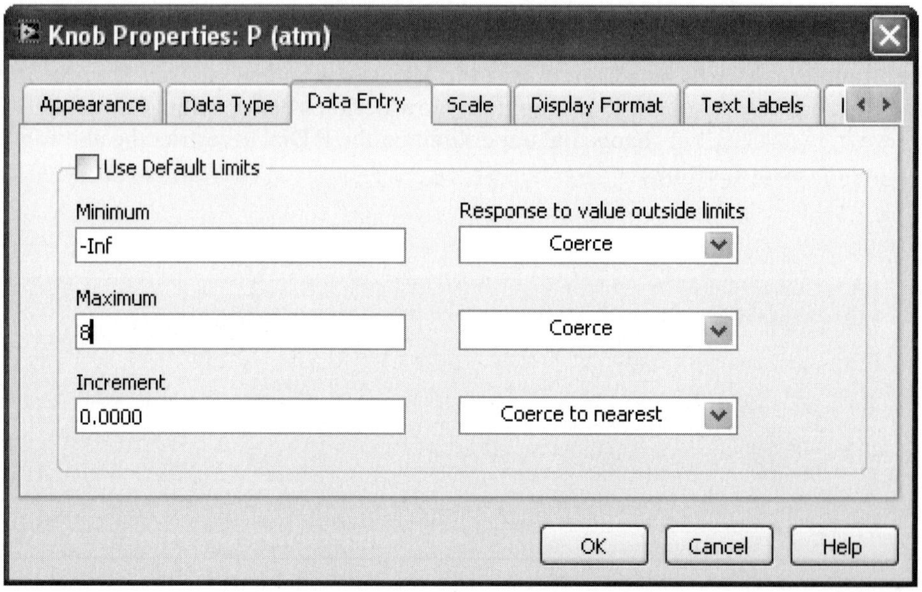

Figure 9.5
Knob Properties dialog, Data Entry panel.

Change the Data Type of the Output Value

The data type of a control should be set to be consistent with the use of the control's output value. By default, numeric controls are set to output double-precision floating point (DBL) values. You can change that by right-clicking on the control and selecting **Representation** and then selecting the desired data type. In Figure 9.6 the Numeric control's output will be 32-bit integer (Long Integer).

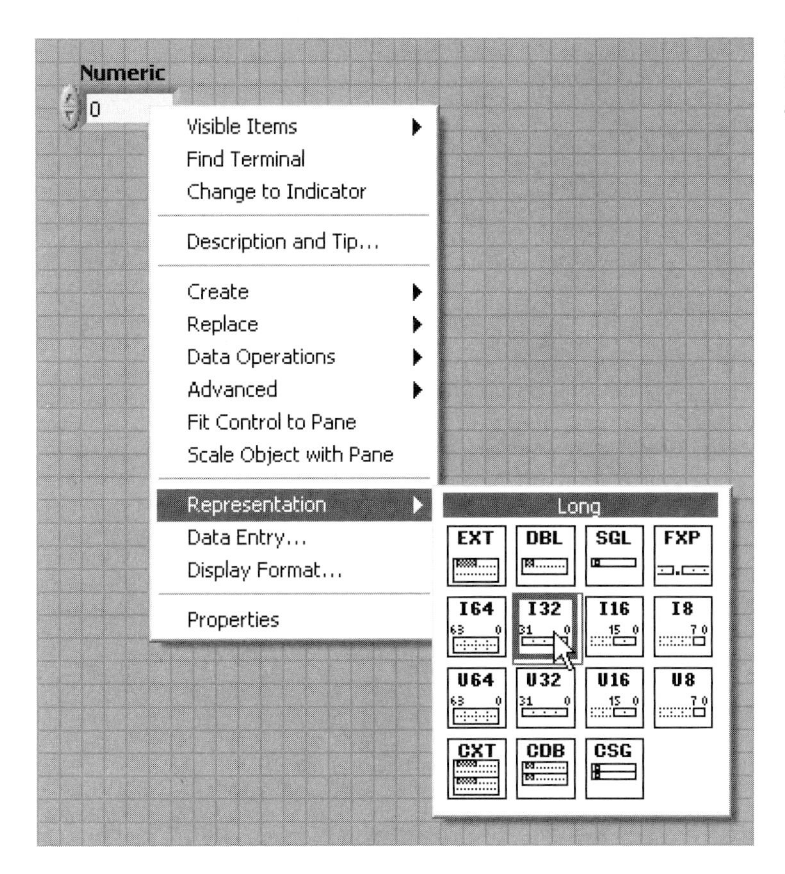

Figure 9.6
Changing the data type of a control's output value.

Use a Logarithm Scale (On Controls with Scales)

By default, scale values on controls with scales are linear. You can switch to a *logarithm scale* as follows (see Figure 9.7):

1. Place the control on the front panel.
2. Change the minimum and maximum scale values (if desired).
3. Right-click on the control and select **Scale / Mapping / Logarithmic**.

Set Default Initial Values

When you open a VI after it has been saved, all controls and indicators are initially displayed with default values (usually zeros). That may not be ideal. For example, if you have invested some time filling a large array with values, and then close LabVIEW and leave to have lunch, it would be nice if that array still displays the same values when you re-open the VI for editing after lunch. It won't unless you save the current value(s) as the default value(s).

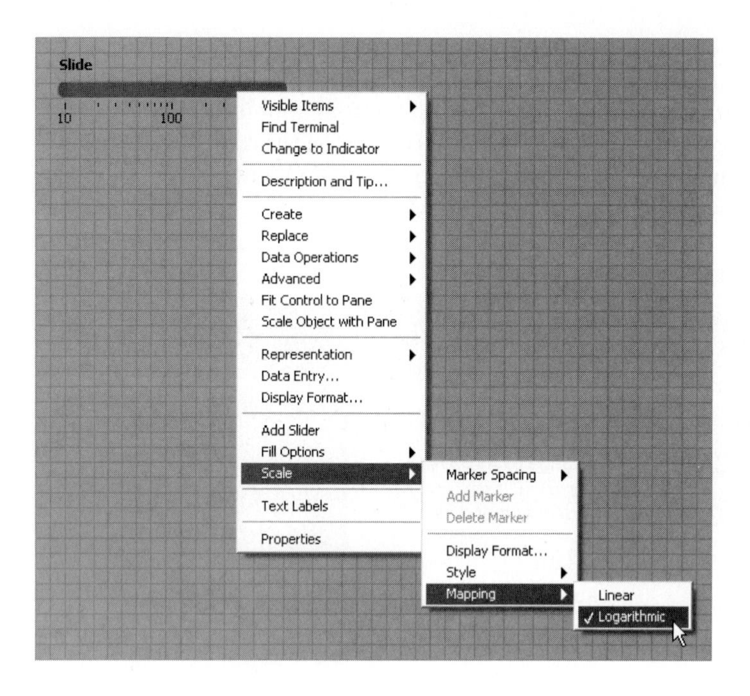

Figure 9.7
Changing to a log scale on a control.

Note: When working with an array or matrix, be sure the entire array or matrix is selected, not just one element. Then, right-click on the control and select **Data Operations / Make Current Value Default**. The current value(s) in the control will be saved with the VI as the default value(s) to load into the control when the VI is opened.

PRACTICE

Setting Default Values on Controls

To observe what happens to array controls when you don't make the current values the default values, build the VI shown in Figure 9.8 and put a few non-zero values in each array.

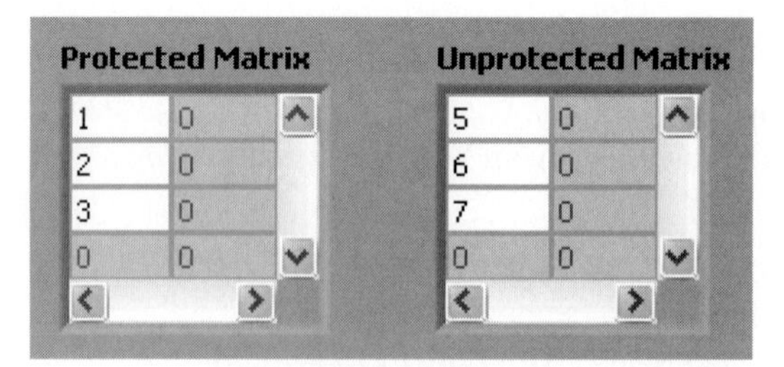

Figure 9.8
Two matrix controls on a VI.

Select the entire matrix labeled "Protected Matrix" and then right-click on the selected matrix and choose **Data Operations / Make Current Value Default** from the pop-up menu. Leave the Unprotected Matrix alone.

- Now: Save the VI and assign it a file name so that it can be re-opened
- Close the VI
- Re-open the VI

When the VI is re-opened (shown in Figure 9.9), LabVIEW fills the matrices with default values. Since the "1, 2, 3" were set as the default values for Protected Matrix, those values are restored. But Unprotected Matrix was initialized with zeros. For small matrices, it is not a big deal to reenter the values before running the VI, but for large matrices you want to be sure to make the entered values the default values before closing the VI.

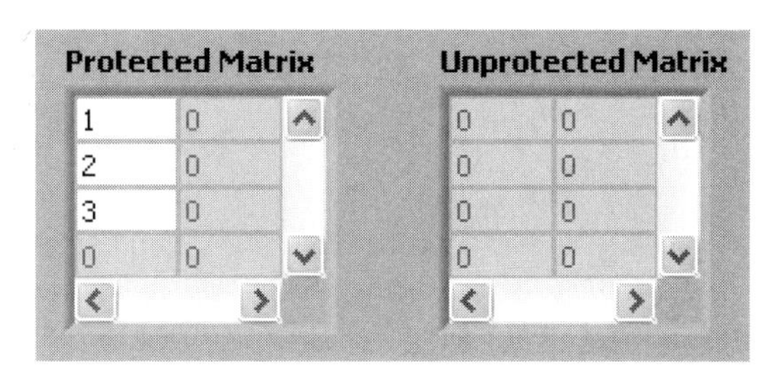

Figure 9.9
The two matrix controls on a VI, reinitialized.

9.2.2 Block Diagram: Nodes, Terminals, and Wires

The block diagram is where the majority of the graphical programming takes place by wiring the nodes (blocks) of the various objects (controls, indicators, functions) together to create a working VI. There are a few things you can do to minimize the time spent working on block diagrams:

- Start on the block diagram with functions rather than on the front panel with controls and indicators.
- Use *automatic wiring* when you can.
- Use the short-cut [Ctrl B] to remove all broken wires from a block diagram.

Start on the Block Diagram

To minimize wiring, place the function on the block diagram first, and then

- Right-click on the function inputs and select **Create / Control**.
- Right-click on the function outputs and select **Create / Indicator**.

This approach has several benefits:

- The data type of the control or indicator is automatically set to be appropriate for the function's requirements.
- The wiring is automatic.
- The control or indicator is automatically labeled using the label of the function's input or output.

Automatic Wiring

When you are placing a new node on a block diagram, if you locate the new node's input terminal near an unused output terminal, LabVIEW will automatically wire the terminals together. As you move the new node near the existing terminal, the proposed wire will flash briefly to let you know that LabVIEW will automatically wire the terminals.

The down side of automatic wiring is that LabVIEW will sometimes wire some terminals that you don't want to have wired. When you place a new node on a block diagram and a wire automatically appears connected to a terminal, check to make sure the automatic wiring is correct.

Remove All Broken Wires from a Block Diagram [Ctrl B]

When you are making changes to a block diagram, removing one control or indicator can sometimes break wires all over the diagram. It is time consuming and frustrating to select and delete all of those broken wires. It is also unnecessary because LabVIEW will remove all broken wires from a block diagram when you press [Ctrl B].

9.2.3 SubVIs

You can build a VI into another VI, in fact, it's done all of the time. Most of the "functions" on the Functions Palette are actually VIs. When a VI is built into another VI, it is called a *SubVI*. You can create your own SubVIs using LabVIEW. By turning commonly used VIs into SubVIs, they become available for use whenever you need them. If you do the same things routinely, SubVIs can save you a lot of time.

One of the tasks that can get tedious is the bundling and array building required to plot multiple curves on an XY Graph. We can create a SubVI that will allow you to wire your X and Y arrays into the SubVI and it will take care of the bundling and array building.

The first step is to get a VI working without using a SubVI. In this example, we create an XY Graph with two curves from four small arrays (X1, Y1, and X2, Y2). The block diagram is shown in Figure 9.10, and the front panel in Figure 9.11.

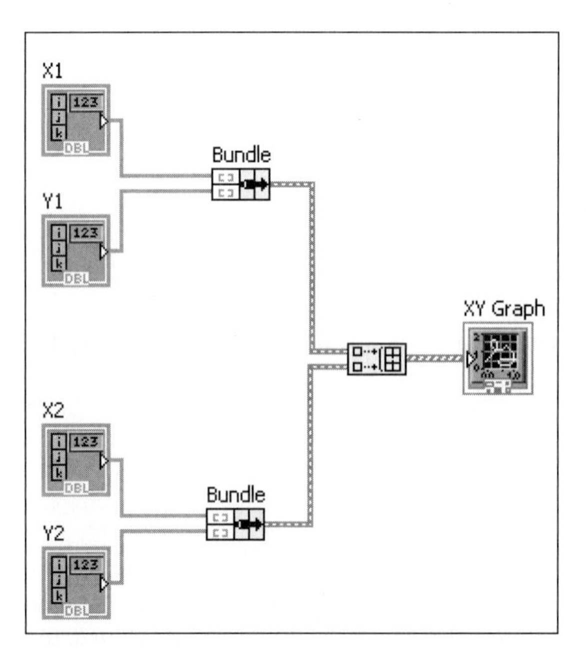

Figure 9.10
A VI that generates an XY Graph with two curves.

Next, to create a SubVI from a portion of an existing VI, you select the part that will become the SubVI, as shown in Figure 9.12, and then use menu

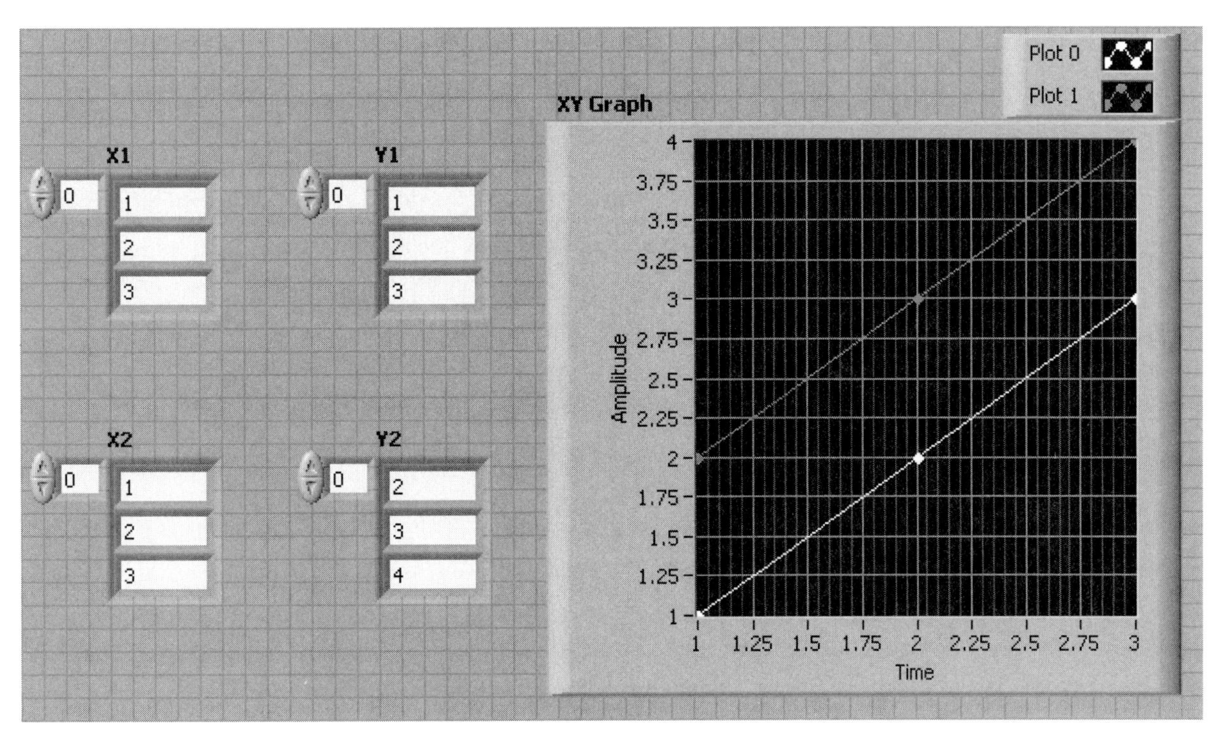

Figure 9.11
Front panel showing simple arrays and XY Graph.

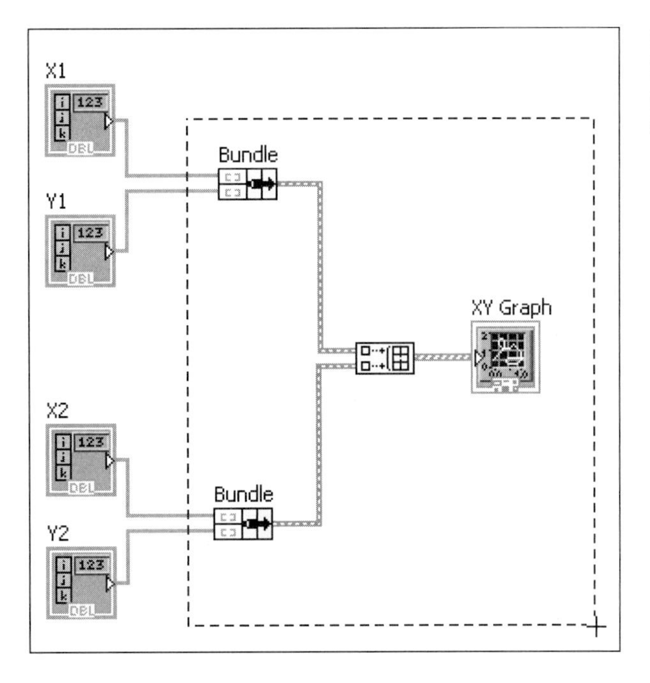

Figure 9.12
Select the portion of the existing VI that will become the SubVI.

options **Edit / Create SubVI**. The resulting SubVI (with generic icon) is shown in Figure 9.13.

Notice that LabVIEW left the XY Graph control out of the SubVI; it can't be included, but LabVIEW put the rest of the functions into the SubVI.

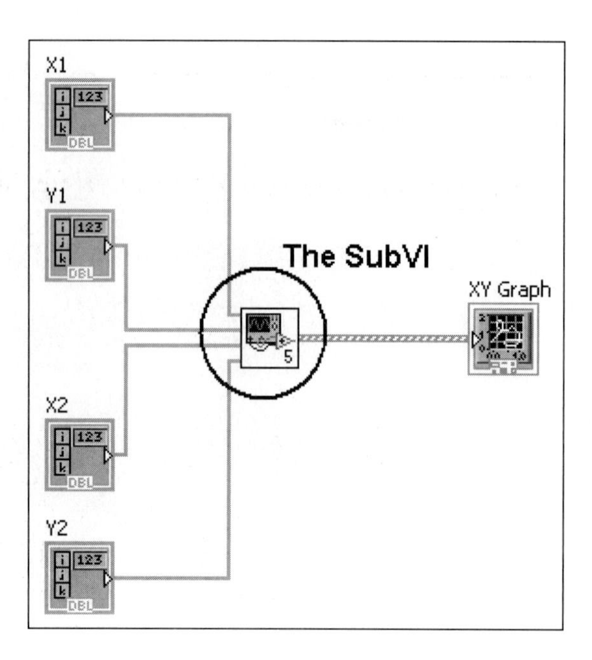

Figure 9.13
The SubVI with generic icon.

There are a few more steps to go through to finish the SubVI:

1. Change the icon.
2. Save the SubVI.
3. Review connections on the connector pane.

Step 1. Change the SubVI's Icon

To change the icon, double-click on the SubVI. LabVIEW will display a front panel for the SubVI—it will look just like the front panel for the VI used to create the SubVI, except for the title, which will have (SubVI) in it, and the number on the icon in the top-right corner of the front panel.

Right-click on the icon in the top-right corner and select **Edit Icon . . .** from the pop-up menu. The Icon Editor will open as shown in Figure 9.14

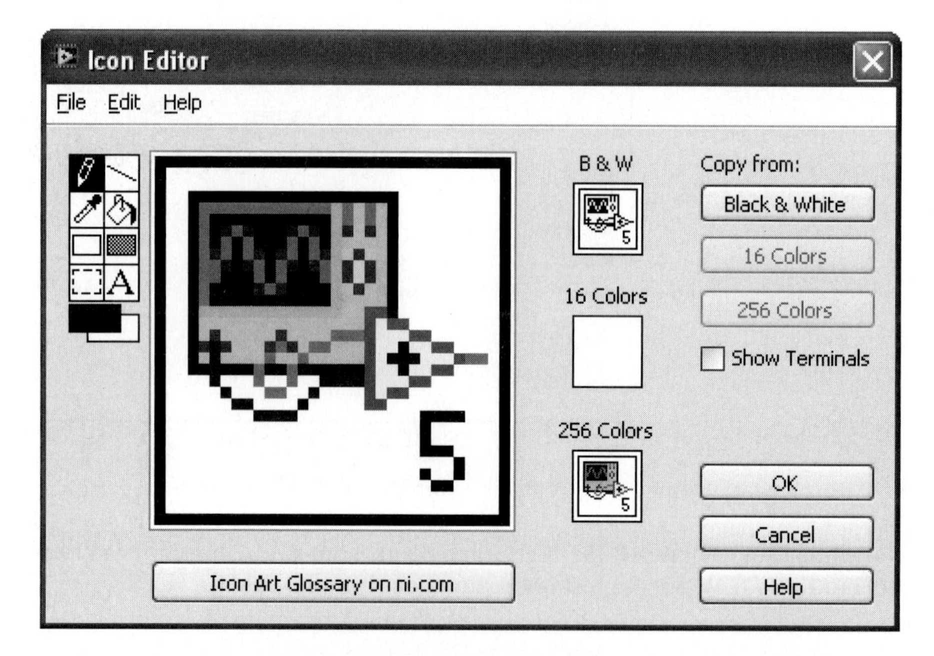

Figure 9.14
LabVIEW 8.5's Icon Editor.

(LabVIEW 8.5) or Figure 9.15 (LabVIEW 2009). The drawing tools are arranged on the left in the old Icon Editor, and on the right in the new one. The new icon is shown in Figure 9.16 (top-right corner).

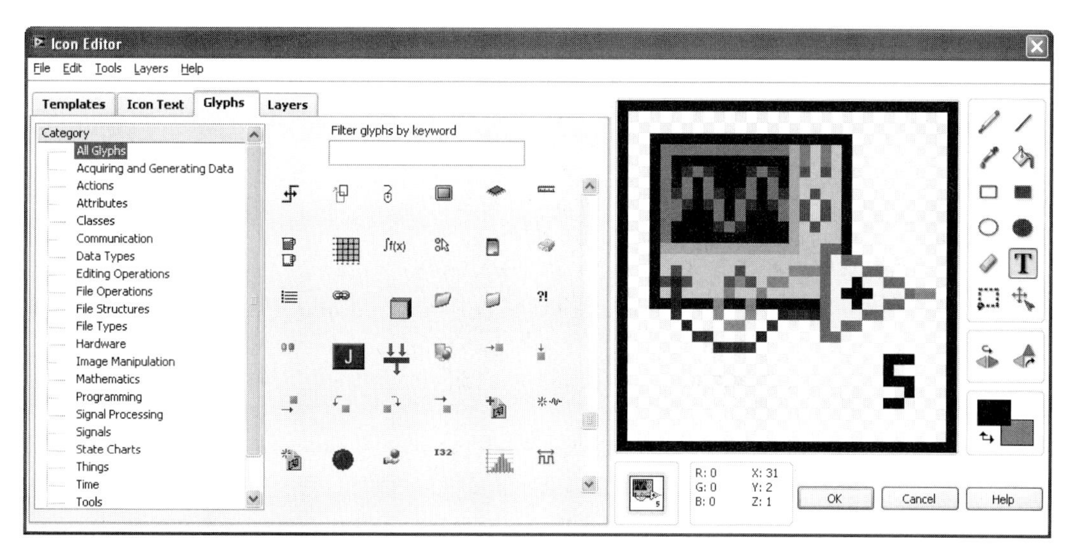

Figure 9.15
LabVIEW 2009's Icon Editor.

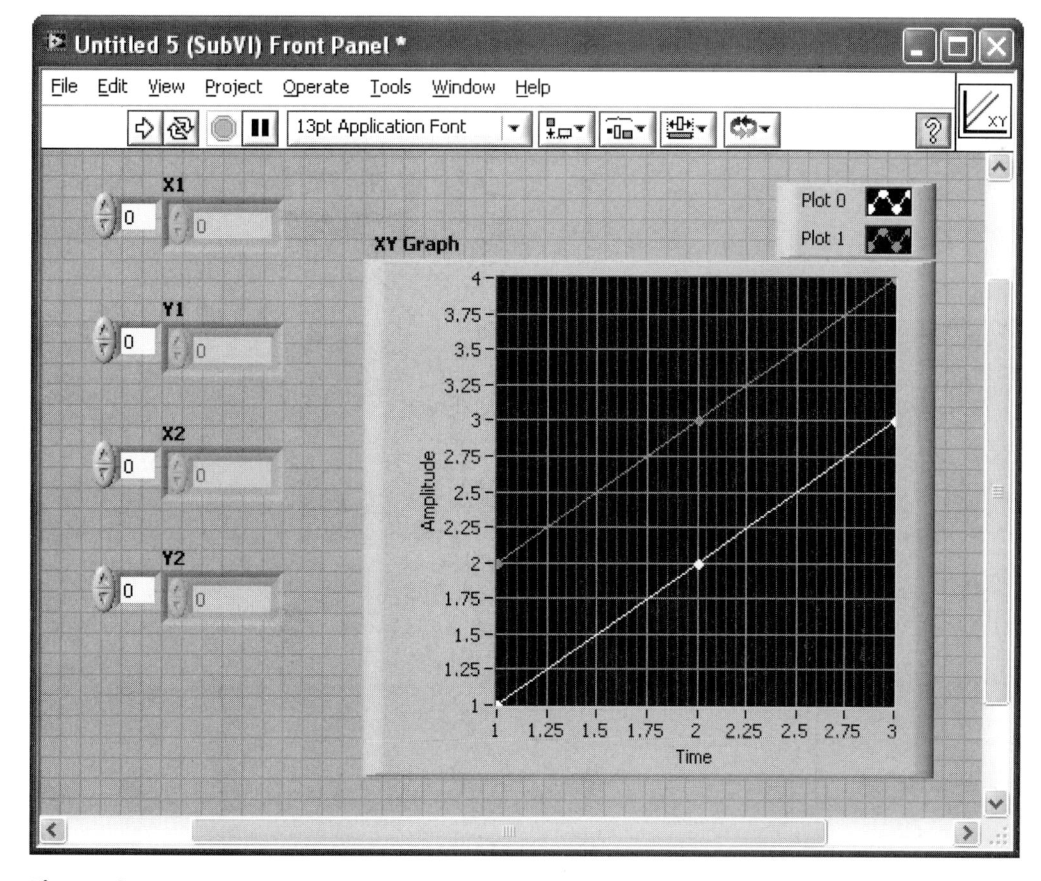

Figure 9.16
Front panel for the untitled SubVI.

Step 2. Save the SubVI

When you close the Icon Editor, you will be returned to the SubVI's front panel; "(SubVI)" should appear in the title bar as shown in Figure 9.16.

Save the SubVI using menu options **File / Save**. You will be asked to specify a name. In this example the SubVI was named "Two Curve XY Graph.VI". Once the SubVI has been assigned a name, the name becomes the label on the icon or Expandable Node, as illustrated in Figure 9.17.

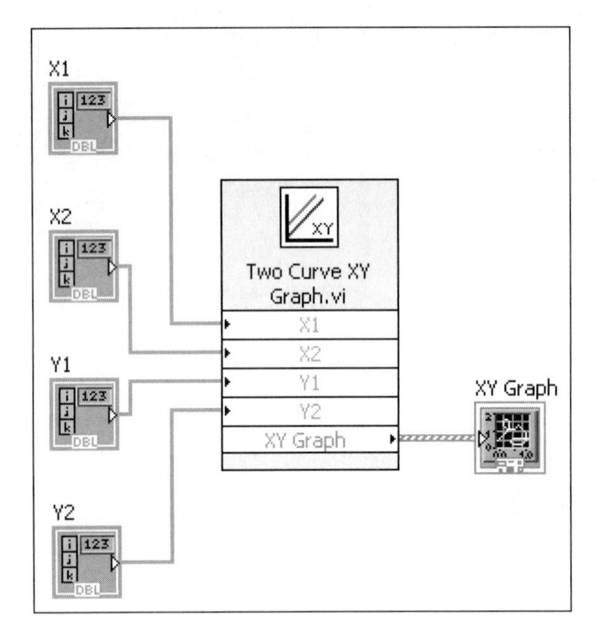

Figure 9.17 The SubVI can be viewed as an Expandable Node which identifies the inputs and outputs.

The power of SubVIs is that they can be used in other VIs whenever you need them. In Figure 9.18 (block diagram) and Figure 9.19 (front panel), the Two Curve XY Graph VI was used to create a plot of time and temperature values.

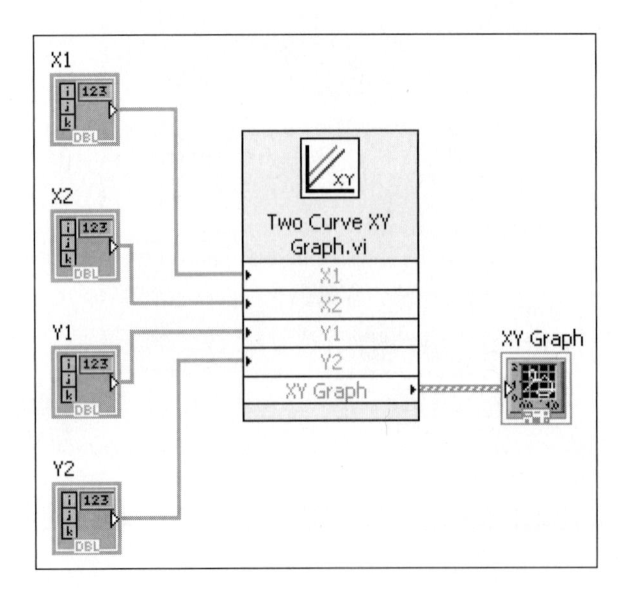

Figure 9.18 Using the Two Curve XY Graph SubVI.

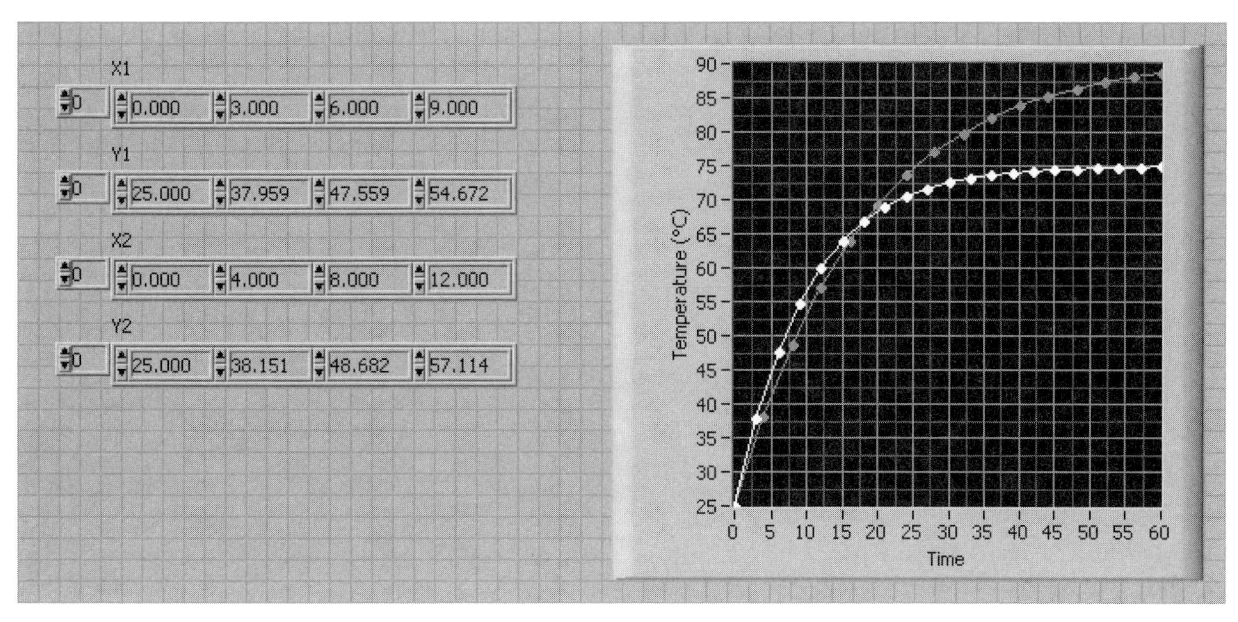

Figure 9.19
Front panel showing the plot created using the Two Curve XY Graph SubVI.

Admittedly, the block diagram looks just the same, but the arrays are totally new, even if they have the same names.

Step 3. Review Connections on Connector Pane
LabVIEW automatically selects a *connector pane* layout with enough connections to handle all of the SubVIs inputs and outputs (up to 14 of each). To see (or modify) how LabVIEW has laid out the connector pane for the SubVI, right-click on the SubVI's icon at the top-right corner of the edit window as illustrated in Figure 9.20. Then select **Show Connector** from the pop-up menu.

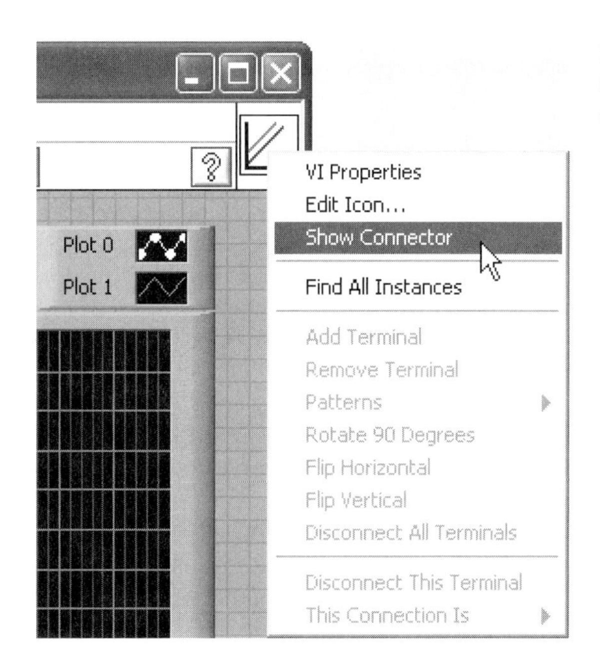

Figure 9.20
Showing the SubVI's connector pane.

The SubVI's icon will be replaced by the SubVI's connector pane as shown in Figure 9.21.

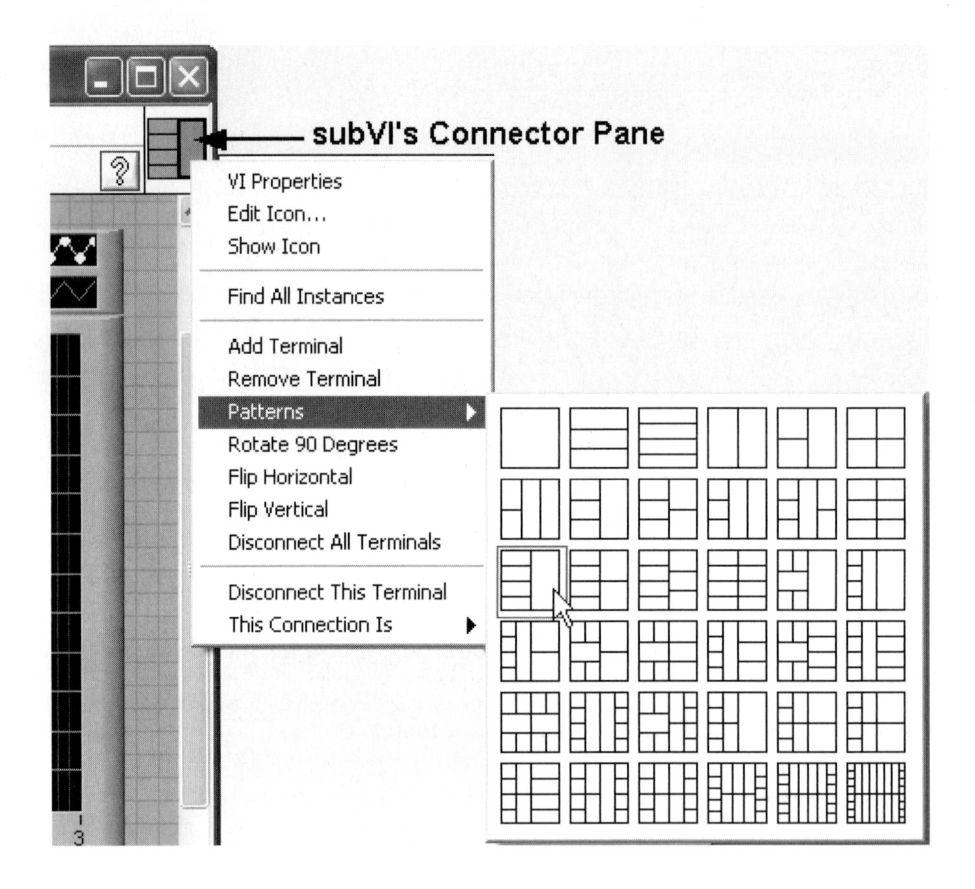

Figure 9.21
The SubVI's connector pane, and options menu.

You can right-click on the SubVI's connector pane to see a menu of options. In Figure 9.21 a display of connector pane options has been displayed. You can select a different arrangement of connectors from this display.
Using the Options menu shown in Figure 9.21, you can also

- Add a terminal to the connector pane.
- Remove a terminal from the connector pane.
- Disconnect all terminals.

If you click on an individual terminal on the connector pane, it will be highlighted to indicate that it is selected, and the control wired to the terminal will also be selected. This is illustrated in Figure 9.22. By clicking on the terminals on the connector pane, it is easy to see how the SubVI connector is wired.

If you need to change how a connector is wired, first unwire the terminal (right-click on connector pane, choose **Disconnect This Terminal** from the options menu). The disconnected terminal is indicated as shown in Figure 9.23. To wire the disconnected terminal, move the mouse over the disconnected terminal (the mouse icon will change to a wiring spool) and click once on the disconnected terminal, and then once on the control that should be wired to the terminal (X2 in this example). After wiring, the connector pane will show that the terminal is connected.

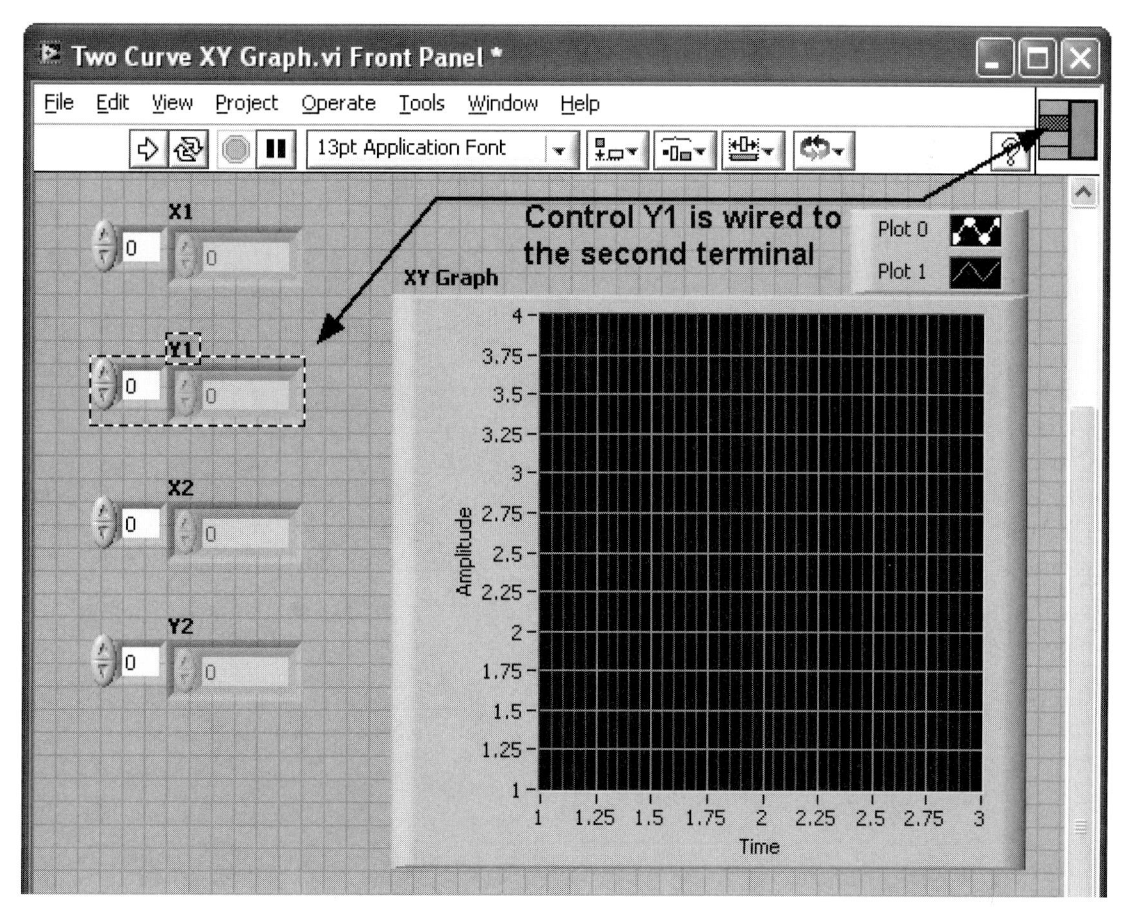

Figure 9.22
Click a terminal on the connector pane to see the control that is wired to that terminal.

PRACTICE

Create a SubVI that Reads a Column from a Text File

Because many LabVIEW graphing controls require 1D arrays, reading one column from a text file is a common task. You can speed up the process by creating a Read Text Column VI.

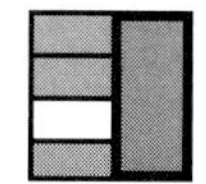

Figure 9.23
The connector pane showing a disconnected input terminal.

The block diagram of a VI that reads one column and sends the values to an indicator is shown in Figure 9.24. The portion that will become the SubVI is indicated with a dashed line.

Select the portion of the VI indicated in Figure 9.24, and then use menu options **Edit / Create SubVI**. LabVIEW will package the Read From Spreadsheet File VI and the Boolean constant connected to the transpose terminal (causes a column to be read). The result is shown in Figure 9.25. The SubVI was assigned the temporary identifier "7"—this will vary depending on how many VIs are open on your desktop.

At this point you should change the icon (optional) and save the SubVI (essential). Then it can be used in other VIs. Since SubVIs don't appear on the Functions

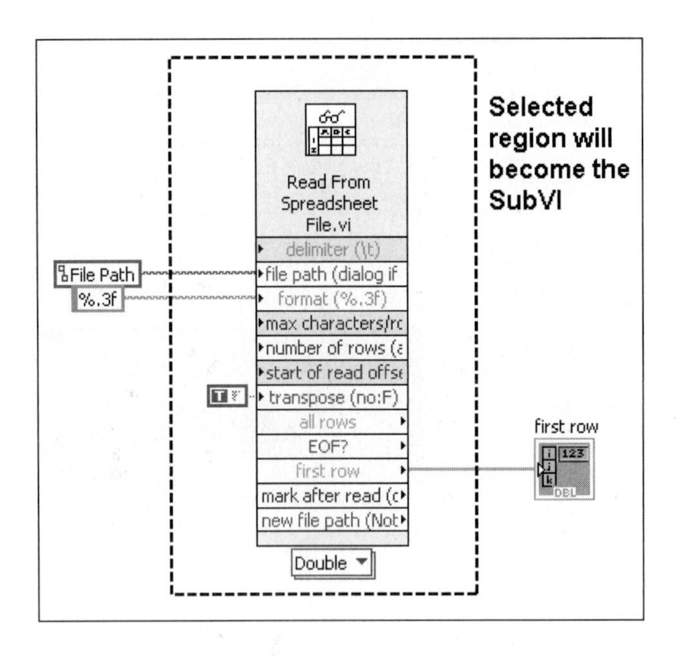

Figure 9.24
Block diagram of the VI before creating the SubVI.

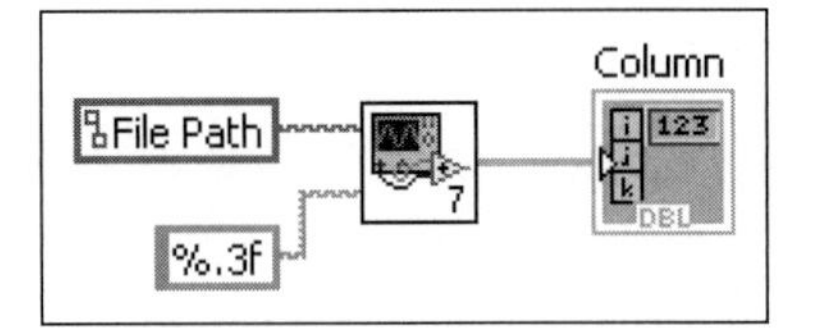

Figure 9.25
The Read Text Column SubVI has been created, temporarily called "7".

Palette, to place a SubVI on a block diagram select **Select A VI . . .** from the bottom of the Functions Palette, then browse for the SubVI file.

In Figure 9.26 the Read Text Column VI has been used twice to read two .txt files. One (InvoluteX.txt) contains a column of X values and the other (InvoluteY.txt) contains a column of Y values. The SubVI can be displayed as an icon (as used to read the X values) or in expanded form (as used to read the Y values). The path string specifying the file path could be omitted; LabVIEW would ask you to select the files when the VI was run.

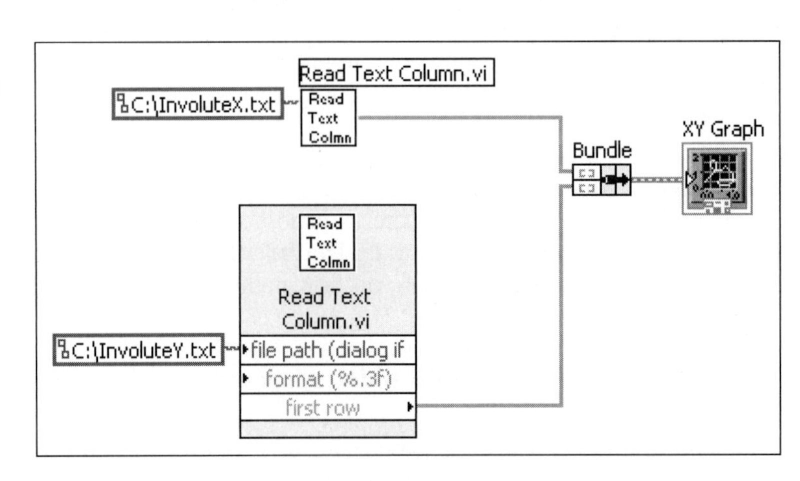

Figure 9.26
Using the Read Text Column VI to read X and Y data columns for plotting.

The involute graph is shown in Figure 9.27.

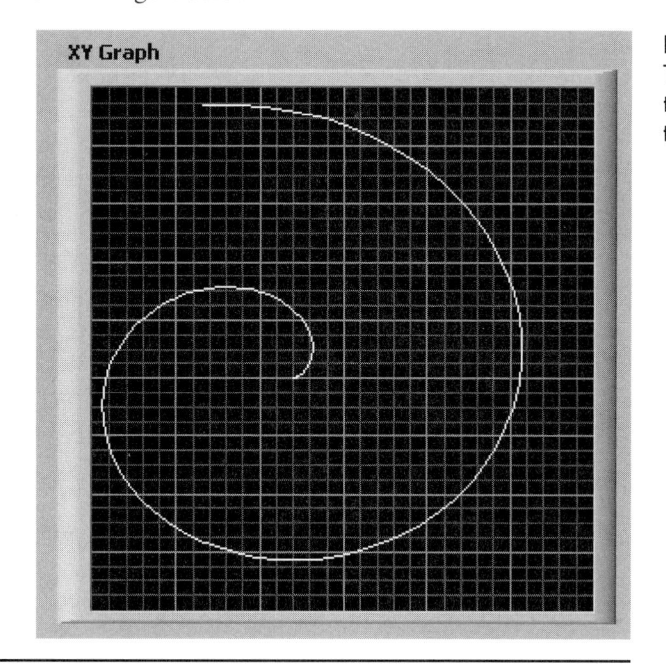

Figure 9.27
The XY Graph crated using the data values read from the files.

9.2.4 LabVIEW Projects

A *project* is a collection of all of the VIs and SubVIs needed to make an application work. As your VIs become more and more complex, keeping track of all the required code is essential. Projects not only collect all the required VIs and SubVIs, they help you keep track of who is editing what (access control) and what changes have been made to which VIs, and what other VIs may be impacted (revision control). Then, when the time comes to compile a stand-alone version of the program, you compile it from the project so that all of the pieces required to make the program work are built together.

For example, the VI created in the last example needed the Two Curve XY Graph SubVI. We could create a project that would keep the two files together.

To create a project, use menu options **Project / New Project . . .**

If you have open VIs, LabVIEW will ask if you want them included in the new project. Click the **Add** button. The Project Explorer (see Figure 9.28) will open

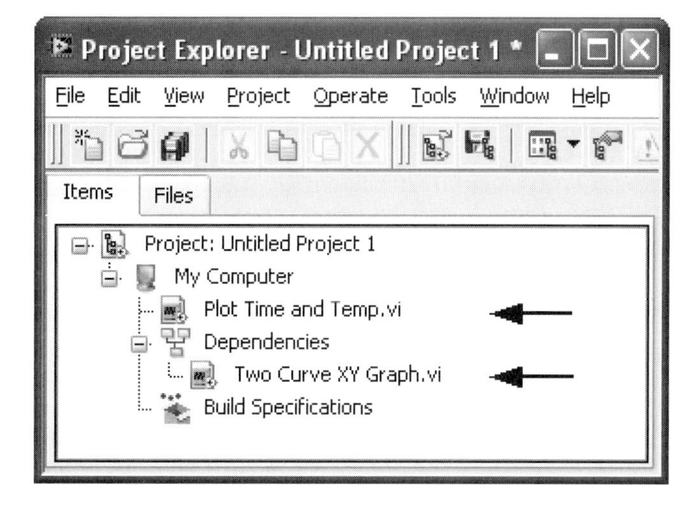

Figure 9.28
The Project Explorer, Items view.

showing the VI that was written to plot the time and temperature values (it was named Plot Time and Temp.vi) and the VI that it depends on, Two Curve XY Graph.vi.

If other SubVIs are added to the Plot Time and Temp VI, they will appear in the Dependencies list as well.

The Project Explorer provides two views, the view by Items (shown in Figure 9.28) shows how the VIs and SubVIs are related. The Files view (see Figure 9.29) shows how the VIs are arranged on disk.

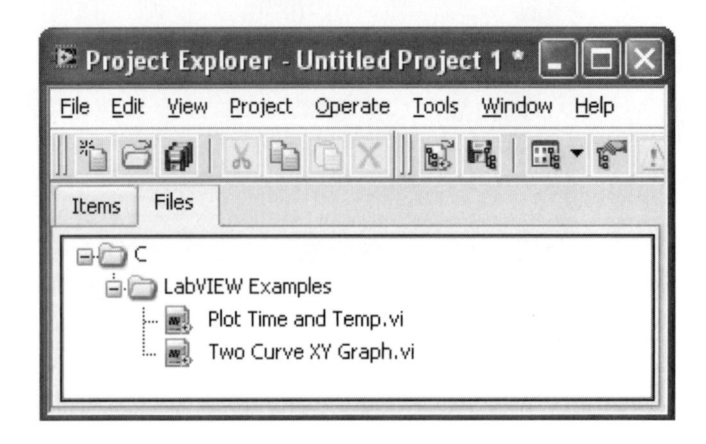

Figure 9.29 The Project Explorer, Files view.

The project shown in Figure 9.29 has not been saved. To save the project, use the menu options **File / Save** from the Project Explorer window.

Projects are very helpful when developing LabVIEW applications. They keep all of the required pieces together so that you can easily access all of the VIs in an application. The Files / Save All provides a quick and easy way to ensure that all of the changes made in an editing session to all of the open VIs are saved before closing the project. Projects are required before you can build an application.

9.3 STRUCTURES

The Programming Group on the Functions Palette provides access to a number of LabVIEW programming *structures*, and these structures give LabVIEW a lot of power as a programming language. The programming structures are relisted here, and then used in the following sections:

- Loop Structures
  - While Loop
  - For Loop
- Case Structure
- Sequence Structures
  - Flat
  - Stacked

9.3.1 While Loop

The *While Loop* keeps looping until a stop condition is satisfied. The While Loop is commonly used to keep a VI running until a STOP button is clicked, but there are many other uses. The example in Figure 9.30 uses a While Loop to keep generating random numbers. The While Loop stops when the random value is exactly 50.

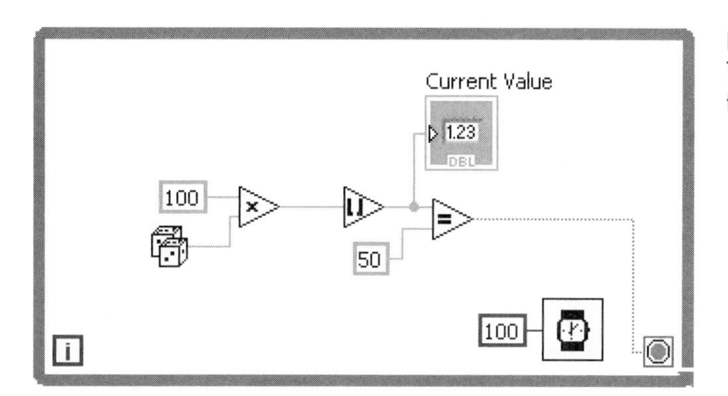

Figure 9.30
The While Loop cycles until the random number is 50.

Controlling the Timing of Loops

In Figure 9.30 a *Wait function* was used to slow the While Loop iterations to make it possible to read the Current Value indicator while the VI was running. The required input for the Wait function is the duration (time to wait). In Figure 9.31 the wait period was set to 100 ms.

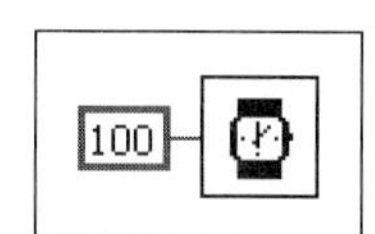

Figure 9.31
The duration of the Wait function set to 100 ms.

When the Wait function executes, all other programming activities stop for the duration of the wait period. When a Wait function is included inside a While loop, the Wait function will execute once for iteration of the While Loop, and the loop must do nothing until the wait period is completed. This has the effect of slowing the While Loop down to approximately one iteration every X milliseconds, where X is the duration value input to the Wait function.

Note: This assumes that the Wait function is the only slow operation inside the While Loop. If there are other slow operations, the loop will cycle even more slowly.

PRACTICE

Using the Wait Function

Create a VI with the block diagram shown in Figure 9.32. This VI simply runs until the STOP button is clicked, loops once every 1000 ms because of the Wait function, and turns the Even LED on each time the remainder of division by two equals 0, and off when the remainder is 1. (The LED turns on or off every second.)

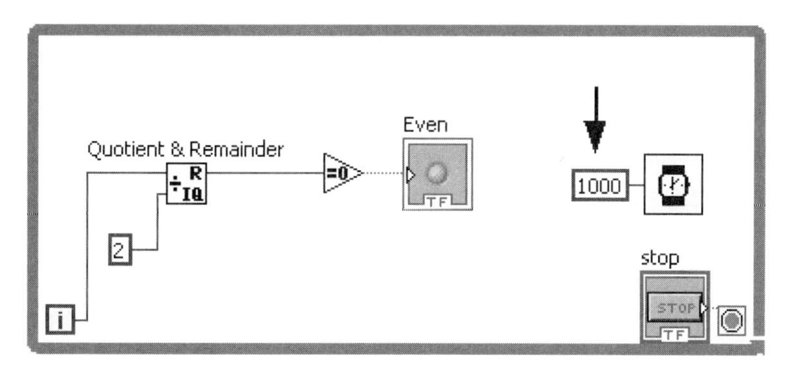

Figure 9.32
A While Loop with an LED that turns on and off every 1000 ms.

Change the value of the constant that inputs the delay time to the Wait function to see how the VIs performance changes.

Reducing the "1000" to smaller values causes the LED to blink faster. You can control the speed of the While Loop by changing the value input to the Wait function.

An alternative to the Wait function is the *Wait Until Next ms Multiple function*, which has a metronome icon as shown in Figure 9.33. The required input is the millisecond multiple value.

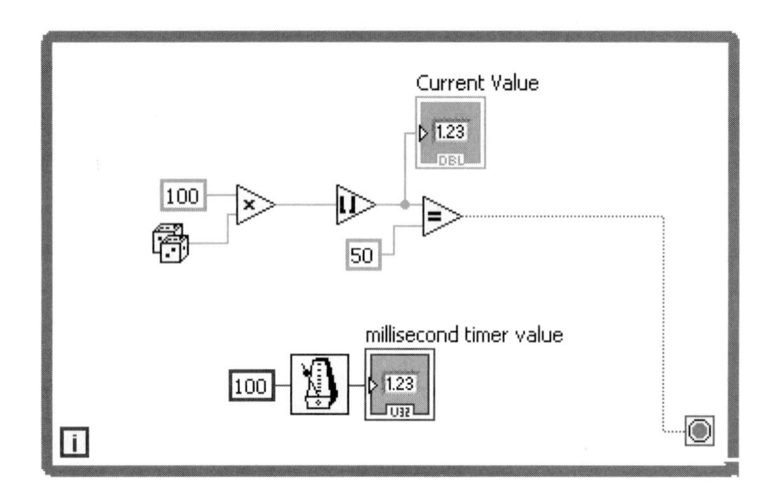

Figure 9.33
Using the Wait Until Next ms Multiple function to control loop timing.

The differences between the Wait function and the Wait Until Next ms Multiple function are as follows:

- If the **Wait function** duration is set to 100 ms, it will always add 100 ms to the time required to complete all other programming steps in the loop. If the other programming steps in the loop take 20 ms, then the loop will cycle once every 120 ms.
- The **Wait Until Next ms Multiple function** will wait whatever time is necessary to get to the next even multiple of the millisecond multiple value input to the function. If the millisecond multiple value is set to 100 ms, then the Wait Until Next ms Multiple function will wait until the millisecond timer has values like 3214500, 3214600, 3214700, and so on. If the other programming steps in the loop take 20 ms, then the Wait Until Next ms Multiple function will add approximately 80 ms per cycle, and the loop will cycle once every 100 ms.

If the other programming steps will be nearly instantaneous, then the Wait function and the Wait Until Next ms Multiple function will both cause the While Loop to iterate at approximately the same rate. But if you are trying to set the loop iteration time, the Wait Until Next ms Multiple function will give you more control.

9.3.2 For Loop

A *For Loop* iterates a specified number of times. An example of a For Loop is shown in Figure 9.34. In a For Loop, the "*i*" is the *iteration terminal* (also called the *iteration counter*) and the "*N*" is the *count terminal*. The value of *i* changes each time through the loop from 0 to $N - 1$. For the example in Figure 9.34, i will take on values from 0 to 9.

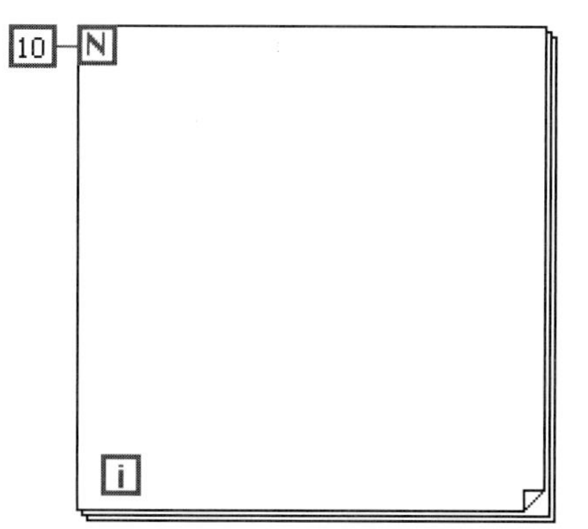

Figure 9.34
For Loop, set to loop ten times ($i = 0 \ldots 9$).

To see the value of i change as the For Loop iterates, we can connect the iteration terminal to an indicator, as shown in Figure 9.35. The Wait function has been added to slow the loop to allow the changes in i to be observed.

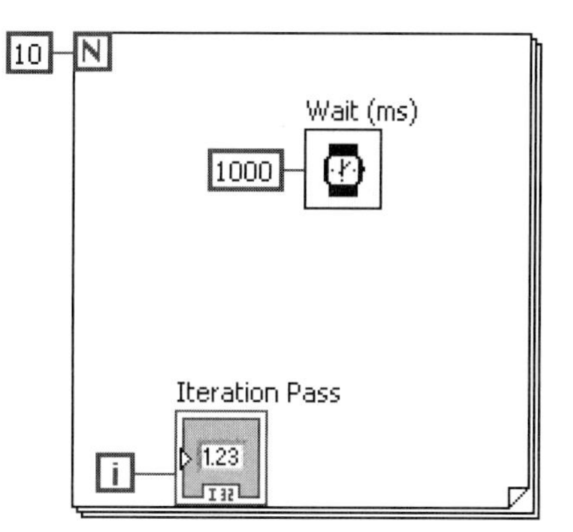

Figure 9.35
For Loop with Iteration Pass indicator added to monitor iteration counter, i.

For Loops are commonly used to build arrays. If the i value is wired to an array indicator outside the loop, as shown in Figure 9.36, and when the VI is run, the array is filled with values from 0 to 9, as shown in Figure 9.37.

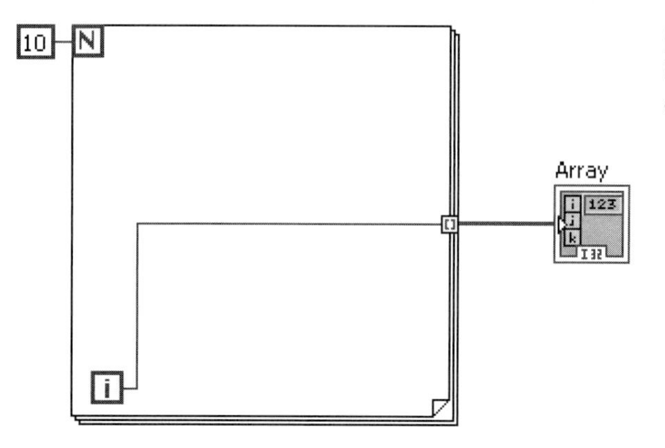

Figure 9.36
Using the For Loop iteration counter to build an array.

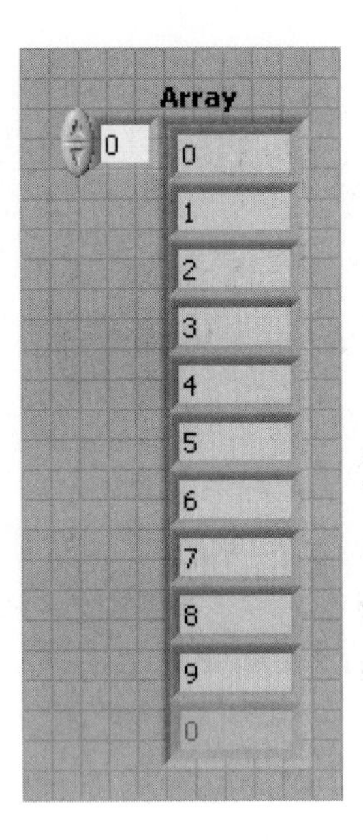

Figure 9.37
The array created by the VI shown in Figure 9.36.

By using math functions with the iteration counter, it is possible to calculate a wide range of array values. For example, the VI in Figure 9.38 creates an array of X values between 0 and 2π, and an array of sin(X) values. The values are clustered then sent into a XY Graph control for plotting. The result is shown in Figure 9.39.

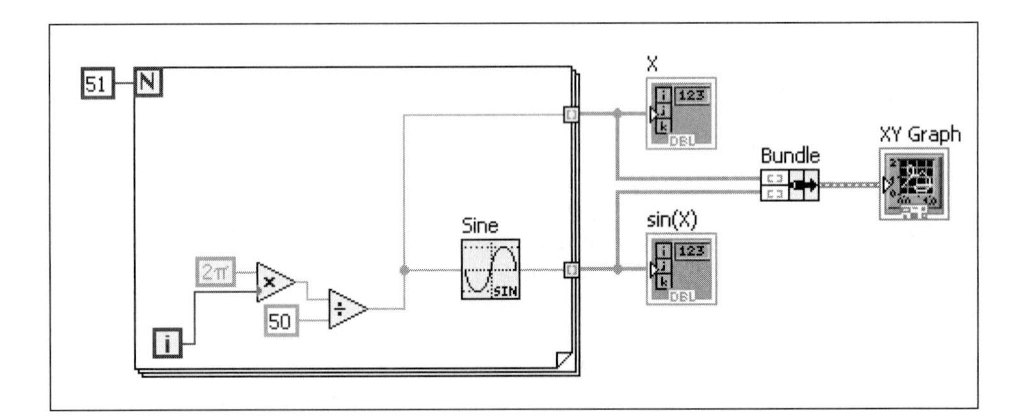

Figure 9.38
For Loop used to create X and sin(X) arrays.

PRACTICE

Using For Loops to Build Arrays

Modify the block diagram shown in Figure 9.38 to build both sin(X) and cos(X) arrays, and send those to an XY Graph.

The modified VI is shown in Figure 9.40, and the resulting XY Graph is shown in Figure 9.41.

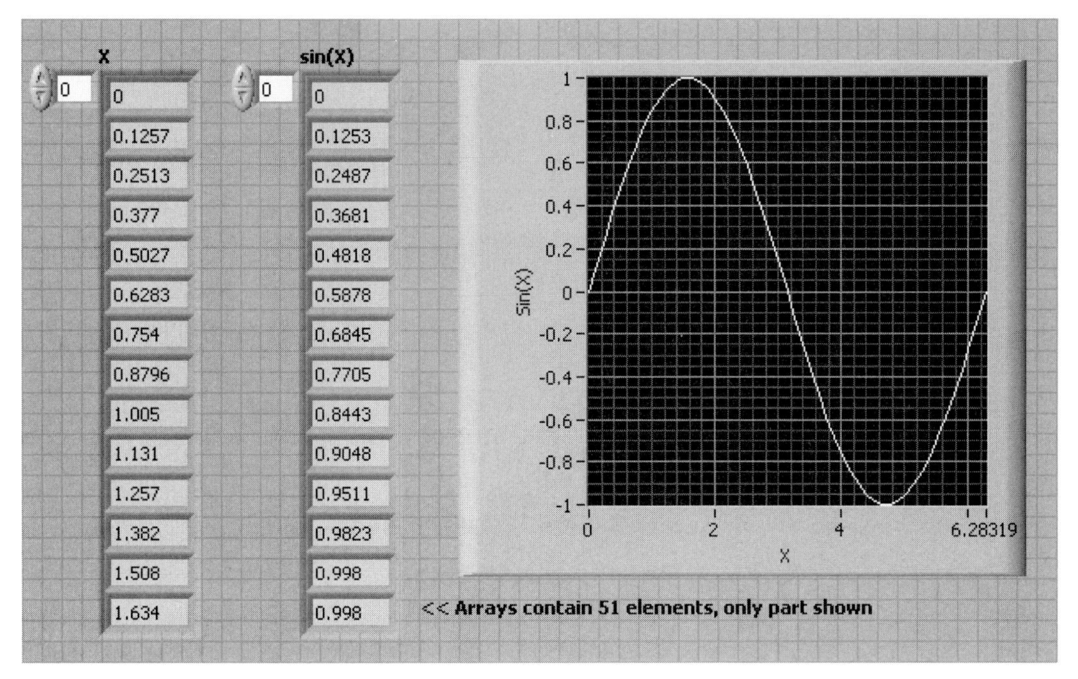

Figure 9.39
The *X* and sin(*X*) arrays created using a For Loop are plotted.

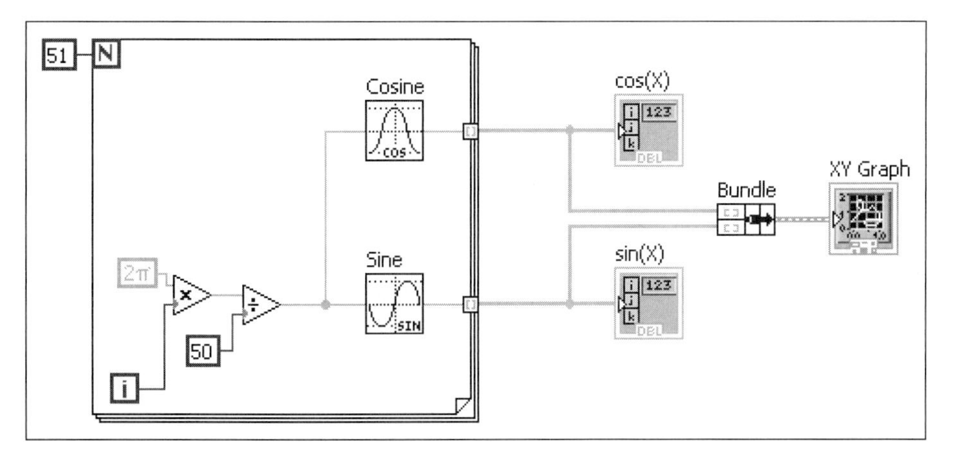

Figure 9.40
Modified VI to create sin(*X*) and cos(*X*) arrays for plotting.

Tunneling Into and Out of Loops

LabVIEW provides a way to get information across a loop boundary, called a *tunnel*. Tunnels were used in the last example to send the array information out of the For Loop to create the arrays. Tunnels can have *indexing* enabled or disabled. Tunnels on For Loops have indexing enabled by default.

- When indexing is enabled, the calculated value in a wire is sent through the tunnel with each loop iteration.
- When indexing is disabled, the calculated value in a wire is sent through the tunnel only when the loop iteration is finished.

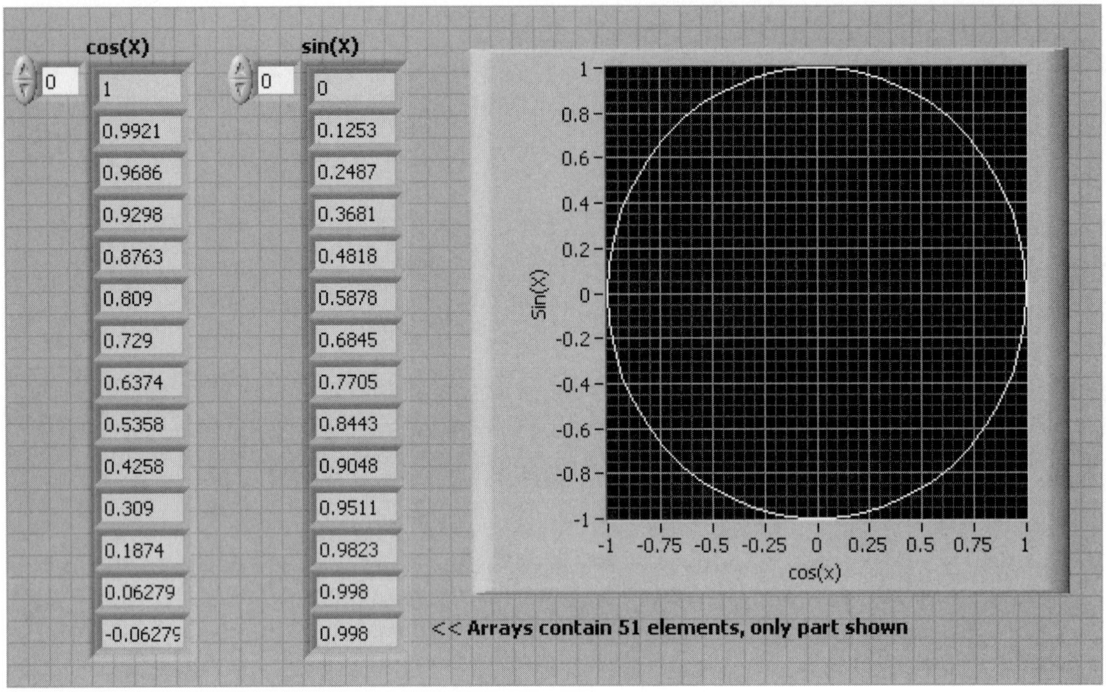

Figure 9.41
The cos(X) and sin(X) arrays created using a For Loop are plotted.

The VI illustrated in Figure 9.42 creates an array of X and sin(X) values from the top For Loop, but only the final X value and sin(X) value leave the bottom For Loop. The resulting front panel is illustrated in Figure 9.43.

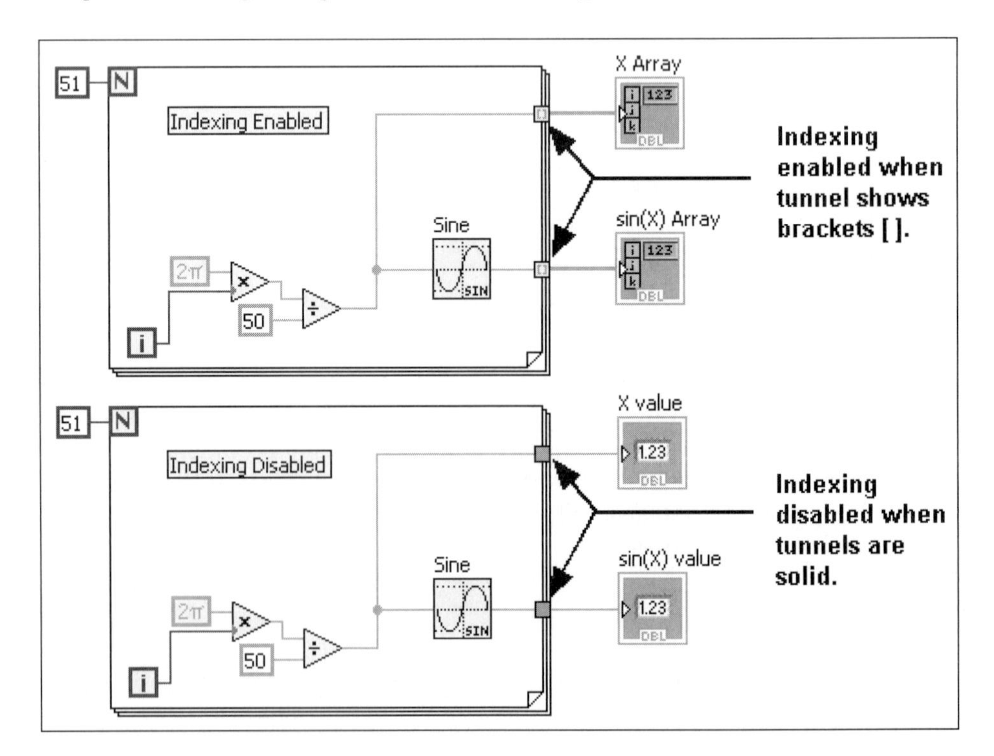

Figure 9.42
Same For Loop, with and without indexing enabled on the tunnels.

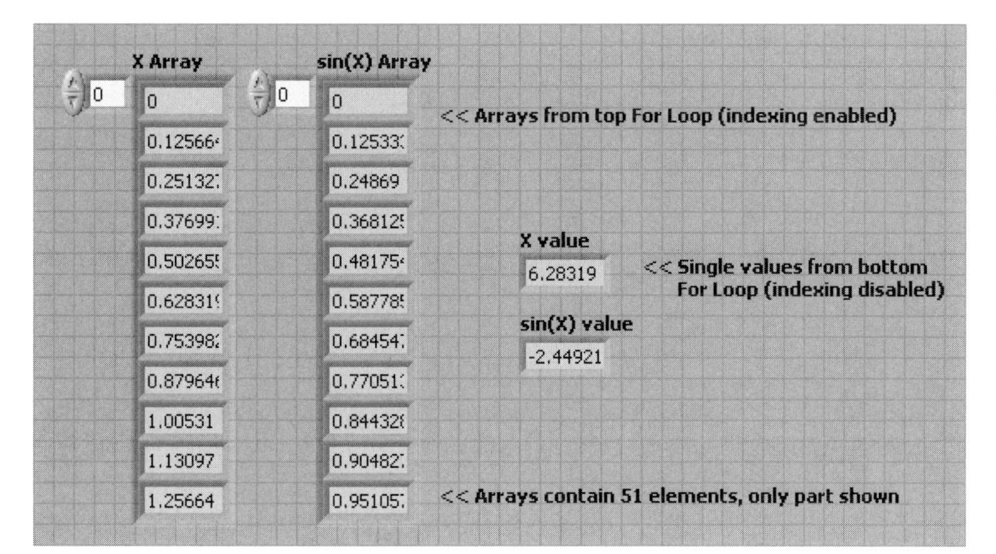

Figure 9.43
Front panel after running the For Loops shown in Figure 9.42.

Auto-Indexing a For Loop

When working with arrays, it is often useful to work with each element in the array, one at a time, one after the other. LabVIEW will do this automatically whenever you wire an array to an input tunnel in a For Loop (see Figure 9.44)—if the tunnel is enabled for indexing.

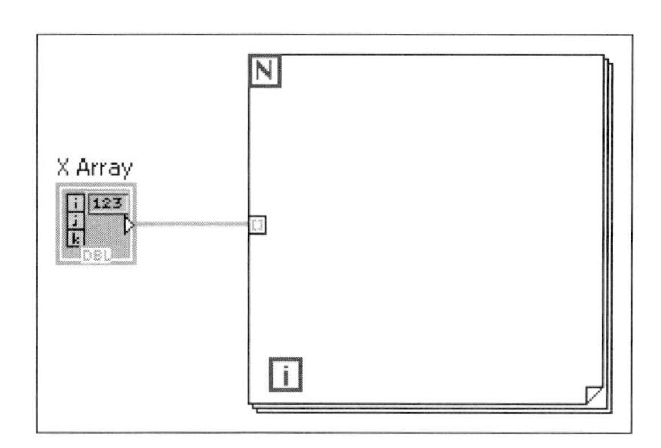

Figure 9.44
Connecting an array to a For Loop tunnel (as input) causes auto-indexing if indexing is enabled.

This is called *auto-indexing* the For Loop.

- When indexing is enabled on the input tunnel, the value of N is set to equal the number of elements in the array, and one array element enters the For Loop each time the loop cycles.
- When indexing is disabled on the input tunnel, the value of N is not set (this generates an error if the value of N is not set in some other way) and the entire array is sent into the For Loop immediately.

Note: Arrays can be wired to While Loops input tunnels too, but the While Loop is not auto-indexed.

In the example VI shown in Figure 9.45, in each cycle of the For Loop, one element of X Array enters the loop. The sine of that value is calculated, and then one-twentieth of the current value of the iteration counter is added to the sine value.

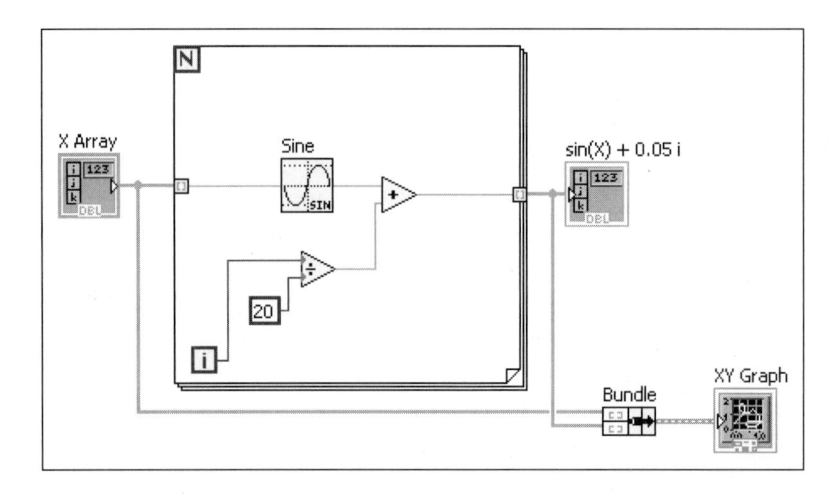

Figure 9.45
The *X* Array elements enter the For Loop one at a time (indexing enabled).

The result exits the For Loop and is added to the $\sin(X) + 0.05\,i$ array. The result is a tilted sine wave, as shown in Figure 9.46.

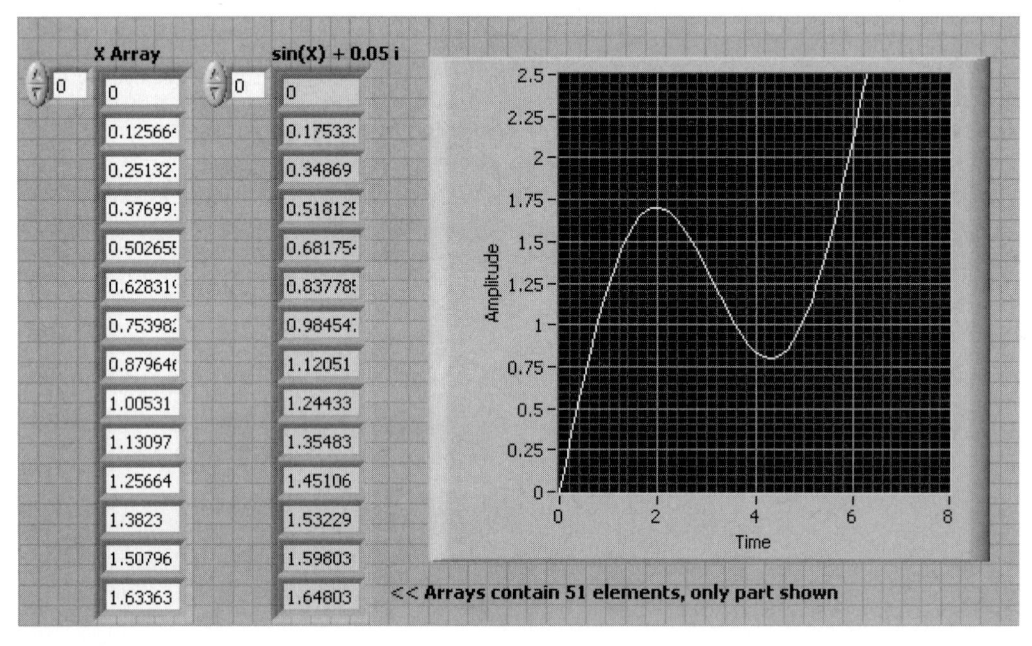

Figure 9.46
The tilted sine wave.

EXAMPLE 9.1

Creating a 3D Parametric Graph of a Torus

To help define terms, consider the torus shown in Figure 9.47.

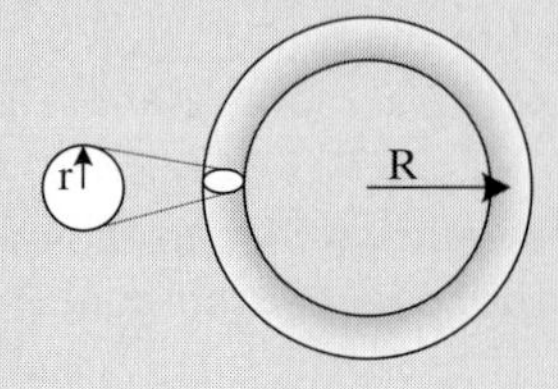

Figure 9.47
Defining the radius values for the torus.

The surface of a torus can be described with the following equations:

$$x = (R + r\cos(u))\cos(v)$$

$$y = (R + r\cos(u))\sin(v)$$

$$z = r\sin(v)$$

Variables u and v are working variables that each range between 0 and 2π radians. The number of increments used for u and v is arbitrary, but a smaller steps size creates a smoother surface when the torus is plotted. In this example we will use 30 steps for both u and v. Variables x, y, and z will all be 2D arrays; they can be created using two nested For Loops as shown in Figure 9.48. The plotted result is shown in Figure 9.49.

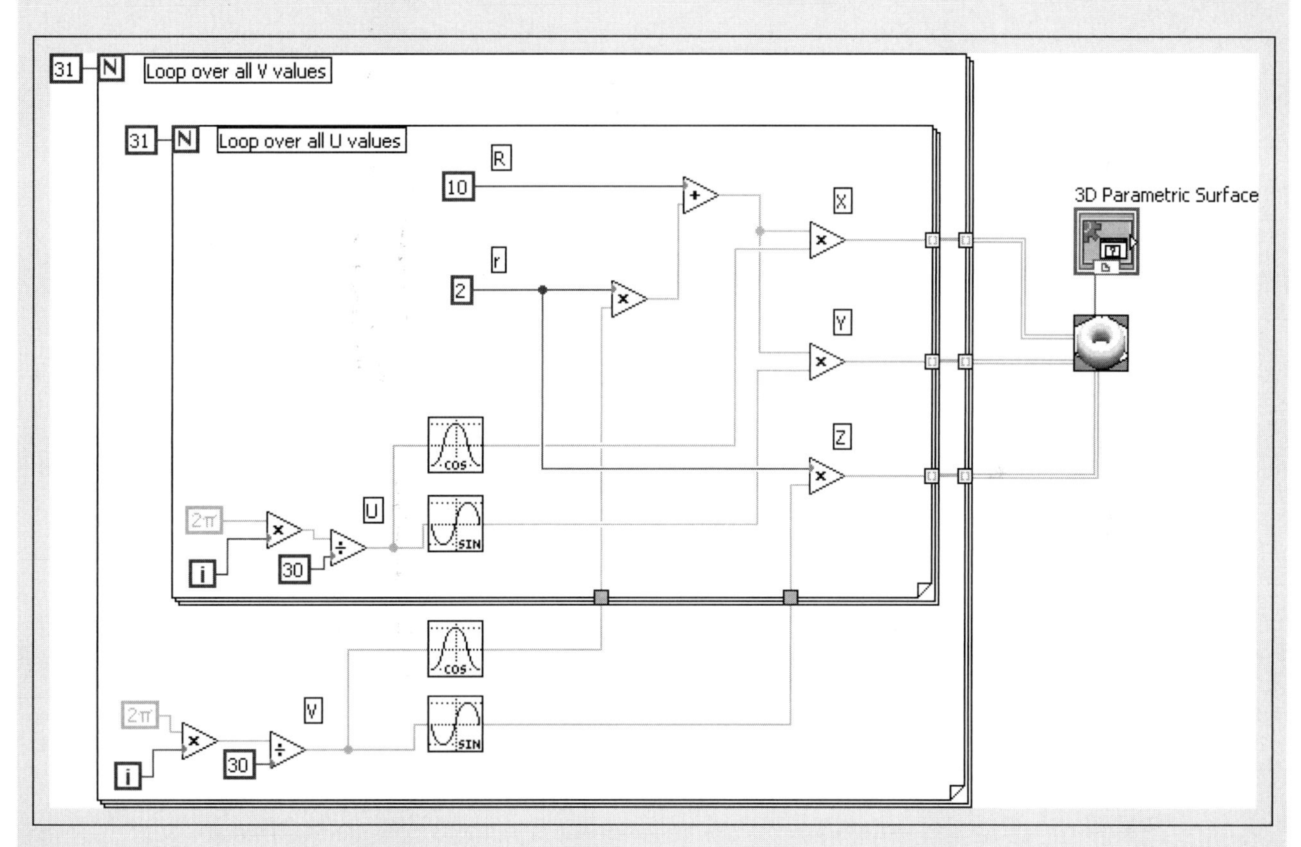

Figure 9.48
Nested For Loops used to generate 2D arrays x, y, and z for plotting.

9.3.3 Shift Registers—Accessing Values from the Previous Loop Iteration

There are times when the current calculation needs to include a value from the previous iteration cycle. *Shift registers* give you that ability in LabVIEW. A shift register sends the calculated value at the end of one iteration into the beginning of the next iteration.

You add a shift register to a loop by right-clicking on the loop boundary and selecting **Add Shift Register** from the pop-up menu. When a shift register is added, indicators appear on both sides of the loop as shown in Figure 9.50.

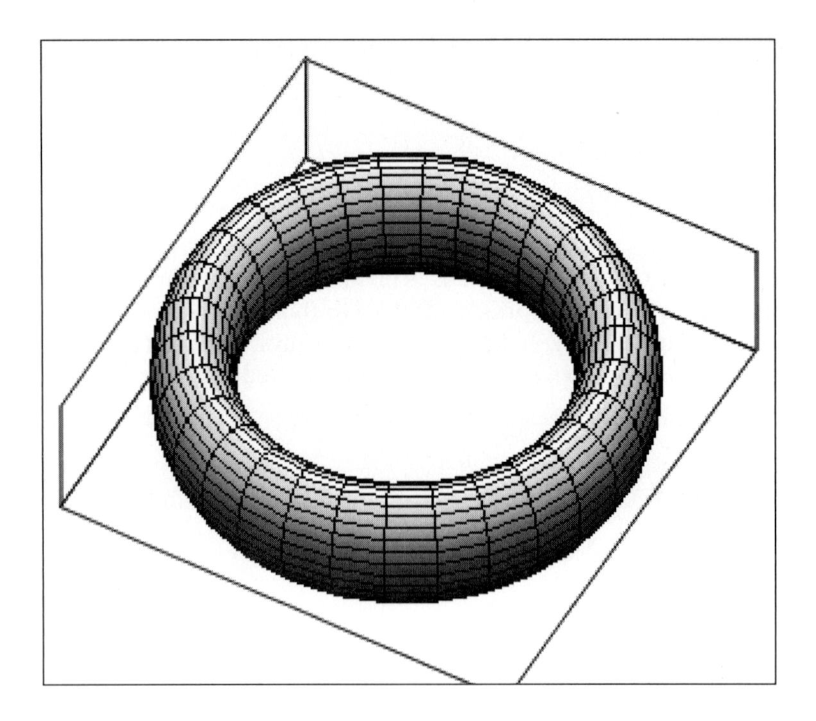

Figure 9.49
The 3D parametric plot of the torus.

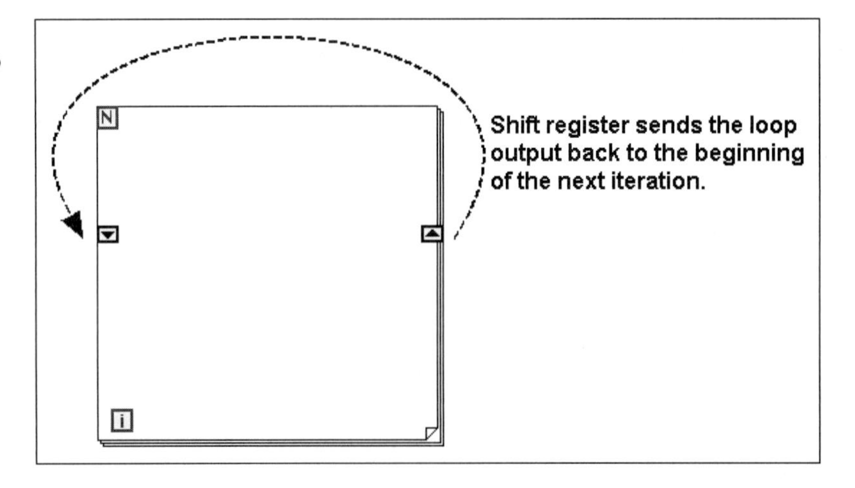

Figure 9.50
A shift register allows you to use previous loop values.

Shift registers can be given initial values or left uninitialized.

- An **uninitialized shift register** gets its starting value from the last output value at the end of the previous execution (the last time the same VI was run). You can use an uninitialized shift register to start the current loop where it left off last time.
- An initialized shift register is assigned a value before the loop begins to iterate. The shift register in Figure 9.51 has been initialized with the value zero. The first time the For Loop cycles, the shift register will have the value zero.

The data type associated with a shift register is assigned when the first connection to any shift register terminal is made. In Figure 9.51 the data type of the constant was double-precision floating point (DBL), so all connections to that shift register must be compatible with the DBL data type.

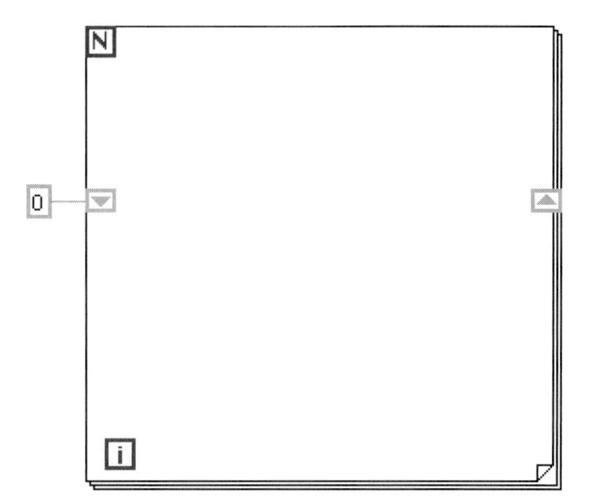

Figure 9.51
Initializing a shift register.

In Figure 9.52 the shift register value is used in a calculation. The For Loop cycles five times because of the "5" wired to the *N* terminal (loop count terminal).

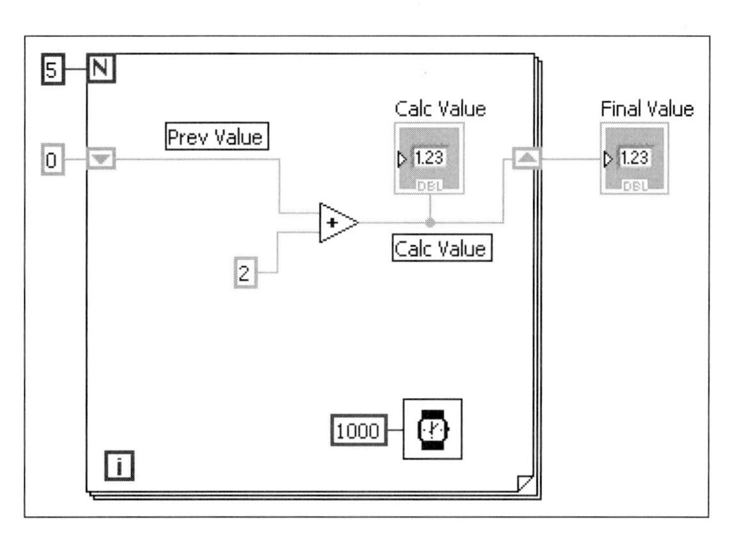

Figure 9.52
Using a shift register in a calculation.

1. The For Loop begins the first iteration.
   a. The shift register receives the initial value, 0.
   b. The 0 and 2 are added; the Calc Value is 2.
   c. The shift register is assigned the Calc Value, 2.
   d. The For Loop ends the first iteration.
2. The For Loop begins the second iteration.
   a. The shift register still contains the value 2 from the end of the last loop.
   b. The 2 (shift register) and 2 (constant) are added; the Calc Value is 4.
   c. The shift register is assigned the Calc Value 4.
   d. The For Loop ends the second iteration.

The process continues. In sum, the shift register receives the values 0, 2, 4, 6, and 8. The Calc Value is 2, 4, 6, 8, and 10, and the Final Value has no value until the For Loop terminates, and then it is assigned the final value of Calc Value, which was 10.

This example wasn't good for much, except (hopefully) to demonstrate how a shift register works. In the next example we'll use a shift register to accomplish something useful.

PRACTICE

Using a Shift Register—Exponential Growth

The VI shown in Figure 9.53 uses a shift register to demonstrate exponential growth. At time zero, there is one cell input to the For Loop through the upper shift register. The 60 ms Wait represents 60 minutes of real time, which is approximately the time required for many types of microbial cells to divide. The For Loop runs for 48 cycles, representing 2 days.

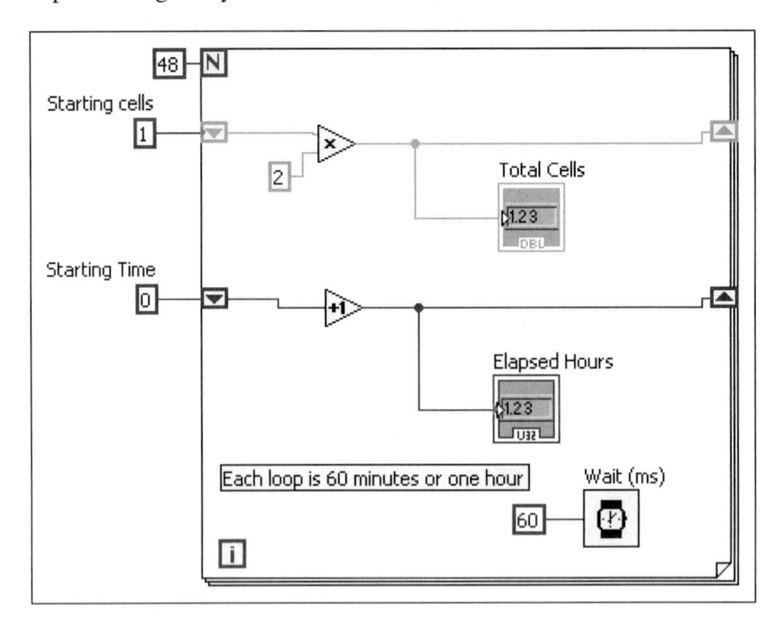

Figure 9.53
Shift registers used to demonstrate exponential growth.

When the VI is run, the time and cell count are continuously updated for the 48 cycles through the For Loop. The final results are shown in Figure 9.54.

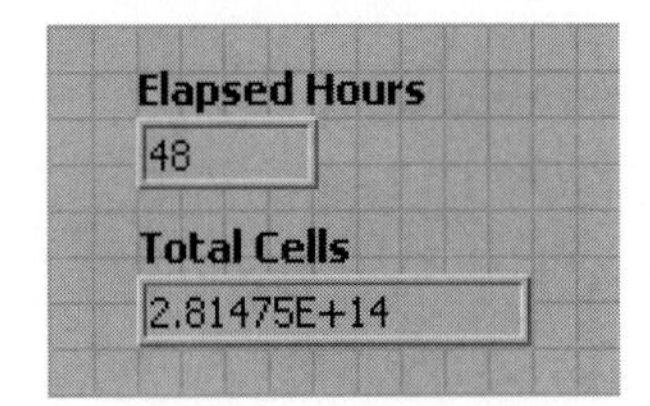

Figure 9.54
Final cell count after 48 hours of growth.

EXAMPLE 9.2

Calculating Slopes from Arrays of X and Y Values

Slope can be approximated as

$$slope = \frac{\Delta y}{\Delta x} = \frac{y_i - y_{i-1}}{x_i - x_{i-1}}$$

When we send an array into a For Loop through an indexed tunnel, the array elements enter the loop one at a time. Therefore, y_i and x_i will be available to use in a calculation, but we also need y_{i-1} and x_{i-1}. Two shift registers will make the previous y and x values available for the slope calculation.

The block diagram for the Slope VI is shown in Figure 9.55.

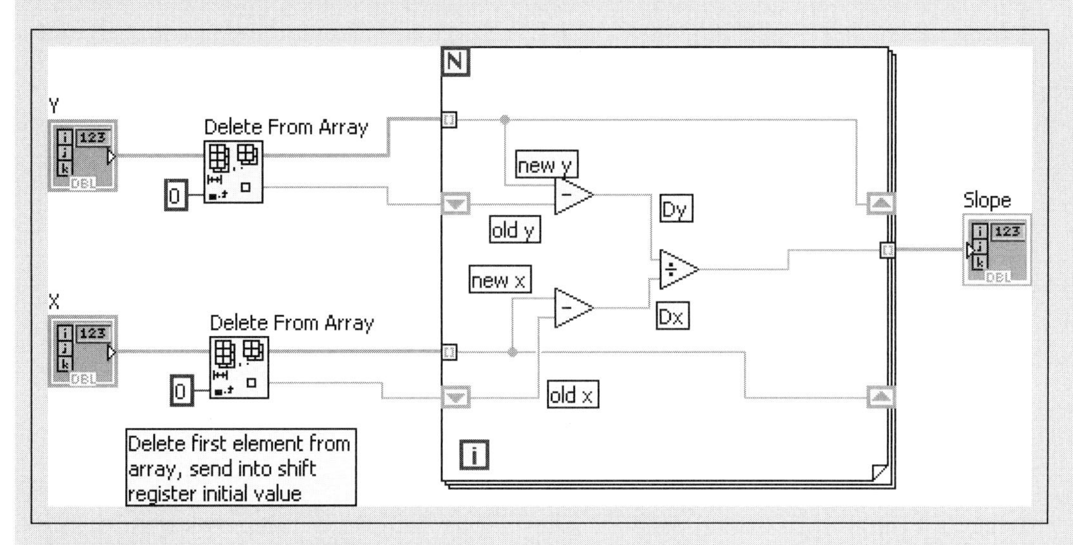

Figure 9.55
Using shift registers to calculate slope values.

The Delete From Array functions are used to strip the first element from the Y and X arrays and use those values to initialize the shift registers. With the first element of each array taken off (and sent in through the shift register), the second element enters the For Loop as the "new" value and the "old" values comes in via the shift register. Old values are subtracted from new values to calculate the Δy and Δx values, which are divided to estimate the slope. The new y and new x values at the end of the cycle are assigned to the shift register and become the "old" values for the next loop. The process continues until all array values have been processed.

Stacking Shift Registers

One shift register provides access to the value of a variable at the end of the previous loop. If you need to go back more than one loop, you can stack shift registers on the left (input) boundary of the For Loop.

To turn a shift register into a stacked shift register, right-click on the shift register on the left boundary of the For Loop and select **Add Element** from the pop-up menu. Each time you select Add Element, another stacked register will be added. In Figure 9.56, the left boundary of the For Loop contains a stack of three shift registers, providing access to values from the past three iterations.

Notice in Figure 9.56 that the shift registers stack on the input side of the For Loop only, the output side is only used to assign the most recent calculated value.

9.3.4 Case Structures

Case structures are common features of modern programming languages; they allow certain programming actions to take place depending on a variable's value. The

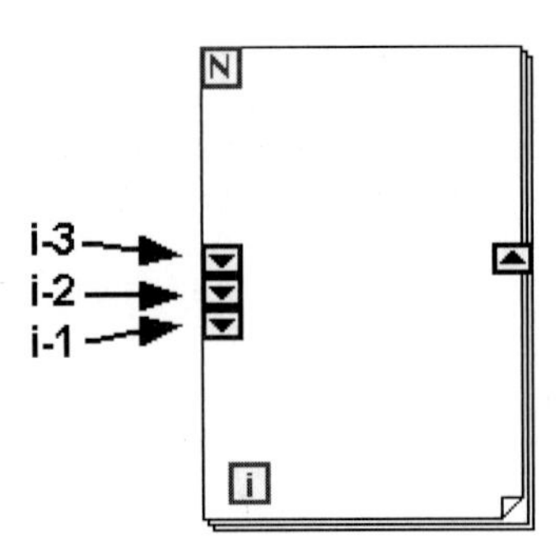

Figure 9.56
Three stacked shift registers provide access to values from past three cycles.

Case Structure is located on the Programming Group on the Functions Palette. Drag the Case Structure icon to the block diagram as illustrated in Figure 9.57.

By default the Case Structure has two cases: **True** and **False**. These can be changed as needed.

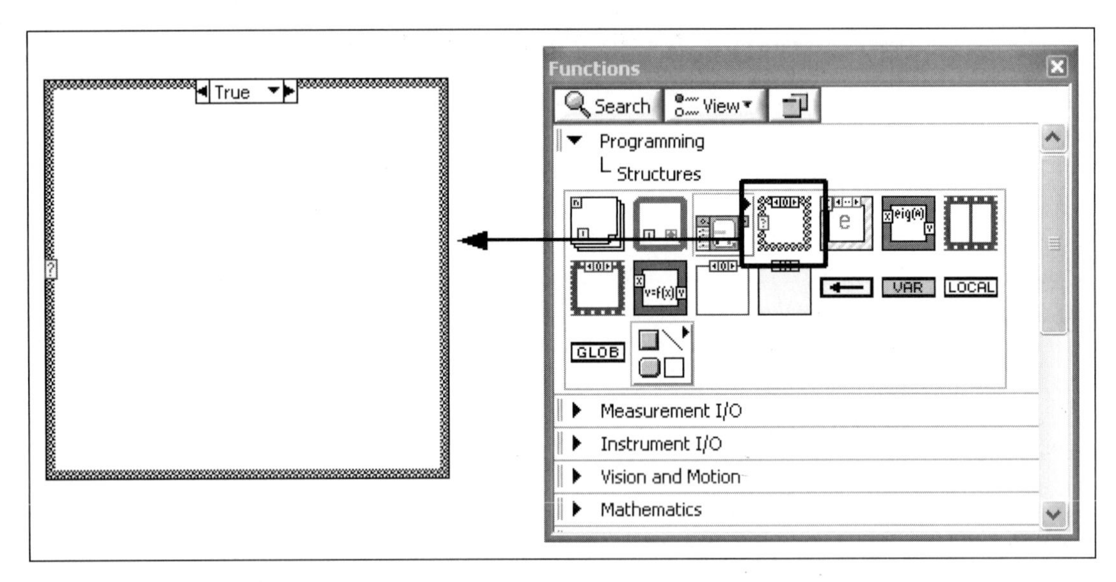

Figure 9.57
Placing a Case Structure on the block diagram.

EXAMPLE 9.3

Checking for Improper Operand

In this example we check to see if the user is attempting to take the logarithm of a non-positive number. The user runs the VI (run continuous) and enters a value for X. If the value of X is greater than 0, then the comparison generates a TRUE, which causes the True case to be selected. The logarithm of X is calculated and displayed (see Figure 9.58) along with a message indicating that there is no problem.

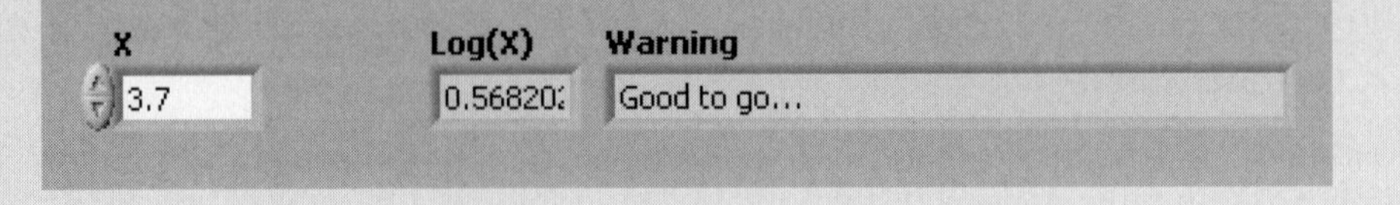

Figure 9.58
Front panel showing results returned by the True case.

But when the user enters a non-positive value, the greater than zero comparison generates a FALSE, which causes the False case to be selected. The False case does not calculate the logarithm, but instead sends a message that there is a problem (see Figure 9.59).

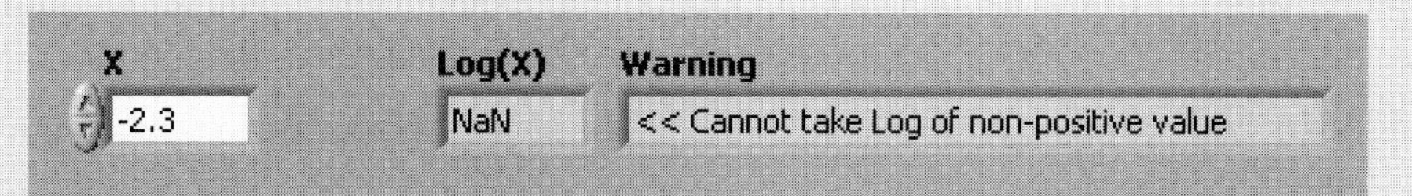

Figure 9.59
Front panel showing results returned by the False case.

The block diagram must be shown in two parts because only one case is visible at a time. Figure 9.60 shows the True case, and Figure 9.61 shows the False case. Click the down arrow on the right side of the Selector Label to choose a different case.

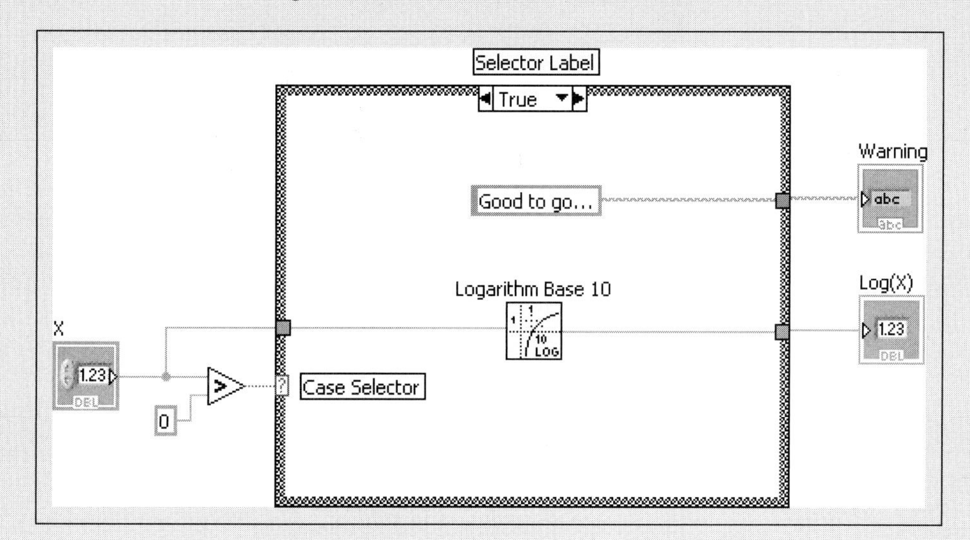

Figure 9.60
Block diagram for the True case.

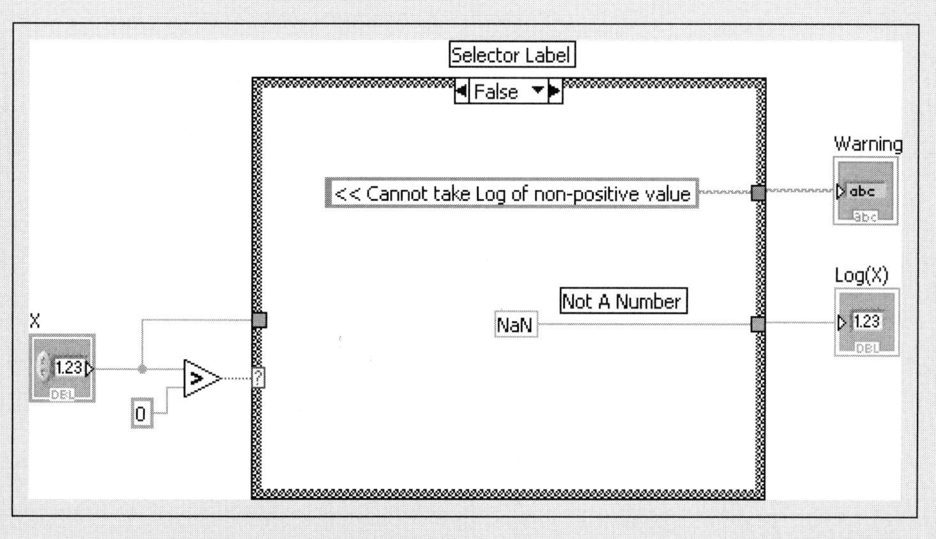

Figure 9.61
Block diagram for the False case.

Notice that in the False case, the value of X is not used. Unused inputs are allowed on Case boundaries, but all output tunnels must be wired. In this case NAN (not a number) to the Log(X) indicator.

PRACTICE

Using Case Structures

Create a VI that multiplies two numbers or divides two numbers depending on whether a switch is open or closed. Use a Case structure that selects the multiply or divide operation depending on the Boolean output from the switch.

A VI that accomplishes this is shown in Figure 9.62 (front panel) and Figure 9.63 (block diagram).

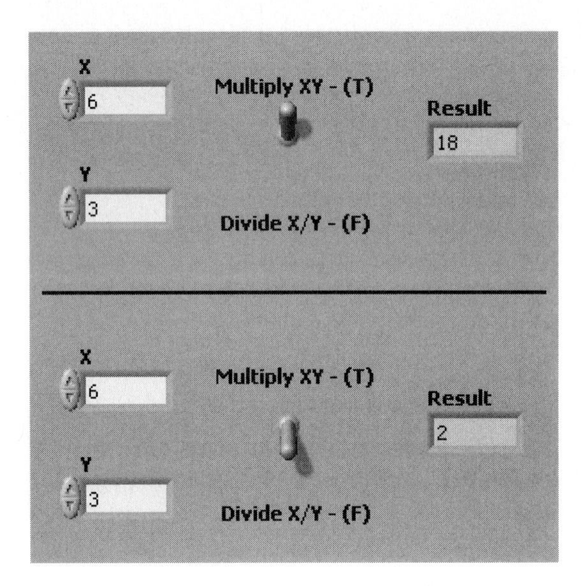

Figure 9.62
Front panel showing how switch changes VI behavior.

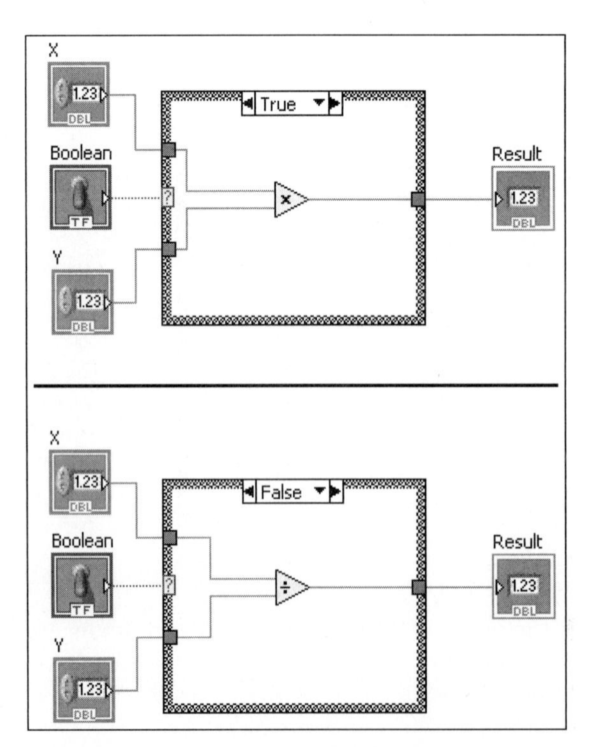

Figure 9.63
Block diagram showing the two cases.

EXAMPLE 9.4

Setting a Control Property Using an Enumerated Control

In this example the user is asked to choose the color of the "fluid" in the Tank control by selecting one of three options on an *Enum (enumerated) control*: Blue, Red, or Green. The front panel of the finished VI is illustrated in Figure 9.64.

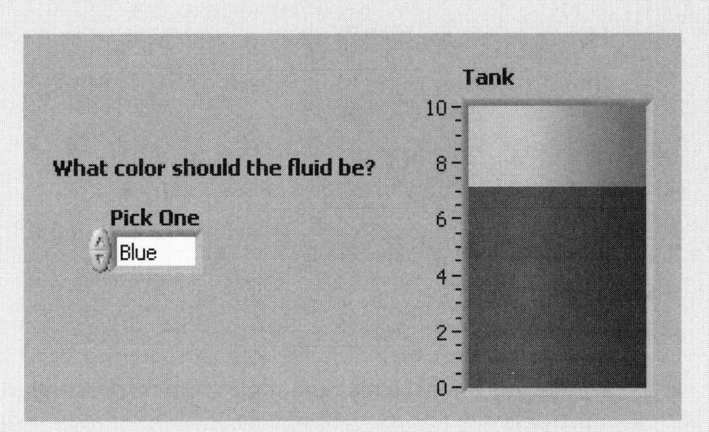

Figure 9.64
Using an Enum (enumerated) control to select a case.

To complete this VI, we will need to work through a few steps:

1. Create a property node for the Tank control's fill color.
2. Determine color codes for blue, red, and green.
3. Add an Enum Control to the front panel.
4. Create a Case Structure with three cases: "Blue", "Red" and "Green".
5. Wire the Controls.
6. Run the VI.

Step 1. Create a Property Node for the Tank control's fill color.
A *Property Node* is a node on the block diagram that can be used to determine or set an object's property value.

To create a property node for a Tank control, first place the Tank control on the front panel. The Tank control is located in either the Modern or Express Groups on the Controls Palette:

Controls Palette / Modern Group / Numeric Group / Tank

Controls Palette / Express Group / Numeric Indicators / Tank

On the block diagram, right-click on the Tank node and select *Create / Property Node/Fill Color* from the pop-up menu. A Property Node for the Tank fill color will be placed on the block diagram (see Figure 9.65). By default, the Node will be set to output the fill color; that's exactly what we want (for now.)

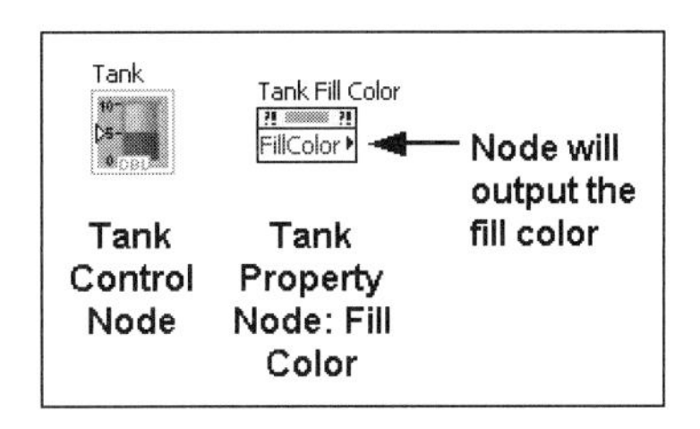

Figure 9.65
A Property Node has been created on the block diagram.

Step 2. Determine color codes for blue, red, and green.
Right-click the Tank Fill Color output terminal and select **Create / Indicator** from the pop-up menu. Now, when you run the VI, the code number for the current tank fill color will be displayed in an indicator on the front panel (see Figure 9.66).

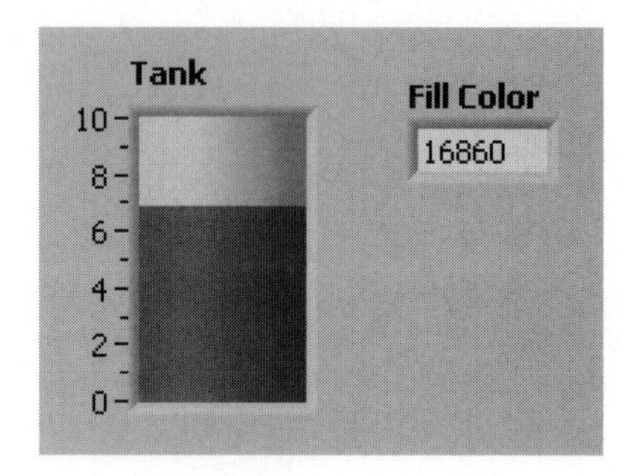

Figure 9.66
Run the VI to determine the color code for the current tank fill color (Blue = 16860).

We can use the Tank control's properties dialog to change the fill color to red and then run the VI to determine the color code (Red = 16711680, but it depends on the selected shade of red). Repeat the process to get the color code for green (Green = 6618880).

So far we have used the property node to output the current fill color; now we want to use it to set the color code to change the fill color. First, delete the Fill Color indicator, then right-click on the property node and select **Change to Write** from the pop-up menu. The property node now displays an input, as shown in Figure 9.67. If we send one of the color codes to the property node input, we can change the Tank control's fill color.

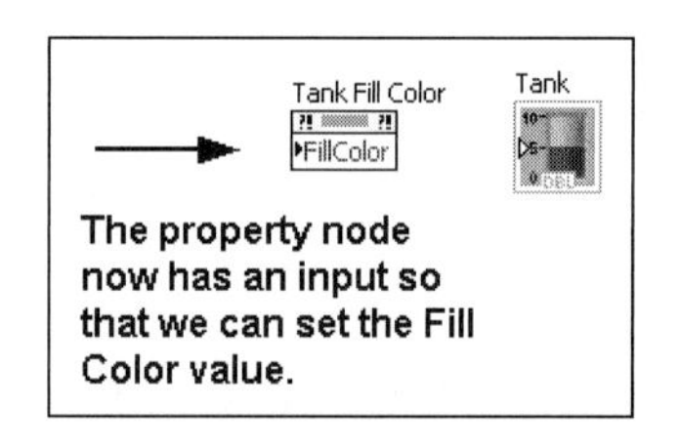

Figure 9.67
Property node changed from Read to Write.

Step 3. Add an Enum Control to the front panel
An Enum (short for enumerated) control allows the user to select between options. In this example, the options are blue, red, and green. Enum controls are found in the Modern Group:

Controls Palette / Modern Group / Ring and Enum Group / Enum

Place the Enum control on the front panel.

The next step is to create the options for the Enum control to display. Right-click on the Enum control and select **Add Item After** from the pop-up menu. A vertical cursor will appear in the Enum control display, indicating that you can type the option (or item) title into the field. Type "Blue" (without the quotes), and then click outside the control to terminate text entry. Repeat the process to add "Red" and "Green" options. The result is illustrated in Figure 9.68.

Note: Use the Increment/Decrement buttons to go through all of the options on the Enum control. If there is a blank item, it must be deleted. Select the blank item, then right-click and select **Remove Item** from the pop-up menu.

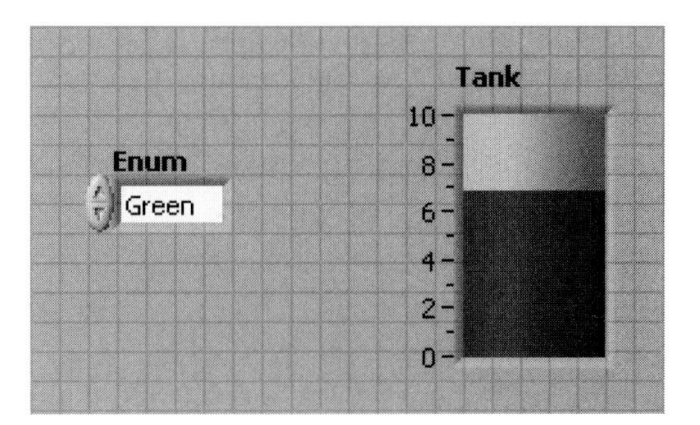

Figure 9.68 The completed Enum control.

Step 4. Create a Case Structure with three cases: "Blue", "Red" and "Green". To create the three-case structure, first place a generic Case Structure on the block diagram. The True and False cases will be included by default.

- Right-click on the Selector Label and select **Add Case After** from the pop-up menu.
- Type in the Case name (e.g., Blue)—you do not need to include quotes, LabVIEW will add them automatically.

The result is shown in Figure 9.69.

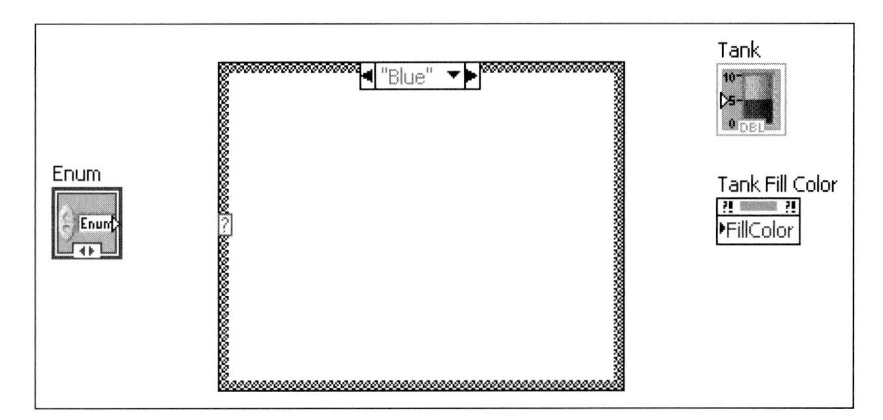

Figure 9.69 Creating the "Blue" case.

Repeat the steps to create the Red and Green cases.

After adding the Blue, Red, and Green cases, there are five cases in the structure; the True and False cases (the defaults) are still present. Click the down arrow on the right side of the Selector Label and choose the True case. Right-click on the Selector Label and select **Delete This Case** from the pop-up menu to delete the True case. Repeat to delete the False case.

Step 5. Wire the controls

All the major pieces are now in place, we just need to do the following:

1. Wire the Enum control output to the *Case Selector* on the left border of the Case structure.
2. Place a numeric constant inside the boundary and give the color code corresponding to the appropriate case:
   - Blue: 16860
   - Red: 16711680
   - Green: 6618880

3. Wire the output of the numeric constant to the right boundary of the Case structure. A tunnel will be created automatically.
4. Wire the tunnel output to the Tank Fill Color input.

Repeat steps 2 and 3 for the other two cases.

Step 6. Run the VI

When you run the VI, you can select a color on the Enum control, and the Tank fill color changes to that color (see Figure 9.70). It doesn't look like much in a black and white text, but you can download the Tank Fill Color VI from the text's website (www.chbe.montana.edu/LabVIEW) to try it on your computer.

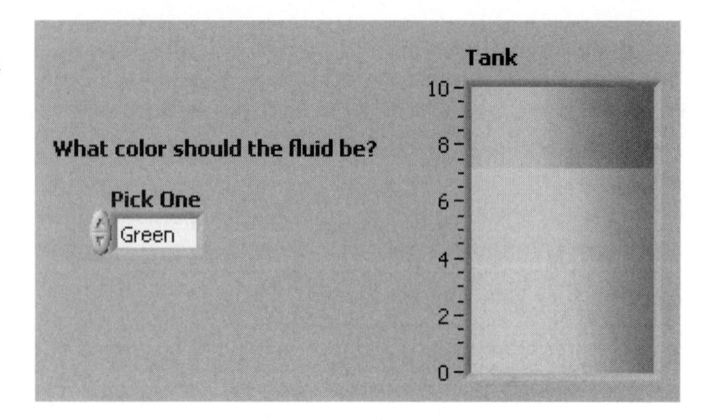

Figure 9.70 When the user picks a color, the Tank fluid color is changed.

9.3.5 Sequence Structures

A *Sequence Structure* is used when something must be done in sequential order. LabVIEW's dataflow programming means that you cannot always control the order of execution. As long as one calculation depends upon the result of a previous calculation (calculations in series), you can be sure that the calculations will occur in the correct order. But when calculations are in parallel (and unconnected), you cannot control the order in which the calculations take place. A sequence structure gives you the ability to force calculations to take place in a defined sequence. Sequence structures are available in the Programming Group:

Functions Palette / Programming Group / Structures Group / Flat Sequence Structure

Functions Palette / Programming Group / Structures Group / Stacked Sequence Structure

When you are tempted to use a sequence structure, first consider whether or not the order of calculation matters. If it does, then see if there is a way to control the order using data flow. In many situations sequence structures are not necessary, but they are available for those instances when things have to happen in the right order.

EXAMPLE 9.5

Checking VI Timing

Programmers often want to know how long it takes to perform a certain set of calculations. A sequence structure can be used for this task as shown in Figure 9.71. The computational task is placed in the center of a three-frame sequence structure. In this example, the task is to read a .txt file, generate two 1D arrays, and create an XY Graph. A Wait function has been included to generate a measurable time lapse.

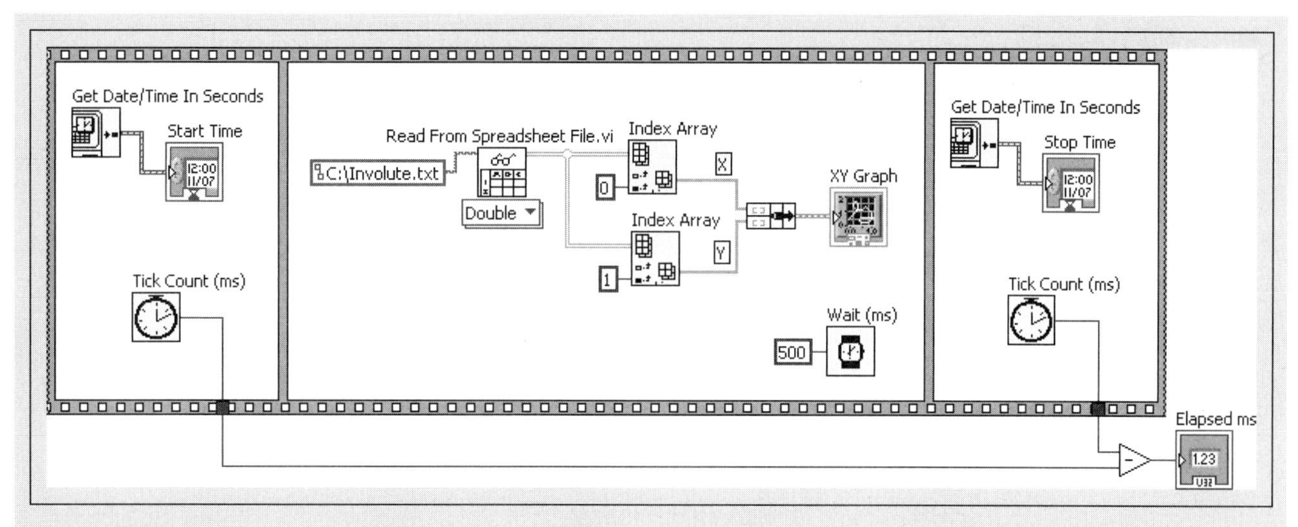

Figure 9.71
Using a Sequence Structure to check program timing.

The frames before and after the computational frame are used to get before and after times in two formats:

- Date/Time—showing the actual date and time
- Tick Count—showing the millisecond count on the clock

When the VI runs, the front panel (Figure 9.72) shows the start time, stop time, and time difference in milliseconds. Even with the file access, the elapsed time for the run was exactly equal to the Wait time: 500 ms.

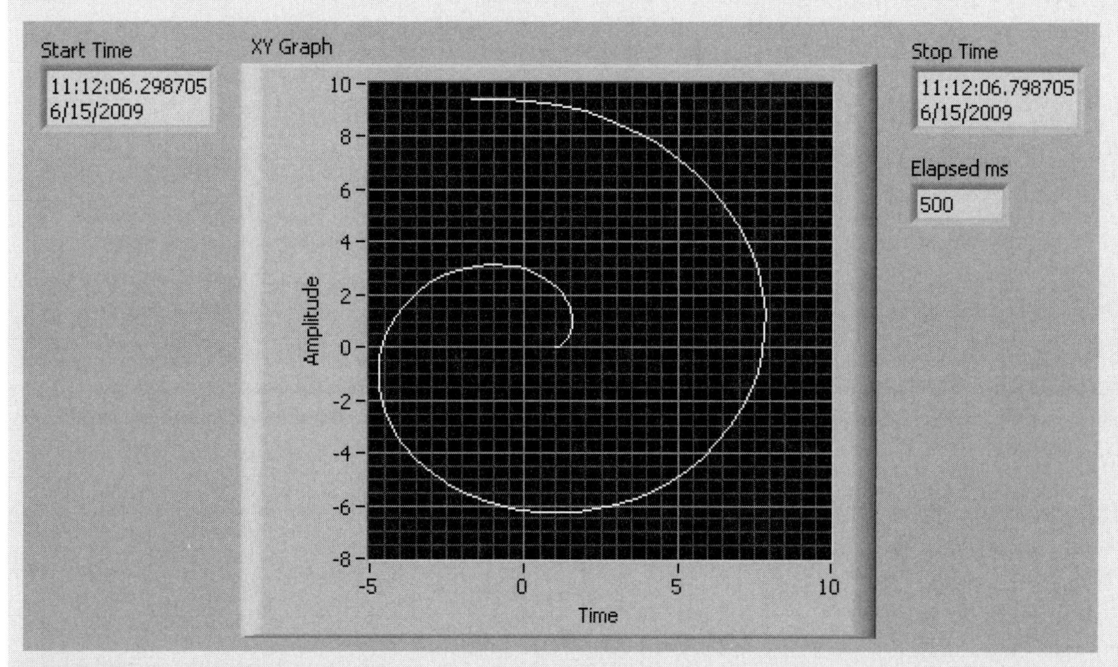

Figure 9.72
Front panel of timing check VI.

APPLICATION

Arithmetic Flash Cards

We want to write a LabVIEW VI that presents two random integers to be added, waits 2 seconds, shows the answer for 1 second, and then repeats the whole thing until the student pushes the STOP button.

Showing random numbers and answers is easy; keeping the answer hidden for 2 seconds is a little trickier. The solution shown here uses a sequence structure, but some enterprising individual may find a way to accomplish the same task without the structure.

Two random integers between 1 and 10 are generated. The integers are displayed in indicators A and B.

- In the first frame of the sequence structure, the Visible property node for the answer indicator (A + B) is set to FALSE so that the answer is not visible. A Wait of 2000 ms is in the first frame as well.
- When the first frame has completed execution (including the 2000 ms wait), execution passes to the second frame of the sequence structure. The Visible property is set to TRUE to display the answer. The answer is displayed for 1000 ms.

When the second frame has completed execution, the While Loop cycles and the whole process repeats.

The block diagram for the first solution is shown in Figure 9.73.

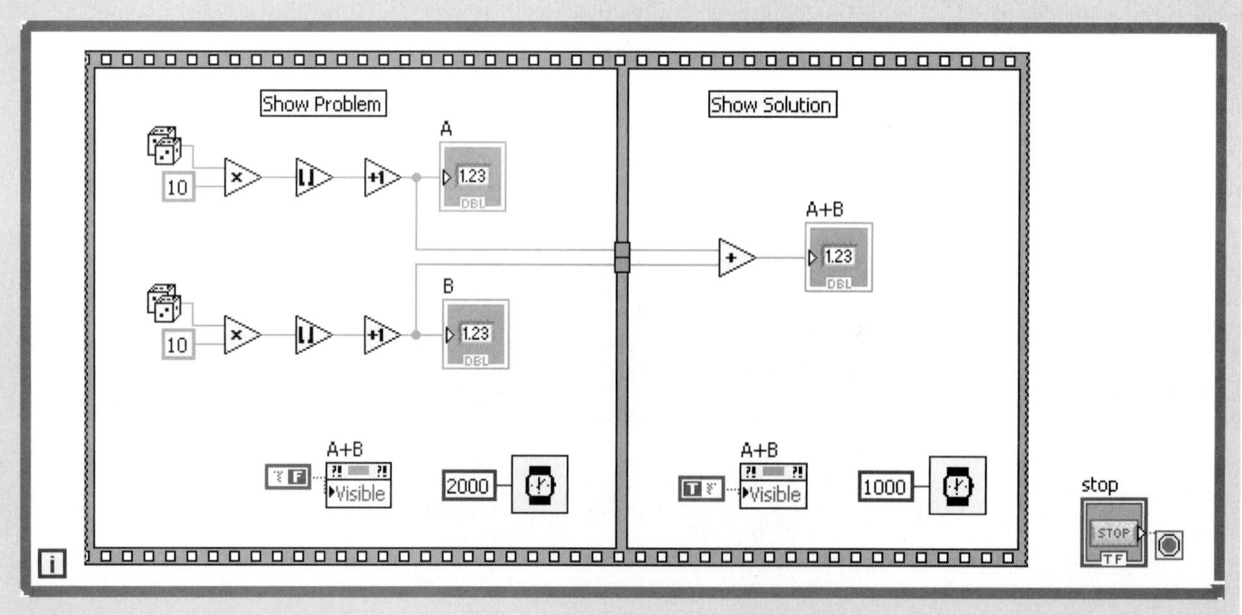

Figure 9.73
First solution to the Flash Cards problem.

Figure 9.74
The problem is displayed for 2 seconds.

When this VI is run, the problem is shown for 2 seconds (see Figure 9.74). Then the answer (see Figure 9.75) is displayed for 1 second.

The first solution works, but there is a slightly simpler solution. The block diagram of the second solution is shown in Figure 9.76.

In the second solution, the calculation and display of the numbers is outside of the sequence structure. All the sequence structure does is control timing and visibility of the answer (A + B) indicator.

In this example the Flat Sequence structure was used. The Flat Sequence structure shows all of the frames (steps) in the block diagram, but can take up a lot of space if there are several frames. The alternative is the Stacked Sequence structure. With the Stacked Sequence structure, the frames are still sequential, but they are stacked one on top of the other (like the Case structure). You use the Selector Label to choose the displayed frame.

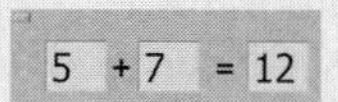

Figure 9.75
The answer is displayed for 1 second.

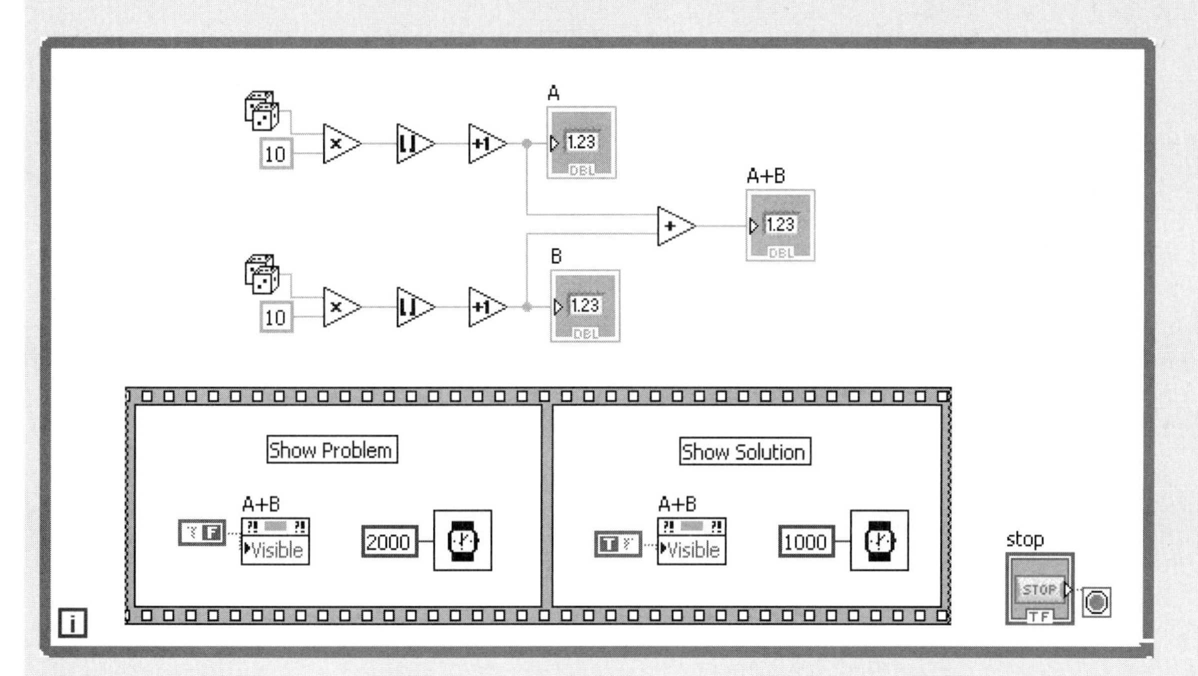

Figure 9.76
Second solution to the Flash Cards problem.

9.3.6 Formula Node

The *Formula Node* is a structure that allows the programmer to perform a series of calculations using sequential statements that are similar to the C programming language. The Formula Node is available in the Programming Group:

> **Functions Palette / Programming Group / Structures Group / Formula Node**

In Figure 9.77, a For Loop has been used to send array elements into the Formula Node one at a time. Inside the Formula Node, the *X* input is used as the input to a polynomial and the result (scalar) is output from the Formula Node to the For Loop. The For Loop reassembles the final *Y* array. The calculated results are shown in the front panel in Figure 9.78.

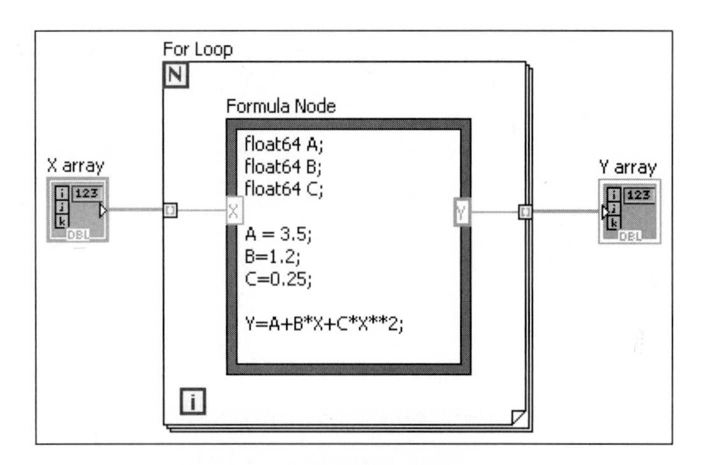

Figure 9.77
Formula Node used to evaluate a polynomial.

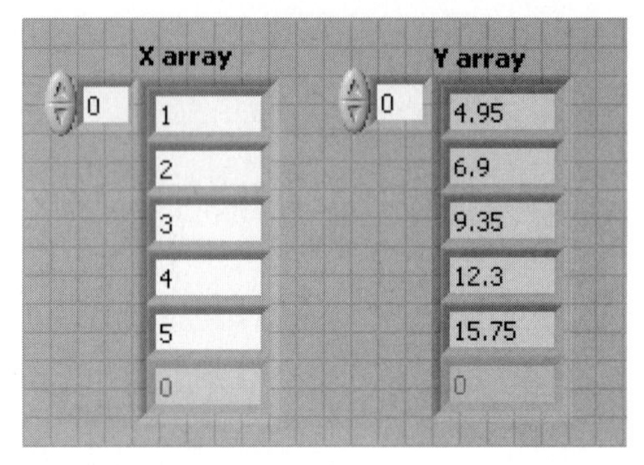

Figure 9.78
Results from the Formula Node inside the For Loop.

9.3.7 MathScript

A *MathScript Node* is a structure that allows the programmer to perform a series of calculations using sequential Matlab-style statements. The MathScript Node is available in the Programming Group:

> **Functions Palette / Programming Group / Structures Group / MathScript Node**

Note: The MathScript Node is not available in 64-bit LabVIEW 2009.

LabVIEW MathScript also includes the *MathScript Window*, which can be opened from LabVIEW using menu options: **Tools / MathScript Window . . .** The MathScript Window is shown in Figure 9.79.

The MathScript Window is designed to operate very much like Matlab®. You can define variables and enter commands in the Command Window, and see the results in the Output Window. The right side of the MathScript Window includes a script editor. In Figure 9.79 a function named polyScript has been created and saved as polyScript.m. Like Matlab, user-written functions must be saved with file names that match the function name, plus the .m extension, called an m-file.

The polySolve function:

function y = polySolve(x)
$A = 3.5;$
$B = 1.2;$
$C = 0.25;$
$y = A + B * X + C * X.\wedge 2;$

Once you save a user-written function, it can be used in a MathScript node inside a LabVIEW VI—as long as LabVIEW knows where to find your m-files. To tell LabVIEW where you are storing your m-files, use the LabVIEW menu options

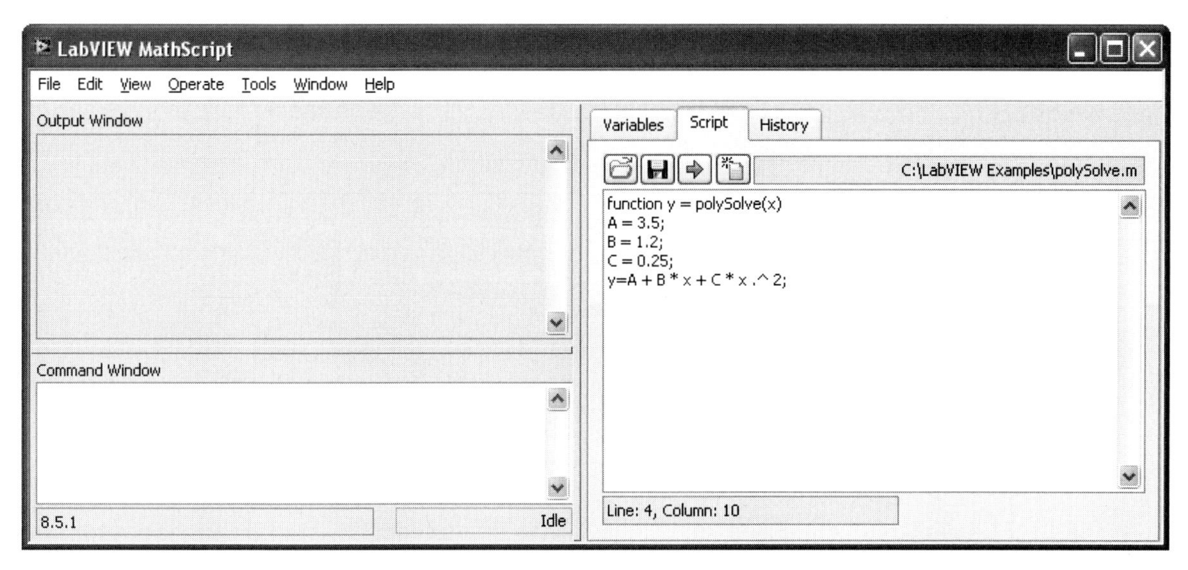

Figure 9.79
The MathScript Window with function polyScript displayed.

Tools / Options . . . Then select the category **MathScript: Search Paths** as shown in Figure 9.80. If the folder in which you are storing your m-files is not included in the Search path list, add it.

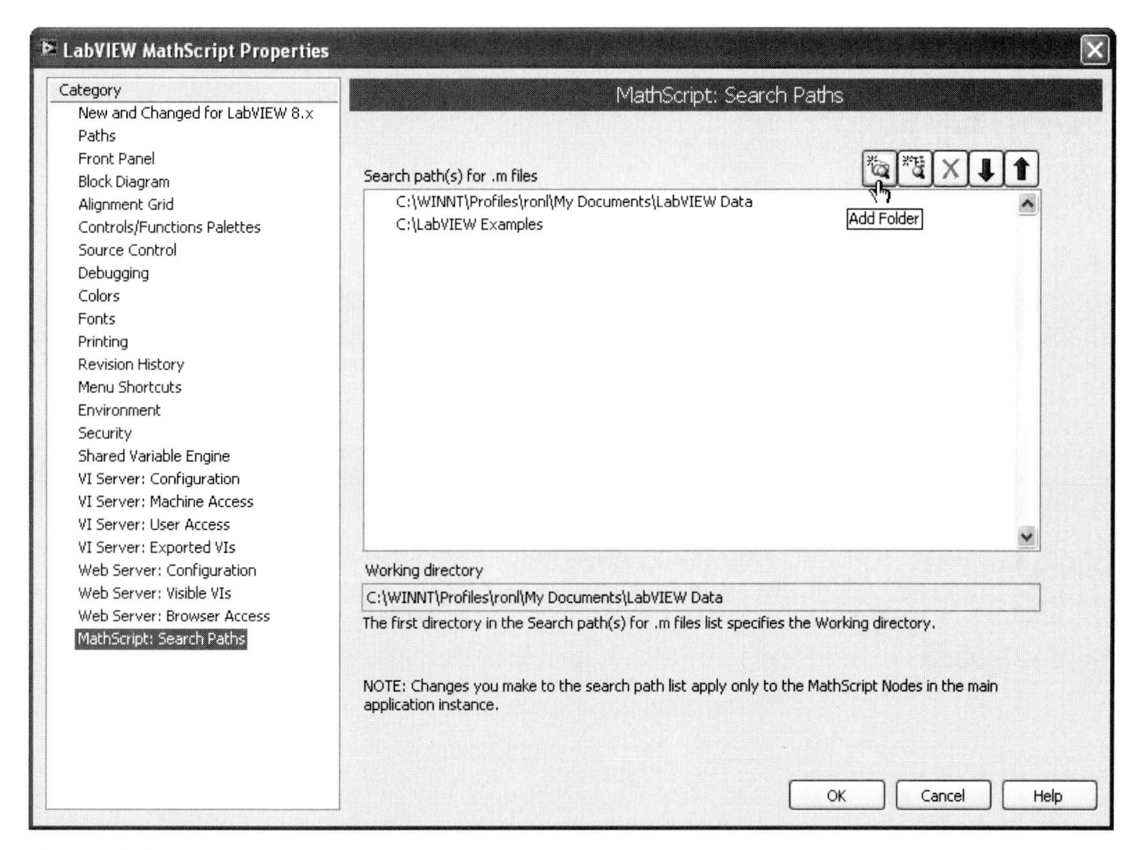

Figure 9.80
Telling LabVIEW where your m-files are stored.

Once LabVIEW can find your user-written m-files, any of your m-files can be used inside of a MathScript node inside a LabVIEW VI. One example is shown in Figure 9.81. The front panel after the VI is run is shown in Figure 9.82.

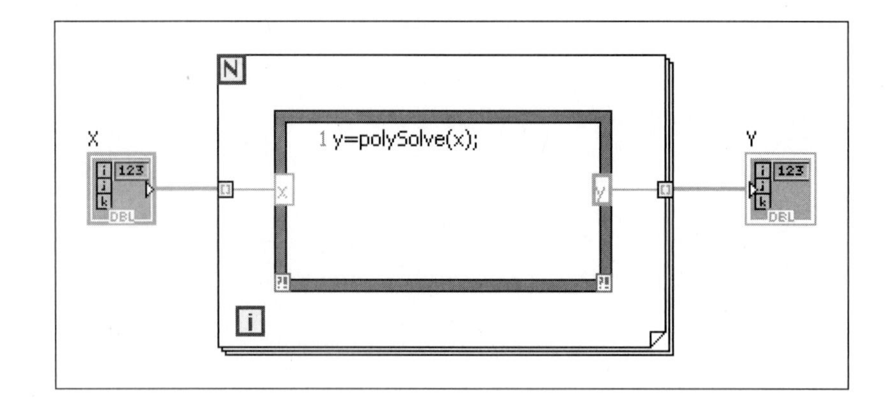

Figure 9.81 Using a user-written MathScript function inside a MathScript node.

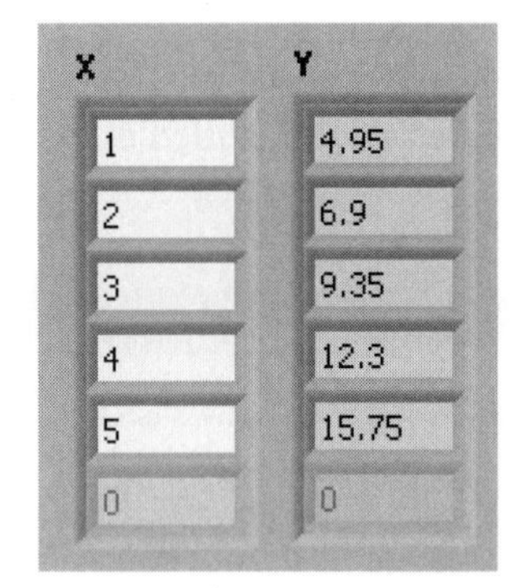

Figure 9.82 The results from the MathScript node (array *Y*).

In the VI shown in Figure 9.81, LabVIEW's For Loop feeds in one *x* value each time the loop cycles, and the polySolve function calculates one *y* value. Alternatively, we can eliminate the LabVIEW For Loop and use MathScript programming statements to loop through the arrays, as shown in Figure 9.83.

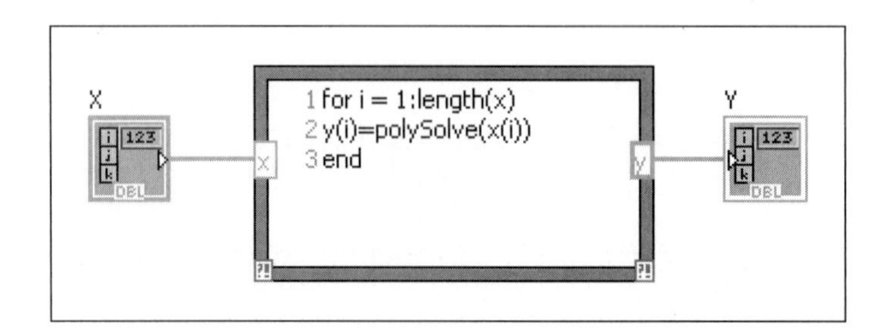

Figure 9.83 Using MathScript commands to step through the *X* array and create the *Y* array.

Using MathScript within LabVIEW:

MathScript adds another approach to solving problems using LabVIEW, but there are a couple of things to watch out for:

- Notice that MathScript array indexing starts at 1, while LabVIEW array indexing starts at 0.
- The y output on the MathScript node (in the right border) had to be explicitly declared to be a 1D array of doubles. To do this, right-click on the variable in the border and select **Choose Data Type** from the pop-up menu.

APPLICATION

Using Loops and Case Structure to Create a Fractal

A fractal is an image created mathematically that has a high degree of repetition, or self-similarity. The fractal generated in this Application is called the fern fractal, and was initially created by M. Barnsley.[1] The fractal is created using a lot of iteration and a series of four attractors, called *fern functions*:

Case 1 ($r \geq 0.93$)

$$x = -0.15x + 0.28y$$
$$y = 0.26x + 0.24y + 0.44$$

Case 2 ($r < 0.93$)

$$x = 0.20x - 0.26y$$
$$y = 0.23x + 0.22y + 1.60$$

Case 3 ($r < 0.86$)

$$x = 0.85x + 0.04y$$
$$y = -0.04x + 0.85y + 1.60$$

Case 4 ($r < 0.05$)

$$x = 0$$
$$y = 0.16y$$

The output of the Fern Fractal VI is shown in Figure 9.84. The block diagram is large and shown in two images in Figures 9.85 and 9.86. The portion in Figure 9.85 generates the fractal, and the rest (Figure 9.86) just prepares the fractal information for graphing.

[1]Barnsley, M. *Fractals Everywhere*, 2nd ed., Boston, MA, Academic Press, 1993, pp. 86, 90, 102 and Plate 2, 193. As reported by Weisstein, Eric W., "Barnsley's Fern." From *MathWorld*—A Wolfram Web Resource, http://mathworld.wolfram.com/BarnsleysFern.html, accessed October 15, 2009.

This fractal VI incorporates the following programming features:

- Two nested For Loops
- Formula Node
- Case Structure
- Shift registers
- Auto-indexed tunnels
- Loop tunnels (auto-indexing disabled)
- Intensity Graph Control

Figure 9.84
The Fern Fractal.

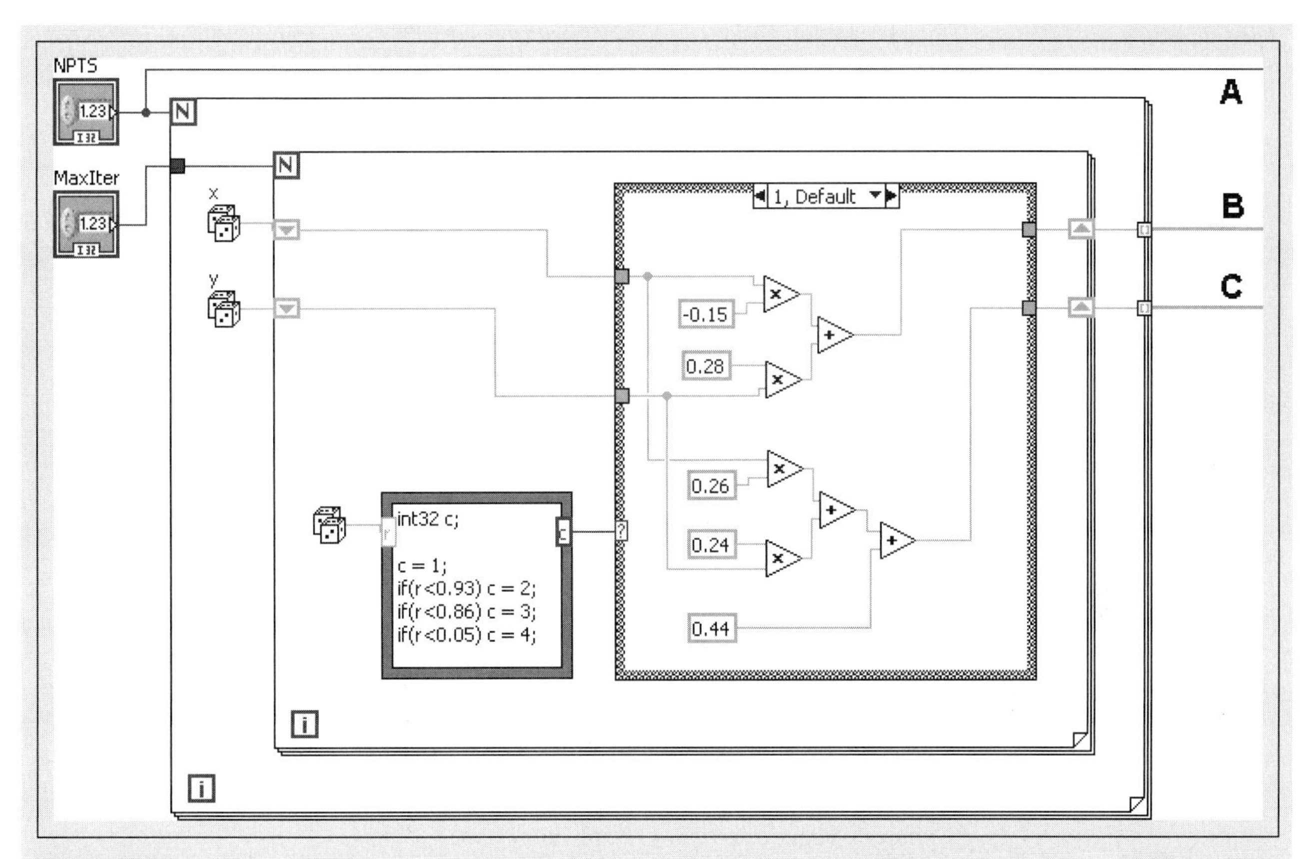

Figure 9.85
Fern Fractal block diagram, part 1.

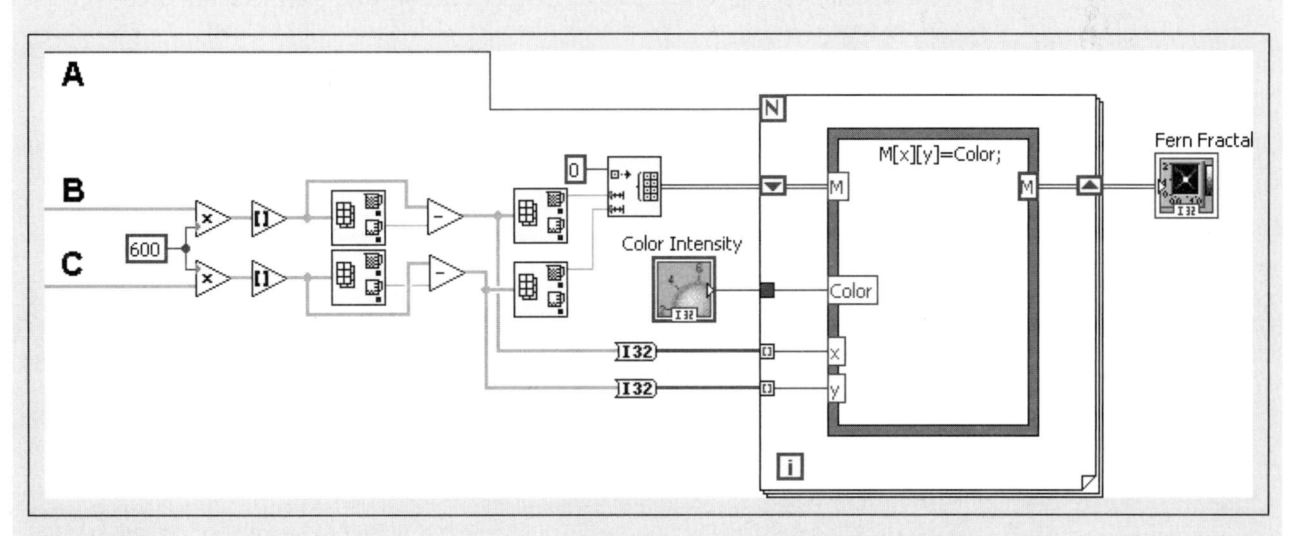

Figure 9.86
Fern Fractal block diagram, part 2.

KEY TERMS

(loop) count terminal
(loop) iteration terminal
auto-indexing (For Loop)
automatic wiring
Case Selector
Case Structure
enumerated control
For Loop
Formula Node
indexing (tunnel, enabled or disabled)
iteration counter
logarithm scale
MathScript Node
MathScript Window
project
property node
Sequence Structure
shift register
Structure
SubVI
tunnel
Wait function
Wait Until Next ms
Multiple function
While Loop

SUMMARY

More LabVIEW Basics

Adjust the display format of displayed values

- Right-click on the control and select **Display Format . . .** from the pop-up menu.
- Numeric Properties dialog, Display Format panel.
  - Select floating point or scientific notation.
  - Number of displayed Digits (six is the default).
  - Change the **Precision Type** between "Significant digits" and "Digits of precision".
  - Show or hide trailing zeroes.

Restrict allowed data entry values

- Right-click on the control and select **Data Entry . . .** from the pop-up menu.
- Numeric Properties dialog, Data Entry panel.

Change the data type associated with a control

- Right-click on the control, select **Representation**, and then select desired data type.

Use a logarithm scale

1. Place the control on the front panel.
2. Change the minimum and maximum scale values (if desired).
3. Right-click on the control and select **Scale / Mapping / Logarithmic**.

Set default initial values

1. Place the control on the front panel.
2. Enter desire values.
3. Right-click on the control and select **Data Operations / Make Current Value Default**.

Minimize Wiring

Place the function on the block diagram first, and then

- Right-click on the function inputs and select **Create / Control**.
- Right-click on the function outputs and select **Create / Indicator**.

Benefits

- Data type set correctly
- The wiring is automatic
- Automatic labeling

Remove broken wires

- **Edit / Remove Broken Wires**, or [Ctrl-B]

Creating SubVIs

1. Get a VI working without using a SubVI.
2. Select the portion of the VI that will become the SubVI.
3. Use menu options **Edit / Create SubVI**.
4. Double-click default SubVI icon to open SubVI for editing.
5. Edit SubVI icon.
   (a) Right-click on the icon in the top-right corner of the SubVI and select **Edit Icon . . .** from the pop-up menu.
   (b) Change the icon as desired.
   (c) Click OK when finished to close the Icon Editor.
6. Save the SubVI—the file name will be used to label the SubVI on block diagrams.

LabVIEW Projects
A project is a collection of all of the VIs and SubVIs needed to make an application work.

- To create a project, use menu options **Project / New Project . . .**

Benefits

- Organization—collect all the required VIs and SubVIs
- Access control—projects help you keep track of who is editing what
- Revision control—projects help you keep track of changes to VIs
- Compilation—projects are used to compile stand-alone program

Programming Structures

While Loop

- Loops until a condition is met.
- **Functions Palette / Programming Group / Structures Group / While Loop**
- **Functions Palette / Express Group / Execution Control Group / While Loop**
- Loop condition can be set. Right-click on Loop Condition indicator to change.
  - Stop if True (default).
  - Continue if True.

For Loop

- Loops a specified number of times.
- **Functions Palette / Programming Group / Structures Group / For Loop**
- Iteration terminal "*i*" provides the iteration number (starts at 0).
- Count terminal "*N*" receives the number of desired iterations.
- Auto-Indexing—when an array output is wired to For Loop input tunnel with indexing enabled, the For Loop will cycle once for each value in array.

Controlling the Timing of Loops

- **Wait** function—adds a defined wait period
- **Wait Until Next ms Multiple** function—causes VI execution to wait for the remainder of a specified period

Tunnels

- Wires crossing loop boundaries create tunnels.
- Tunnels can have *indexing* enabled or disabled:
  - Input tunnel, indexing disabled—value (or all values, if array) passes into loop when loop starts
  - Input tunnel, indexing enabled—one value of array enters loop each cycle—For Loop is *autoindexed*; loop cycles once for each element in array
  - Output tunnel, indexing disabled—value (or all values, if array) passes out of loop when loop terminates
  - Output tunnel, indexing enabled—values built into array at tunnel; array gets released when loop terminates

Shift Registers

- Sends the calculated value at the end of one iteration into the beginning of the next iteration.
- Right-click on the loop boundary and select **Add Shift Register** from the pop-up menu.
- An **uninitialized shift register** gets its starting value from the last output value at the end of the previous execution (the last time the same VI was run). You can use an uninitialized shift register to start the current loop where it left off last time.
- An initialized shift register is assigned a value before the loop begins to iterate.
- Use stacked shift registers to store results from earlier iterations:
  - One shift register—access to i - 1 iteration values, where i is current iteration number
  - Two shift registers—access to i - 1 and i - 2 iteration values.
  - Three shift registers—access to i - 1, i - 2, and i - 3 iteration values

Case Structures

- Allow program actions to vary depending on a variable's value.
- **Functions Palette / Programming Group / Structures Group / Case Structure**
- Default: Two cases; True and False—right-click on case selector to change

Sequence Structures

- Used to force operations to be done in sequential order.
- **Functions Palette / Programming Group / Structures Group / Flat Sequence Structure**

  Functions Palette / Programming Group / Structures Group / Stacked Sequence Structure

Formula Node

- Allows the programmer to perform a series of calculations using sequential statements that are similar to the C programming language.
- **Functions Palette / Programming Group / Structures Group / Formula Node**
- Right-click on left boundary to add an input; type variable name in input field.
- Right-click on right boundary to add an output; type variable name in output field.

MathScript Node

- Allows the programmer to perform a series of calculations using sequential Matlab-style statements. [not available in 64-bit LabVIEW]
- **Functions Palette / Programming Group / Structures Group / MathScript Node**
- Right-click on left boundary to add an input; type variable name in input field.
- Right-click on right boundary to add an output; type variable name in output field.

MathScript Window

- Open from LabVIEW using the menu options **Tools / MathScript Window . . .**
- Define variables and enter commands in the Command Window.
- View results in the Output Window.
- Create scripts and functions using the script editor.
- Functions must be saved with file names that match the function name, plus the .m extension.

SELF-ASSESSMENT

1. How do you change the display format used to display numbers on the front panel?
   ANS: Right-click on the control and select **Display Format . . .** from the pop-up menu. This opens a properties dialog to the Display Format tab.
2. How do you restrict the values that a user can enter into a control?
   ANS: Right-click on the control and select **Data Entry . . .** from the pop-up menu. This opens a properties dialog to the Display Format tab. Set the minimum and/or maximum allowed values.
3. On controls that show a scale, how do you specify that the scale should be linear or logarithmic?
   ANS: Right-click on the control and select **Scale / Mapping / Linear** (or **Logarithmic**).
4. Most people first place controls and indicators on the front panel, and then go to the block diagram to wire them. What are the benefits of right-clicking on terminals and creating controls and indicators on the block diagram instead?
   ANS:
   - The correct data type for the terminal is automatically set.
   - The wiring is automatic.
   - The control or indicator is automatically labeled.
5. How can you remove all broken wires from a block diagram?
   ANS: Press [Ctrl B]
6. What is a SubVI?
   A SubVI is a VI that has inputs and or outputs, and has been saved separately so that it can be used inside another VI. SubVIs act like the functions used in other programming languages.
7. What are the benefits of using LabVIEW projects?
   ANS: They keep all files needed for a particular application together so that they are easy to find and work with.
8. What type of loop structure is used to keep a VI running until a **STOP** button is pressed?
   ANS: A While Loop
9. How can you slow down a While Loop?
   ANS: Use a Wait function or a Wait Until Next ms Multiple function.
10. What kind of loop structure is used to populate arrays with values?
    ANS: For Loop
11. What is a "tunnel" in LabVIEW?
    A tunnel allows information to flow into or out of a loop structure.

12. Tunnels can have auto-indexing enabled or disabled. What's the difference?
ANS: Auto-indexing (when enabled) causes array values to be passed through the tunnel one at a time, one for each cycle through the loop structure. At an input, this has the effect of passing array elements into the loop one at a time for processing. At an output, this has the effect of building up an array, element by element, each time the loop cycles.

    When indexing is disabled on an input tunnel wired to an array, the entire array is passed through the loop boundary (through the tunnel) at one time. When indexing is disabled on an output tunnel, only the final value of a calculation is passed through the tunnel.
13. What does a shift register do?
ANS: A shift register sends a value calculated within a loop back to the beginning of the next loop. Shift registers make the results of previous calculations available for the next iteration.
14. Why are shift registers sometimes stacked?
ANS: Stacked shift registers provide access to results calculated more than one loop cycle earlier.
15. How does a Case Structure work?
ANS: A Case Structure causes the program to respond differently depending on the value of the Case selector.
16. What are Sequence Structures used for?
ANS: Sequence structures are used to force LabVIEW to carry out calculations in sequence.
17. What is a Formula Node?
ANS: A Formula Node is a programming structure that allows you to build some sequential programming statements into a LabVIEW program. The required syntax is somewhat like the C programming language.
18. What is a MathScript Node?
ANS: A MathScript Node is a programming structure that allows you to use Matlab-style statements in a LabVIEW program.

PROBLEMS

1. Some programs have a feature that allows you to quickly prepare a plot of a function over a standard range, such as -10 to $+10$. This can be handy when you need to select a function for some purpose such as modeling or curve fitting. We can create a VI in LabVIEW that does the same thing because LabVIEW allows you to replace one function with another without breaking all the wires.

    Create a Function Plot VI for the Sine function that uses a For Loop to evaluate the function over the range -10 to $+10$, with at least 200 points (for creating a smooth plot). Send the X and $\sin(X)$ values to an XY Graph control. Your VI should look something like Figure 9.87 (front panel) and Figure 9.88 (block diagram). Once your VI is working with the Sine function, replace the Sine function with the following:
    a. Sin(X)/X function—what is the minimum value in the range -10 to $+10$?
    b. Asec(X) function—how does the VI respond to undefined values, such as Asec(0)?
    c. Bessel Function jn—why is half the plot missing?
2. Use a For Loop to create an array containing 100 elements between 0 and 500.

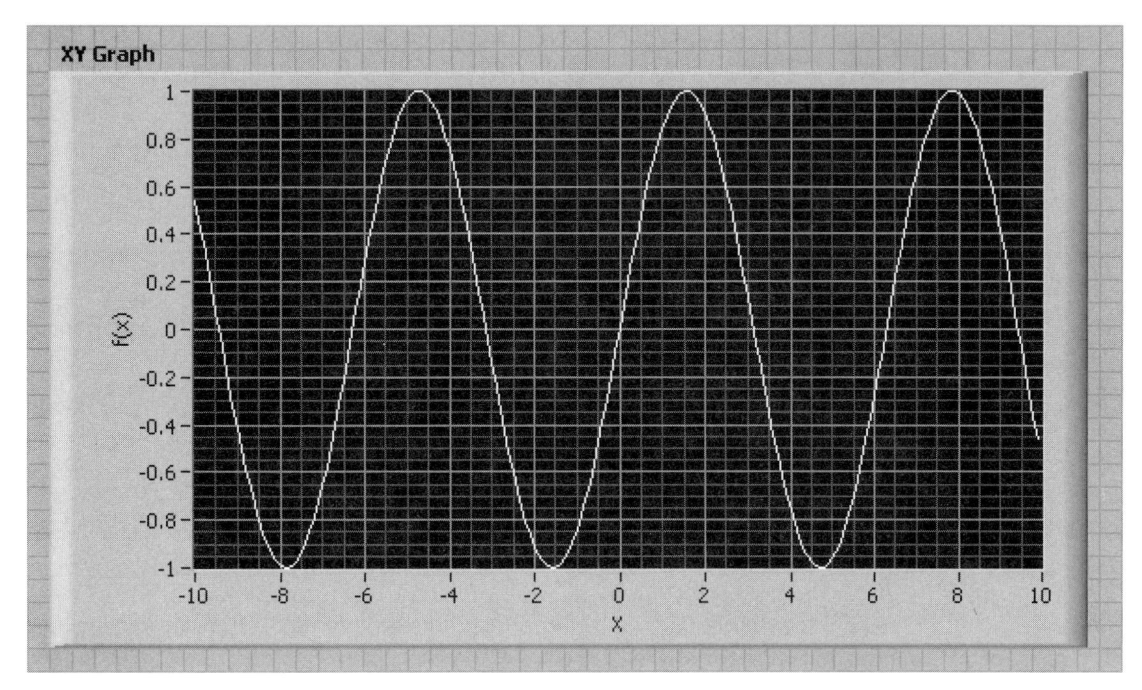

Figure 9.87
Function Plot VI, front panel.

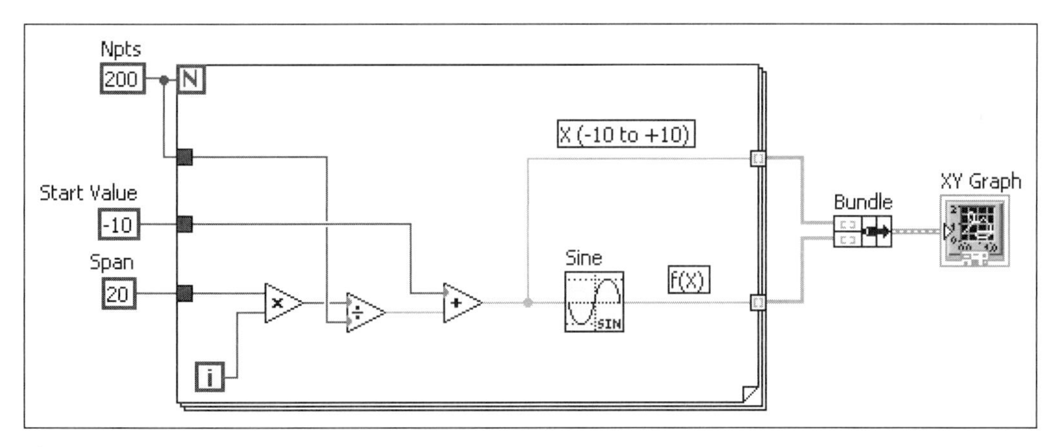

Figure 9.88
Function Plot VI, block diagram.

3. Use a For Loop to create an array of 50 elements ranging between -20 and 80.
4. Use a For Loop to create arrays for plotting a cosine wave between $-\pi$ and $+\pi$ using 300 points. Send the X and $\cos(X)$ arrays to an XY Graph control for plotting.
5. Extend the previous problem by using a Case Structure and Enum control to select between sine, cosine, and tangent plots. Add your own case that creates a different function plot.
6. In solving problems in fluid mechanics, the friction factor is a necessary value for determining pressure losses in piping systems due to fluid friction. The equation used to calculate friction factor depends on the type of fluid flow.

(The two types of flow are called "laminar" and "turbulent" flow, but you don't need to know that to solve this problem.) The Reynolds number is used to determine if the flow is laminar or not; if Reynolds number is less than 2100, the flow must be laminar. If not, we will assume the flow is turbulent.
The Reynolds number for a pipe flow is defined as

$$Re = \frac{DV_{avg}\rho}{\mu}$$

Where

D is the pipe diameter (m)
V_{avg} is the flow velocity (m/s)
ρ is the fluid density (kg/m^3)
μ is the fluid viscosity (kg/m s)

If the flow is laminar ($Re < 2100$), the friction factor is calculated as

$$f = \frac{64}{Re}$$

But if the flow is turbulent, the friction factor is calculated in a variety of ways depending on the Reynolds number and the type of pipe. One way to calculate friction factor for low-Re turbulent flow in smooth pipes is

$$f = 0.184\ Re^{-0.2}$$

Create a VI that accepts D, V_{avg}, ρ, and μ as inputs, calculates Reynolds number, and then uses a Case Structure to solve for friction factor using the appropriate equation.
Test your VI with these values (slow, room temperature water flow):

$D = 0.02$ m
$V_{avg} = 0.1$ m/s
$\rho = 1000$ kg/m^3
$\mu = 0.001$ kg/m s
$Re = 2000$
$f = 0.032$

Then, use your VI to find the friction factor for the following flows:

a. Fast room temperature water flow

$D = 0.02$ m
$V_{avg} = 2$ m/s
$\rho = 1000$ kg/m^3
$\mu = 0.001$ kg/ms

b. Flowing honey

$D = 0.02$ m
$V_{avg} = -0.01$ m/s
$\rho = 1400$ kg/m^3
$\mu = 7$ kg/m s

7. A classic example of the use of a While Loop is calculating factorials. The factorial of X (an integer) is the product of X and all smaller integers, down to 1. The factorial of 4 is

$$4 \times 3 \times 2 \times 1 = 24$$

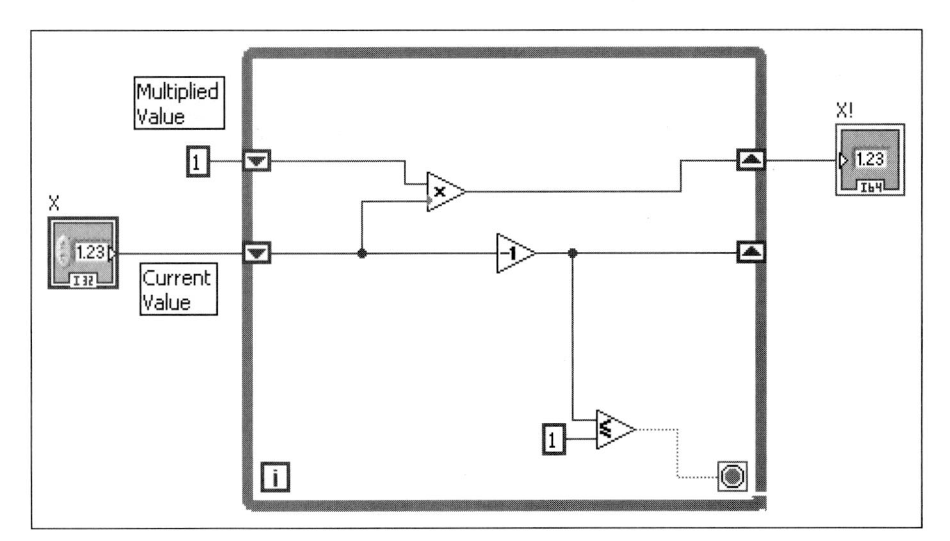

Figure 9.89
A block diagram for calculating a factorial.

To calculate a factorial using a While Loop (see Figure 9.89), you

- Send the initial value of X into a shift register that keeps track of the Current Value.
- Use a second shift register to keep track of the increasing product value (initialize with a value of 1).
- Multiply by the current X value and the product from the previous cycle
- Decrement the current X value.
- Loop until the current X value is less than or equal to 1.

Write a VI that calculates factorials using a While Loop. Set the Representation of the controls, indicators, and constants to integer (I32, for example).

a. What is the maximum initial value of X that will work with I32 controls and indicators?

b. What is the maximum initial value of X that will work with I64 controls and indicators?

c. What happens if someone tries to use an initial value of X that is too large?

d. Modify the control for the initial value of X to restrict the allowable values that the user can enter to integers between 1 and the maximum possible value from part b.

8. The PRACTICE problem in Section 8.3.3 used a For Loop to demonstrate exponential growth. Modify the VI from that PRACTICE problem to use a While Loop to determine how long it takes for the number of cells to exceed 1×10^5 if single cell divides once each hour.

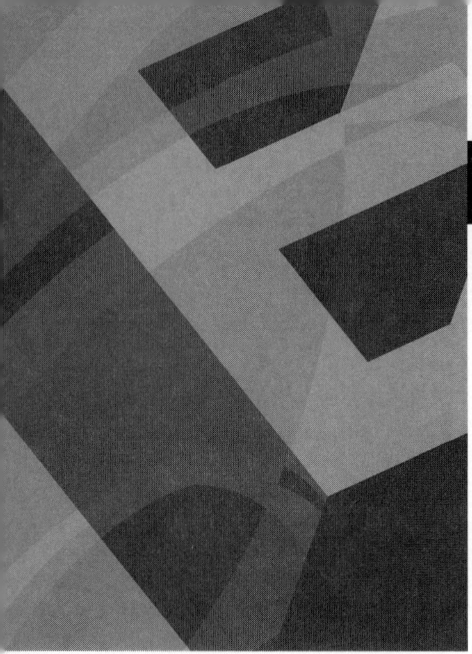

CHAPTER 10

Looking Forward: Advanced Math Using LabVIEW VIs

Objectives

This chapter is intended to give you a preview of more advanced LabVIEW applications, including:

- how to use LabVIEW functions to work with polynomials (to find roots, for example)
- how to use LabVIEW's T-Test VI to perform hypothesis testing on a data set
- how to perform numerical differentiation on data set values
- how to integrate data set values
- how to use LabVIEW's ODE Runge-Kutta 4th Order VI to integrate sets of differential equations
- how to create an exponential filter VI using LabVIEW
- how to perform a spectral analysis using a LabVIEW VI
- how to create a Monte Carlo simulation using LabVIEW
- how to use a PID controller with a LabVIEW data acquisition system

10.1 INTRODUCTION

The target audience for this text is freshmen and sophomore engineering students, and the topics in this chapter are well beyond the expected skill level of that audience. But freshmen engineering students turn into juniors and seniors, and LabVIEW has capabilities that may be useful for upper division engineering students too.

The goal of this chapter is to let you know about some of the more advanced math capabilities that are available in LabVIEW, and hopefully pique your interest in what lies ahead.

Note: Most of the features presented in this chapter are not available in LabVIEW's base package; the student, full, or professional packages are required.

In this chapter we present some VIs built to demonstrate LabVIEW's abilities in the following areas:

- Working with Polynomials
- Statistics: Hypothesis Testing
- Integration
- Differentiation
- Runge–Kutta Integration of Differential Equations
- Exponential Filter
- Spectral Analysis
- Monte Carlo Simulation
- PID Controller

10.2 WORKING WITH POLYNOMIALS

LabVIEW provides a wide range of polynomial functions in the Mathematics Group:

Functions Palette / Mathematics Group / Polynomial Group

The polynomial functions include the following:

- Order of Polynomial
- Polynomial Evaluation
- Polynomial Plot
- Polynomial Roots
- Roots Classification
- Remove Zero Coefficients
- Polynomial Real Zeros Counter
- Polynomial Eigenvalues and Vectors
- Add Polynomials
- Multiply Polynomials
- Divide Polynomials
- Partial Fraction Expansion (PFE)
- Create Polynomial From PFE
- Create Polynomial From Roots
- GCD of P(x) and Q(x)
- LCM of P(x) and Q(x)
- nth Derivative of Polynomial
- Indefinite Integral of Polynomial
- Integral of Polynomial over [a,b]

The VI shown in Figure 10.1 uses several of these functions.

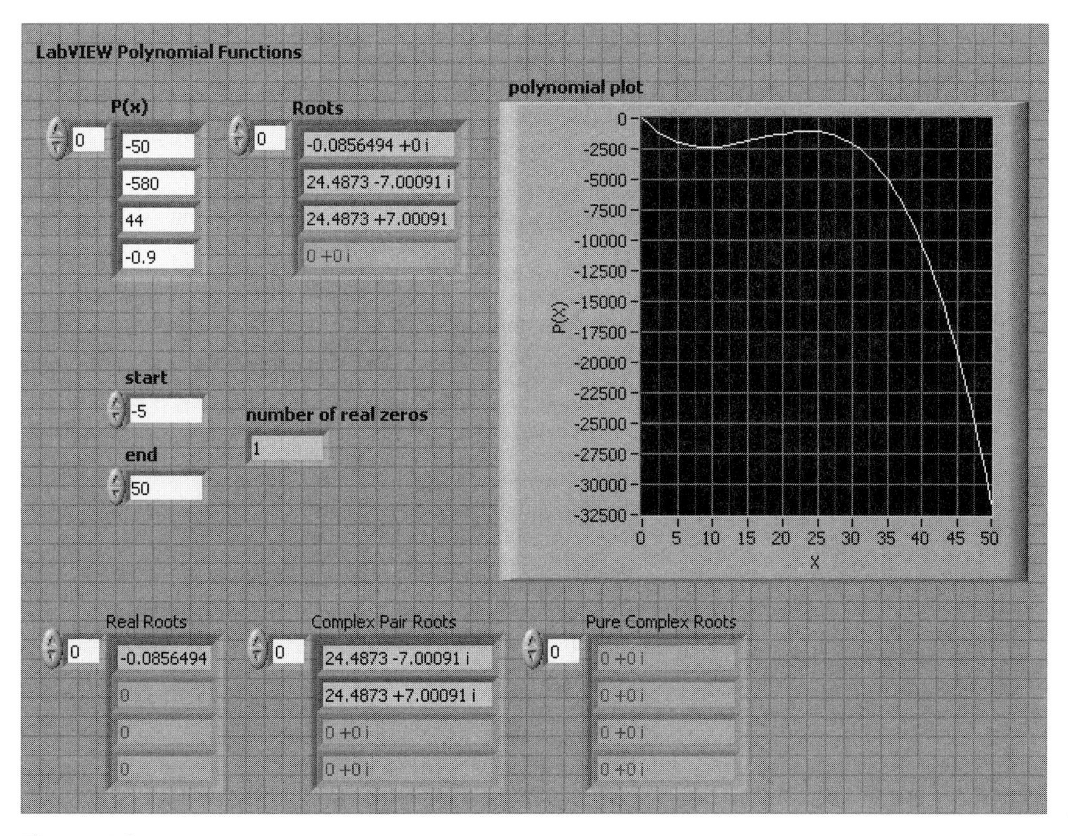

Figure 10.1
Working with polynomials.

- Polynomial Plot
- Polynomial Roots
- Polynomial Real Zeros Counter
- Roots Classification

The polynomial is described by the coefficient array P(x), with coefficients in ascending order of power.

- The **Polynomial Roots** function was used to find all roots (real and imaginary), and file the **Roots** array.
- The **Polynomial Real Zeros Counter** function was used to determine the number of roots in the specified range (–5, 50 in Figure 10.1).
- The **Roots Classification** function was used to sort the roots into Real Roots, Complex Pair roots, and Pure Complex Roots.
- The **Polynomial Plot** function was used to prepare the data arrays for plotting.

The block diagram is shown in Figure 10.2.

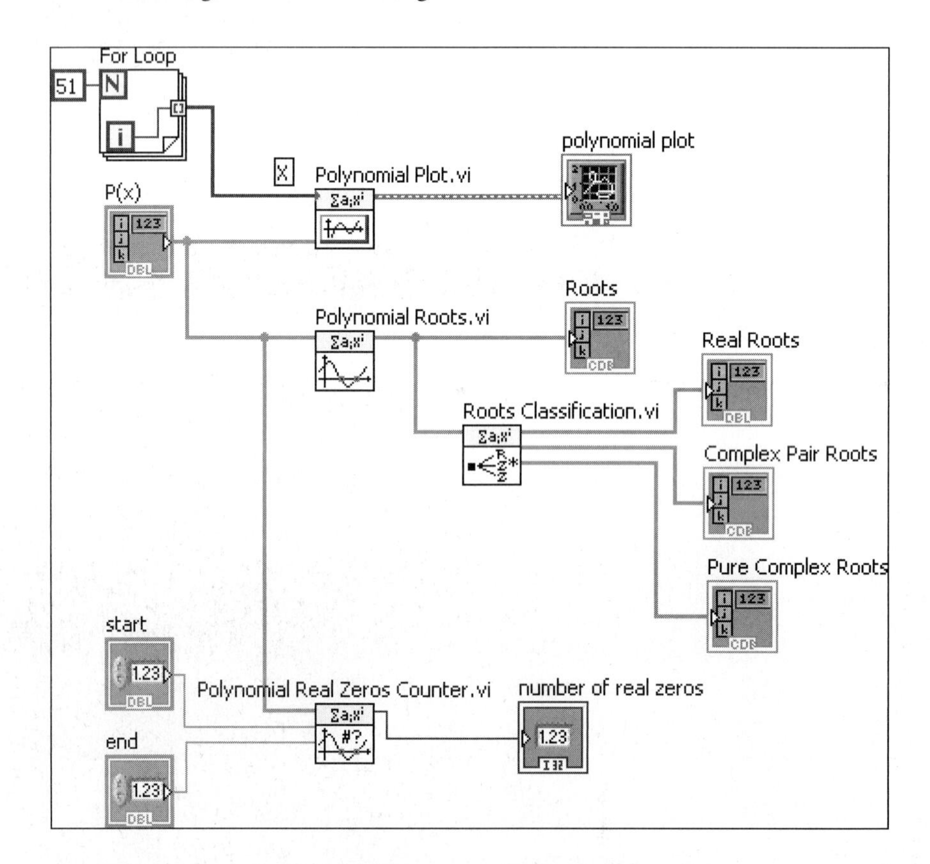

Figure 10.2 Using the polynomial functions.

10.3 STATISTICS: HYPOTHESIS TESTING

LabVIEW provides a good collection of functions for ANOVA (analysis of variance) and hypothesis testing. The functions include the following:

- T-Test
- Z-Test
- Correlation Test
- 1D, 2D, 3D ANOVA

The functions are located in the Mathematics Group:

> **Functions Palette / Mathematics Group / Probability & Statistics / Hypothesis Testing**
>
> **Functions Palette / Mathematics Group / Probability & Statistics / Analysis of Variance**

As an example of the use of these functions, some circumference and diameter data values were collected. These values were used to calculate π, and the experimental value of π was tested against the known value. The test failed; the experimental value of π was deemed to not be equal to the known value.

The T-Test results are shown in Figure 10.3. The block diagram is shown in Figure 10.4.

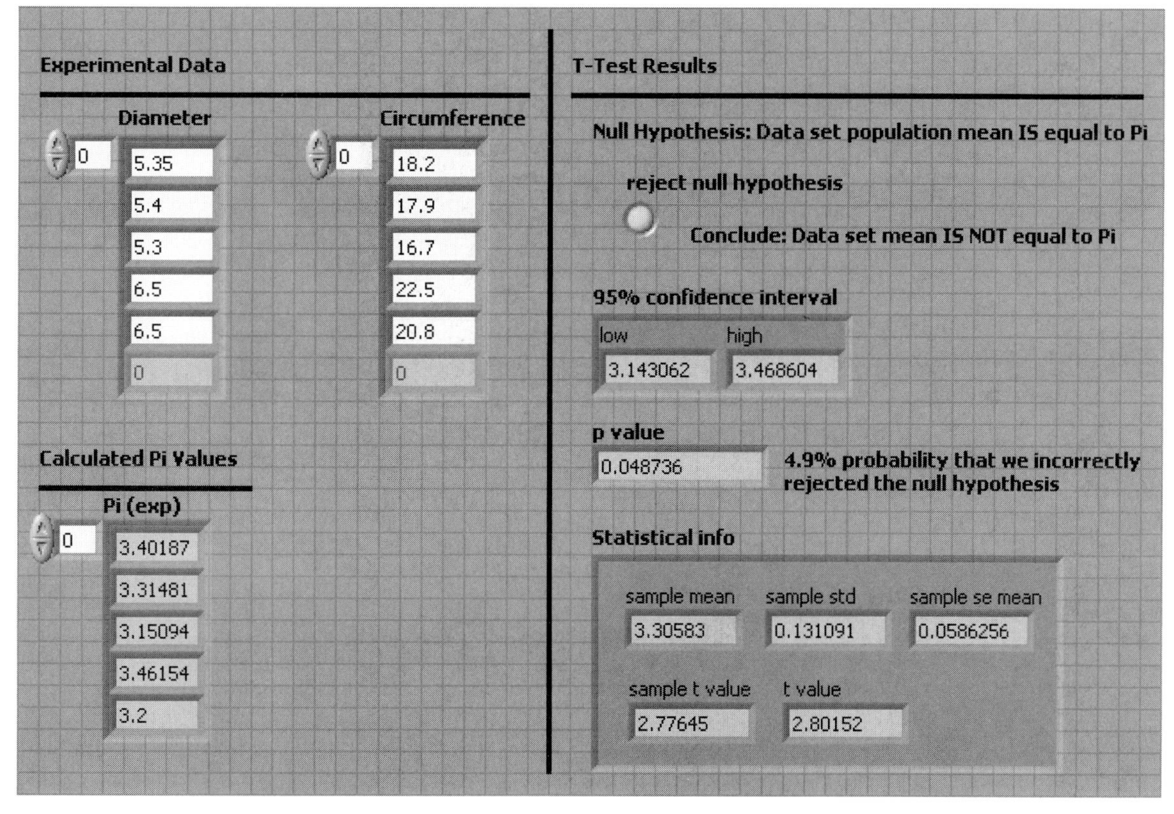

Figure 10.3
Results of a T-Test.

10.4 DIFFERENTIATION

LabVIEW provides one function for performing numerical differentiation of a data set, Derivative x(t).VI. The function is located in the Mathematics Group:

> **Functions Palette / Mathematics Group / Integration & Differentiation / Derivative x(t)**

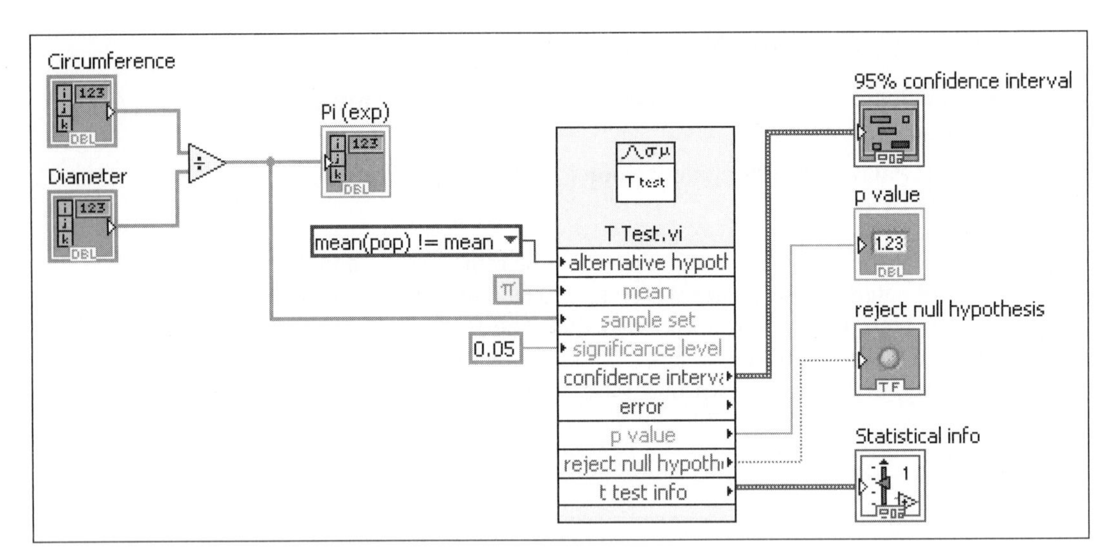

Figure 10.4
T-Test VI.

The Derivative x(t) function will approximate a derivative using four methods:

- Second-order central difference
- Fourth-order central difference
- Forward difference
- Backward difference

In this example, a sin(x) signal was numerically differentiated using the second-order central difference. Since the derivative of sin(x) is cos(x), the numerical derivative of sin(x) was plotted with cos(x) in the front panel shown in Figure 10.5. The results are very similar except right at the boundaries.

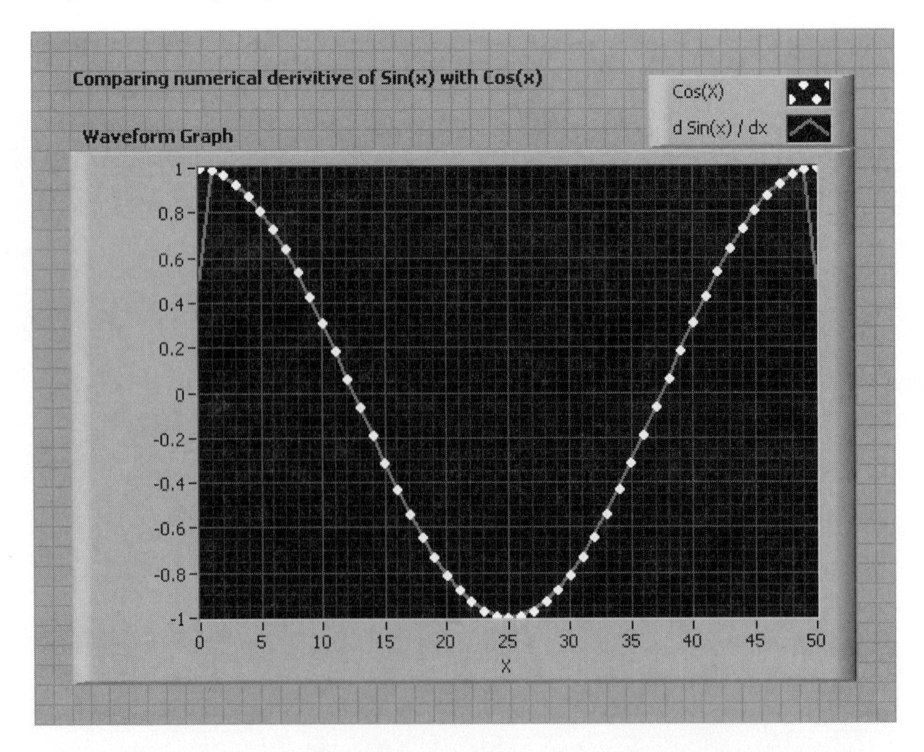

Figure 10.5
Comparing numerical differentiation result (line) with analytical result (points).

The VI used to perform the numerical integration is shown in Figure 10.6.

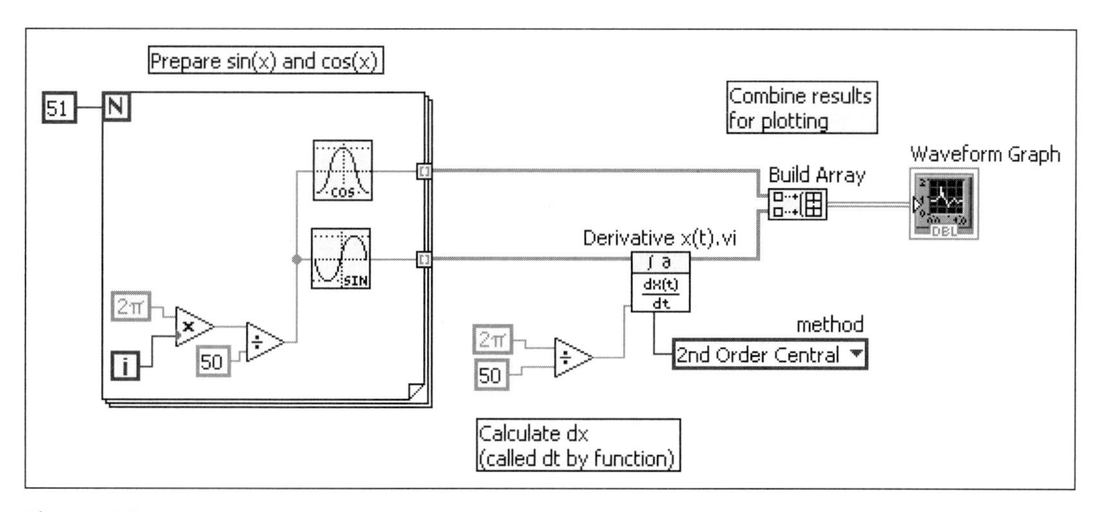

Figure 10.6
Numerical differentiation VI.

10.5 INTEGRATION

LabVIEW provides a number of functions for numerical integration. These functions are located in the Mathematics Group:

Functions Palette / Mathematics Group / Integration & Differentiation

- Integral x(t)
- Numeric Integration
- Quadrature
- Uneven Numeric Integration
- Time Domain Math

The Integral x(t) function is similar in design to the Derivative x(t) function used in Figure 10.6. We should be able to integrate cos(x) to get sin(x). The numerical integration was carried out by the Integral x(t) function in the block diagram shown in Figure 10.7. The result is shown in Figure 10.8. There is a significant lag in

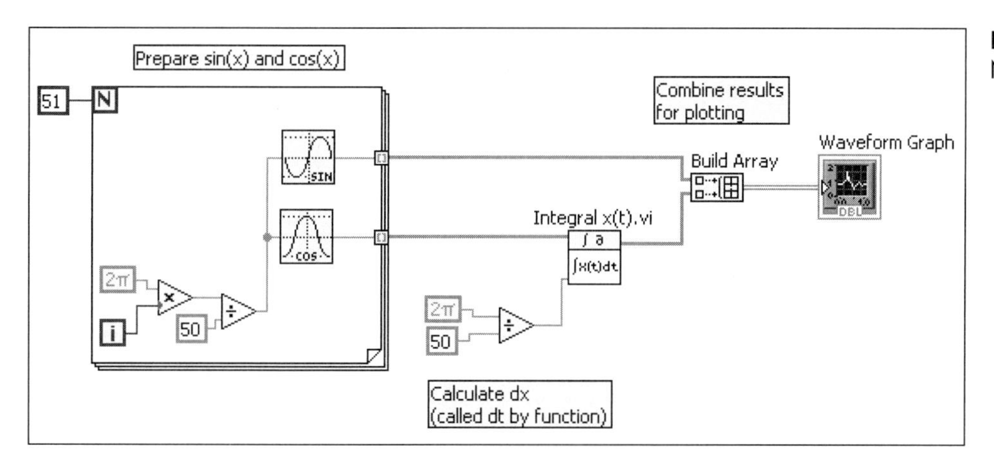

Figure 10.7
Numerical Integration VI.

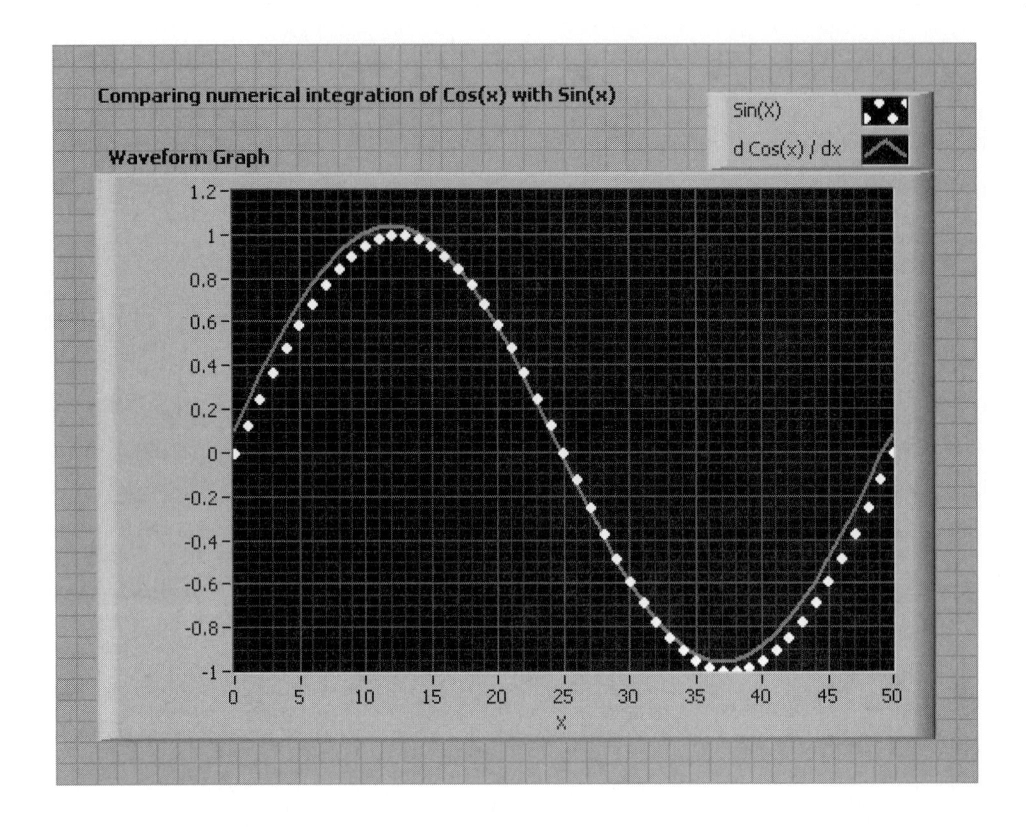

Figure 10.8
Comparing the numerical integration of cos(x) with analog result.

the numerical result caused by the relatively small number of data points in the array sent to the integration function. With more closely spaced data points, the lag is less noticeable.

The other type of numerical integration is solving for the area beneath a curve. The result is a single value, the area. The Numerical Integration function performs this type of integration.

We know that the area under a sine curve between 0 and 2π radians should be 0 because the positive portion is counterbalanced by the negative portion. The block diagram shown in Figure 10.9 performs this integration. The result is shown in Figure 10.10.

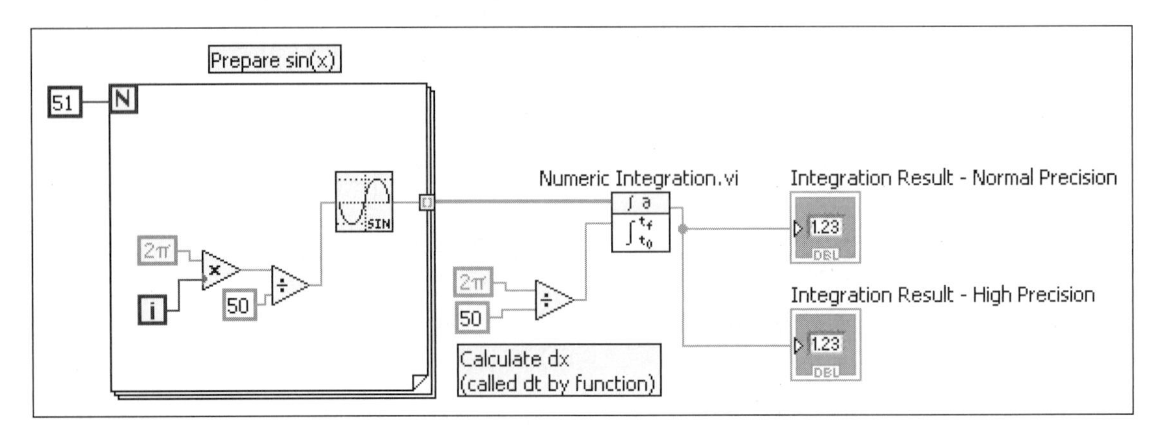

Figure 10.9
Numerical integration for area beneath a sine curve.

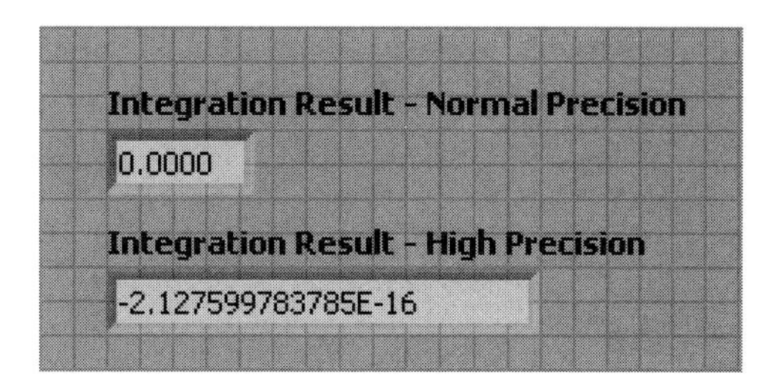

Figure 10.10
Results of sine integration.

It's not a very interesting result, except that the zero value was so "exact" that it was suspect. Showing the result to a lot of decimal places confirmed the area is really close to zero.

To see a non-zero result, let's modify the VI to integrate from 0 to π. The modified VI is shown in Figure 10.11 and the results in Figure 10.12.

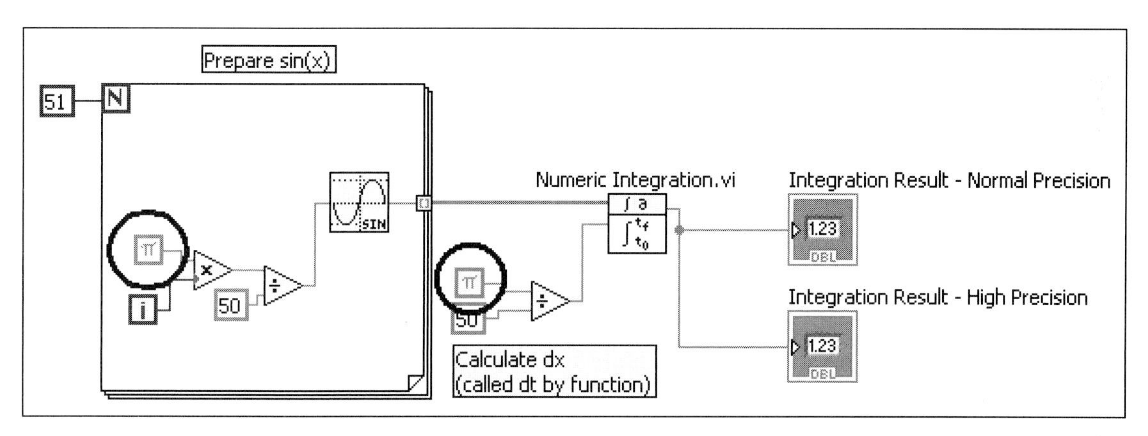

Figure 10.11
Numerical integration for area beneath a sine curve between 0 and π.

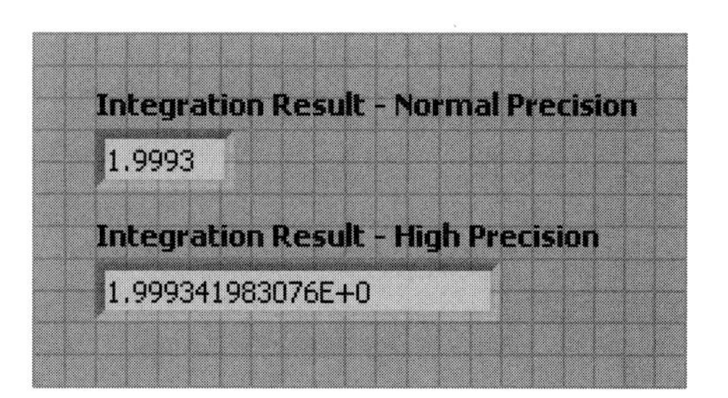

Figure 10.12
Results of sine integration between 0 and π.

The analytic result is 2. The numerical result is close at 1.9993, but not perfect.

10.6 RUNGE–KUTTA INTEGRATION

A commonly used set of routines for integrating sets of linear, first-order ordinary differential equations (ODEs) is the Runge–Kutta methods. The fourth-order method is the most common, and it is available using LabVIEW's **ODE Runge Kutta 4th Order.VI**.

One set of linear ODEs can be developed from the equation for a harmonic oscillator:

$$\frac{d^2x}{dt^2} = -\frac{k}{m}x$$

This one, second-order ODE can be rewritten as two, first-order ODEs as

$$\frac{dx}{dt} = y$$

$$\frac{dy}{dt} = -\frac{k}{m}x$$

To integrate, we need initial conditions or starting values for each variable; x and y at $t = 0$. Variable x represents the location of the harmonic oscillator; $x = 0$ initially just puts the oscillator at a point when the oscillations begin. Variable y represents the velocity of the harmonic oscillator; assigning an initial velocity value (say, $y = 0.5$) is like giving the harmonic oscillator a kick in the positive direction (up). Assigning a negative value ($y = -0.5$ is used in this example) is like giving the harmonic oscillator a kick down.

The system parameters are k, the spring constant, and m, the mass of the oscillator. We will choose arbitrary values of ($k = 1, m = 5$) for these variables. The VI used to solve the ODEs is shown in Figure 10.13 (front panel) and Figure 10.14 (block diagram).

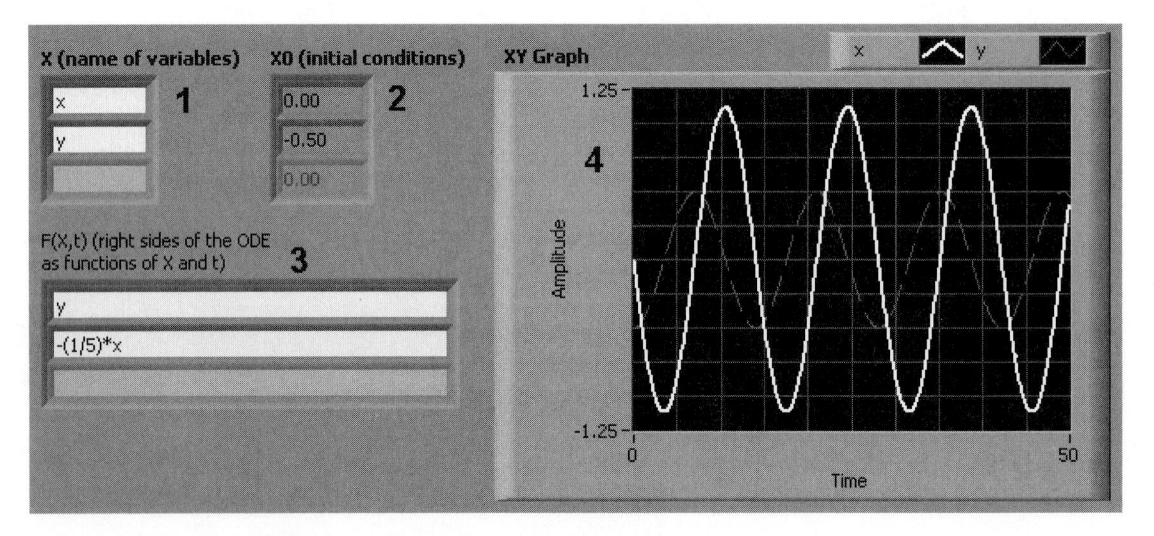

Figure 10.13
Front panel of Runge–Kutta VI.

1. The variable names that will be used in the ODE definitions are entered.
2. The initial values of each variable are entered.
3. The right-hand sides of each ODE are entered.
4. The solutions to the integrated ODEs are plotted when the VI is run.

What is interesting about this VI is not the solution—this is a complicated way to plot sine waves—but the flexibility of LabVIEW's approach to integrating ODEs using the ODE Runge Kutta 4th Order.VI. By using arrays of variable names and

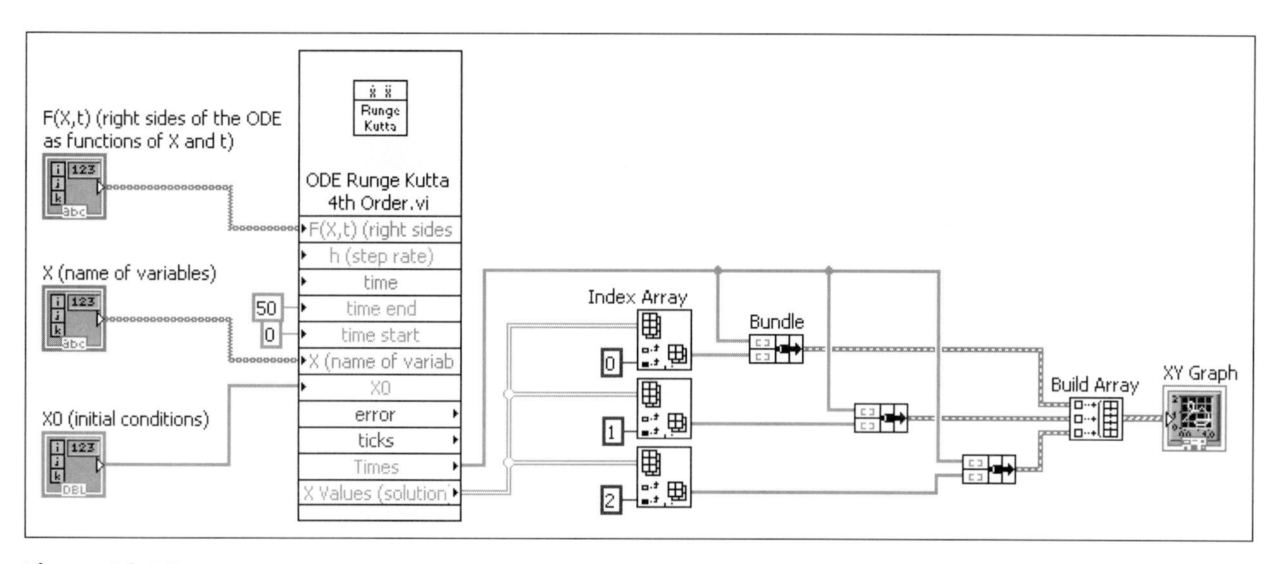

Figure 10.14
Block diagram of the Runge–Kutta VI.

defining the equations on the front panel, this VI can be used for a wide range of integration problems.

Note: The variable names must be single-character, lower case.

10.7 EXPONENTIAL FILTER

The VI in this example (Figure 10.15) uses an **Exponential Filter SubVI** to filter a composite signal made up of a sine wave and random noise.

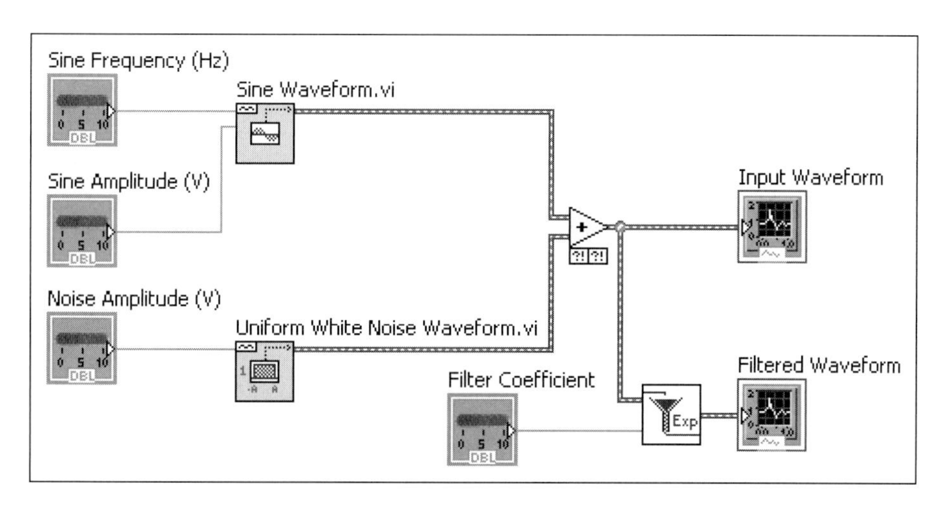

Figure 10.15
Filtering a composite waveform.

An exponential filter uses the current signal value and the previous signal value to reduce signal noise. The equation is

$$x_{\text{filtered}} = (1 - \alpha) \cdot x_{\text{previous}} + \alpha \cdot x_{\text{current}}$$

Where α is the filter coefficient, which is reduced to increase the extent of filtering. The block diagram of the Exponential Filter SubVI is sown in Figure 10.16.

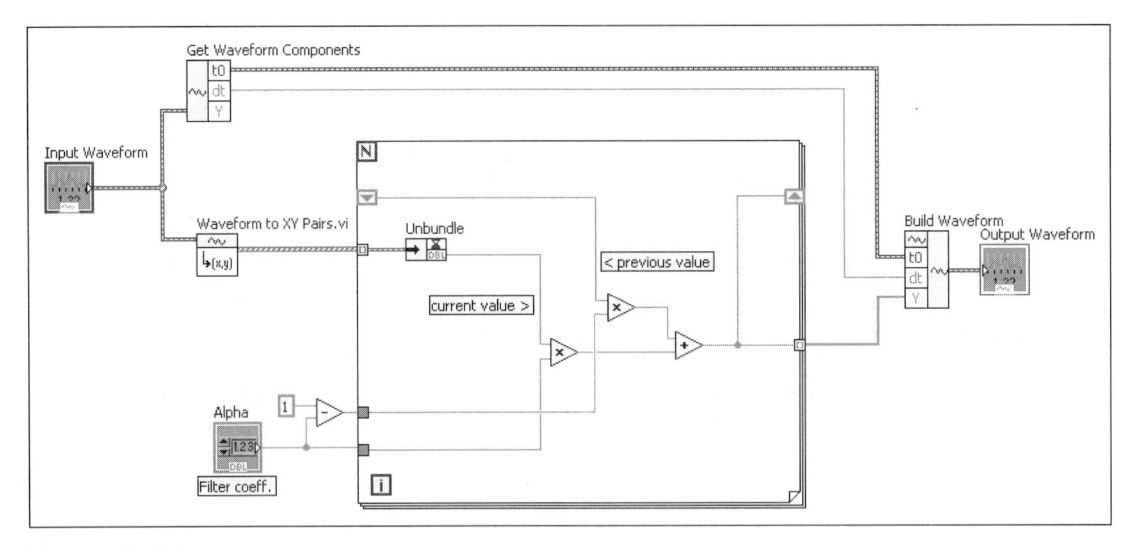

Figure 10.16
Exponential Filter SubVI.

The use of the filter is shown in the front panel graphs in Figure 10.17.

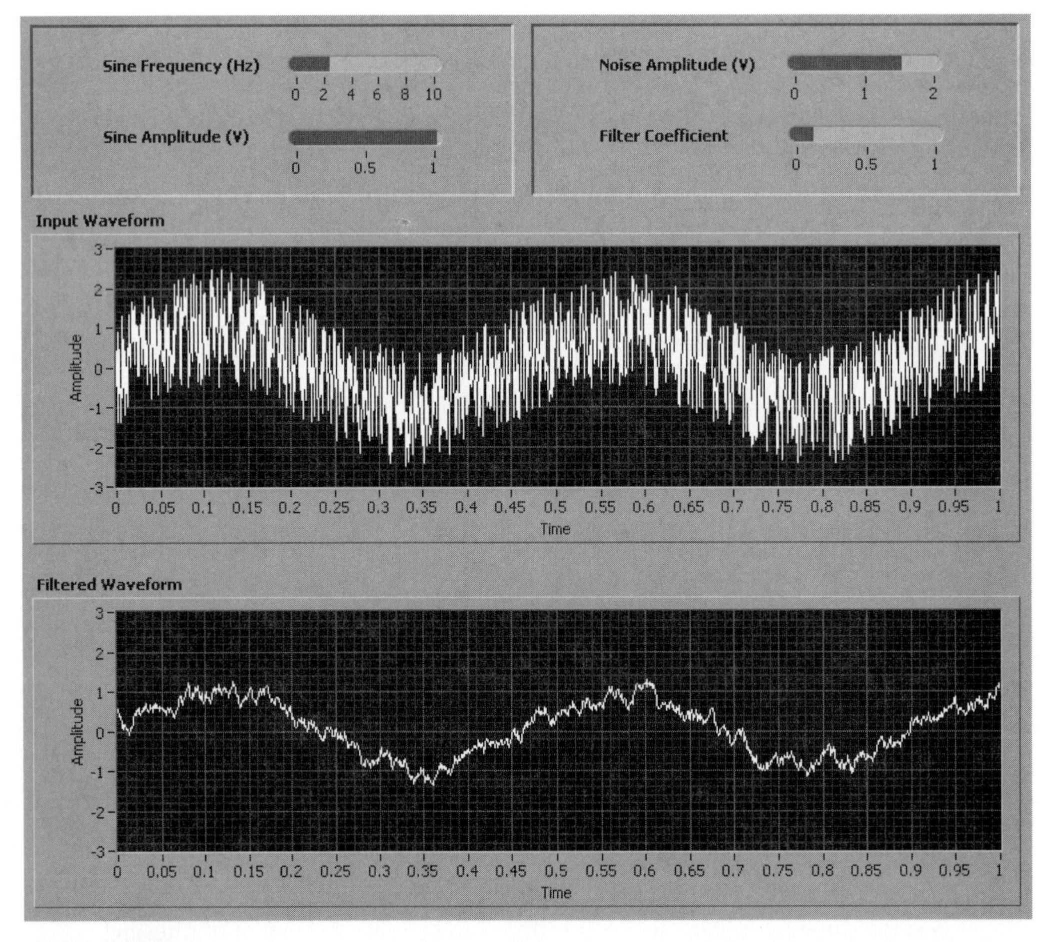

Figure 10.17
Exponential Filter VI.

10.8 SPECTRAL ANALYSIS

LabVIEW offers a number of tools for analyzing waveforms. In this example, LabVIEW's Amplitude and Phase Spectrum.vi has been used to identify the major frequency components of a waveform. In Figure 10.18, the input waveform has frequency components at 20 Hz and approximately 150 Hz, with a low noise level.

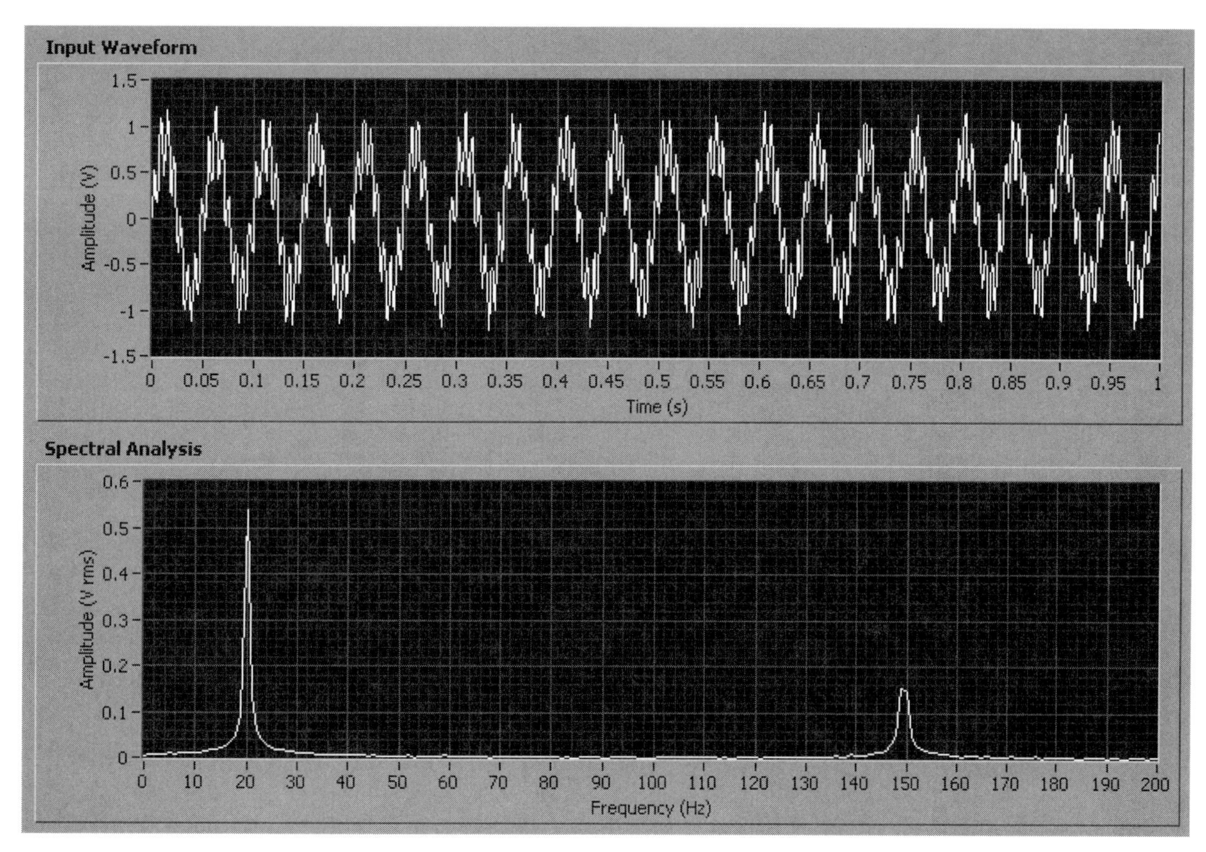

Figure 10.18
Spectral Analysis VI.

The block diagram for the spectral analysis portion of the VI is shown in Figure 10.19.

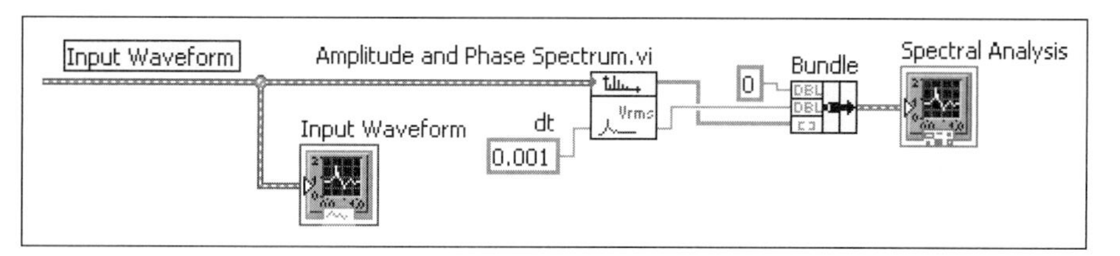

Figure 10.19
The spectral analysis portion of the VI.

This VI actually generates the input waveform too. The waveform generation portion of the VI is shown in Figure 10.20 (front panel) and Figure 10.21 (block diagram).

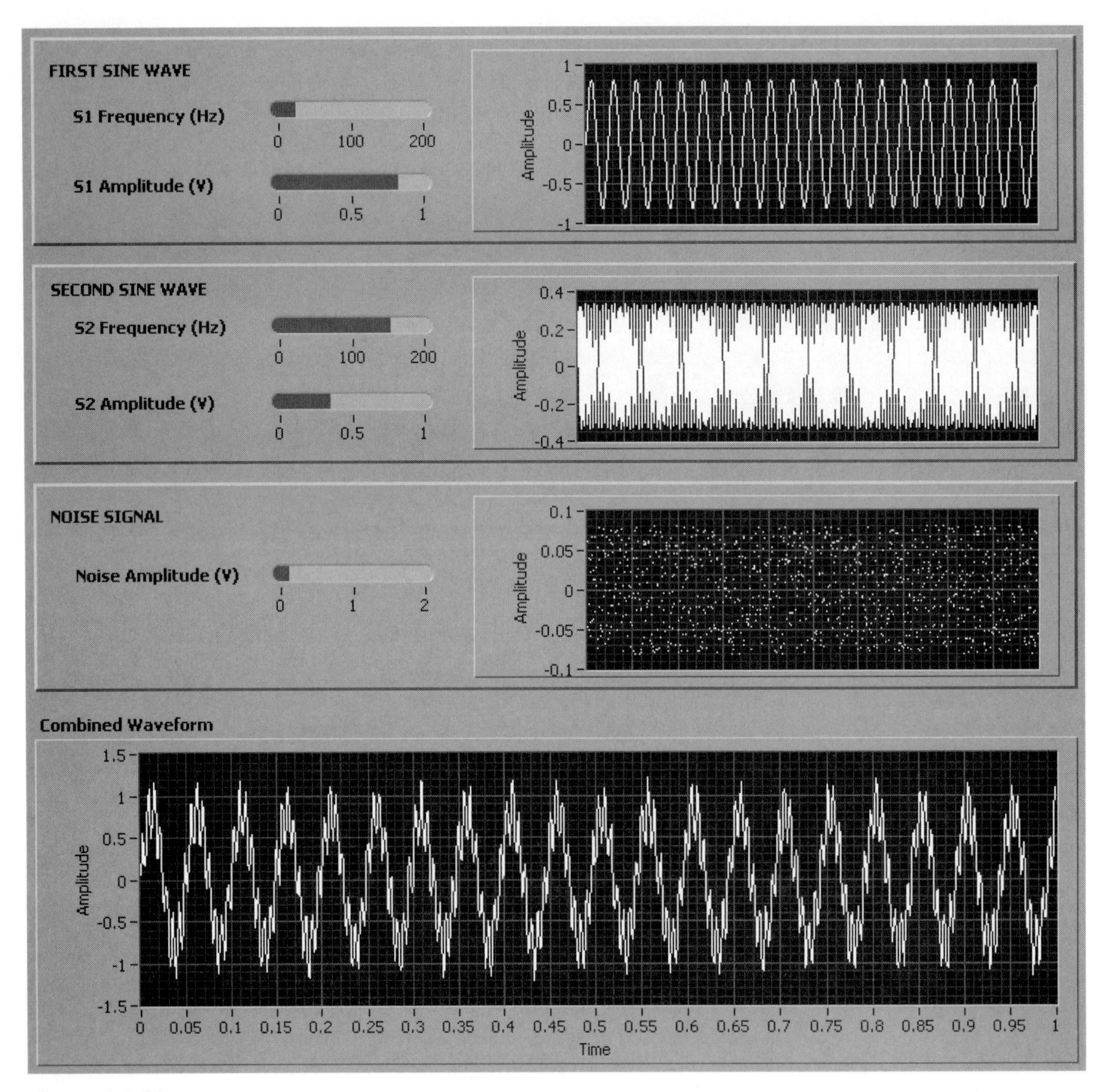

Figure 10.20
Waveform generation portion of Spectral Analysis VI (front panel).

10.9 MONTE CARLO SIMULATION

Monte Carlo simulations are used to determine the likelihood of certain events when the events have a certain extent of randomness associated with them. A common example is rolling dice. There is a randomness associated with each roll, but there are certain outcomes that are more probable than others. For example, with three dice there are several ways to roll a 9 and only one way to roll a 3—the probability of rolling a 9 is higher than the probability of rolling a 3. A Monte Carlo simulation of the dice throws would build the dice throws (based on a random number generator) into a loop and simulate throwing dice a few thousand times. Then, the results would be sorted out to see which outcomes occur most frequently.

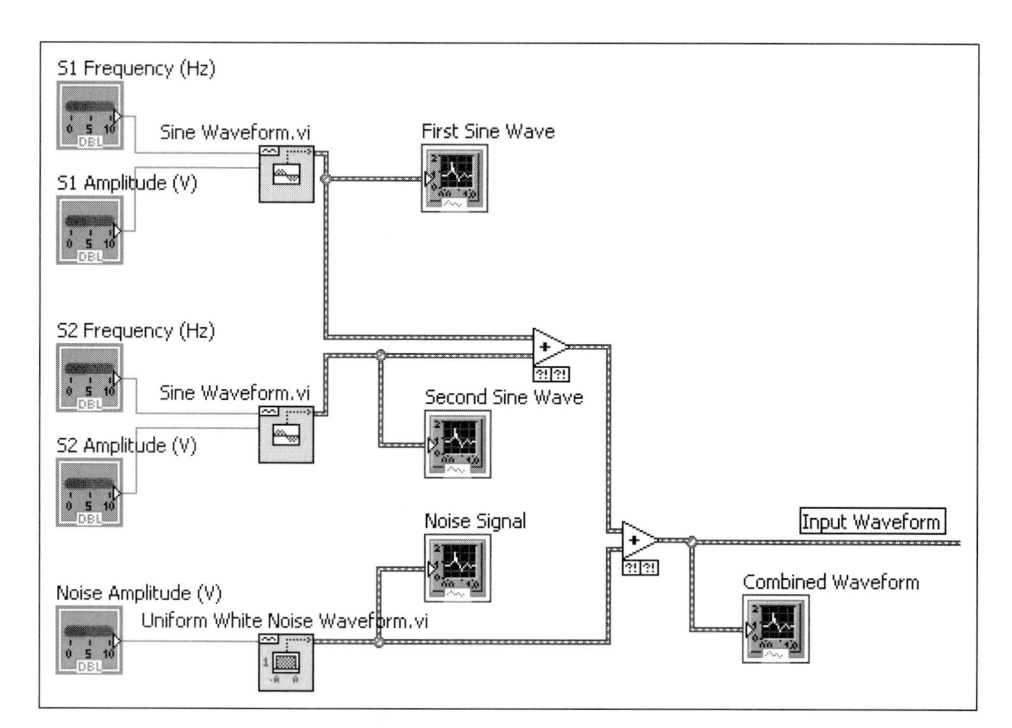

Figure 10.21 Waveform generation portion of Spectral Analysis VI (block diagram).

This example does not simulate throwing dice—that's been done elsewhere. Just for fun, we'll look at the probabilities of how long it takes to tour Yellowstone Park. The author lives just north of Yellowstone and often takes visitors through the park. How long will the trip take? With experience, we have found that it takes about four hours to drive across the park if there are no stops. Beyond the driving time, the additional trip time depends on several factors:

- **Elk**—Elk don't slow the traffic too greatly, because they are very common. On any given trip there is a 90% chance that you will spend 20 minutes watching elk and a 10% chance you will spend 45 minutes.
- **Bears**—Bears are not common, so there is an 85% chance that they will not slow you down at all. However, there is a 10% chance that you will see one, and it will slow you down by an hour. There is a 5% chance that you will see more than one and be slowed down by two hours.
- **Bison**—Bison (a.k.a. buffalo) are very common, and they have a habit of walking down the road that you are trying to drive down. On any given trip there is a 60% chance you will spend 20 minutes watching bison and a 40% chance that you will spend one hour waiting for them to get out of the way.
- **Wolves**—Wolves are rarely seen (50% chance of not seeing one), and when they are spotted, they are often far off (25% chance of spending 15 minutes watching). Occasionally, you see a wolf up close doing something interesting (15% chance of watching for 50 minutes).
- **Old Faithful Geyser**—It takes about 10 minutes to watch Old faithful erupt, plus some random fraction of an hour waiting for the eruption to start.
- **Lower Falls of the Yellowstone River**—With most trips (80%) it takes about 30 minutes to visit one of the viewing platforms. Less frequently (20%) the visitor wants to hike to the brink of the falls, which takes 2.5 hours.
- **Mammoth Hot Springs**—Most of the time (80%) a quick stop is enough (20 minutes); sometimes (20%) the visitor wants to take a lot of pictures (one hour).

The VI shown in Figures 10.22 and 10.23 builds these random factors into a Monte Carlo simulation of the likely time required to tour Yellowstone. About

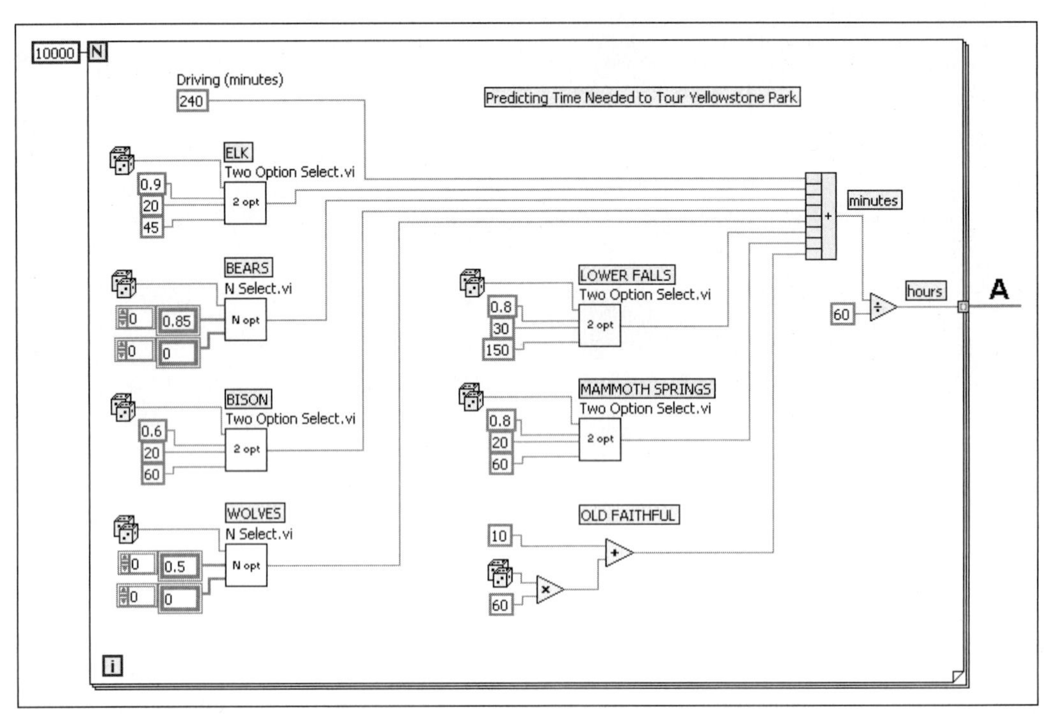

Figure 10.22
Monte Carlo simulation—calculating times.

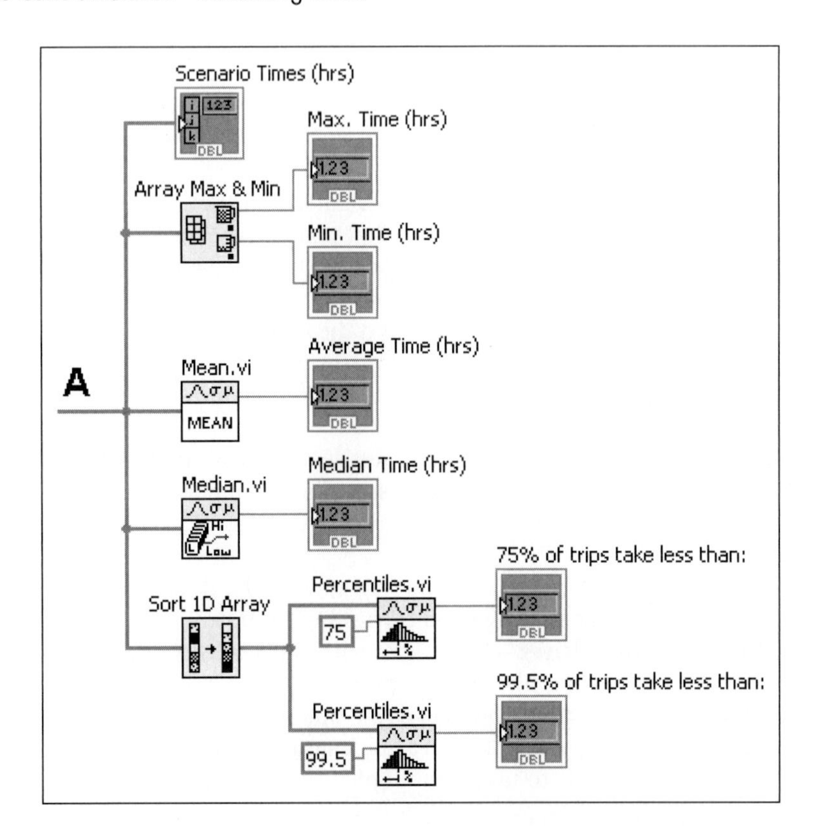

Figure 10.23
Monte Carlo simulation—analyzing results.

10,000 simulations are performed before the results are determined. The results are shown on the front panel, in Figure 10.24.

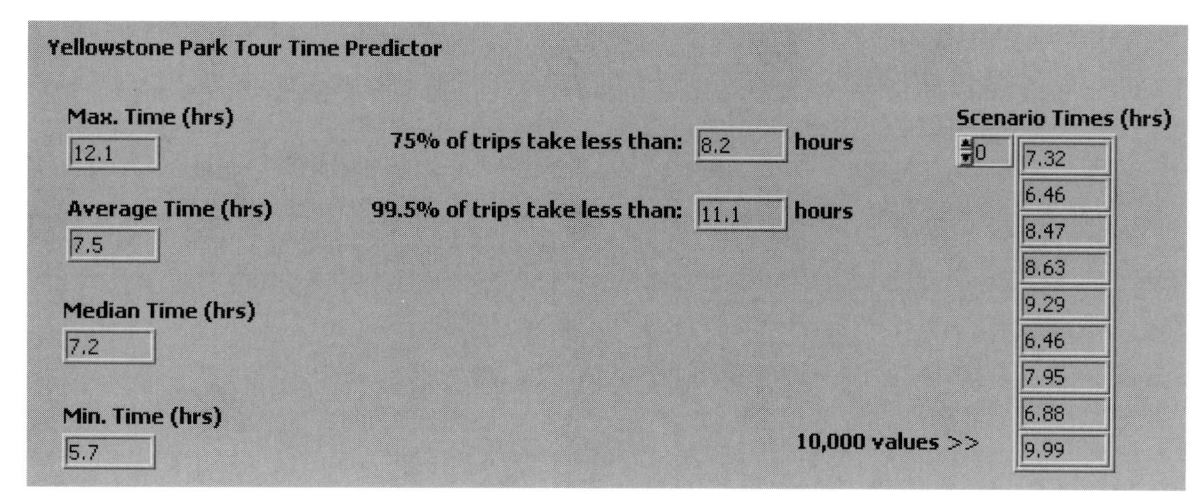

Figure 10.24
Monte Carlo simulation—results.

The results indicate that the minimum time required is 5.7 hours (the 4 hours of straight driving, plus at least 1.7 hours of sightseeing). The average trip takes 7.5 hours, and the maximum time (due to multiple bear sightings, a visitor who wants to hike, and bison on the road) is 12.1 hours. About 75% of trips take less than 8.2 hours.

This is a silly little example, but the surprising outcome is that the results agree well our experience taking many visitors through Yellowstone Park.

The block diagram shown in Figure 10.22 makes us of two SubVIs. The **Two Option Select SubVI** (Figure 10.25) chooses between two possible event durations depending on whether or not a random number (between 0 and 1) is less than or greater than the probability assigned to event A.

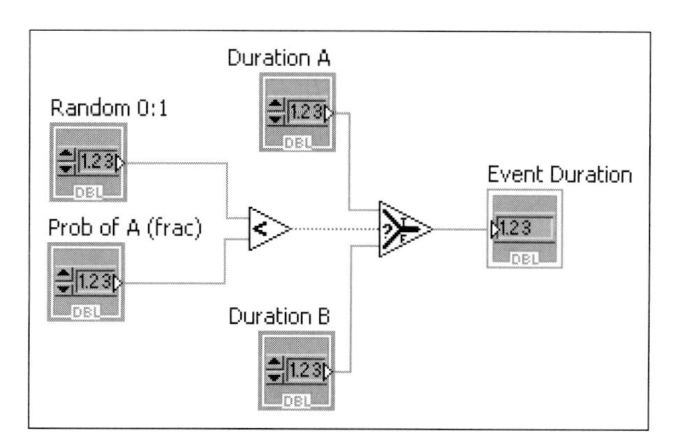

Figure 10.25
Two Option Select SubVI.

The **N Option Select SubVI** shown in Figure 10.26 uses a Formula Node to determine the event duration from arrays of event probabilities and durations.

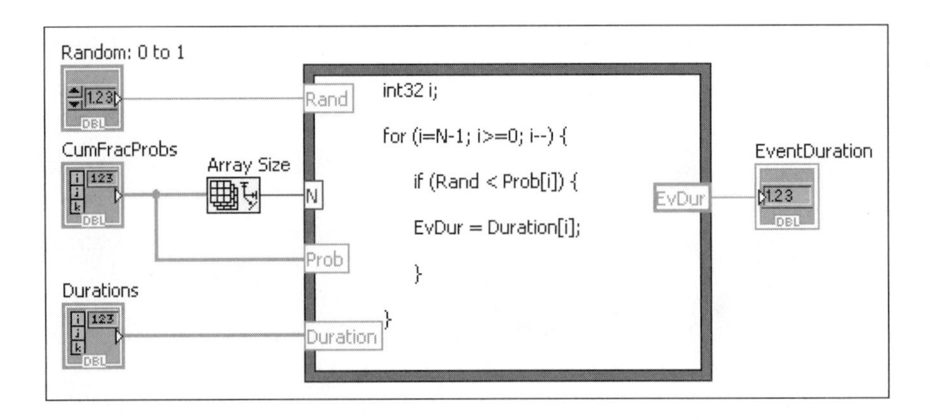

Figure 10.26
N Option Select SubVI.

10.10 PID CONTROLLER

We conclude this chapter with a VI that can be used to add PID (Proportional, Integral, and Derivative) control to a signal obtained via data acquisition. This example (Figure 10.27) simulates a heat control situation in which the controller is trying to keep the process measurement constant at 20°C. Notice that when something caused the temperature to drop suddenly, the controller responded by quickly increasing the controller output to add energy to the process (by opening a valve carrying a heat transfer fluid).

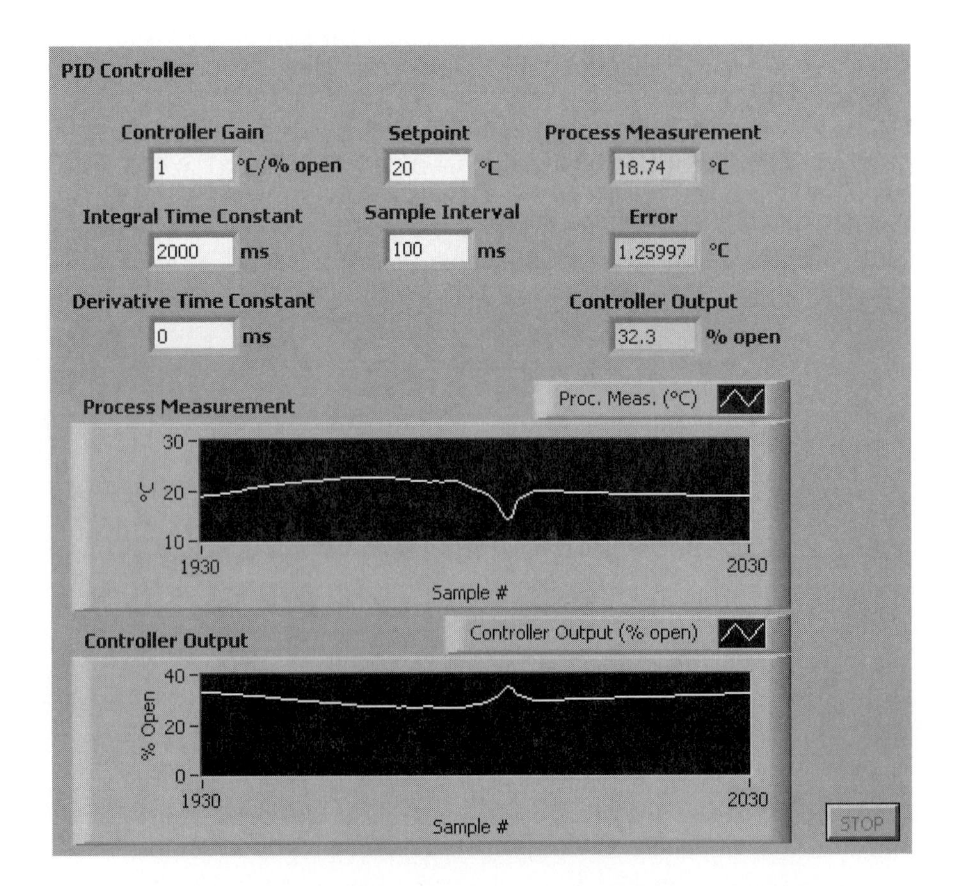

Figure 10.27
PID Controller VI.

The PID Controller VI block diagram is shown in Figure 10.28. It uses a velocity form of the PID algorithm:

$$\mathrm{CO}_{\mathrm{new}} = \mathrm{CO}_{\mathrm{old}} + K_C\left[(E_i - E_{i-1}) + \frac{\Delta t}{\tau_1}E_i - \frac{\tau_D}{\Delta t}(PM_i - 2PM_{i-1} + PM_{i-2})\right]$$

where

CO is the controller output
E is the error (i is current time step, $i-1$ is previous time step)
Δt is the sample interval
PM is the process measurement
K_C is the controller gain
τ_I is the integral time constant
τ_D is the derivative time constant

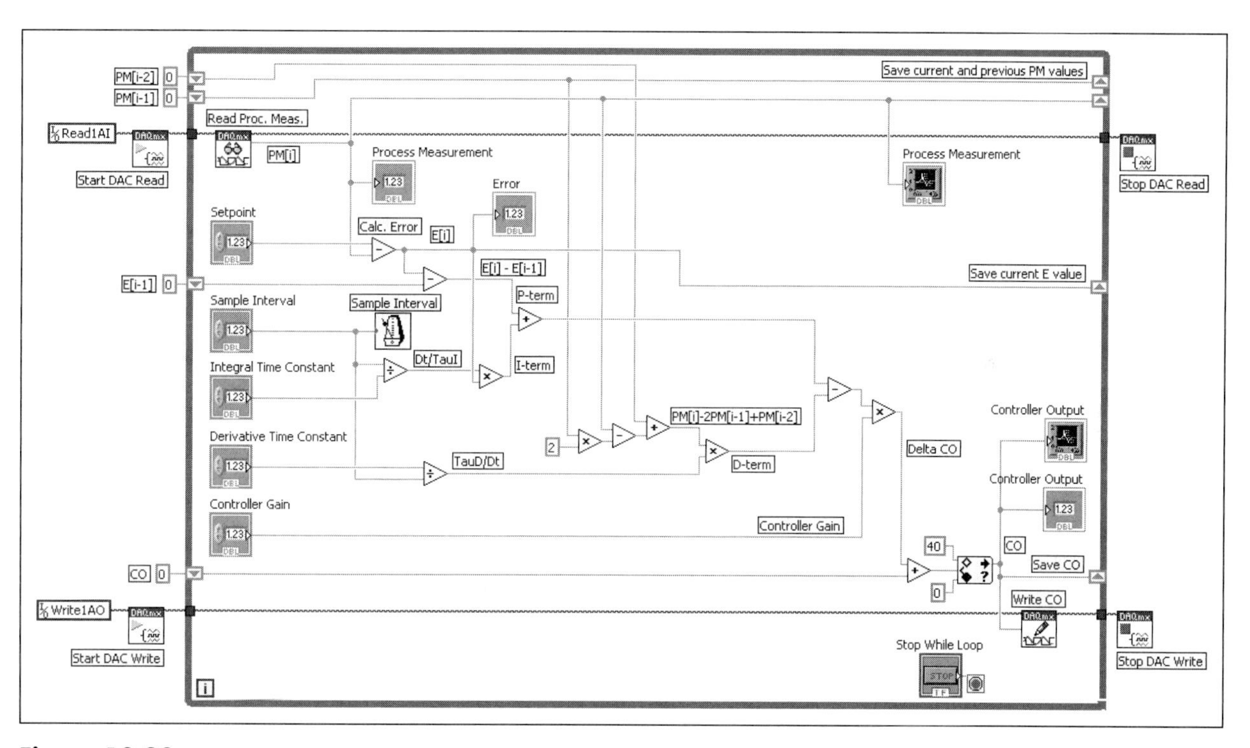

Figure 10.28
PID Controller VI, block diagram.

Appendix: Printing VIs

In a classical programming language, printing the program meant printing a listing of the programming statements. Because LabVIEW is a graphical programming environment, printing the program is a little different, and generally involves printing some combination of the following:

- The front panel
- The block diagram
- Information about the controls on the block diagram
- Names of any SubVIs (none have been used in any examples yet)

If you just want a printout of the current window (either the front panel or the block diagram), use menu options **File / Print Window . . .** This approach does not allow you to select options, but it is the quickest way to get a picture of your block diagram or front panel to a printer.

Opening the Print dialog by selecting Print . . . from the File menu (from either the front panel or the block diagram) gives you a lot of control over:

- what is printed
- in what format
- to what destination

For most situations the **File / Print Window . . .** approach is adequate.

1.1 USING THE PRINT DIALOG

To choose from the many print options involved in printing a LabVIEW program, selecting Print . . . from the File menu (from either the front panel or the block diagram):

File / Print . . .

These menu options open the Print dialog shown in Figure 1.1.

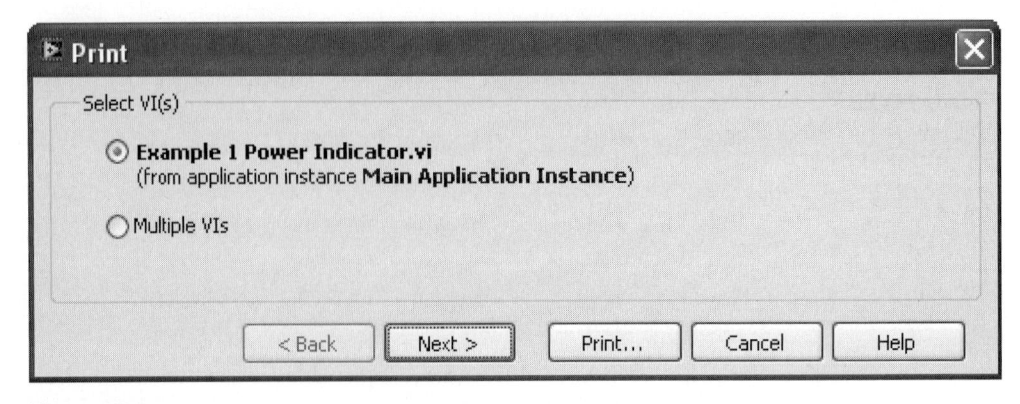

Figure 1.1
Print dialog, page 1.

The first page of the Print dialog collects information on which VI(s) you want to print. The first option (selected by default) is the VI from which the Print dialog was opened ("Example 1 Power Indicator.VI in Figure 1.1).

The buttons along the bottom of the Print dialog provide some options:

- **< Back**—Returns to the previous page of the Print dialog to allow you to make changes.
- **Next >**—Moves to the next page of the Print dialog.
- **Print . . .**—Skips the remaining pages of the Print dialog and allows you to select a printer.
  Printing without going through all of the pages of the Print dialog causes LabVIEW to use the output options specified the previous time the Print dialog was used. This can be very handy if you make a small change and simply want to reprint a VI.
- **Cancel**—Closes the Print dialog without printing.
- **Help**—a—Opens the LabVIEW Help system to information on the Print Dialog Box.

The most common printing need is to print the VI from which the Print dialog is opened. This is the default selection on the first page of the Print dialog.

Click **Next >** to move to the second page of the Print dialog, shown in Figure 1.2.

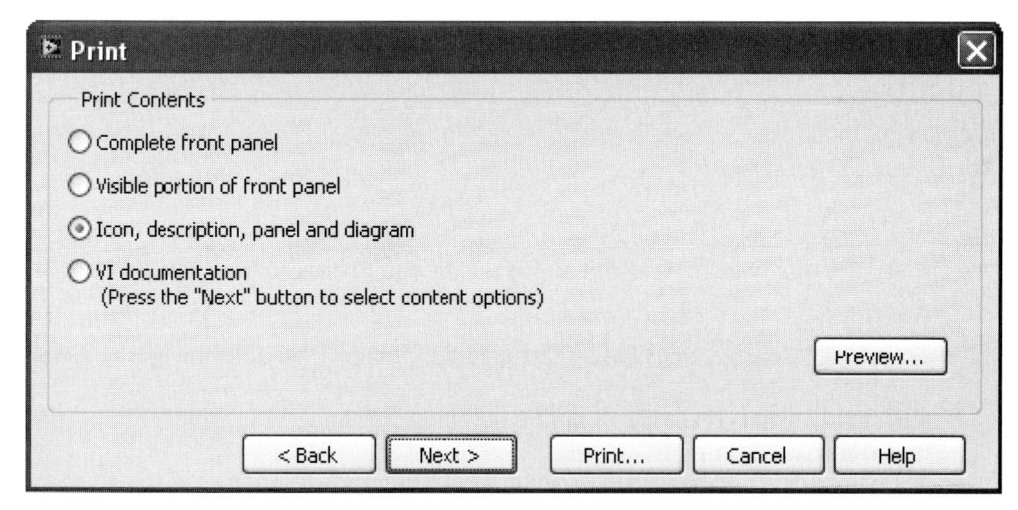

Figure 1.2
Print dialog, page 2: Print Contents.

The second page of the print dialog collects information on what you want to print. The options on the second page are as follows:

- **Complete front panel**—even if only a portion is visible when the Print dialog is opened. A large front panel may be scaled down significantly when printed.
- **Visible portion of front panel**—prints exactly what is visible when the Print dialog is opened. This allows printing a small portion of the front panel in a larger image.

- **Icon, description, panel and diagram**—prints both the front panel and the block diagram. (This is a commonly used option.)
- **VI documentation**—allows you to select exactly what you want to have printed.

The **Preview . . .** button will allow to see what the printout is going to look like.
What happens when you click the Next > button depends on which option you have selected on page 2:

- If you have selected **VI documentation**, the next page allows you to select exactly what you want to have printed.
- If you select any of the other options, the next page allows you to specify the destination for the printout, as illustrated in Figure 1.3.

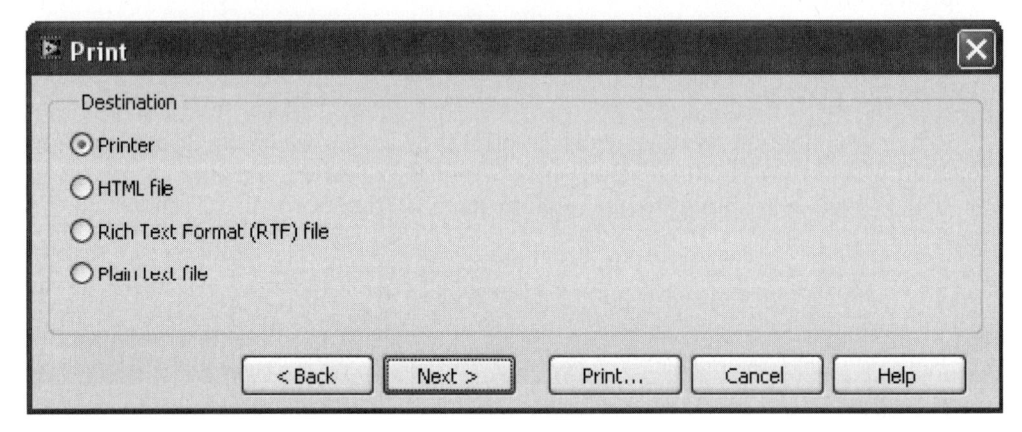

Figure 1.3
Print dialog, page 3: Destination.

The Destination page of the Print dialog allows you to choose from several options:

- **Printer**—send the printout to a printer.
- **HTML file**—create an HTML file (including images) suitable for use as a web page.
- **Rich Text Format (RTF) file**—create a file (including images) suitable for importing into a word processor.
- **Plain text file**—create a text file (no images) describing the VI.

The first three options create virtually identical outputs in various formats. Because the **Plain text file** option cannot include images, it provides much less information about the VI.

Note: You can type in a description of a VI as one of the VI's properties (**File / VI Properties**). If you have not provided a description of the VI, the **Plain text file** option provides little beyond the VI's title and file location.

In this example we will select the Printer option and click **Next >** to move to the Print dialog's Page Setup page shown in Figure 1.4.

The Page Setup page of the Print dialog allows you to set margins and indicate whether or not to include a Print Header. The defaults are shown in Figure 1.4. By

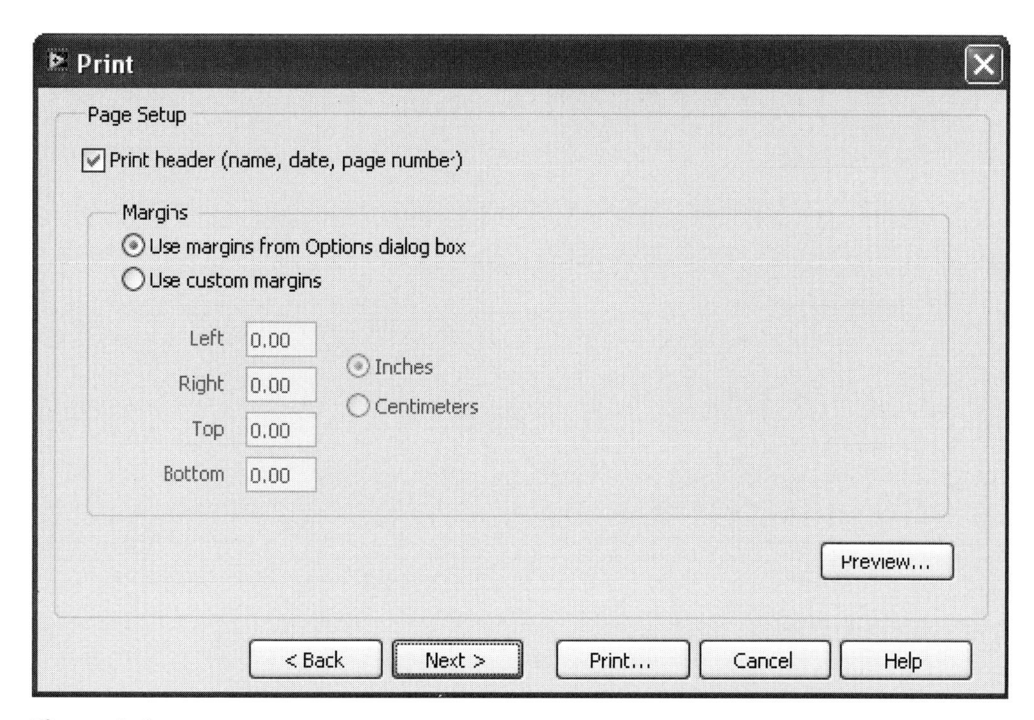

Figure 1.4
Print dialog, page 4: Page Setup.

default, no margins are included on the printouts. This maximizes the area available for graphics, but margins may be useful. You can change the margins for the current printing on the Page Setup page of the Print dialog. However, if you want to change margins for all of your printouts, change the LabVIEW options using the following menu options:

Tools / Options / Category: Printing → set Margins

Click **Next >** to move to the Print dialog's Printer options page, shown in Figure 1.5.
The Print dialog's Printer options page allows the following options:

- **Scale front panel to fit**—if not checked, a large front panel will be printed on multiple pages.
- **Scale block diagram to fit**—if not checked, a large block diagram will be printed on multiple pages.
- **Page breaks between sections**—when checked, each section (heading, front panel, block diagram) will start on a separate page.
- **Print section headers**—when checked, LabVIEW will draw a line between each section and add headings (e.g., "Front Panel", "block diagram") to each section.
- **Surround panel with border**—when checked, LabVIEW includes a thin line border around the image of the front panel.

The **Printer Setup** button allows you to access your computer's Printer Setup dialog to select a printer and set printer properties (such as printing landscape, or double-sided printing).

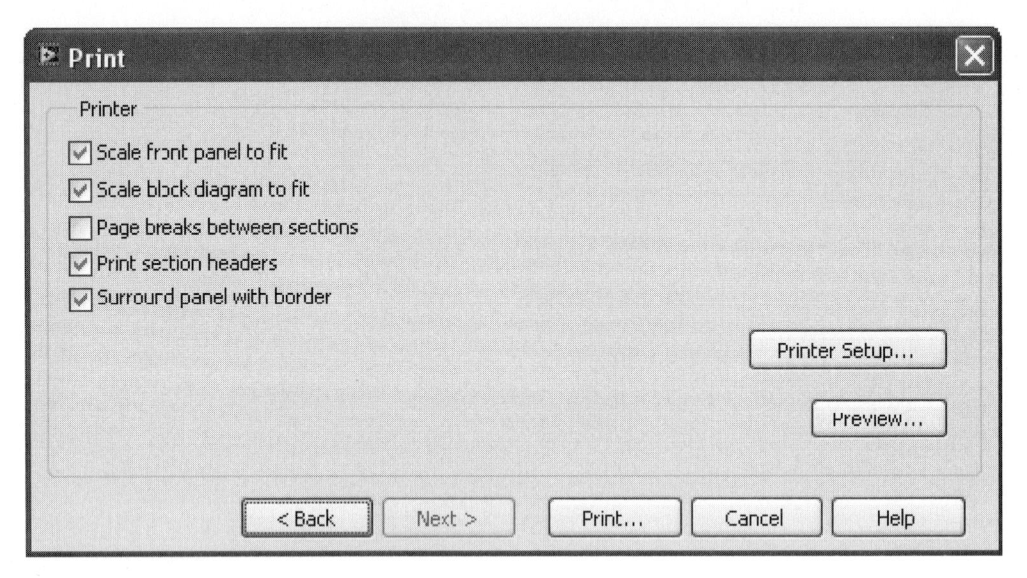

Figure 1.5
Print dialog, page 5: Printer options.

The **Next >** button is not enabled on the Print dialog's Printer options page because there are no more pages. Instead, click the **Print . . .** button to open your computer's Print dialog (Figure 1.6).

Use the Print dialog to select a printer, and then click the **Print** button to (finally) send the printout to the printer.

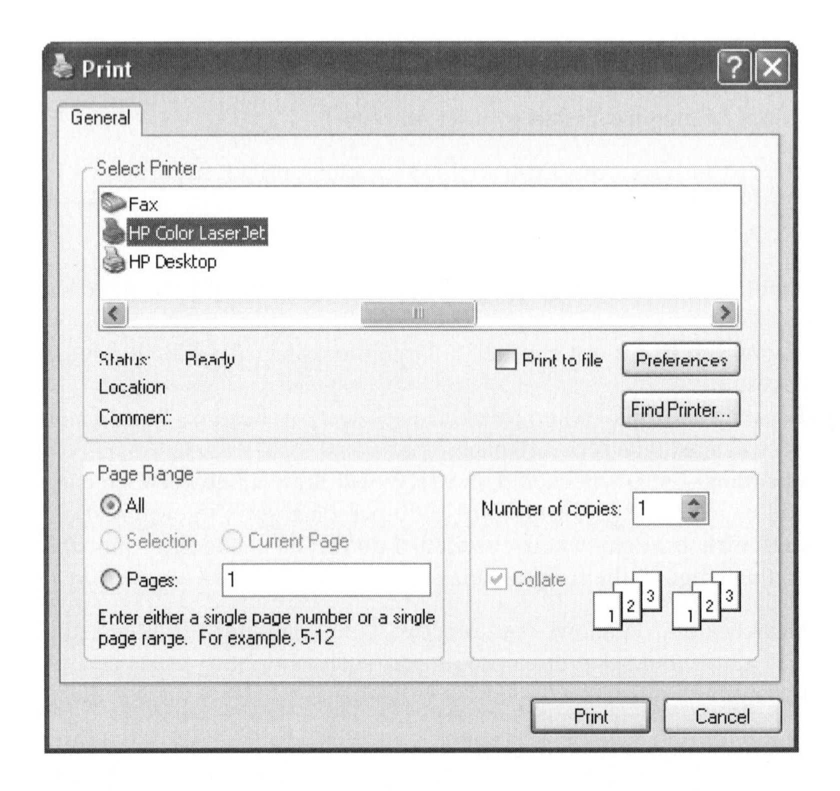

Figure 1.6
The Print dialog is used to select a printer and print the document.

There were a lot of steps involved in printing a VI. Fortunately, the defaults can be used on most of the dialog pages to speed up the printing process. Also, once you are familiar with LabVIEW's Print dialog pages, you can use the **Print . . .** button (see the bottom of Figure 1.5) to skip pages and get your printout to the printer more quickly.

Index

D

G

H

I

Q

R

S

X

Z